Operations Research Proceedings 1987

DGOR/NSOR

Papers of the 16th Annual Meeting of DGOR
in Cooperation with NSOR
Vorträge der 16. Jahrestagung der DGOR
zusammen mit der NSOR

Edited by/Herausgegeben von
H. Schellhaas P. van Beek H. Isermann
R. Schmidt M. Zijlstra

With 140 Figures/Mit 140 Abbildungen

Springer-Verlag
Berlin Heidelberg New York
London Paris Tokyo

Prof. Dr. Helmut Schellhaas

Technische Hochschule Darmstadt
Fachbereich Mathematik
Schloßgartenstraße 7, D-6100 Darmstadt

Prof. Dr. Paul van Beek

Agricultural University Wageningen
Department of Mathematics
De Dreijen 8, NL-6703 BC Wageningen

Prof. Dr. Heinz Isermann

Johann Wolfgang Goethe-Universität Frankfurt
Fachbereich Wirtschaftswissenschaft
Mertonstraße 17, D-6000 Frankfurt

Prof. Dr. Reinhart Schmidt

Christian-Albrechts-Universität Kiel
Institut für Betriebswirtschaftslehre
Olshausenstraße 40, D-2300 Kiel

Drs. Mynt Zijlstra

Nederlandse Philips Bedrijven B.V.
Centre for Quantitative Methods
P.O. Box 218, NL-5600 MD Eindhoven

ISBN 978-3-540-19365-4 ISBN 978-3-642-73778-7 (eBook)
DOI 10.1007/978-3-642-73778-7

2142/7130-543210

Vorwort

Die „Operations Research Proceedings 1987" geben eine vollständige Übersicht über die Vorträge, die auf der sechzehnten Jahrestagung der Deutschen Gesellschaft für Operations Research zusammen mit der Nederlandse Stichting voor Operations Research in der Zeit vom 23. bis 25. September 1987 im Kongreßzentrum Koningshof in Veldhoven bei Eindhoven gehalten wurden. Mehr als 350 Teilnehmer waren der Einladung der DGOR und der NSOR gefolgt, neue methodische Entwicklungen und aktuelle Anwendungen aus dem breiten Spektrum des Operations Research miteinander zu diskutieren. In bewährter Weise war das Programm nach thematischen Schwerpunkten in Sektionen gegliedert. Themen von breitem Interesse wurden in Plenarveranstaltungen behandelt. Ich stelle mit Freude fest, daß sämtliche Vorträge der Plenarveranstaltungen in diesen Proceedings in ausführlicher Fassung enthalten sind. Von den 166 Fachvorträgen in den Sektionen erscheinen 50 Vorträge in den Proceedings in Langfassung, die restlichen Vorträge in Kurzfassung. Die Langfassungen haben in der Regel 8 Seiten, umfangreicher sind teilweise die Übersichtsvorträge. Dabei erscheinen Übersichtsvorträge als erste innerhalb der jeweiligen Sektion. Im Inhaltsverzeichnis sind Übersichtsvorträge durch (ÜV) gekennzeichnet. Die Kurzfassungen haben überwiegend 1 Seite. Dem Leser, der sich aufgrund der Lektüre einer Kurzfassung für die behandelte Thematik besonders interessiert, bietet das Verzeichnis der Autoren und Referenten am Ende des Buches die Möglichkeit, sich mit dem jeweiligen Verfasser unmittelbar in Verbindung zu setzen.

Auf der Tagung wurden einige Themen schwerpunktmäßig diskutiert. Eines dieser Themen ist *Flexible Fertigungssysteme.* Der Eröffnungsvortrag und ein Plenarvortrag waren diesem Gebiet gewidmet, die neu in das Programm aufgenommene Sektion *Flexible Fertigungssysteme* (12 Vorträge) fand große Resonanz. Die Sektionen *Produktionsplanung und Lagerhaltung* (13 Vorträge) sowie *Anwendungen aus der industriellen Praxis* (8 Vorträge) waren traditionell stark vertreten. Dies gilt auch für die Sektionen *Decision Support- und Expertensysteme* (14 Vorträge) und *Organisations- und Entscheidungstheorie* (10 Vorträge). Einen weiteren Schwerpunkt bildete die Sektion *Logistik und Verkehr* (20 Vorträge). *OR in Banken und Versicherungen* wurde in einem Plenarvortrag und einer Sektion (7 Vorträge) diskutiert. Wiederum einen Schwerpunkt bildeten die Sektionen aus der Stochastik: *Zuverlässigkeits- und Bedienungstheorie* (14 Vorträge) und *Stochastische Entscheidungsprozesse* (9 Vorträge). Bemerkenswert war die Entwicklung der Sektionen aus dem Bereich der Optimierung. Mit den Sektionen *Mathematische Optimierung* (9 Vorträge) und *Kombinatorische Optimierung* (16 Vorträge) sowie mit 2 Plenarvorträgen war die Optimierung schwerpunktmäßig vertreten. Im Studentenwettbewerb für Diplomarbeiten wurden 3 Arbeiten prämiert, die in einer eigenen Sektion vorgestellt wurden. Entsprechend finden sich in den Proceedings diese Arbeiten in einem separaten Abschnitt, bei der obigen Themenübersicht wurden diese Arbeiten jedoch bei den fachlich zugehörigen Sektionen mitgezählt. Insgesamt läßt sich feststellen, daß neue methodische Entwicklungen und aktuelle Anwendungen gleichermaßen Gegenstand der Tagung waren.

Die Tagung wurde durch den Vorsitzenden der Deutschen Gesellschaft für Operations Research e. V., Herrn Prof. Dr. REINHART SCHMIDT, und den Vorsitzenden der Nederlandse Stichting voor Operations Research, Herrn Prof. Dr. PAUL van BEEK eröffnet. Herr Ir. M. KUILMAN, Vice-President N.V. Philips' Gloeilampenfabrieken, Eindhoven, hielt den Eröffnungsvortrag über *Flexible Manufacturing – a Requirement of the Future.* Im Rahmen der Tagung wurden vier Plenarvorträge gehalten, die bei den Teilnehmern auf lebhaftes Interesse stießen. Den ersten Plenarvortrag hielt Herr Prof. Dr. JAN KAREL LENSTRA über *Algorithmics and Heuristics in Combinatorial Optimization.* Im zweiten Plenarvortrag sprach Herr Prof. Dr. Ir. LUDO F. GELDERS über *Planning in Flexible Manufacturing Systems.* Im dritten Plenarvortrag referierte Herr Prof. Dr. WOLFGANG BÜHLER über *Bewertung und Management festverzinslicher Wertpapiere,* während im vierten Plenarvortrag Herr Prof. Dr. KLAUS RITTER Probleme und Lösungswege zu *Numerical Methods for Nonlinear Programming Problems* darlegte.

Zum Gelingen dieser Tagung trugen viele bei. Referenten, Sitzungsleitern, Sektionsleitern und der Jury des Studentenwettbewerbs sei herzlich gedankt. Besonderer Dank gilt Herrn Prof. Dr. PAUL van BEEK, Herrn Prof. Dr. C. BERNHARD TILANUS und Herrn Prof. Dr. JAAP WESSELS sowie ihren Mitarbeiterinnen und Mitarbeitern für ihre hervorragende Organisation. Auch haben sie mit einem attraktiven Rahmenprogramm die Teilnehmer begeistert. Dank auch an Frau ASTRID SCHNEEWEISS in der Geschäftsstelle der DGOR für ihre organisatorische Hilfe. Hoch einzuschätzen ist der Beitrag einer Vielzahl von Spendern, die die Tagung finanziell und durch Sachmittel gefördert haben. Sie sind vor dem Inhaltsverzeichnis benannt.

Herr Dipl.-Math. EGBERT FALKENBERG unterstützte mich bei der Vorbereitung der Tagung und der Proceedings, Herr Dipl.-Math. RAINER WEILBÄCHER betreute die Datenverwaltung. Frau MAGDALENE TABBERT übernahm – unterstützt durch Frau TRAUDEL RIDDER – die umfangreichen Sekretariatsarbeiten. Ihnen allen möchte ich für die gute Zusammenarbeit herzlich danken.

Dem Springer-Verlag spreche ich meinen Dank für die gute Kooperation aus.

Darmstadt, im Februar 1988 H. SCHELLHAAS

Programmausschuß

H. Schellhaas, Darmstadt (Vorsitzender); P. van Beek, Wageningen; H. Isermann, Frankfurt; R. Schmidt, Kiel; M. Zijlstra, Eindhoven

Sektionsleiter	**Sektion**
S. Albers, Lüneburg	Marketing
G. Altrogge, Hamburg	Studentenwettbewerb
K.-H. Borgwardt, Augsburg	Mathematische Optimierung
H. Braun, Bochum	Berichte aus den DGOR-Arbeitsgruppen
W. Gaul, Karlsruhe	Datenanalyse und Prognoseverfahren
H. van Gelder, Heiloo	OR in Banken und Versicherungen
J.P.C. Kleijnen, Tilburg	Simulation
H. Laux, Frankfurt	Organisations- und Entscheidungstheorie
J.K. Lenstra, Amsterdam	Kombinatorische Optimierung
M. Meyer, Nürnberg	OR im Gesundheitswesen
J.A.E.E. van Nunen, Rotterdam	Decision Support- und Expertensysteme
W. Popp, Bern	Strategische Planung
M. Schäl, Bonn	Stochastische Entscheidungsprozesse
Ch. Schneeweiß, Mannheim	Produktionsplanung und Lagerhaltung
S. Stöppler, Bremen	Kontrolltheorie
H.C. Tijms, Amsterdam	Zuverlässigkeits- und Bedienungstheorie
C.B. Tilanus, Eindhoven	Logistik und Verkehr
R. Vahrenkamp, Kassel	OR im Personal Computing
M. Zijlstra, Eindhoven	Anwendungsbereiche aus der industriellen Praxis
W.H.M. Zijm, Eindhoven	Flexible Fertigungssysteme

Jury des Studentenwettbewerbs

DGOR

G. Altrogge, Hamburg (Vorsitzender); K.-P. Schuster, Hamburg; H. Steckhan, Regensburg

NSOR

H. van Gelder, Heiloo (Vorsitzender); J.P.M. de Kroon, Eindhoven; A.T. Langevelo, Amsterdam

Wir danken den folgenden Firmen und Institutionen, die die Tagung finanziell und durch Sachmittel unterstützt haben:

COVAM SYSTEMEN, Eindhoven
DELTA LLOYD VERZEKERINGSGROEP N.V., Amsterdam
DIRECTIE LANDBOUWKUNDIG ONDERZOEK MINISTERIE VAN LANDBOUW EN VISSERIJ, s'-Gravenhage
DMV CAMPINA B.V., Veghel
EOS – ISAR, München
EURO III FOUNDATION, Wageningen
F. FLICK FÖRDERUNGSSTIFTUNG, Düsseldorf
HENDRIX INTERNATIONAL B.V., Boxmeer
HOOGOVENS IJMUIDEN, IJmuiden
IBM NEDERLAND N.V., Amsterdam
LANDBOUWUNIVERSITEIT, Wageningen
NMB BANK, Amsterdam
PARENCO B.V., Renkum
PHILIPS, Eindhoven
PREUSSAG AG, Hannover
RABO BANK, Utrecht
RWE, Essen
SCHERING AG, Berlin
SHELL INTERNATIONAL PETROLEUM COMPANY LIMITED, London
STORK BRABANT B.V., Boxmeer
TECHNISCHE UNIVERSITEIT, Eindhoven
TNO, s'-Gravenhage
UNILEVER RESEARCH LABORATORIUM, Vlaardingen
VAN DOORNE'S BEDRIJFSWAGENFABRIEK DAF B.V., Eindhoven
VVS, Eindhoven

Inhaltsverzeichnis

Eröffnungsvortrag

Plenarvorträge

Studentenwettbewerb

Produktionsplanung und Lagerhaltung

Flexible Fertigungssysteme

Organisations- und Entscheidungstheorie

Decision Support- und Expertensysteme

Strategische Planung

Logistik und Verkehr

Marketing

Anwendungsbereiche aus der industriellen Praxis

OR in Banken und Versicherungen

OR im Gesundheitswesen

Datenanalyse und Prognoseverfahren

OR im Personal Computing

Simulation

Zuverlässigkeits- und Bedienungstheorie

Stochastische Entscheidungsprozesse

Kontrolltheorie

Mathematische Optimierung

Kombinatorische Optimierung

Berichte aus den DGOR-Arbeitsgruppen

FLEXIBLE MANUFACTURING: A REQUIREMENT OF THE FUTURE

Ir M. Kuilman,
vice-president N.V. Philips' Gloeilampenfabrieken,
Eindhoven, The Netherlands

Mr Chairman,
Ladies and Gentlemen,

Imagine, if you will, an engineer sitting at a computer terminal punching in data for the design of a new product and sketching freely with a lightpen on the screen before him. Happy with the design, he presses a button and the details are passed electronically to another computer running software that checks to see whether the design's stresses and strains are within prescribed limits. The information then zips along to a third computer which generates instructions that command the tools in the workshop to machine, assemble and store the engineer's product ready for distribution - all done automatically, without hassle, delay or hefty manhandling, and all before the morning's coffee break. One more satisfied customer.

Welcome to the "factory of the future"!

This utopian picture of the factory of the future is described in the introduction to a recent article in The Economist[1]. Utopian as it may be, the story reflects a trend in industrial production, or rather in the interaction between the market and industrial production, which is inescapable.

The name of the game in modern manufacturing is not simply "product quality" or "low costs", but <u>flexibility</u>, the quest to supply the customer with his or her personal choice or design, to supply it <u>quickly</u> yet at the same price as a mass-produced item. A "personalised" Porsche for the price of a standard Volkswagen, to be supplied within one day of ordering.

Exactly _what_ the factory of the future will look like is still far from certain. Besides, you can take it from me that _the_ factory of the future does not exist. Just think of the different demands made on a factory which assembles consumer goods in large numbers, a factory which produces components in very large quantities, and a factory which builds professional installations one by one. Certainly, however, we have not only to deal with the considerable influence of changing market demands on the technical equipment used in the production processes. There is also going to be a revolution in the _organisation_ of production, purchasing and distribution. Indeed, this revolution has already begun; let me remind you of the immense effort made by industry over recent years to achieve integrated logistic planning, a reduction of stocks and the amount of work-in-process, and shorter lead times. Think of the many quality campaigns too.

In order to be able to guide this revolution in the right direction, it is essential to have a good understanding of the underlying processes and of the parameters which control them. Quantitative models and analyses can be applied exceptionally well here, and in this field mathematics, particularly Operations Research, has a tradition to uphold. Let me immediately add a brief comment. In the introduction I have already described to you how much industrial production is subject to change - how it is on the brink of a revolution. The world is more dynamic than ever before; numerous things which used to be seen as established facts, now no longer seem to apply. If Operations Research wants to retain its credibility in the future, then its models and analyses will have to represent this dynamism adequately. I shall come back to this in greater detail later.

Before doing so, however, I want to outline briefly those changes which are occurring in industrial production - as a result of changed markets, alterations in cost ratios and technical developments. These changes constitute a true revolution - "the factory is being reinvented from scratch". We are miles away from the traditional industrial production of the past. What were the characteristics of this? What line of thought was it based on?

A Brief Review

To answer this question, we will have to go a long way back, as far as the end of the 18th Century. The first industrial revolution at that time brought about a drastic shift from small-scale traditional and individual production to capitalist, market-oriented production. Until that time a product was manufactured from start to finish by one person. Later, partly owing to the influence of Adam Smith, work was divided into smaller and smaller tasks which could be carried out separately, each by a different person. Smith used as an example the manufacture of pins: by carefully dividing up the work into different tasks, each to be carried out by a different person, he achieved a production level which was 240 times greater than if each worker were to produce one complete pin. This principle of the division of labour ultimately culminates in the theory of labour organisation which is now known as Scientific Management (Taylor, Gilbreth[2]).

In addition to this, the striving for increased efficiency leads to mass production; economies of scale are the main theme. Factories are based on "repetitive manufacturing", on short-cycle work. There is hardly any variety in the product range; the production managers at that time believed that greater product diversity would have an adverse effect on efficiency and, in view of the technological capabilities in those days, this was not such a stupid belief. Remember the

fig. 1

answer Henry Ford gave when asked in what colours his new T-Ford would be supplied: "Any colour, as long as it's black." That was the heyday of the dedicated assembly line. And make no mistake, in about 1920 the total lead time for the production of the Model T Ford was only four days, from casting the steel to when the car rolled off the production line.[3]

At that time the Philips factories were also organised like this. In 1930 Henry Ford paid a visit to Anton Philips in Eindhoven (fig. 1) and was clearly impressed by the efficient,large-scale method of production (fig. 2).

fig. 2

After the Second World War, during the period of recovery, this system initially proved to be quite adequate. The aim was to produce large quantities: everything that was made could be sold; this was said to be a seller's market. (fig. 3) However, prosperity increased, so the demand for product variety also rose sharply. This market demand had to be satisfied; at the same time, however, efficiency based on the division of labour and economies of scale continued to be the order of the day. The answer was found to be batch-type production methods. Different product models are often made up of different parts and therefore require different processing methods (fig. 4). Machines have to be reset. Economies of scale call for a functional arrangement of the machines; in Operations Research literature this is referred to as the "Job Shop". This functional arrangement, combined with the often long changeover times, leads to large batches and therefore large stocks of work-in-process

fig. 3

and very long lead times (fig. 5). Ratios of 1:25 or even 1:100 of actual job content to lead time are the rule rather than the exception.

fig. 4

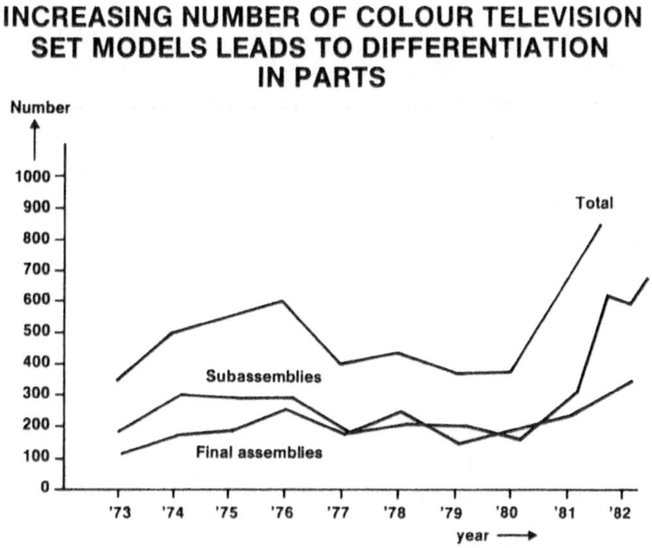

fig. 5

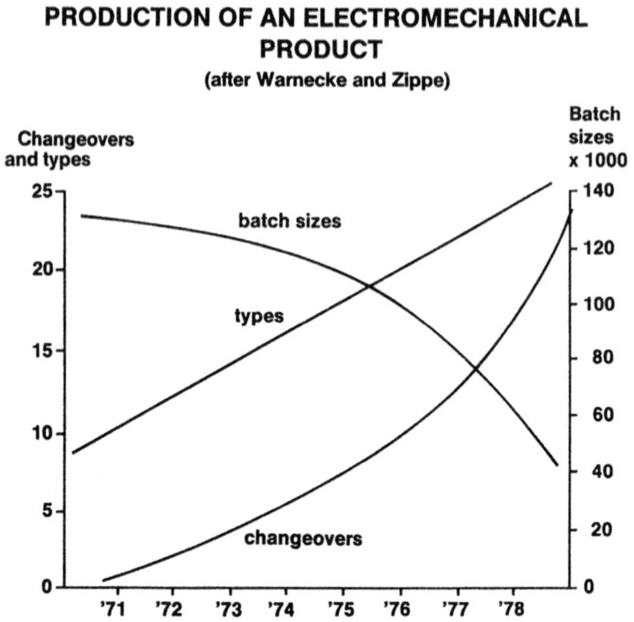

The foundation of the EEC is another reason for increasing the scale of operations - the old formula for success still applies. Philips concentrates production in IPCs (International Production Centres). Added to the internal lead times in the factory are the external ones: the supply of materials and international transport between various factories push the length of the total logistic chain up even higher (fig. 6). Then there's the time required for processing data - central planning departments have to convert market data into plans. Thus the lead times become longer and this causes stocks of work-in-process to rise. What is more, these long lead times force us to maintain large commercial stocks (stocks of finished products) in order to be able to provide a reasonable service to the public. I do not need to explain to you the relationshipbetween lead time and safety stocks.

However, the cost of stocks is still relatively low because interest rates are low. Products still have a reasonable commercial life, which means the risk of obsolete stock is practically zero. We can still plan for the long term.

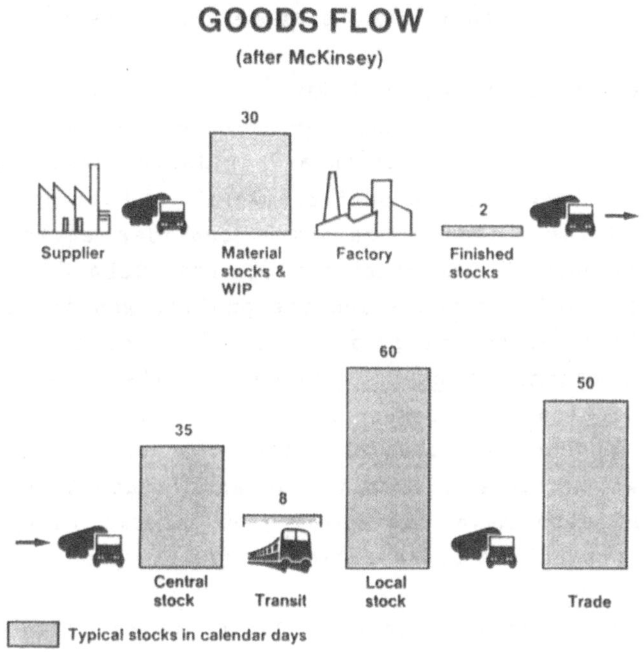

fig. 6

Let us turn now to your specialist field. Mathematical theory on production and stocks flourish in the sixties. This was the time of the Camp/Wilson[5] formula for the calculation of the economic lot size in the case of constant demand, the Wagner and Whitin[6] algorithm for determining the optimum production series in the case of variable demand; it was also the time of statistical ordering rules such as the well-known ($\underline{s}$,$\underline{Q}$) and ($\underline{s}$,$\underline{S}$) strategies[7]. These are also applied within Philips. However, the analyses relate mainly to the "single-phase" production stock models; the time of the multi-echelon models in which chain reactions are studied, is yet to come. I shall make an exception for the work of Forrester[8] on "Industrial Dynamics" in which, as early as the late 50's, he discusses the cyclical variation of stocks in longer production-distribution chains. It was only later that this work was fully appreciated, even by industry.

Furthermore, I think it can be said that traditional Operations Research is focused above all on the characteristics and management of fairly stable processes. Indeed, it is considerably more difficult to formulate good strategies for non-stationary than for stationary processes. What is more important, however, is that up to the end of the 60's these assumptions were not unrealistic - the world was indeed changing, but still only at a relatively slow rate.

But then comes the oil crisis.

A combination of stagnation and inflation causes a recession such as has been unknown to the post-war world. Fuel prices rise sharply, government intervention increases - consider environmental legislation, for example. Competitors from cheap-labour countries flood our markets with their products and the public, who are by now accustomed to a high level of spending, demand more and more variation; the seller's market becomes a buyer's market. The life cycle of consumer goods in particular becomes shorter (in some cases less than one year). The functional, batch-oriented production method now results in stocks which can not be paid for and in addition to this, shorter life cycles also make the holding of stocks risky. Fundamental changes to production organisation become inevitable.

And these changes do come. Japan, forced like no other country to stretch its very scarce raw materials, flourishes in this particular situation which itself helped create. At Toyota, automation is used on a large scale to cut changeover times, sometimes by a factor of 50 or more. Just-In-Time production is carried out using Kanban systems. Even Deming's[9] ideas on quality control are well received in Japan. Europe and America are confronted with a flood of products from the Far East which are very competitive in terms of price and quality. The West must react. And we do react. Let me outline how, by using Philips' response as an example.

Philips' Reaction

Philips has implemented a strategy in response to the developments described above. This is based on a number of points:

- acquisitions, to increase our market share (Bauknecht, Westinghouse lamp division)
- establishments closer to the market (Signetics, Magnavox)

- R&D
  (Megaproject)

- company-wide quality improvement

- improved logistic organisation and rationalisation of the goods flow

- reorientation of the production processes
  (flow production)

The key word in all these points is "flexibility".

Flexibility

Flexibility means: reacting quickly to the market. In particular, short lead times when developing, producing and distributing high-quality products for the market. With really short lead times, production can be order-driven instead of forecast-driven. Thus the risk of redundant stock is avoided.

Therefore, flexibility is not only a case of installing automated production systems on a large scale, it also has an important role to play in development and distribution. Within Philips we make a distinction between innovation flexibility, mix flexibility and volume flexibility.

Innovation flexibility concerns the speed at which new products can be brought onto the market. It is mainly determined by the development lead time and the duration of the starting phase in production.
Mix flexibility refers to the speed at which alterations can be made to the composition of the product range to be supplied. The lead times in production and distribution are of importance here.
Finally, volume flexibility concerns the extent to which fluctuations in volume can be dealt with. Of importance here are the potential excess capacity of the production system and the speed with which components or semi-finished articles can be obtained.

As you can see: flexibility comes into play at many levels in the organisation. I shall now go on to explain some of them.[4]

Production

In our historical review we first looked at the traditional assembly line and then the Job Shop. We saw that the traditional assembly line was characterised by very high productivity but was also particularly inflexible as far as variations in output were concerned. In addition to this, the assembly line system proved to be particularly sensitive to disturbances. The Job Shop, on the other hand, made it possible to supply a fairly varied range of products, although this meant very high intermediate stocks and extremely long lead times compared with the actual job content.[10]

We are aiming for Flexible Manufacturing Systems and Flexible Assembly Systems. The use of these is really intended to combine the advantages of both of the previously-mentioned production systems, i.e. a high level of efficiency and great diversity with short lead times.[10] Short production lead times have their advantages also with regard to quality: faults are detected earlier and information about these can therefore be fed back more quickly. High intermediate stock levels, on the other hand, only help to conceal faults.

To show you that all this is not just beautiful theory, let me mention a few examples. Our car radio factory in Rambouillet, France, is now able to produce and skip orders within five days, starting with the assembly of the printed circuit boards. Large parts of the manufacturing process are completely automated. This example will also be presented in a paper this afternoon.

Our metalware factory in Leeuwarden is implementing a new shop floor control methodology; at this moment significant improvements have already been made such as:

- a reduction of structural backlog from four weeks to one week,
- a reduction of throughput times and work-in-progress inventory by 40%,
- a reduction of finished products stocks by 20%.

Another fine example of what can be achieved is shown by the implementation of our LOPIC-TOOPIC program at ELCOMA, our components division. LOPIC (Logistics, Planning and Information Control in ELCOMA) was first implemented in the group capacitors (about 15 000 type numbers) involving seven factories. Figures 7 and 8 show some of the key results obtained up to now.

fig. 7

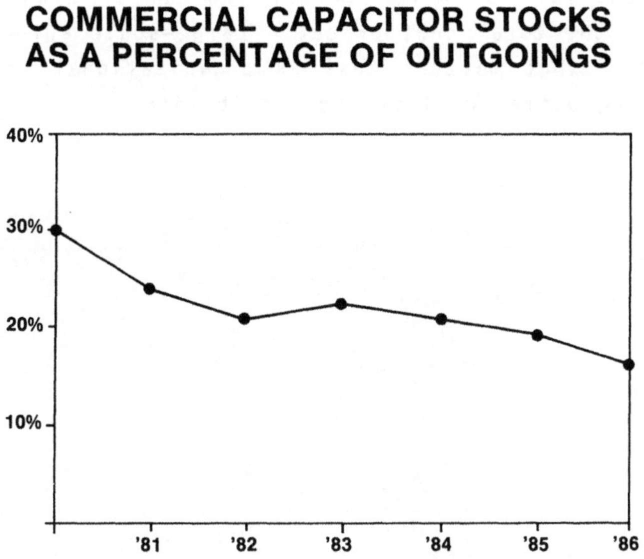

fig. 8

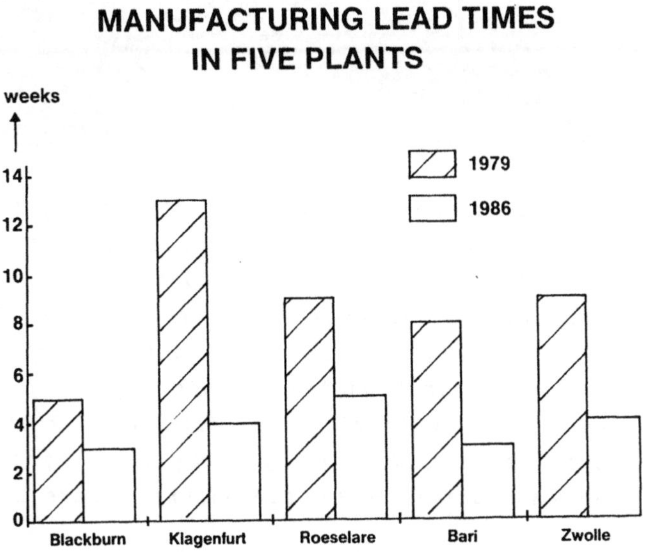

Furthermore, flexible manufacturing will have a different effect and manifest itself in a different way in each factory. Let me explain this with regard to the situation at Philips. In assembly activities we are nearly always dealing with three areas: the manufacture of components, followed by subassemblies and then final assembly.

It has been shown that the labour costs, and thus the added value, of the final assembly of many products no longer constitute such a large factor in the total cost price. Large-scale automation is not necessary here - what matters most is organisational changes with a view to achieving better quality and flexibility.

fig. 9

In the field of subassemblies, such as cathode ray tubes, motors and mounted printed circuit boards, a higher level of automation and robotisation lies ahead. Figure 9 shows robotised handling of cathode ray tubes - robotisation is necessary here because of the weight of the components. In other areas the components are so small that robotisation is necessary for handling reasons. (fig. 10) Sometimes process conditions allow so little room for error that people are unable to carry out the work accurately enough. The same is often true also for the manufacturing of components, where automation is furthest advanced.

fig. 10

But, automation or no automation, orientation of the production process to Just-In-Time production plays a key role in every case.

fig.11

This means, flow production instead of batch production and organisation based on the product axis instead of the functional arrangement which we were accustomed to. This latter point definitely does not mean a return to the dedicated assembly line - it means systems which are capable of producing in a short time a large number of variations more or less in whatever order required within a given product range, and with negligible changeover times. In automated systems this was made possible by the advent of Computer Numerically Controlled (CNC) machines, whilst in our consumer electronics factories advanced transport systems, for example the automatic guided vehicles (AGV, fig. 11) or the variable transport system (VTS), ensure immediate transport of individual products between the various production units.

Product Development

Products should also be developed according to their manufacturability. Product development and process development should be coor-

DESIGN FOR ASSEMBLY

fig. 12

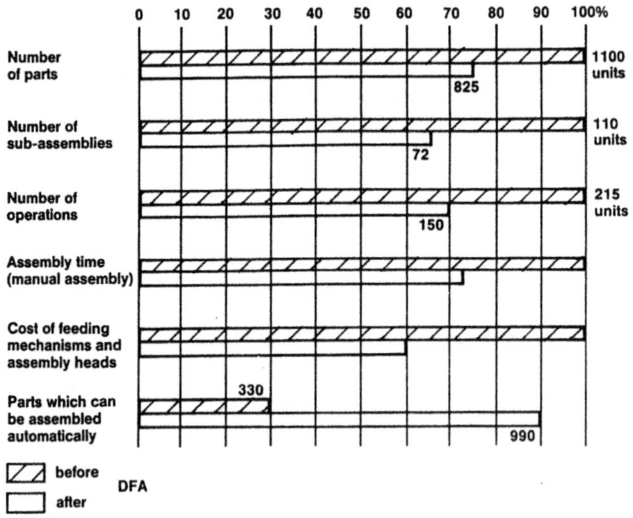

dinated and not run separately. Within Philips Design For Assembly (DFA) techniques are being introduced (fig. 12). This combination of product and process development is essential in order to bring a sufficiently developed product of superior quality onto the market on time.

At the same time we are trying to make our products from an ever-decreasing number of components or parts, by incorporating in each component a greater number of functions. An optical fibre cable can perform the same task as hundreds of conventional copper cables, a modern chip contains tens of thousands of circuits.

Finally, products which are produced in many different finished versions should be made so that wherever possiblethe specific features are added at the end of the production chain. The logistic advantages will be clear to you: stocks of a uniform semi-finished product will be smaller than the total stocks of specific finished products. In Operations Research literature, several recent publications are devoted to this subject.

Supply and Distribution

Ladies and gentlemen, there is very little point in making factories highly flexible if this flexibility does not extend over the entire logistic process - from first supplier to ultimate buyer. Our suppliers must also deliver Just-In-Time : smaller batches at more frequent intervals and with short delivery times. A closer relationship between Philips and its suppliers then becomes necessary; we are aiming very much for co-development and co-makership.

The same applies to management of the flow of materials between factories and the goods flow from production to the market. Large quantities of accurate and literally "up-to-date" data on our stock levels and goods flow movements are necessary in order for us to be able to plan or replan rapidly and on a sound basis. Philips invests a great deal of time and money in implementing information systems and planning systems in order to make various activities possible. To this end, Philips has set up its own world-wide data network called CODAN. Short lead times and smaller batches force us towards high resolution: Our ultimate goal is control of the goods flow with daily resolution. A lot of work has still to be done before this can be achieved. At the moment control of the goods flow often is carried out with monthly resolution. In Operations Research terms : we wish to progress from periodic review planning systems to continuous review systems.

Implications for Operations Research

Ladies and gentlemen - in achieving greater flexibility the main issue is the reduction of lead times. Markets, technology and other factors in our environment are changing so quickly that we can only survive if we react appropriately (and quickly) to these changes. Dealing with unexpected change has become one of the most critical aspects of management. Flexibility is a must for management. What are the implications of this situation for Operations Research?

The answer to this question will ultimately have to come from you and not from me, since I am less familiar than you are with your specialist field. I will gladly leave it to you to determine which models should be constructed and analysed. I presume these models will be a frequent subject of discussion at this conference, but may I be so bold as to propose some main ideas which may serve to guide your discussions.

1. As said before, I am afraid that most traditional Operations Research is focused on the properties and management of stable processes. And I know that without the assumptions about the long-term regularity of a process, it is difficult to formulate simple policies and test their impact on performance measures, much less to identify an optimal policy. I also know that it is difficult to model unexpected changes (changes that can be forecast are not unexpected), but it must be possible to find out for whichparameter settings of a system it is at least possible to react adequately to these unexpected changes. For example, how should production systems be designed, and how should they be controlled, in order to attain maximum flexibility? And what is the effect of more modularity in our product designs?

2. In close connection with the preceding remarks: never accept

the structure of a given situation, or the values of most of its parameters, as given facts. If you do so, your job will be reduced to optimizing (or simulating) <u>within</u> this structure. But <u>parameter values are not given values</u>: they are <u>ultimately the product of managerial action</u>. Let me give you an example. In the Western world we generally tend to accept a production changeover (set-up) time as a fixed constraint and use it to calculate the optimal lot size. This lot size, in turn, determines the average amount of work-in-process inventory. The Japanese, on the other hand, have concluded that work-in-process is bad because it represents "waste" (unused resources), because it lengthens lead times and reduces the speed of feedback about processing problems. Therefore they have set about reducing the production changeover times; often they were able to reduce them to one-tenth or even one-fiftieth of their original levels!

The conclusion is not difficult: instead of optimizing within a given structure, ask the responsible manager what is the desired performance and let the Operations Research expert indicate under what parameter settings this performance can be achieved.

3. I would like to see more attention paid to the phenomenon of learning. There has recently been a growing interest in examining the properties and the implications of the "learning curve". Unfortunately, we tend to accept the curve and its parameters once again as given and use them as the basis for calculating some optimal decision or policy. However, different factories "learn" differently: their learning curves neither follow smooth patterns over time nor seem to be governed by similar parameters. Again, management makes an immense difference. This implies that we should spend more time understanding the learning process and what causes it - and then <u>work on those causes</u> - instead of accepting the theoretical structure as fixed. The job of operations managers is not to achieve steady-state efficiency, but to grope their way toward a long-term dynamic optimality in an evolving competitive environment.[11]

I could continue, but I dare say the message is clear. Your models need to be as flexible as the environment demands our systems to be.

In Conclusion

Ladies and gentlemen, I have started to describe for you today a, perhaps somewhat extreme, picture of the factory of the future. Whether we shall ever see this factory or whether it should come, I cannot tell. The idea on which it is based - Flexible Manufacturing - is, however, something we cannot avoid. I have shown you, using history as an illustration, just what a tremendous change this means, and how far-reaching the effects of this are in all parts of our organisation. Finally, I have tried to indicate just what a pertinent contribution Operations Research can make to solving the many problems which still lie ahead of us.

I am especially pleased with the increased readiness of Operations Research to devote attention to complex industrial problems. Many sections of this conference are directly related to the problems I have just described. This willingness is also evident in other areas. The Eindhoven University of Technology has recently introduced a number of post-doctoral designer courses; in the course on "Designing logistic control systems" in particular, Operations Research plays an important role. Philips wholeheartedly supports these developments with material and manpower. In addition to this, I am happy to say that right here within Philips there is a group at work, the Centre for Quantitative Methods, which works mainly on the problems which arise within the company and gears its mathematical activities to these problems.

Ladies and Gentlemen - I hope and trust that this conference will stimulate and encourage you to approach the problems I have mentioned with fresh energy and enthusiasm. Here I realise that in some cases mathematical elegance will have to give way to the practical relevance of the models. I hope that bringing your mathematical models ever closer to the complexity of the real world will enhance your pleasure in working in this field.

Thank you.

References

[1] Factory of the Future, The Economist, May 30, 1987.

[2] Frederick W. Taylor, Frank and Lillian Gilbreth introduced what has become known as "Scientific Management". They perfected work-study techniques so that the worker's task could be standardized.

[3] T. Kumpe, Fabrieken met een toekomst (in Dutch), Inaugural Address, Katholieke Universiteit Brabant, February 1986.

[4] P.T. Bolwijn et al., Flexible Manufacturing, Elsevier, Amsterdam 1986.

[5,6,7] H.M. Wagner, Principles of Operations Research, 2nd. ed., Prentice-Hall 1975.

[8] Jay W. Forrester, Industrial Dynamics - a Major Breakthrough for Decision Makers, Harvard Business Review **36**, No. 4, pp. 37-66, July-August 1958.

[9] W.E. Deming, Quality, Productivity and Competitive Position, MIT-Press, Cambridge, MA, 1982.

[10] W.H.M. Zijm, Flexible Manufacturing Systems: background, examples and models, DGOR/NSOR Conference, Veldhoven 1987.

[11] R.H. Hayes and K.B. Clark, Explaining Observed Productivity Differentials Between Plants: Implications for Operations Research, Interfaces **15** : 6, November-December 1985.

BEWERTUNG UND MANAGEMENT FESTVERZINSLICHER WERTPAPIERE

Wolfgang Bühler, Dortmund

Zusammenfassung: In den letzten 15 Jahren wurde auf den Primär- und Sekundärmärkten für festverzinsliche Wertpapiere eine Vielzahl neuer Instrumente zur Begrenzung und Umverteilung des Zinsänderungsrisikos entwickelt. Die Mehrzahl dieser Innovationen lassen sich auf die einfachen Konstruktionsprinzipien

- Stripping und
- Replicating

zurückführen. Diese Konstruktionsprinzipien können auch zur Bewertung komplexer finanzieller Instrumente und zur Systematisierung von Strategien des Managements von Anleihenportefeuilles herangezogen werden.

Summary: The primary and secondary markets for interest-bearing instruments experienced in the recent 15 years a large number of innovations to limit and redistribute interest rate risk. Most of these innovations can be traced back to two simple principles:

- Stripping and
- Replicating.

These principles are also useful to value complex financial instruments and to systematize strategies of bond portfolio management.

I. FINANZ-CHEMIE

Vier Stichworte prägen die Entwicklung der vergangenen 15 Jahre auf den Märkten für festverzinsliche Titel: Zunehmende **Volatilität der Zinssätze, Liberalisierung, Globalisierung** und **Securitization**. Zur Begrenzung und zur Umverteilung des Zinsänderungsrisikos wurden auf den Primär- und Sekundärmärkten eine Vielzahl neuer Instrumente entwickelt. Zinsswaps, Caps und Zinswarrants stellen Beispiele für Innovationen auf den Primärmärkten dar. Mit Futures, Optionen auf Futures sowie Forward Rate Agreements auf festverzinsliche Instrumente sind die wichtigsten Innovationen auf den Sekundärmärkten benannt.

Verbunden mit der Einführung neuer Instrumente ist die Entstehung einer Begriffsvielfalt, die den Blick auf die elementaren Bausteine und

Operations Research Proceedings 1987

die Konstruktionsprinzipien von Finanzinnovationen außerordentlich erschwert. Dabei beruhen die Instrumente zur Kontrolle und Steuerung von Zinsänderungsrisiken nur auf drei **Basiselementen**:

- **Zero-Bonds,**
- **Terminkontrakte** und
- **Optionen**.

Aus theoretischer Sicht lassen sich zudem Optionen durch eine permanente Umschichtung von Portefeuilles auf die beiden anderen Bausteine zurückführen, so daß insgesamt nur zwei Bausteine benötigt werden.

Die Risikoprofile der drei Basiselemente sind für einen festen zukünftigen Zeitpunkt, den Fälligkeitstag T, in der nachstehenden Abbildung dargestellt. Dabei sind auf der Ordinate der Wert des Instrumentes am Fälligkeitstag und auf der Abszisse die ungewisse, diesen Wert beeinflussende Größe Z abgetragen. Im einfachsten Fall repräsentiert Z einen Zinssatz oder den Kurs eines festverzinslichen Titels am Fälligkeitstag T.

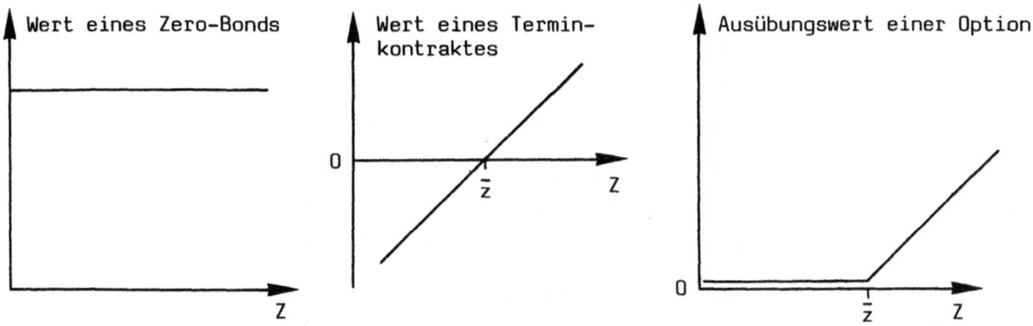

Abb. 1: Risikoprofile der Basiselemente

Das Risikoprofil des Zero-Bonds, eine kuponlose Anleihe, ist unabhangig von dem zukünftigen Zinsniveau. Er repräsentiert die risikolose Anlage- und Verschuldungsmöglichkeit. Der Wert des Terminkontraktes mit dem Terminkurs (Terminzinssatz) $\bar{z}$ wächst oder fällt streng monoton mit Z. Der "geknickte" Verlauf des Risikoprofils von Optionen beruht auf dem für Optionen typischen Recht, den Kontrakt nur dann zu erfüllen, wenn dies für den Kontraktinhaber vorteilhaft ist.

Wie in der Chemie, lassen sich nun zwei grundsätzliche Prinzipien zur Erzeugung neuer Risikoprofile unterscheiden:

- **Stripping** und
- **Replicating**.

Beim **Stripping** werden in Analogie zur chemischen **Analyse** Risikopositionen in zwei oder mehrere Teilpositionen **zerlegt**. Die umgekehrte Vorgehensweise liegt dem **Replicating** zugrunde. Wie bei der **Synthese** werden hier aus den elementaren Bausteinen durch Portefeuillebildung neue Risikopositionen erzeugt. Die nachstehenden Beispiele sollen diese beiden Konstruktionsprinzipien verdeutlichen.

II. STRIPPING

1. Kupon-Stripping

Die Zerlegung des Zahlungsstromes einer gesamtfälligen Anleihe in die einzelnen Kupon-Zahlungen und den Tilgungsbetrag wird als Kupon-Stripping bezeichnet. Dieses, in erster Linie von großen US-amerikanischen Investmenthäusern wie Merrill Lynch oder Salomon Brothers seit 1982 betriebene Geschäft der Emission von Zero-Bonds auf der Grundlage von Staatsanleihen (Treasury Bonds) besteht aus vier Schritten:

- Das Investmenthaus erwirbt Staatsanleihen in einem größeren Umfang.
- Diese Staatsanleihen werden bei einer Verwahrbank als Sicherheit hinterlegt. Diese Bank übernimmt auch die Funktion eines Treuhänders.
- Die Verwahrbank oder das Investmenthaus emittieren Zero-Bonds, deren Fälligkeit und Rückzahlungsbeträge mit den Kuponzahlungen und dem Tilgungsbetrag der Staatsanleihe übereinstimmen.
- Die Zinszahlungen und der Tilgungsbetrag werden von der Verwahrbank unmittelbar an die Inhaber von Zero-Bonds bei deren Fälligkeit weitergeleitet.

In der nachfolgenden Tabelle ist die formale Struktur des Kupon-Stripping aus Sicht des Investmenthauses unter Vernachlässigung der Verwahrbank dargestellt:

t_0	t_1	t_2	....	t_N	
$-K_0$	c	c		c+RZ	Staatsanleihe
$c \cdot B_1$	-c				Zero-Bond 1
$c \cdot B_2$		-c			Zero-Bond 2
⋮					⋮
$c \cdot B_N$				-c	Zero-Bond N
$RZ \cdot B_N$				-RZ	Zero-Bond N+1

Tab. 1: Kupon-Stripping

Für das Investmenthaus ist ein Kupon-Stripping dann vorteilhaft, wenn der Emissionserlös aus den N+1 Zero-Bonds mit den Kursen $B_1,...,B_N$ abzüglich der Emissionskosten größer als der Kaufpreis K_0 der Staatsanleihe ist. Die Rückzahlungsverpflichtungen sind durch die Kuponzahlungen c und den Tilgungsbetrg RZ gedeckt.

Als Motive für den Erwerb von Zero-Bonds können

- die Ausschaltung des Wiederanlagerisikos und
- die steuerliche Sonderbehandlung von Kursgewinnen in bestimmten Ländern (z.B. in Japan)

genannt werden.

2. Harmless Warrants

Warrants (Optionsscheine) auf Anleihen verbriefen das Recht, während einer bestimmten Bezugsfrist Anleihen eines festgelegten Typs zu einem vereinbarten Preis zu beziehen. Harmless Warrants werden zusammen mit einer durch den Emittenten **kündbaren** Anleihe begeben; sie besitzen die Besonderheit, daß in ihrer einfachsten Ausgestaltung durch die Ausübung des Warrants das Kündigungsrecht des Schuldners rückgängig gemacht wird.

Ein Beispiel hierzu findet sich in den Anleihebedingungen der 6 1/2 % DM-Anleihe von 1986/96 der Dai-Ichi Kangyo Bank NV, die zusammen mit einem Harmless Warrant emittiert wurde. Vor dem 1. Kündigungstermin am 1.4.91 führt die Ausübung des Warrants zu einem Tausch von kündbaren gegen unkündbare, sonst identische Anleihestücke.

Aus Sicht des Inhabers von Teilschuldverschreibungen kann die Emission einer kündbaren Anleihe mit einem Harmless Warrant als das Stripping der Risikoposition einer unkündbaren Anleihe in eine verkaufte Kaufoption (Stillhalterposition des Gläubigers beim Kündigungsrecht) und eine gekaufte Kaufoption (Harmless Warrant) aufgefaßt werden. Diese Form des Strippings ist in der nachstehenden Abbildung für den Fall genau eines Kündigungstermines dargestellt. Hierbei werden auf den Abszissen der Kurs der unkündbaren Anleihe am Kündigungstermin und auf den Ordinaten die Risikopositionen der unkündbaren, der kündbaren und der kündbaren Anleihe zusammen mit dem Harmless Warrant aufgetragen. KK bezeichnet den Kurs, zu dem die Anleihe vorzeitig gekündigt werden kann.

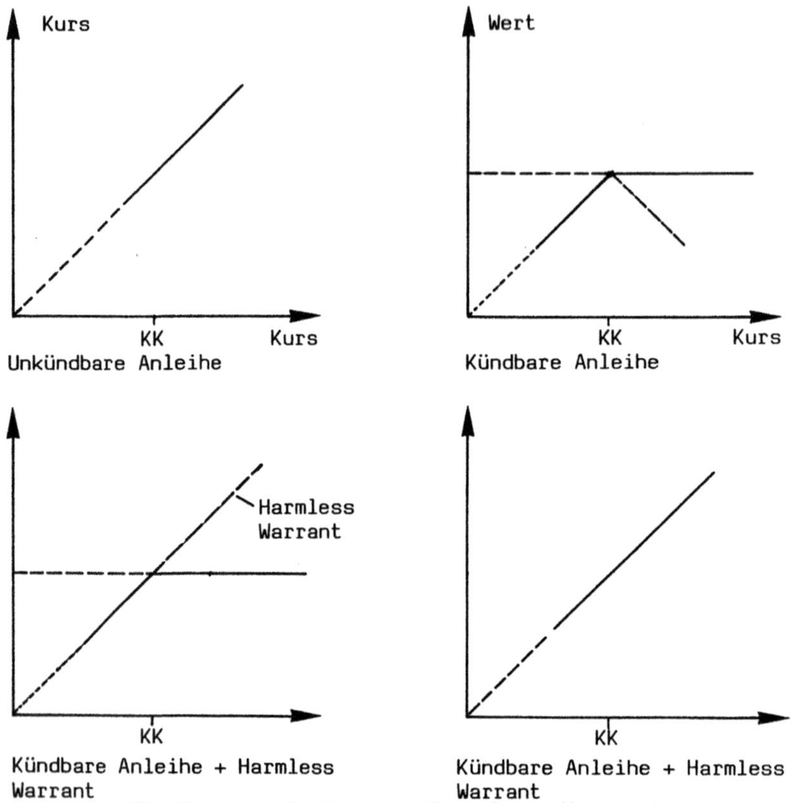

Abb. 2: Kündbare Anleihe und Harmless Warrant

Für den Schuldner besitzt die Emission einer kündbaren Anleihe dann einen Vorteil, wenn der Abschlag beim Emissionskurs der kündbaren Anleihe aufgrund des Kündigungsrechtes geringer als der Emissionserlös für den Harmless Warrant ausfällt.

3. Bull- und Bear Anleihen

Anleihen, bei denen der Kupon und/oder der Rückzahlungsbetrag an einen Index gekoppelt sind, besitzen eine vergleichsweise lange Historie. Als Beispiele können die Bindung der Zahlungsreihe an den Preis für Gold, Silber, Öl, Roggen oder einen Wechselkurs genannt werden. Bull- und Bear Anleihen zeichen sich gegenüber herkömmlichen Indexanleihen durch folgende Besonderheiten aus:

- Die gesamte Anleihe wird in zwei gleich große Tranchen zerlegt.
- Die Indizierung der beiden Tranchen erfolgt in der Weise, daß für den Emittenten eine risikolose Position entsteht.

Eine Bull- und Bear Anleihe läßt sich somit auch als die Zerlegung einer risikolosen Position in zwei risikobehaftete Positionen auffassen. Das Prinzip dieser speziellen Form des Strippings ist in Abbildung 3 dargestellt:

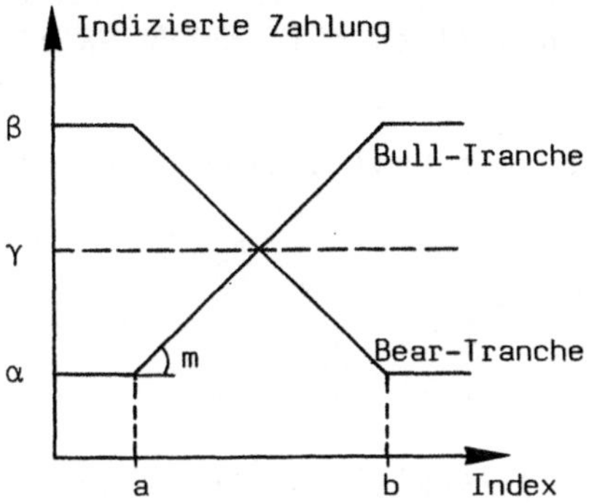

Abb. 3: Bull- und Bear Anleihe

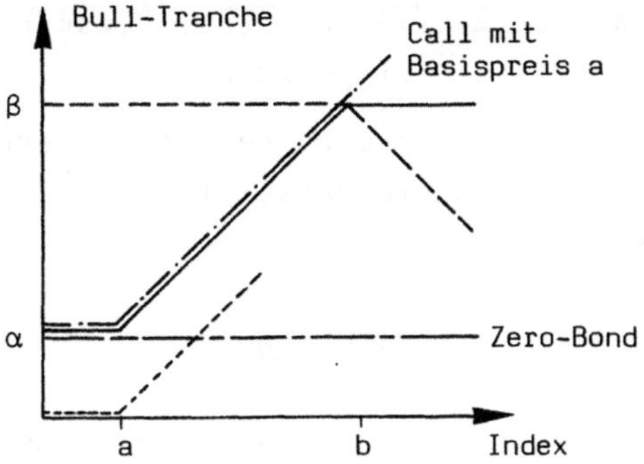

Abb. 4: Stripping der Bull-Tranche

Bei der Bull-Tranche wächst die indizierte Zahlung in dem Intervall [a,b] linear mit dem Index, bei der Bear-Tranche nimmt sie mit derselben betragsmäßigen Steigung ab. Die Summe der Zahlungen aus beiden Tranchen ist unabhängig vom Index. Die bekannteste Bull- und Bear Anleihe am DM-Euromarkt wurde 1986 von der Deutschen Bank emittiert. Ihr Rückzahlungsbetrag im Jahr 1991 ist an die prozentuale Änderung des FAZ-Aktienindex' gekoppelt. Die Risikopositionen dieser Anleihe lassen sich mit den Bezeichnungen in Abbildung 3 durch die Parameterwerte $a=0$; $b=230{,}85$; $m=0{,}94$; $\alpha=0$; $\beta=217$; $\gamma=108{,}5$ charakterisieren.

Aus Abbildung 4 wird deutlich, daß die Bull-Tranche (analog die Bear-Tranche) sich weiter in einen Zero-Bond mit Rückzahlungsbetrag α, eine gekaufte Kaufoption auf den Index mit Basispreis a und eine verkaufte Kaufoption mit Basispreis b zerlegen und damit auf die in Abschnitt I dargestellten Bausteine zurückführen läßt.

III. REPLICATING

Die Erzeugung einer gewünschten zukünftigen Risikoposition durch die Bildung eines Portefeuilles aus verfügbaren Bausteinen tritt in einer Vielzahl von Variationen auf:

- Duplizieren einer am Markt verfügbaren Risikoposition zur Erzielung eines risikofreien Arbitragegewinns.
- Duplizieren der Risikoposition einer Innovation zur Ermittlung des theoretischen heutigen Wertes des neuen Instrumentes (Abschnitt IV).

- **Hedging** durch Erzeugung einer Gegenposition zu einer vorhandenen Risikoposition.
- Approximative Erzeugung der gewünschten Risikoposition in **genau einem** zukünftigen Zeitpunkt.
- Approximative Erzeugung gewünschter Risikopositionen in **mehreren** zukünftigen Zeitpunkten.

Einige Beispiele zu den verschiedenen Formen des Replicating sind nachstehend zusammengestellt.

1. Cash Flow Matching

Das **Hedging** zukünftiger, bekannter Auszahlungsverpflichtungen gegenüber dem Zinsänderungsrisiko mit Hilfe eines kursminimalen Portefeuilles aus Anleihen wird als Cash Flow Matching bezeichnet. Stehen n Anleihen mit den Kupon- und Tilgungszahlungen

$$b_{i1},\ b_{i2},\ldots,b_{iT} \qquad (i=1,\ldots,n)$$

in den Zeitpunkten t=1,...,T und den aktuellen Kursen

$$K_1,\ K_2,\ldots,K_n$$

zur Verfügung, dann läßt sich das Problem der kursminimalen Abdeckung der zukünftigen Zahlungsverpflichtungen

$$a_1,\ a_2,\ldots,a_T$$

durch das folgende lineare Programm beschreiben:

$$\left.\begin{array}{ll} \min\ K_1x_1 + K_2x_2 + \ldots + K_nx_n & \\ b_{1t}x_1 + b_{2t}x_2 + \ldots + b_{nt}x_n = a_t & (t=1,\ldots,T) \\ x_i \geq 0 \qquad i=1,\ldots,n & \end{array}\right\} \quad (1)$$

x_i (i=1,...,n) bezeichnet hierbei die Anzahl der Anleihestücke des Typs i. Besitzt (1) eine optimale Lösung, dann sind die Auszahlungen a_t gedeckt, sofern diese

- hinsichtlich des zeitlichen Anfalls und ihrer Höhe sicher und sofern
- die Anleihen im Portefeuille bonitätsrisikofrei sind und nicht durch den Emittent gekündigt werden können.

Reicht das Spektrum der verfügbaren Anleihen nicht für eine zulässige Lösung von (1) aus, dann müssen die Gleichheitszeichen durch "≥"-Zeichen ersetzt und zudem einperiodige Reinvestitionsmöglichkeiten mit einem "sicheren" Reinvestitionszinssatz eingeführt werden.[1] Eine zweite, allerdings nicht zu einer sicheren Abdeckung der Zahlungsverpflichtungen führende Möglichkeit bildet die auf der Duration basierende "Multiple Liability Immunization."[2]

2. Zero-Bonds

Die synthetische Erzeugung eines Zero-Bonds aus Kupon-Anleihen ist für

- die theoretisch exakte Berechnung von Zinsstrukturkurven und für
- die Absicherung einer einzelnen zukünftigen Auszahlungsverpflichtung (Sonderfall des Cash Flow Matching Problems) von Bedeutung.

Soll ein Zero-Bond mit einer Laufzeit von T Jahren generiert werden, dann besitzt diese Aufgabe eine exakte Lösung, falls zwei in T gesamtfällige Anleihen mit verschiedenen Kupons existieren. Mit Hilfe des folgenden Beispiels kann diese Aussage verdeutlicht werden:[3]

Beispiel 1:

Zur Berechnung der Zinsstrukturkurve wird ein Zero-Bond mit 8-jähriger Laufzeit benötigt. Am Markt werden zwei 8-jährige Anleihen mit einem Kupon von 5 % (Anleihe A) und 10 % (Anleihe B) notiert. Das Portefeuille mit zwei Teilschuldverschreibungen der Anleihe A und einer leerverkauften Teilschuldverschreibung der Anleihe B erzeugt den gewünschten Zero-Bond:

Anleihe	Kurs	Zins- und Tilgungszahlung				Interner Zinsfuß
A	83,95	5	5	....	105	7,77 %
B	113,90	10	10	....	110	7,62 %
2*A-B	54,00	0	0	....	100	8,01 %

Tab. 2: Synthetische Erzeugung eines Zero-Bonds

Am deutschen Rentenmarkt stehen in der Regel nicht genügend geeignete Anleihen zur exakten Erzeugung von Zero-Bonds der gewünschten Restlaufzeit zur Verfügung. Durch Portefeuillebildung kann deshalb nur näherungsweise die Risikoposition eines Zero-Bonds erzeugt werden. Als Hilfsmittel hierzu dient die **Duration** einer Kupon-Anleihe. Diese kann mit Hilfe des internen Zinsfußes y sowie der Zins- und Tilgungszahlungen b_t der Anleihe definiert werden als[4]

1) Fußnoten sind am Ende der Arbeit zusammengestellt.

$$D(y) = [\sum_{t=1}^{T} t \cdot b_t(1+y)^{-t}] / \sum_{t=1}^{T} b_t(1+y)^{-t}. \qquad (2)$$

Die Duration eines Portefeuilles aus zwei Anleihen A, B mit den wertmäßigen Anteilen x und (1-x) am Portefeuille ergibt sich als Konvexkombination der beiden Anleiheduration D_A und D_B:

$$D_x(y) = x \cdot D_A(y) + (1-x) \cdot D_B(y) \qquad (3)$$

Für die näherungsweise synthetische Erzeugung eines Zero-Bonds mit Fälligkeit in T Jahren sind nun zwei Anleihen A,B mit $D_A < T$ und $D_B > T$ im dem Umfang in ein Portefeuille aufzunehmen, daß dessen Duration $D_x(y)$ mit der Restlaufzeit T des Zero-Bonds übereinstimmt. Bei einem sich ändernden Zinsniveau wird dann näherungsweise die Zunahme (Abnahme) der Zinserträge auf reinvestierte Zins- und Tilgungszahlungen durch die Abnahme (Zunahme) des Kurswertes des Portefeuilles im Zeitpunkt T ausgeglichen.

Beispiel 2:

Es liege eine flache Zinsstrukturkurve mit 8 % vor. Eine Anlage von 300.000 DM in einen (nicht existierenden) Zero-Bond mit zweijähriger Restlaufzeit würde risikolos auf 349.920 DM anwachsen. Zur Verfügung stehen die in Tabelle 3 durch ihre Zahlungsreihen charakterisierte 8 %-Tilgungsanleihe (A) und die 8 % gesamtfällige Anleihe (B). Tabelle 4 zeigt für den Zeitpunkt T=2 die Konsequenzen einer Änderung des Zinsniveaus auf die Ertragskomponenten der beiden Anleihen und des Portefeuilles mit der Duration D_x=2.

	t=1	t=2	t=3	Duration
A	124.000	116.000	108.000	1,902
B	24.000	24.000	324.000	2,783
0,89A+0,11B	112.876	105.766	132.027	2,000

Tab. 3: Anlage der Mittel in Anleihen A, B oder ein Portefeuille aus A und B

Bemerkenswert ist in Tabelle 4, daß der Wert des Portefeuilles in T unabhängig von der Zinsänderung nicht unterhalb des konstanten Wertes des Zero-Bonds liegt. Diese Dominanz des Zero-Bonds durch das betrachtete Portefeuille steht im Widerspruch zur Arbitragefreiheit von hochorganisierten Finanzmärkten.[5)]

	6 %		8 %		10 %	
	Änderung Wiederanl.	Kurs	Zins	Kurs	Änderung Wiederanl.	Kurs
A	-2.480	1.887	249.920	100.000	2.480	-1.818
B	- 480	5.660	49.920	300.000	480	-5.455
0,89 A +0,11 B	349.969		349.920		349.955	

Tab. 4: Näherungsweise synthetische Erzeugung eines Zero-Bonds

3. Optionen

Optionen auf Anleihen lassen sich unter speziellen Annahmen über die Kursentwicklung durch ein Portefeuille aus einer geeigneten Anzahl von Anleihen zusammen mit einer vom Zinsänderungsrisiko nicht betroffenen Verschuldungsmaßnahme synthetisch erzeugen. Dieses für die Bewertung von Aktienoptionen von Black und Scholes[6)] entwickelte Prinzip läßt sich am eindruckvollsten mit Hilfe einer auf Sharpe zurückgehenden Modifikation für vollständige Märkte demonstrieren:

Beispiel 3:

Eine 10 %-Anleihe notiert sechs Monate vor dem nächsten Zinstermin zu 106,75. In sechs Monaten soll ihr Kurs entweder 109,00 oder 100,00 betragen können. Eine Europäische Kaufoption mit Ausübungstag in 6 Monaten und einem Basispreis in Höhe von 106 besitzt dann die möglichen Ausübungswerte 3,00 oder 0. Aus Tabelle 5 wird deutlich, daß das Portefeuille aus einer Anleihe und einer Termingeldaufnahme zu 5 % im Umfang von 107,32 in sechs Monaten dieselbe Risikoposition erzeugt wie drei Kaufoptionen:

	Aktuelles Datum	Sechs Monate später Anleihekurs 100	Anleihekurs 109
Anleihe	-106,75	100,00	109,00
Stückzinsen	- 5,00	10,00	10,00
Termingeld	+107,32	-110,00	-110,00
Portefeuille	- 4,43	0	9,00
3 Kaufoptionen	-3*C	0	9,00

Tab. 5: Synthetische Erzeugung von Kaufoptionen

IV. BEWERTUNG DURCH REPLICATING

Zwei Portefeuilles, die in der Zukunft zu identischen Risikopositionen führen, müssen auf arbitragefreien Märkten in der Gegenwart denselben Wert besitzen. Überträgt man dieses Prinzip auf das Beispiel aus Tabelle 5, so folgt, daß der Wert C der dort betrachteten Kaufoption zum aktuellen Datum C = 4,43/3 = 1,48 sein muß. Auf dieser Überlegung basieren die auf Black und Scholes zurückgehenden Ansätze einer zeitstetigen, arbitragefreien Bewertung von Optionen. Äquivalent hierzu ist die Bedingung, daß das Portefeuille aus Anleihe, Termingeldaufnahme und der Stillhalterposition in drei Kaufoptionen mit den zukünftigen Nettorückflüssen der Höhe 0 auch einen Gegenwartswert von 0 haben muß.

Im folgenden wird durch Replicating eine Bewertungsbeziehung für Kaufoptionen auf Anleihen abgeleitet.[7)] Ausgangspunkt ist hierbei die Modellierung des Anleihenkurses als Diffusionsprozess mit Hilfe einer stochastischen Differentialgleichung. Gegenüber der in der Optionspreistheorie typischen Modellierung von Aktienkursverläufen sind dabei folgende Besonderheiten zu beachten:

- Der Kurs einer Anleihe am Ende ihrer Laufzeit ist bei Ausschluß des Bonitätsrisikos mit Wahrscheinlichkeit 1 gleich ihrem Rückzahlungskurs.
- Anleihenkurse sind beim Ausschluß negativer Renditen nach oben durch die Summe aus dem Rückzahlungskurs und den zukünftigen Stückzinsen beschränkt.
- Die Kursvolatilität nimmt im Zeitablauf ab.

Ferner ist zu berücksichtigen, daß

- der Zinssatz für eine risikolose Anlage- oder Verschuldungsmöglichkeit zum Ausschluß negativer Terminzinssätze nicht unabhängig von der Anleihenrendite sein kann und daß
- beim Kauf und Verkauf einer Anleihe Stückzinsen anfallen.

Charakterisiert man die Kursentwicklung einer Anleihe durch eine stochastische Differentialgleichung bzgl. der Ableitung des Standard-Wienerprozesses $\tilde{Z}(t)$

$$d\tilde{B}_t = \mu_B(B_t,t)dt + \sigma_B(B_t,t)d\tilde{Z}_t, \qquad (4)$$

so müssen die oben dargestellten Besonderheiten in der Drift $\mu_B(B_t,t)$

und der Momentanstandardabweichung $\sigma_B(B_t,t)$ erfaßt werden. Die Drift beschreibt die Kursentwicklung bei konstanten Zinsverhältnissen. Für einen Rückzahlungskurs von 100 und einer Fälligkeit der Anleihe im Zeitpunkt T_B wird sie als

$$\mu_B(B_t,t) = [-B_t \cdot \ln B_t] \;/\; (T_B - t) \tag{5}$$

definiert. Die Momentanstandardabweichung wird als Produkt

$$\sigma_B(B_t,t) = f(B_t,t) \cdot \text{Duration}\,(B_t,t) \tag{6}$$

dargestellt. Hierbei ist die Funktion $f(B_t,t)$ für $B_t=0$ und für den maximalen Anleihenkurs $B_{max}(t)$ im Zeitpunkt t gleich 0. Auf diese Weise wird über (5) und (6) erreicht, daß B_t nicht negativ und nicht größer als $B_{max}(t)$ werden kann. Der zweite Produktterm, die von den gesamten Ausstattungsmerkmalen der Anleihe abhängige Duration, führt zu einer streng monotonen Abnahme von $\sigma_B(B_t,t)$ in t.

Hinsichtlich des unbekannten Wertes $C(B_t,t)$ einer Kaufoption mit dem Basispreis E und dem letzten Ausübungstag T wird unterstellt, daß dieser zweimal stetig nach B_t und einmal stetig nach t differenzierbar ist. Die Wertentwicklung von $C(B_t,t)$ kann dann ebenfalls durch eine stochastische Differentialgleichung

$$dC_t = \mu_C dt + \sigma_C d\tilde{Z}_t \tag{7}$$

beschrieben werden. μ_C und σ_C lassen sich mit Hilfe des Itôschen Lemmas[8] durch μ_B, σ_B und durch die partiellen Ableitungen $\partial C/\partial B$, $\partial^2 C/\partial B^2$ und $\partial C/\partial t$ darstellen.

In jedem Zeitpunkt t wird aus der Anleihe, der Kaufoption und einer Kreditaufnahme mit der risikolosen Momentanverzinsung r ein Portefeuille (X_B, X_C, X_K) mit folgenden lokalen Eigenschaften zusammengestellt:

Kein Kapitaleinsatz:	$(B_t+c_t)X_B + C_tX_C + X_K$	$= 0$
Kein Risiko:	$\sigma_B X_B + \sigma_C X_C$	$= 0$
Kein Ertrag:	$(\mu_B + c \cdot dt)X_B + \mu_C X_C + rX_k$	$= 0$

Die beim Erwerb der Anleihe im Zeitpunkt t zu entrichtenden Stückzinsen sind dabei mit c_t bezeichnet und der Kupon mit c. Aus diesen drei Gleichungen ergibt sich die folgende Beziehung zwischen den relativen Risikoprämien der Anleihe und der Kaufoption:

$$[\mu_B + cdt - r\cdot(B_t + c_t)] / \sigma_B = [\mu_c - r\cdot C_t] / \sigma_c \quad (8)$$

Werden μ_c und σ_c in (8) mit Hilfe des Lemmas von Itô durch μ_B, σ_B und die partiellen Ableitungen von C nach B und t ersetzt, ergibt sich die folgende partielle Differentialgleichung des parabolischen Typs für C:

$$1/2\sigma_B^2 \cdot \frac{\partial^2 C}{\partial B^2} + [r\cdot(B+c_t) - c] \cdot \frac{\partial C}{\partial B} + \frac{\partial C}{\partial t} - r\cdot C = 0 \quad (9)$$

Endbedingung: $C(B,T) = \max(0,B-E)$

Randbedingungen: $C(0,t) = 0$

$C(B_{max},t) = B_{max}(T)-E$ für eine Europäische Kaufoption

$C(B_{max},t) = B_{max}(t)-E$ für eine Amerikanische Kaufoption.

Ferner ist für eine Amerikanische Kaufoption die Ausübungsbedingung $C(B_t,t) \geqq \max(B_t-E,0)$ zu berücksichtigen.

Als Beispiel für die Bewertung von Optionen durch **Replicating** sind in Abbildung 5 die Werte Amerikanischer und Europäischer Kaufoptionen als Funktion der Optionsfrist T wiedergegeben. Die Optionen berechtigen zum Bezug einer Anleihe mit dem aktuellen Kurs B_0=100, einem Kupon von 6,5 % und einer Restlaufzeit von 10 Jahren. Der Basispreis beträgt 100 und der aktuelle Zinssatz für eine Kreditaufnahme 5 %. Die Kursvolatilität σ der 10-jährigen Anleihe zu Beginn der Optionsfrist wurde 10 % gesetzt.

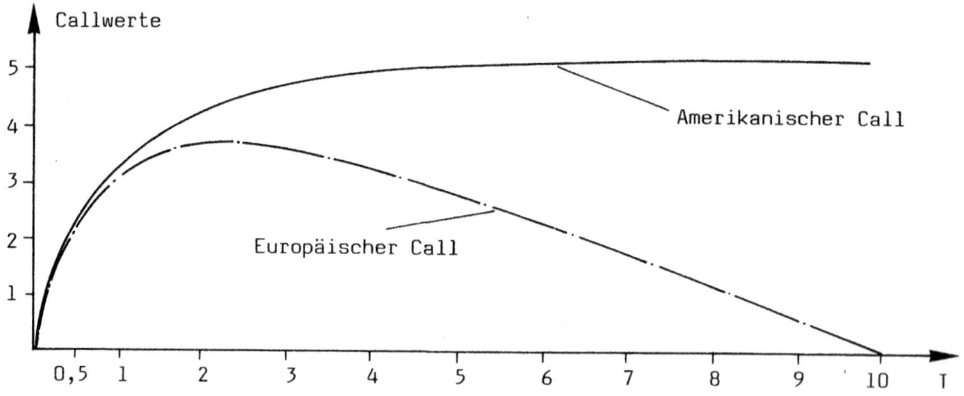

Abb. 5: Werte Amerikanischer und Europäischer Kaufoptionen in Abhängigkeit von der Optionsfrist

Im Gegensatz zu dem Werteverlauf von Kaufoptionen auf **Aktien** nimmt der Wert Europäischer Kaufoptionen für Optionsfristen von mehr als ca. 25 % der Restlaufzeit der Anleihe aufgrund der abnehmenden Kursvolatilität ebenfalls ab. Aus demselben Grund erzielen Amerikanische Kaufoptionen für Optionsfristen von mehr als 40 % der Anleihenlaufzeit keinen bemerkenswerten Wertzuwachs.

V. PORTEFEUILLEMANAGEMENT UND REPLICATING

Aus der Vielzahl denkbarer und praktizierter Portefeuillestrategien zur Erzielung einer gewünschten zukünftigen Risikoposition durch Replicating werden im folgenden zwei Strategien,

- die Immunisierung eines zukünftigen Endvermögens gegen das Zinsänderungsrisiko und
- die Portefeuille Versicherung

zusammen mit empirischen Simulationsergebnissen für den deutschen Rentenmarkt dargestellt.

1. Immunisierung

Strategien zur Immunisierung eines in festverzinslichen Wertpapieren angelegten Vermögens werden in vielfältigen operativen Varianten realisiert. Diese lassen sich als praktische Umsetzungen des in Abschnitt III.2 dargestellten Prinzips der synthetischen Erzeugung der Risikoposition eines Zero-Bonds in einem zukünftigen Zeitpunkt auffassen. Grundsätzlich wird dabei in der Weise vorgegangen, daß zu Beginn des Planungszeitraumes ein Portefeuille aus Kuponanleihen zusammengestellt wird, dessen Duration mit dem zeitlichen Abstand T zum Planungshorizont übereinstimmt. Dieses Portefeuille wird dann während des Planungszeitraumes laufend derart umgeschichtet, daß die Bedingung

Duration = zeitlicher Abstand zum Planungshorizont (10)

erhalten bleibt.

Bei der Realisierung dieser Portefeuillestrategie bestehen erhebliche Freiheitsgrade, da

- die Bedingung (10) die Zusammensetzung des Portefeuilles nicht eindeutig festlegt und

- die Umschichtungsauslöser unterschiedlich gewählt werden können.

Darüber hinaus stellt sich die Frage, ob unter den praktischen Bedingungen des Portefeuillemanagements eine Ausschaltung des Zinsänderungsrisikos möglich ist. Als Ursachen für eine mögliche Abweichung vom theoretisch erzielbaren Endvermögen kommen in Frage

- Zinsänderungen, die mit der verwendeten Durationformel nicht verträglich sind,
- Transaktionskosten sowie
- Zuflüsse zum und Abflüsse aus dem Portefeuille.

Fisher und Weil berichten in der ersten empirischen Untersuchung zu diesem Problemkreis von einer deutlichen Reduktion des Zinsänderungsrisikos durch Immunisierungsstrategien.[9)] Diese Ergebnisse wurden von anderen Autoren bestätigt,[10)] von Gultkein/Rogalski[11)] und insbesondere von Ingersoll[12)] jedoch relativiert.

In einer noch nicht vollständig abgeschlossenen empirischen Studie des Verfassers wird diese Problematik für den deutschen Rentenmarkt wiederaufgegriffen. Diese Studie ist folgendermaßen aufgebaut:

- **Untersuchte Portefeuillestrategien**
  - Maturity-Strategie: Alle Mittel werden in diejenige Anleihe angelegt, deren Fälligkeitszeitpunkt am nächsten am Planungshorizont liegt.
  - Barbell-Strategie: Das Portefeuille wird aus den beiden Anleihen mit der größten und der kleinsten Duration gebildet.
  - Bullet-Strategie: Das Portefeuille wird aus den beiden Anleihen zusammengestellt, deren Duration D_A, D_B am wenigsten vom Planungshorizont T abweichen und für die $D_A < T$, $D_B > T$ gilt.
  - Ladder-Strategie: Es wird ein Portefeuille aus 10 Anleihen in der Weise gebildet, daß die Restlaufzeit der i-ten Anleihe mindestens i und höchstens i+1 Jahre beträgt (0=1,...,9).
- **Planungszeiträume:** 36, 48, 60 und 72 Monate.
- **Untersuchungszeitraum:** 1972 - 1985.
- **Anleihengesamtheit:** Alle gesamtfälligen Anleihen von Bund, Bahn und Post.
- **Umschichtungshäufigkeit:** Monatlich und halbjährlich.
- **Transaktionskosten:** 0 % und 0,35 %.

Die Ergebnisse sind in den Abbildungen 6 und 7 auf der folgenden Seite zusammengefaßt. Sie beruhen für einen Planungszeitraum von k Jahre auf (14-k)*12 im Monatsabstand durchgeführten Einzelsimulationen. Die Rendite eines Zero-Bonds wurde aus den mit Hilfe des kubischen Spline-Verfahrens von McCulloch[13)] gewonnenen Zinsstrukturkurven ermittelt und zur Berechnung des "risikolosen" Vergleichsvermögens verwandt. Auf der Abszisse ist das Verhältnis von dem durchschnittlichen mit einer Immunisierungsstrategie erzielten Endvermögen und dem Vergleichsvermögen, auf der Ordinate die Semistandardabweichung der prozentualen Abweichung vom Vergleichsvermögen abgetragen. Die neben den einzelnen Punkten eingetragenen Zahlen i/j bezeichnen die Länge des Planungszeitraumes i in Monaten und den zeitlichen Abstand j zwischen zwei Umschichtungsterminen.

Abbildung 6 zeigt, daß ohne Transaktionskosten die untersuchten Strategien hinsichtlich des Zieles, ein Endvermögen in Höhe des Vergleichsvermögens zu sichern, in folgende Reihenfolge gebracht werden können:

$$\textbf{Bullet} \succ \textbf{Maturity} \succ \textbf{Ladder} \succ \textbf{Barbell} \qquad (11)$$

Unter Berücksichtigung von Transaktionskosten ändert sich die Reihenfolge (11) erheblich:

$$\textbf{Maturity} \succ \textbf{Ladder} \succ \textbf{Bullet} \succ \textbf{Barbell} \qquad (12)$$

Bemerkenswert ist hier insbesondere, daß die einfache Maturity-Strategie, deren Duration stets kleiner als der zeitliche Abstand zum Planungshorizont ist, am besten abschneidet. Zudem führen, insbesondere bei längeren Planungszeiträumen, Bullet- und Barbell-Strategien zu erheblichen Abweichungen vom Vergleichsvermögen. Strategien mit durchschnittlichen Abweichungen von mehr als 3 % sind in Abbildung 7 nicht mehr eingetragen.

2. Portefeuille Versicherung

Existiert eine Europäische Verkaufsoption (Put) mit Verfallstag T und Basispreis E auf eine Anleihe, so kann der Kurswert einer Anlage von Mitteln in diese Anleihe nach unten abgesichert werden, indem zusätzlich dieser Put gekauft wird. Wie in Abbildung 8 dargestellt, besitzt das Portefeuille aus Anleihe und Put bei Wahrung der Kurssteigerungschancen einen Mindestkurswert in Höhe des Basispreises. In analoger Weise kann bei der Verfügbarkeit eines geeigneten Puts der Kurswert

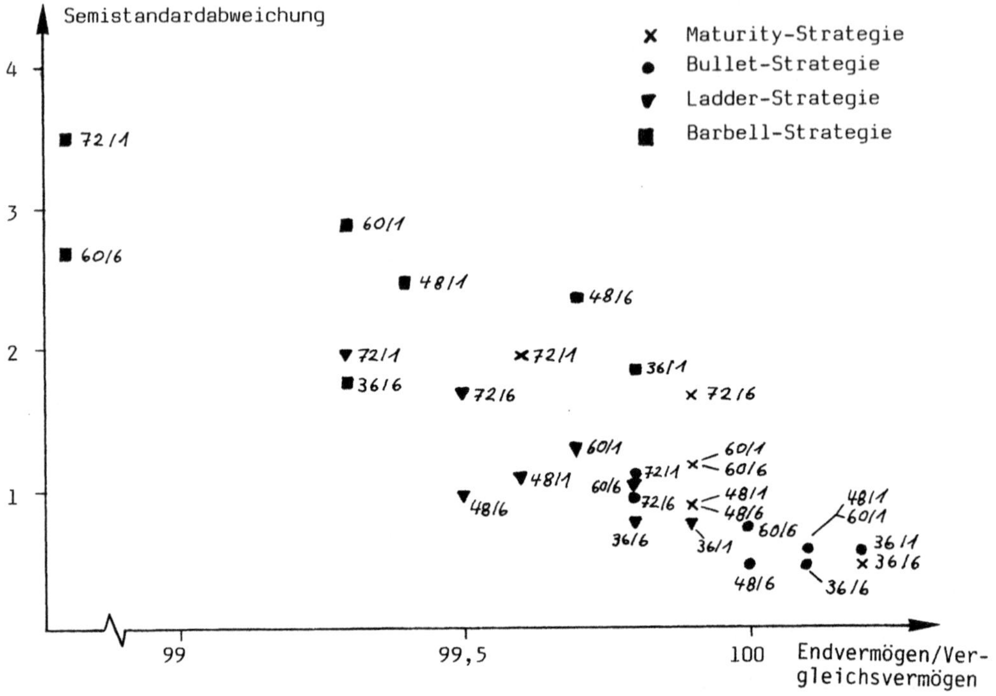

Abb. 6: Ergebnisse der Immunisierungsstrategie ohne Transaktionskosten

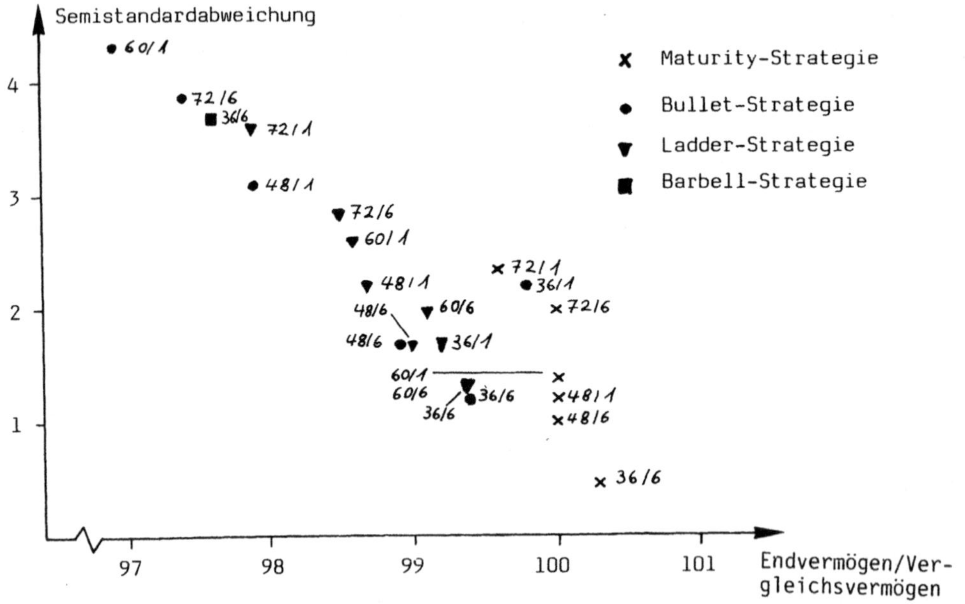

Abb. 7: Ergebnisse der Immunisierungsstrategien mit Transaktionskosten

eines **Portefeuilles** aus Anleihen abgesichert werden.

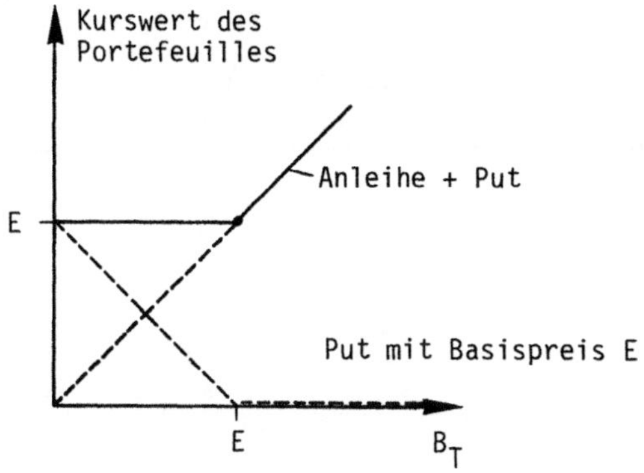

Abb. 8: Portefeuille Versicherung

Die Prämie für eine Portefeuille Versicherung der beschriebenen Art besteht aus der Anzahl der erworbenen Puts multipliziert mit dem Putpreis. Nur der verbleibende Teil des verfügbaren Budgets - verglichen mit dem gesamten Budget bei einem Verzicht auf eine Absicherung - kann im zinstragenden Titel angelegt werden. Der abgesicherte Betrag in DM ist deshalb (ohne Berücksichtigung der Zinszahlungen während der Optionsfrist) stets kleiner als das anfängliche Budget.

Für die praktische Umsetzung der Portefeuille Versicherung ist von Bedeutung, daß in der Regel Europäische Puts mit der gewünschten Optionsfrist nicht gehandelt werden. Es liegt deshalb nahe, die in den Abschnitten III.3 und IV dargestellten Zusammenhänge zwischen Anleihen (Portefeuilles aus Anleihen), Termingeldern und Optionen auszunutzen und den benötigten Put synthetisch durch Replicating zu erzeugen. Hierzu ist in Analogie zu den dort für Calls abgeleiteten Beziehungen in jedem Zeitpunkt ein Portefeuille aus **leerverkauften** Anleihen und einer risikolosen Mittelanlage MA zu bilden und in Abhängigkeit von der Kursentwicklung derart umzuschichten, daß am Planungshorizont das Auszahlungsdiagramm eines Puts entsteht.[14)]

Die Anzahl der zur synthetischen Erzeugung **eines** Puts leer zu verkaufenden Anleihen ist gleich dem

Put-Delta,

einer für die Portefeuille Versicherung zentralen Größe.[15)] Zur Erläuterung dieser Größe ist in Abbildung 9 der Putwert als Funktion des **aktuellen** Kurswertes einer Anleihe (eines Portefeuilles aus Anleihen) dargestellt. Der Betrag der Tangentensteigung in einem Punkt $(B_t, P(B_t))$

ist gleich dem Put-Delta.

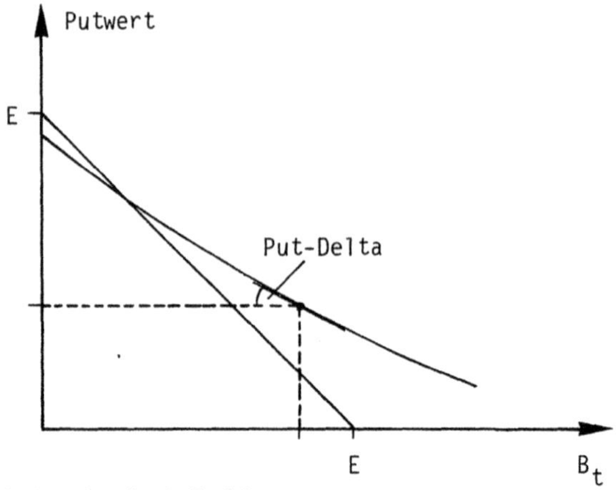

Abb. 9: Put-Delta

Das Put-Delta δ_P besitzt folgende Eigenschaften:

- $0 \leq \delta_P \leq 1$.
- δ_P fällt streng monoton in B_t.
- Am Verfallstag der Put-Option ist $\delta_P \in \{0;1\}$ für $B_t \neq E$.

Für die Berechnung des Put-Deltas ist ein Optionsbewertungsmodell erforderlich. In der weiter unten dargestellten empirischen Untersuchung wurde das Bewertungsmodell des Abschnittes IV eingesetzt.

Bei der praktischen Durchführung oder der Simulation einer Portefeuille Versicherung sind nun die folgenden Schritte durchzuführen:

- Festlegung der risikobehafteten Anlagemöglichkeiten und des Planungshorizontes T.
- Bestimmung des abzusichernden Kursniveaus im Zeitpunkt T.
- Aufteilung des verfügbaren Budgets auf x_0 Einheiten der riskanten Anlage und x_0 synthetische Puts mit dem Wert P_0^S. Die riskante Anlage erfordert pro Einheit Mittel im Umfang des aktuellen Kurses B_0 zuzüglich der Stückzinsen c_0. Der synthetische Put P_0^S besteht aus einer risikolosen Anlage des noch zu bestimmenden Umfangs $\delta_p^0 \cdot MA_0$ und aus δ_P^0 leerverkauften riskanten Instrumenten. Das Portefeuille besitzt damit folgende Struktur:

$$x_0 \cdot (B_0 + c_0; P_0^S) = x_0 \cdot (B_0 + c_0; \delta_P^0(B_0) \cdot (MA_0 - (B_0 + c_0))$$

$$= x_0 \cdot ((1 - \delta_P^0) \cdot (B_0 + c_0); \delta_P^0 MA_0). \qquad (13)$$

MA_o ist derart festzulegen, daß $x_o \cdot ((1-\delta_p^o)(B_o+c_o)+\delta_p^o MA)$ durch das verfügbare Budget gedeckt wird.

- Definition eines Umschichtungsauslösers.
- In jedem Umschichtungszeitpunkt t wird unter Berücksichtigung des aktuellen Kurses B_t, der Stückzinsen c_t und des aktuellen Put-Deltas δ_p^t das Portefeuille gemäß der Beziehung (13) an das aktuelle Kursniveau angepaßt. Steigende (fallende) Kurse führen hierbei zu steigenden (fallenden) Anteilen des risikobehafteten Instrumentes, am Portefeuillewert.

In der nachfolgenden Tabelle sind exemplarisch einige Ergebnisse einer Portefeuille Versicherung am deutschen Rentenmarkt zusammengestellt. Da der Absicherungscharakter der synthetischen Option nur bei steigenden Zinsen wirksam werden kann, werden Planungszeiträume (1. Spalte) ausgewählt, in denen die Umlaufsrendite nahezu ausschließlich gestiegen ist (2. Spalte). Abgesichert wurde die Kursentwicklung derjenigen Anleihe, die zu Beginn des Planungszeitraumes die höchste Kurssensitivität gegen Zinsänderungen aufwies. Kupon, Restlaufzeit (RLZ) und Kursnotiz (B_o) dieser Anleihe sind in den Spalten 3, 4 und 5 angegeben.

Planungszeitraum	Umlaufsrendite	Kupon	RLZ	B_o	BP	PW_{abs}	PW_T	B_T+c_T
7.1.72-27.7.73	7,5 ↑ 10,0	7,75	10,8	100,25	100,82	100,34	99,72	92,6
7.1.72- 2.8.74	7,5 ↑ 10,7	7,75	10,8	100,25	99,79	100,19	101,27	88,1
31.7.70- 2.8.74	8,4 ↑ 10,7	6	10,5	83,90	87,76	100,94	102,33	94,7
31.3.78- 4.4.80	5,2 ↑ 9,8	6	14,9	102,10	102,74	94,72	92,35	73,8
5.9.80- 4.9.81	8,0 ↑ 11,4	6	12,5	85,70	85,36	97,11	97,80	87,9
7.1.83- 6.1.84	7,6 ↑ 8,2	6	10,2	92,40	88,46	94,82	95,75	95,2

Tab. 6: Ergebnisse der Portefeuille Versicherung für steigende Zinssätze

Als **Basispreis** (BP) **(abzusichernder Kurswert)** wurde derjenige Kurswert der Anleihe verwendet, der sich einstellen würde, wenn die zu Beginn des Planungszeitraumes bestehende Zinsstrukturkurve bis zum Planungshorizont unverändert bliebe (Spalte 6). Der abzusichernde Portefeuillewert PW_{abs} (Spalte 7) ergibt sich bei einem auf 100 normierten Anfangsbudget mit Hilfe des theoretischen Putwertes P_o^S aus den beiden Beziehungen

$$x_0 \cdot (B_0 + c_0) + x_0 \cdot P_0^S = 100 \qquad \text{(Budgetbedingung)}$$

$$PW_{abs} = x_0 \cdot (BP + c_T).$$

c_T bezeichnet die Stückzinsen am Planungshorizont. Im Planungszeitraum fällige Kuponzahlungen werden weder in PW_{abs} noch bei der Ermittlung des tatsächlich erzielten Portefeuillewertes P_T (Spalte 8) berücksichtigt.

Bei der Ermittlung von P_T wurde das Portefeuille wöchentlich unter Vernachlässigung von Transaktionskosten umgeschichtet. Eine Berücksichtigung von Transaktionskosten führt zu geringfügig niedrigeren Werten. Als "risikolose" Anlagemöglichkeit wurde unter Berücksichtigung der Ergebnisse des vorhergehenden Abschnittes 1 diejenige Anleihe definiert, deren Fälligkeitstermin am engsten am Planungshorizont liegt.

Der Vergleich von PW_{abs} mit PW_T zeigt, daß der abzusichernde Portefeuillewert mit Ausnahme von einem Planungszeitraum auch weitestgehend durch die Portefeuille Versicherung erreicht wird. Dies trifft nicht für den am 31.3.1978 beginnenden Zeitraum zu. Bei der sehr starken Zunahme der Umlaufsrendite von 5,2 auf 9,8 % wird PW_{abs} um 2,37 Einheiten oder 2,5 % verfehlt. Die starre wöchentliche Anpassung des Portefeuilles an die geänderten Kurse erweist sich in diesem Planungszeitraum als zu grob. Verglichen mit einem Portefeuillewert von 73,80 bei einem Verzicht auf die Absicherung (Spalte 9) wird dieser Mißerfolg jedoch etwas relativiert.

Fußnoten:

1) Vgl. hierzu auch Fong, G. und Fabozzi, F. (1985), S. 156-161.
2) Vgl. Fong, G. und Fabozzi, F. (1985), S. 138-144; Bierwag, G., Kaufman, G. und Toevs, A. (1983a), S. 113-123
3) Vgl. Rosinski, B. (1984), S. 50f und Caks, J. (1977), S. 105f.
4) Vgl. zu einer Übersicht Rudolph, B. (1979), S. 190f. und Bühler, W. (1983), S. 89-92.
5) Auf diesen Widerspruch haben erstmals Ingersoll, Skelton und Weil hingewiesen. Vgl. Ingersoll, J.E., Skelton, J. und Weil, R.: (1978), S. 634f.
6) Vgl. Black, F. und Scholes, M. (1973), S. 640-645.
7) Zu Einzelheiten vgl. Bühler, W. (1988), Rationale Bewertung von Optionsrechten auf Anleihen, erscheint in Zeitschrift für betriebswirtschaftliche Forschung, Oktober 1988.
8) Vgl. Arnold, L. (1973), S. 103f.
9) Vgl. Fisher, L. und Weil, R. (1971), S. 413 und 423.
10) Vgl. Bierwag, G., Kaufman, G. und Toevs, A. (1983b), S. 131.
11) Vgl. Gultekin, B. und Rogalski, R. (1984), S. 257ff.

12) Vgl. Ingersoll, J. (1983), S. 172.

13) Vgl. McCulloch, H. (1971), S. 19ff.

14) Zu einer Darstellung der Portefeuille Versicherung für Aktien und Portefeuilles aus Aktien vgl. Rubinstein, M. und Leland, H. (1981), S. 70ff., Rubinstein, M. (1985), S. 42ff., Schwartz, E. (1986/87), S. 9ff.

15) Aus Tabelle 5 ergibt sich für das dort dargestellte Beispiel ein Call-Delta von 1/3.

Literatur:

ARNOLD, L. (1973): Stochastische Differentialgleichungen, München 1973.

BIERWAG, G., KAUFMAN, G. und TOEVS, A. (1983a): Immunization Strategies for Funding Multiple Liabilities, in: Journal of Financial and Quantitative Analysis 18, 1983, S. 113-123.

BIERWAG, G., KAUFMAN, G. und TOEVS, A. (1983b): Recent Developments in Bond Portfolio Immunization Strategies, in: Kaufman, G., Bierwag, G., Toevs, A.: Innovations in Bond Portfolio Management, Greenwich 1983, S. 105-157.

BLACK, F. und SCHOLES, M. (1973): The Pricing of Options and Corporate Liabilities, in: Journal of Political Economy 81, 1973, S. 637-654.

BÜHLER, W. (1983): Anlagestrategien zur Begrenzung des Zinsänderungsrisikos von Portefeuilles aus festverzinslichen Titeln, in: Zeitschrift für betriebswirtschaftliche Forschung, Sonderheft 16, 1983, S. 82-137.

BÜHLER, W. (1988): Rationale Bewertung von Optionsrechten auf Anleihen, erscheint in Zeitschrift für betriebswirtschaftliche Forschung 40, Oktober 1988.

CAKS, J. (1977): The Coupon Effect on Yield to Maturity, in: The Journal of Finance 32, 1977, S. 103-115.

FISHER, L. und WEIL, R. (1971): Coping with the Risk of Interest-Rate Fluctuations: Returns to Bondholders from Naive and Optimal Strategies, in: The Journal of Business 44, 1971, S. 408-431.

FONG, G. und FABOZZI, F. (1985): Fixed Income Portfolio Management, Homewood 1985.

GULTEKIN, B. und ROGALSKI, R. (1984): Alternative Duration Specifications and the Measurement of Basis Risk: Empirical Tests, in: The Journal of Business 57, 1984, S. 241-264.

INGERSOLL, J. (1983): Is Immunization Feasible? in: Kaufman, G., Bierwag, G. und Toevs, A.: Innovations in Bond Portfolio Management, Greenwich 1983, S. 105-157.

INGERSOLL, J., SKELTON, J. und WEIL, R. (1978): Duration Forty Years Later, in: Journal of Financial and Quantitative Analysis 13, 1978, S. 627-650.

McCULLOCH, H. (1971): Measuring the Term Structure of Interest Rates, The Journal of Business 44, 1971, S. 19-31.

ROSINSKI, B. (1984): Zur Zinsstruktur am deutschen Kapitalmarkt, Diss. Berlin 1984.

RUBINSTEIN, M. und LELAND, H. (1981): Replicating Options with Positions in Stock and Cash, in: Financial Analysts Journal, Juli-August 1981, S. 63-72.

RUDOLPH, B. (1979): Zinsänderungsrisiken und die Strategie der durchschnittlichen Selbstliquidationsperiode, in: Kredit und Kapital 12, 1979, S. 181-204.

SCHWARTZ, E. (1986/87): Options and Portfolio Insurance, in: Finanzmarkt und Portfolio Management 1, 1986/87, S. 9-17.

PLANNING IN FLEXIBLE MANUFACTURING SYSTEMS

Ludo F. Gelders, Leuven (B)

1. The Need for Improved Planning

Flexible Manufacturing Systems are defined as computer controlled production systems capable of processing a variety of part types. The main components of such a system are :

- numerically controlled manufacturing machines including the tools to operate these machines,
- an automated material handling system to move the workpieces (e.g. conveyors, tow-lines, automated guided vehicles,...),
- on-line computer control to manage the entire FMS, including the manufacturing and handling under variable part type mix.

Flexible Manufacturing Systems may differ enormously in the extent of automation and the diversity of parts.

FMSs possess characteristics of both transfer lines and job shops. Fixed sequence transfer lines are typically aimed at large volume production, high machine utilization and short lead times. On the other hand, traditional job shops are designed to manufacture small series of a variety of products. This flexibility is usually paid for with a low utilization rate of the production facilities, long lead times and high in-process inventories. The highly integrated FMSs offer the opportunity to combine both the efficiency of transfer lines and the flexiblity of job shops.

A majority of electrical and mechanical products is manufactured in small series. Storage costs are high and the market is characterized by short product life cycles. Therefore, an efficient and flexible production system is of vital importance.

Although FMSs may offer tremendous benefits, like higher machine utilization, lower unit costs, lower lead times, higher quality and quicker response to market changes, these potential advantages are not easy to realize. An FMS is a very complex system consisiting of many interconnected components of hardware and software, and with many limited resources such as pallets, fixtures and tools. This complexity makes the succesful implementation of such a system very hard. In many cases the planning effort necessary before purchasing the equipment has been heavily underestimated.

Operations Research Proceedings 1987

Both technical and organizational problems have been encountered during the installation of FMSs. Technical aspects to which special attention should be given are e.g. chip removal, swarf clearance and retrieval, design, maintenance and control of fixtures, tool management and tool condition monitoring. But, the successful implementation of an FMS will depend as much upon the selection of efficient planning and control policies. Managing production in an FMS is more complex than for transfer lines or for job shops because :

- each machine is quite versatile and capable of performing many different operations,
- the system can manufacture several part types,
- each part type may have alternative routings,
- there is little slack in the system because of the interrelated components and the requirement to operate in real time.

So, in setting up an FMS one is confronted on the one hand with increased capabilities or a larger number of decision variables, and on the other hand with additional constraints. This creates the need to develop new and appropriate planning and control procedures that can take advantage of the system's capabilities for higher production rates.

2. An hierarchical approach to FMS planning

Given the complexity of FMSs, it is clear that a single analytical model cannot be used to solve all planning problems at once. The classical three-level view of the organization (HOLSTEIN, (1)) can be used to provide the hierarchical framework in which several smaller problem subsets can be identified. We distinguish the strategic, operational and tactical levels. Figure 1 represents this hierarchical approach in an FMS context.

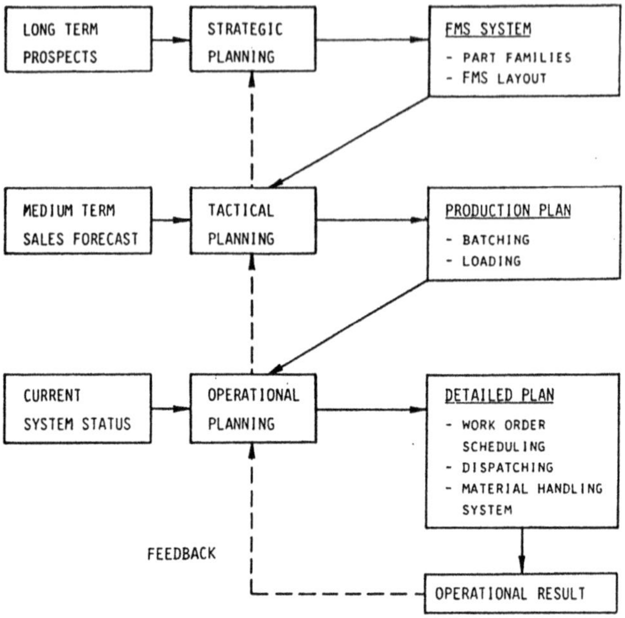

As mentioned before, an FMS is usually only a part of a larger manufacturing environment. In that case, the overall objectives and production targets determined at the corporate level will serve as inputs to the following three levels of operation.

The strategic level is typically the responsibility of top management and deals with long term decision making, i.e. the selection of a parts spectrum and the corresponding FMS layout. These decisions should be based on a long term production plan.

At the tactical level the sales forecasts are translated into a production plan. This plan divides the production requirements into several batches. These batches should be loaded into the system in such a way as to guarantee a good utilization of the production resources and to respect the due dates.

The operational level is concerned with the detailed decision making required for real time operation of the FMS. This includes scheduling and dispatching the workpieces as well as controlling the material handling system. Input parameters are the overall allocation of system resources made at the previous level and the current status of the system.

Any plan may need some adjustment. An unbalanced workload can be corrected by changing the production schedule, or if necessary the production plan. A permanent distortion can lead to a change of the system itself (e.g. add or delete a machining centre, change the parts spectrum).

We now discuss each of these three levels in some detail.

2.1. Strategic level or design phase

The decisions made at this level are crucial and should be made with utmost care.

The first decision concerns the selection of a parts spectrum. Ideally, one should make a specification of what will be manufactured on the system during the whole lifetime. In reality, this means one should try to design a system for a family (or families) of parts, rather than for a number of very specific parts which may be out of the market before the system is installed. Based upon this general specification the machine requirements can be defined. An examination of the technical characteristics of the part types allows to determine which type of machine tool should perform each operation, together with the appropriate cutting tools and cutting conditions. With these data the necessary number of tools of each type can be specified. In most cases, several technical solutions will be available. But, it is not enough to reach technical feasibility. The expensive FMS should also perform well with respect to system utilization and production rates. These criteria will permit a rough preliminary economic evaluation of alternatives by identifying inefficient designs. After this inital screening process only a few alternatives are left. VAN LOOVEREN, et al. (2) illustrate by an example that the design process is not finished when technical feasibility is reached.

Finally, a careful investment analysis should be made. Today, a growing number of managers believe traditional investment criteria (e.g. DCF analysis) are not fully appropriate when justifying new manufacturing techniques. But the amount of capital involved is so huge that justification by faith alone might lead to company suicide. GELDERS (3) gives a review of different approaches available, and concludes that the use of Multiple Criteria Analysis is highly recommended.

2.2. Tactical level or production planning phase

The FMS procuction planning problem consists in organizing the production in such a way as to satisfy the master production plan as well as to obtain an efficient use of the system resources (machines, pallets, fixtures, tools). In an FMS of reasonable size, this planning process is quite complex.

At the tactical level several classifications have been suggested. SURI and WHITNEY (4) distinguish between the batching problem and the balancing problem. First, the required production has to be divided into a set of batches. This is necessary because of the limited set of pallets, fixtures and tools. Then these batches should be balanced so as to guarantee an efficient use of system resources.

STECKE (5) identifies the following decisions :
- the selection of part types which will be produced (part type selection problem)
- the partitioning of machines into machine groups
- determining part mix and routing mix
- the allocation of pallets and fixtures to part types (the resource allocation problem)
- the assignment of operations and tools to machine groups (the loading problem).

These decisions will influence the flexibility and performance of the system. Of course, these problems cannot be separated completely ; e.g. a good part type selection cannot be done without considering the necessary tools and operations which are to be loaded. Even these subproblems are quite complex. The following example illustrates the difficulty of defining an efficient subset of part types for immediate processing. In deciding which part mix will be made, alternative routings should be taken into account.

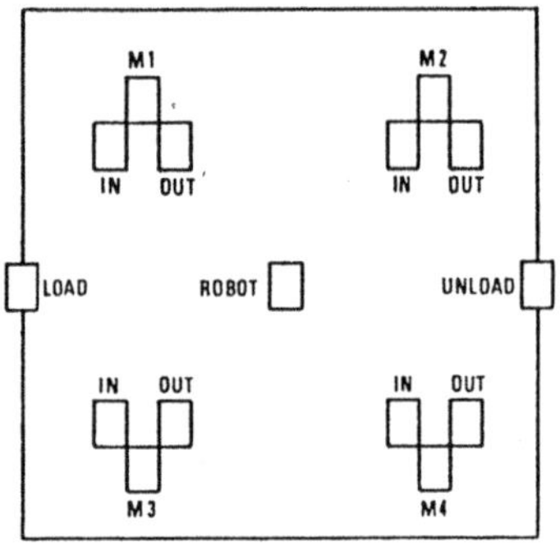

Consider a manufacturing system consisting of four machining centres (M1 to M4) and a robot to move parts between machines (figure 2). The production requirements for the next period are 15 pieces for each part type (A, B, C, D).
The table gives processing times and sequences for each part and for each routing.

Part type	Routing	M1	M2	M3	M4
A	1	1/234	2/1456	3/1785	4/352
	2	1/1438	-	-	2/2387
B	1	2/98	1/1098	-	3/1785
	2	1/762	-	2/1164	3/1023
C	1	3/1785	-	1/2785	2/576
	2	2/3987	1/1078	-	3/167
D	1	-	1/892	-	1/352
	2	3/72	-	2/584	1/598

The effect of alternative routings on system performance is shown in figure 3 which gives completion times for different routing mixes (the same mix is used for each part type). Similar figures can be made for different part mixes. It is clear that completion times can be reduced by a judicious use of alternative routings. Besides part mix and routing mix, many other elements need to be considered in defining a feasible (and efficient) production plan. E.g. one has to check whether sufficient pallets, tools and fixtures are available.

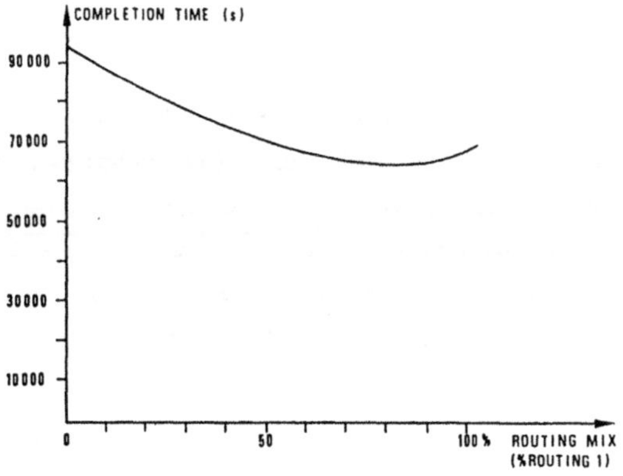

2.3. Operational level

At this level,the decisions include the workpiece release function as well as the workpiece handling control policies.The release policy controls the flow of workpieces into the system and takes as input two sets of parameters. The first set of parameters results from the overall allocation of system resources to the production of each part type decided at the tactical level. The second set of parameters specifies the current status of the system, such as breakdowns, types of pallets and fixtures currently available, deviations from desired production rates, etc. Based upon these inputs it is decided which workpiece should be introduced next and when this should be done.

Once a part is in the system, it has to be routed through that system. This workpiece handling function includes e.g. which machine will be selected in case there are several alternatives ; which buffer will be selected when a piece has to wait ? The control policy of the material handling system should also be considered. It should answer questions such as which carrier or robot should take care of a particular workpiece when several AGVs or robots are available. Finally the workpiece handling task incorporates special procedures in case system components break down.

Although these scheduling problems also occur in traditional job shops, solving them is more complex and critical in an FMS environment. The complexity is explained by the higher (potential) flexibility of the system. The fact that there are multiple part types, parallel machines, alternative routings and so on, boosts the number of alternatives and therefore the number of decisions. Moreover an efficient FMS scheduling system should operate in real time. A fixed, off-line schedule would deny the system's ability to adapt to unexpected changes in the system.

3. Planning Techniques

In this section, we illustrate to which extent some well-known O.R. techniques can be used to solve the aforementioned planning problems. The techniques considered are : waiting line theory (queueing networks), simulation analysis and mathematical programming. It is assumed that the reader is familiar with these techniques. Figure 4 summarizes the use of these techniques in FMS planning. For an extensive literature review we refer to GELDERS and VAN LOOVEREN (25).

PHASE	MODELS
DESIGN	- SIMULATION (layout) - WAITING LINES (capacity computations)
PRODUCTION-PLANNING	- MATHEMATICAL PROGRAMMING (batching) - WAITING LINES (number of pallets)
DETAILED PLANNING	- SIMULATION (alternative routings) - MATHEMATICAL PROGRAMMING (dispatching parts)

3.1. Waiting line models (queueing networks)

The basic theory of queueing networks was developed by JACKSON (6), and later on extended by GORDON and NEWELL (7) and BUZEN (8). Queueing networks can be used as an aid in making strategic or tactical decisions. They model the FMS as a set of stations having one or more servers. A station is a point at which work is done, which means it can be a machine as well as a transportation system. A work queue is allocated to each station. The system manufactures one or more workpieces of which the routing and average processing time are known. For such a system, queueing analysis allows the determination of the production rate, the average queue length at every station and the utilization of every server. Queueing analysis is based on a set of rather restrictive assumptions, including the fact that the processing times are exponentially distributed and that there is no restriction on queue length. However, queueing networks have proved to be very useful in providing a first indication of system performance.

Consider e.g. an FMS consisting of following elements :
- a loading/unloading station
- a series of 10 manufacturing machines (lathes, grinders, etc...)
- a single conveyor linking the different stations.

The queueing network model (represented by figure 5) uses as input :
- the number of machining stations
- the average processing times
- the average arrival frequencies.

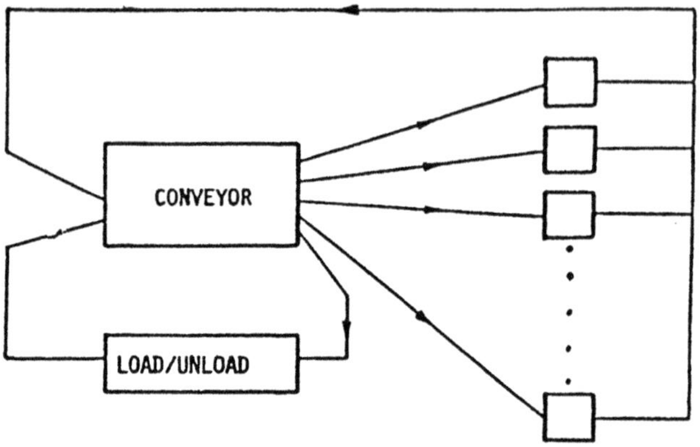

The model then allows for the computation of different system characteristics (see table). For this specific case it is clear that the single conveyor can handle the traffic, but machining station 5 would create major problems.

STATION	UTILIZATION	ITEMS IN QUEUE
1	0.86	5.0
2	0.86	5.0
3	0.86	4.9
4	0.86	4.9
5	0.98	17.7
6	0.43	0.3
7	0.13	0
8	0.43	0.3
9	0.47	0.42
10	0.26	0.1
CONVEYOR	0.63	0
AVERAGE TROUGHPUT TIME/ITEM = 1.76 min.		

It is obvious that this type of analysis can be very useful in screening alternative policies. It is reiterated once more that the previous analysis is based on a number of restrictive assumptions. It is assumed e.g. that the service time of the material handling system is identical for all jobs and all movements irrespective of the closeness of the workstations. Moreover, real FMSs are usually much more complicated. In reality one is confronted with multiple parts, breakdowns, a limited number of pallets, finite queues at the workstations and so on.

For screening purposes a lot of queueing models have been suggested. One of the first models is the CAN-Q model (closed queueing network) developed by SOLBERG (9). This model has been extended by e.g. DUBOIS (10) and SURI and HILDEBRANT (11). An evaluation of the suitability of several queueing models for analyzing the performance of an FMS is given by BUZACOTT (12) and BUZACOTT and SHANTIKUMAR (13).

3.2. Simulation

Simulation analysis is an indispensable tool in designing complex systems. Usually, discrete event simulation is used for modelling (flexible) manufacturing systems (MITTRA (14), BULGREN (15)). These models mimic the detailed operation of a system through a computer program which effectively steps through each event that would occur in the system. As such, simulation permits controlled experimentations on a complex system with little or no disturbance to the actual system. Simulation analysis can be performed at different levels of sophistication and therefore with varying degrees of accuracy and credibility. The price to be paid is in terms of programming time to create the model, time to collect detailed data sets, and computer time for each run. In FMSs, simulation can be used to test the layout of the system, and to study the effects of different control strategies, scheduling priority rules, breakdown scenarios and maintenance schemes.

Simulating large systems with several machine groups, alternative routings, multiple part types, is a very complex matter. The first challenge is to build a model that is as simple as possible and at the same time remains an efficient approximation of the actual system. Next, the simulation experiments have to be designed in such a way as to provide a better understanding of the system. The number of parameters and decision variables will be very high and therefore the possible combinations of control policies will be almost unlimited. A well-thought out setup of the (expensive) test runs is indispensible.
To simulate an FMS one can use a general simulation language or a tailor made FMS-package. Several general purpose simulation languages have been used in modelling FMSs, e.g. GPSS (RATHMILL et al. (16), SLAM (MARTIN and PRITSKER (17)) and Q-Gert (RUSSELL and TANCHOCO (18)). Specific packages have been developed by ELMARAGHY and HO (19), WARNECKE et al. (20), and many others. These packages usually have a modular structure to ease the model building and data input effort. M. MONTAZERI et al. (26) describe a modular general purpose simulator developed specifically for the analysis of FMS.

3.3. Mathematical programming

Mathematical programming concerns the optimal allocation of limited resources among competing activities, under a set of constraints. Since Dantzig developed the simplex technique for solving linear programming problems in 1947, many additional techniques have been developed which relax the assumptions of the linear programming model and broaden the application of the mathematical programming approach (HILLIER and LIEBERMAN (21)). It is clear that these techniques can be very useful in solving several FMS planning problems which are characterized by many scarce resources (e.g. machining time, tools, pallets, fixtures) and numerous constraints (e.g. due dates, routings, part mix).

The following example illustrates the use of linear programming for solving problems at the tactical level. In section 2.2. the problem of selecting a good routing mix was discussed. This problem can be formulated as follows :

$$\min \quad T$$

$$\text{s.t.} \quad \sum_{j=1}^{m} n_{ij} \geq N_i \qquad ,\forall i$$

$$\sum_{i=1}^{n} \sum_{j=1}^{m} n_{ij} \cdot t_{ijk} \leq T \qquad ,\forall k$$

$$n_{ij},\ T \geq 0 \qquad ,\forall i\ ,\ \forall j$$

where i = part type index ($i=1,\ldots,n$)

j = routing index ($j=1,\ldots,m$)

k = machine index

T = production makespan

N_i = number of pieces of type i to be produced during T

n_{ij} = number of pieces of type i that will be produced following route j

t_{ijk} = time for performing the operations on part i that are assigned to machine k in case route j is chosen.

This formulation minimizes the makespan of the required production. Part types can follow different routings. However, once a routing is selected it has to be followed till the end.

This problem was solved for the system specification and production requirements of the example in section 2.2. The "optimal" (rounded) routing mix is :

Part type / Routing	A	B	C	D
1	11	13	10	15
2	4	2	5	0

The production makespan (T) is 49,588 seconds. Of course this is an absolute minimum as transportation times and precedence relations are not taken into account in the LP formulation. A detailed simulation of the system indicated a completion time of 63,780 seconds, which is indeed lower than the results shown in figure 3 (minimum completion time for the considered routing mix combinations : 66,450 s).

This formulation is kept very simple. First of all, it assumes that there are no integer decison variables. Moreover, several constraints are neglected. If for instance one wants to model tooling requirements, one is confronted with nonlinearities. This makes it very hard to solve realistic problems within reasonable computation times.

Another example of LP formulation is provided by AVONTS, GELDERS and VAN WASSENHOVE in (22). They describe a real-life case in which management had to decide which products in what quantities had to be produced on the (new) FMS and/or on the existing conventional jobshop. One formulation maximizes FMS utilisation :

$$\text{Max} \quad \sum_{i=1}^{I} \sum_{j=1}^{J} t_{ij} x_i$$

$$\text{s.t.} \quad \sum_{i=1}^{I} t_{ij} x_i \leq C_j \qquad ,\forall j$$

$$x_i \leq d_i \qquad ,\forall i$$

$$xi \geq 0 \qquad ,\forall i$$

where : i = part type index

j = machine index

x_i = number of parts type i produced on the FMS

t_{ij} = operation time part i on machine j

d_i = demand rate product i

C_j = capacity machine j

4. Conclusions

In this article attention was given to some planning problems that may occur in Flexible Manufacturing Systems. A distinction was made between the strategic, the tactical and the operational planning level. The applicability of several planning techniques was illustrated. Other OR methods may also be useful in this context (e.g. network flow minimization and dynamic programming). At the strategic level there is a strong need for a general framework to evaluate the value of the increased flexibility which such a manufacturing system offers. At the tactical and the operational level one is confronted with a set of isolated and fragmentary approaches. Only integrated models will enable to fully realize the potential benefits of FMSs.

Clearly an FMS increases the complexity of the planning problem. Not only more information, but also more accurate and timely information is needed. Therefore, a successful implementation of any planning and control system will heavily depend upon the efficiency of the information-processing system (YOUNG (23) ; RANKY (24)).

REFERENCES

1. Holstein, W., "Production Planning and Control Integrated", Harv. Bus. Rev., vol. 46, n°3, 1986.

2. Van Looveren, A.J., L.F. Gelders, L.N. Van Wassenhove, "A Review of FMS Planning Models", Modelling and Design of Flexible Manufacturing Systems, A. Kusiak (ed.), Elsevier Science Publishers B.V., Amsterdam, 1986.

3. Gelders, L., "On the Justification of New Manufacturing Technology", 16th ISIR Conference, Sept.30-Oct.2, 1986, Brussels.

4. Suri, R. and Whitney, C., "Designing a Decision Support System for FMS", J. Manuf. Syst., vol.3, n°1, 1984.

5. Stecke, K., "Formulation and Solution of Nonlinear Production Planning Problems in Flexible Manufacturing Systems", Manage. Sci., vol.29, n°3, 1983.

6. Jackson, J., "Networks of Waiting Lines", Oper. Res., vol.5, 1957.

7. Gordon, W. and Newell, G., "Closed Queueing Systems with Exponential Servers", Oper. Res., vol.15, 1967.

8. Buzen, J., "Computational Algorithms for Closed Queueing Networks with Exponential Servers", Commun. ACM, vol.16, 1973.

9. Solberg, J., "A Mathematical Model of Computerized Manufacturing Systems", Proc. 4th Int. Conf. on Prod. Res., Tokyo, 1977.

10. Dubois, D., "A Mathematical Model of a Flexible Manufacturing System with Limited in Process Inventory", 5th Eur. Congress on Oper. Res., Lausanne, 1982.

11. Suri, R. and Hildebrant, R., "Modelling Flexible Manufacturing Systems Using Mean-Value Analysis", J. Manuf. Syst., vol.3, n°1, 1984.

12. Buzacott, J., "Modelling Flexible Manufacturing Systems", Operational Research '84, North-Holland, Amsterdam, 1984.

13. Buzacott, J. and Shanthikumar, J., "Models for Understanding Flexible Manufacturing Systems", AIIE Trans., vol.12, n°4, 1980.

14. Mittra, S., "Discrete System Simulation Concepts", Simulation, Sept. 1984.

15. Bulgren, W., Discrete System Simulation, Prentice Hall, New Jersey, 1982.

16. Rathmill, K., Greenwood, N. and Houshmand, M., "Computer Simulation of FMS", Proc. 2nd Int. Conf. on FMS, Brighton, 1982.

17. Martin, D., and Pritsker Associates, "Simulation in the Life Cycle of Flexible Manufacturing Systems", Proc. 1st ORSA/TIMS Conf. on FMS, Ann Arbor, 1984.

18. Russell, R. and Tanchoco, J., "An Evaluation of Vehicle Dispatching Rules and Their Effect on Shop Performance", Material Flow, vol.1, 1984.

19. ElMaraghy, H. and Ho, N., "A Simulator for Flexible Manufacturing Systems with Graphical Animation", Proc. 2nd Int. Comp. Eng.Conf., San Diego, 1982.

20. Warnecke, H., Zipse, T. and Zeh, K., "Simulation and Computer-aided Planning of FMS", Proc. 3rd Int. Conf. on FMS, Boeblingen, 1984.

21. Hillier, F. and Lieberman, G., Introduction to Operations Research, Holden-Day Inc., San Francisco, 1984.

22. Avonts, L., Van Wassenhove, L. and Gelders, L., "Allocation Work Between an FMS and a Conventional Jobshop. A Case-Study.", EJOR, forthcoming.

23. Young, R., "Software Control Strategies for Use in Implementing Flexible Manufacturing Systems", Ind. Eng., Nov. 1981.

24. Ranky, P;, The Design and Operation of FMS, IFS (Publications) Ltd., Kempston, 1983.

25. Gelders, L. and Van Looveren, A., "Hierarchical Planning and Control in Flexible Manufacturing Systems", Handbook of Technology Management, D.F. Kocaoglu, ed., J. Wiley, forthcoming.

26. Montazeri, M., Gelders, L., Van Wassenhove, L., "A Modular Simulator for Design, Planning an Control of FMS", Advanced Manufacturing Technology, forthcoming.

Algorithmics and Heuristics in Combinatorial Optimization

Jan Karel Lenstra, *Amsterdam*

I would like to start by saying that I am pleased that this annual DGOR meeting is being held in the Netherlands and that I am honored that the program committee has invited me to give a plenary lecture.

There exist long-standing relations between the operations research communities in Germany and the Netherlands, and as far as DGOR is concerned, these relations have been especially close in the area of stochastic operations research. I am gratified that it has been decided to choose a topic in deterministic operations research for this presentation, and I have hopes that this signifies a broadening of the contacts between DGOR and the Dutch OR Society.

The subject of today is combinatorial optimization. In view of my affiliation with the CWI, you will not be surprised to hear that I will talk in particular about some tools from mathematics and computer science that are available to help us to solve problems in combinatorial optimization. I will thus concentrate more on techniques and less on problems.

Mr. Kuilman has spoken about flexible manufacturing, where the use of techniques from operations research in general and combinatorial optimization in particular could be beneficial. Logistics in a broad sense, covering production planning and distribution management, is just one application area of combinatorial optimization. Others are timetabling, marketing, investment planning, health care, network design, and circuit layout. It is fair to say that combinatorial optimization as a research area would not exist without the demand for effective and efficient planning and design methods in a variety of practical situations.

Let me give you an example of optimization in practice. For me, it is a historical example.

Operations Research Proceedings 1987

History

The first time that I encountered the subject of today was in the mid 1950's, in a novel called *Professor Sealingwax and his cuckoo.* Professor Sealingwax was a remarkable operations researcher. Many of the actions he took were based on calculations he carried out in pencil on the back of a sheet of wallpaper.

Once upon a time he found himself on a ship, together with his wife Sweetie and their cuckoo, but without a crew, without coal, and without wind. They wanted to get to a country by the name of Foreveronia. This is how it goes.

He multiplied Foreveronia with Southwind, added Sealingwax, Sweetie, ten stokers, three engineers and forty sailors, extracted the square root, divided by Nip Tune, and then by starboard and the Great Bear, subtracted steam and smoke and added lull. Then he multiplied the result by lifeboat, removed the first and the last digit, changed the middle digit nine into a four, and cried with pleasure: 'Sweetie, I've got it! One bucket of mustard, a pot of pepper, a bag of onions and a bag of garlic in the middle funnel of the ship, and that should do the trick.' 'Sealingwax,' Sweetie said, 'I have always trusted you, and I also trust you this time.'

There is little doubt that this is an algorithm, with components of a real-life situation as input and a decision as output. It is also clear that there is mathematics and computing involved, and the sheet of wallpaper suggests the infinite tape of a Turing machine, so he also used some sort of computer. We could call this operations research.

But Sealingwax calculates with real objects. There is no transformation of a practical problem situation into a mathematical problem type, which is called *modelling*, and there is no translation of a mathematical solution into a practical decision, which is called *implementation*. There is no gap between theory and practice. Abstraction and reality are the same. I will return to this relation later.

When I became seriously interested in combinatorial optimization in the 1960's, it went without saying that we needed mathematics and computers. One question continued to intrigue us in those days: how should we use our computers? More precisely, how do we implement our algorithms efficiently? It came as something of a surprise that a new discipline was emerging that addressed exactly this issue: the art of computer programming.

But we needed more. Some problems could be solved efficiently; other problems escaped us and no truly efficient method for their solution could be found. And the question was: how far can we go? That is, where are the limits to efficiency? It came as an even bigger surprise, in the early 1970's, to find that the difference between easy and hard problems could be explained. This is what computational complexity theory is about.

Complexity

You all know what a graph is: a collection of nodes and a collection of edges, each of which links two nodes together. The graph in Figure 1 is connected, because you can get from each node to any other. The graph in Figure 2 is disconnected. The graph in Figure 3 is Hamiltonian, since it contains a cycle that visits each node exactly once. The graph in Figure 4 is not Hamiltonian; you might want to prove this.

Suppose you are attending a conference and you want to buy a present for the

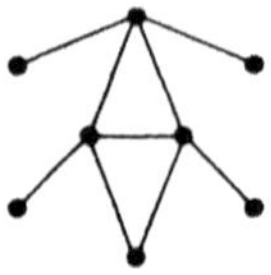

FIGURE 1. A connected graph.

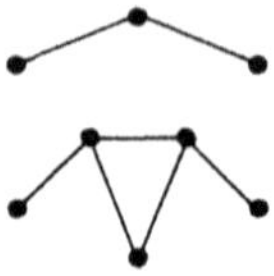

FIGURE 2. A disconnected graph.

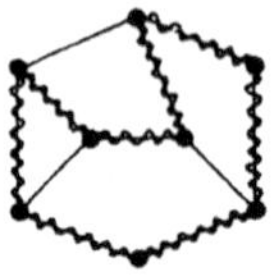

FIGURE 3. A Hamiltonian graph.

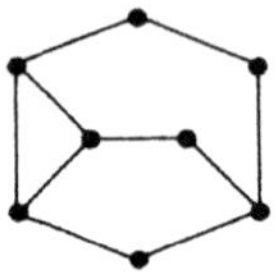

FIGURE 4. A non-Hamiltonian graph.

(a) ? (b) Right. (c) Wrong.
FIGURE 5. Buying a connected graph.

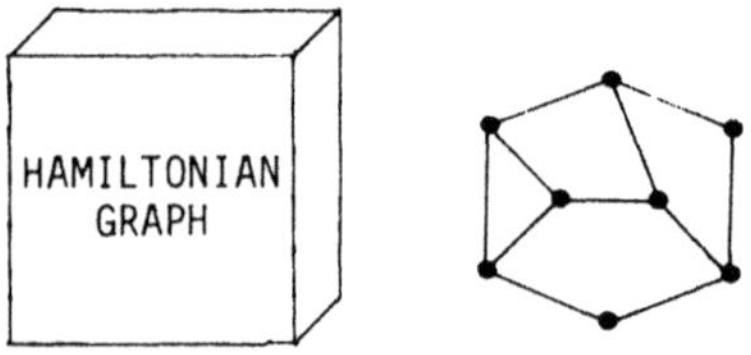

(a) ? (b) ??
FIGURE 6. Buying a Hamiltonian graph.

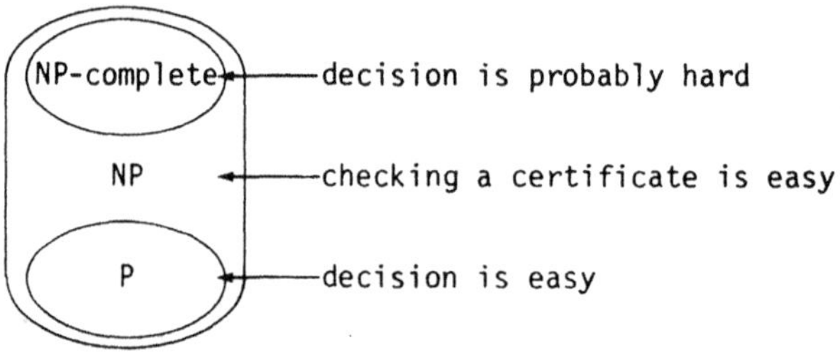

FIGURE 7. A likely map of *NP*.

people at home. You go to a graph store and ask for a connected graph. The shopkeeper puts a box on the counter (Figure 5(a)). You want to check this, open the box, and take out the graph. When it sticks together, it is connected (Figue 5(b)); when it falls apart, it is not (Figure 5(c)). The point I want to make here is that you can easily test a graph for connectivity by yourself; it takes an amount of time proportional to the number of edges.

Now you want to buy something special: a Hamiltonian graph. Again, there is a box (Figure 6(a)). You open it - but now you may find yourself in trouble (Figure 6(b)): there is no fast method available that can test any given graph for Hamiltonicity. Trial and error may work, but not necessarily so. However, the shopkeeper can easily convince you, namely by pointing out a Hamiltonian cycle as in Figure 3; this takes an amount of time proportional to the number of nodes.

This is exactly the difference between the problem classes *P* and *NP*. Both classes contain only decision problems, which require a yes/no answer; I will return to optimization problems shortly. *P* contains all those problems for which one can easily come up with the correct answer. *NP* contains all those problems for which one can easily be convinced of the correctness of the yes answer by checking a given structure: a Hamiltonian cycle in the example, a *certificate* in terms of complexity theory.

These definitions only make sense if the notion of *easiness* is formalized. A computation is easy if its running time is bounded by a polynomial function of the size of the problem under consideration. For a graph on *n* nodes, checking all nodes or all edges takes time polynomial in *n*, but generating all permutations of the node set in the hope of finding a Hamiltonian cycle is superexponential.

What are the virtues of an algorithm when it runs in polynomial time? First of all, its robustness. An algorithm that is polynomial on one machine is polynomial on any other reasonable type of machine, including theoretical models and commercial computers (but excluding parallel machines). Secondly, its asymptotic behavior. Any polynomial function in *n* is ultimately, when *n* is large enough, smaller than any exponential function. In the third place, its practical efficiency. Polynomial algorithms tend to work well in practice. Some polynomial algorithms are pretty bad, but it seems to be the case that once a problem has been shown to belong to *P*, a truly efficient method is found sooner or later. Finally, polynomiality allows us to come to grips with computational complexity in a theoretical sense. It serves to explain why some problems appear to be harder than others. More generally, it has proved to be a fundamental concept in the broad area of computational mathematics.

Any problem in *P* also belongs to *NP*, so *P* is a subclass of *NP*. I have indicated that the connectivity problem is a member of *P* and that Hamiltonicity is in *NP*. If it could be shown that Hamiltonicity is outside *P*, then the problem would have no solution in polynomial time and one would justifiably call it *hard*. Such a proof seems to be beyond the reach of present-day mathematics. However, we can do slightly less. It can be shown that the Hamiltonicity problem is a generalization of any other problem in *NP*. Hamiltonicity is *NP-complete*, i.e., it is representative of the entire class *NP*. If Hamiltonicity would belong to *P*, then all other problems in *NP* would be easy as well and *P* would be equal to *NP*. No one believes this to be true, for the simple reason that *NP* seems to be so much richer than *P*. It follows that the Hamiltonicity problem is unlikely to be easy and therefore *probably hard*. See Figure 7.

Next to Hamiltonicity, many other combinatorial decision problems have been shown to be *NP*-complete. I have not told you how results of this type are obtained. That is of secondary importance here; suffice it to say that it is conceptually a simple affair, although it can be technically very intricate.

What is the use of all this for combinatorial optimization? Complexity theory deals with yes/no problems, we consider optimization problems. If a problem has a polynomial optimization algorithm, then it is said to be *easy* (or well solved, or tractable). If a problem is at least as hard as some *NP*-complete problem, then it is said to be *NP-hard*. This should not be the last word on the problem, but the first. It tells you that you cannot expect to find a guaranteed optimum in worst case polynomial time. You have to give in on either speed or solution quality.

INTRACTABILITY

The next question is, of course: how to solve these hard problems? How to cope with intractability? At this point, we enter the area of the design and analysis of algorithms in combinatorial optimization. There is no way that I could review all that has happened here since the mid 1970's, even if you would give me the time.

The main criteria to be taken into account are *efficiency* and *effectivity*, or speed and quality. How do we make optimization methods more efficient, that is, run faster? How do we make approximation algorithms more effective, that is, achieve better solutions?

The performance of an algorithm can be analyzed in terms of its worst case or average case behavior. Worst case analysis is a pessimistic approach, since it has to take the isolated difficult problem instance into account, but it provides solid performance guarantees. Average case analysis is a complementary approach, which presupposes some probability distribution over the problem instances. I do not want to go into any detail here. Let me just mention one brief example: the simplex method for linear programming. This algorithm requires exponential time in the worst case but performs very satisfactorily in practice. In order to explain this, you have to resort to an analysis of the average case, and significant progress has been made in this direction.

I would like to indicate four recent tools in the design and implementation of combinatorial algorithms. *Geometric* and *randomized* methods are algorithmic approaches of a mathematical nature. *Parallelism* and *interaction* employ new architectures, which have become available due to achievements in computer engineering.

GEOMETRY

Geometric algorithms, and in particular those based on polyhedral combinatorics, represent an extremely important topic, which is worthy of a full plenary address at one of your future meetings. That is all I will say about it.

RANDOMIZATION

Before talking about randomization, I should clarify the role of stochastics in combinatorial optimization. Stochasticity occurs at three levels.

First, there are the stochastic *problem types*, which occur in areas like stochastic programming and queueing theory, but also in routing and scheduling. The model is stochastic in the sense that one has to determine a solution before the data is realized, so as to optimize a global parameter like the expected criterion value. Secondly, the problem type is deterministic but the *problem instances* are random. This refers to the case of probabilistic analysis. For each realized instance, an optimal or approximate solution is defined, and one is looking for a probabilistic characterization of its value. Finally, the problem type and its instances are deterministic, but the *algorithm* is randomized in the sense that it is able to toss a coin at certain points in order to decide how to proceed. In the context of machine scheduling, for example, one may want to minimize the expected makespan at the first level, to find the expected minimum makespan at the second, and to minimize the makespan in a randomized fashion at the third.

Let me give you an example of the use of randomization. Consider the problem of determining whether a given number is prime. It is an open question whether primality can be tested in polynomial time. However, there exist randomized algorithms that run in polynomial time and that behave as follows: if the output is 'no', then the number is definitely composite; if the output is 'yes', then the number is prime with probability at least one half, and repeated trials can reduce the error probability. Such a test gives moral rather than mathematical certainty, but it does so very fast.

Randomized methods in combinatorial optimization are, of course, approximation algorithms. A few randomization schemes are indicated here, with the randomized action in italics:

Monte Carlo:	$[\textit{generate}]^{\text{many}}$;
Monte Carlo & local search:	$[\textit{generate} \rightarrow \text{improve}]^{\text{many}}$;
simulated annealing:	$\textit{generate} \rightarrow \text{improve or } [\textit{deteriorate}]^{\text{prob}\downarrow 0}$;
sampling & clustering:	$[\textit{generate}]^{\text{many}} \rightarrow \text{select} \rightarrow [\text{improve}]^{\text{some}}$;
extension & rotation:	$\textit{extend} \rightleftarrows \textit{rotate}$.

Monte Carlo is an old approach: a number of feasible solutions is generated and the best one is selected. The combination with local search is slightly more sophisticated: each generated solution is subjected to iterative improvement on the basis of neighborhood search. Several variations on this approach have recently been proposed.

Simulated annealing is a technique for iterative improvement, again based on neighborhood search, which accepts deteriorations with a small and decreasing probability in the hope of avoiding bad local optima and getting settled in the global optimum. Sampling & clustering is another variation, which selects only one starting point out of each cluster of sampled solutions that are likely to lead to the same local optimum. Extension & rotation is an altogether different idea.

The investigation of the randomization principle in OR is at a relatively early stage. Some notable successes have been obtained, and more can be expected.

PARALLELISM

What has been achieved in parallel computing?

Architectures. - In a parallel computer, several processors operate in parallel and communicate with each other. We distinguish three classes. In vector machines, the operations are pipelined rather than parallelized. In SIMD (single instruction multiple data) machines, the processors perform at each point in time the same operation on local data. Usually, there is a large number of small processors and a fast interconnection network. Both classes are suitable for regular computations, where many operations of the same type have to be performed in a synchronized fashion. In MIMD (multiple instruction multiple data) machines, the processors can perform different instructions at a time. In practice, there is a moderate number of processors that operate in an asynchronous mode and communicate through a slow network or a shared memory.

Computations. - Parallel computing has provided a new playground for computational OR. Most experience has been obtained with numerical algorithms and nonlinear optimization on vector and SIMD machines. In combinatorial optimization, these types of machines perform well as long as the computational process is regular, as in dynamic programming. MIMD seems to be more suitable if the structure of the computation is not known in advance. An example is branch and bound, where the processors should explore different parts of the search tree and communicate only if the need occurs.

Computational models. - In comparison to sequential computing, there are two problems. First, existing machines are by no means equivalent; implementations tend to be highly machine dependent. Secondly, realistic models are lacking; theoretical analyses have to take account of physical features of the computational environment. Much attention has been paid to the PRAM (parallel random access machine), which allows for unbounded parallelism and unit-time interprocessor communication, but such a model is hardly realistic.

And what are the perspectives?

Computational models. - We need an investigation of severe restrictions on parallelism and communication. In particular, a robust theory for models with at most a linear number of processors that communicate over a bounded degree network would be very useful.

Architectures. - The main obstacle for a breakthrough of parallel computing is not the lack of realistic models but the chaos in the real world of architectures. We need a consensus on a single concept of a flexible MIMD computer in which the user can define the sort of parallelism he desires. Before attempting a hardware realization of this machine, we should build it in software and analyze its performance. This requires a flexible set of tools, including a versatile programming language, which does not bother the user with the internal machine structure.

Computations. - We need a theoretical approach towards the design and analysis of parallel algorithms for hard problems. The fundamental question is how to distribute the computational effort over the processors and how to arrange the communication so as to maximize the speedup. Operations researchers are well positioned to model and solve this complicated design problem.

INTERACTION

Man-machine interaction is a less formal and more fashionable topic than geometry, randomization, or parallelism. I will concentrate on *interactive planning systems*, also known as *decision support systems* (DSS). By this, I mean systems that are designed to support decision making in practical planning situations by the integration of human perception and mechanical algorithmics in an interactive environment.

Where do we place DSS if we look at it from an OR point of view? OR has various sides. The mathematics of OR is a *normative* occupation that intends to develop a theory of models and algorithms. Practical OR is an *empirical* activity in which quantitative tools are put to use in actual problem situations in a heuristic fashion. DSS is then nothing but a novel approach towards practical OR, made possible by advances in information technology. DSS merges the areas of OR and information systems. From this point of view, one should not expect a formal theory of DSS. One should, however, expect the influence of other disciplines, including database theory and software engineering, computer graphics and computational geometry, and even pattern recognition and psychology.

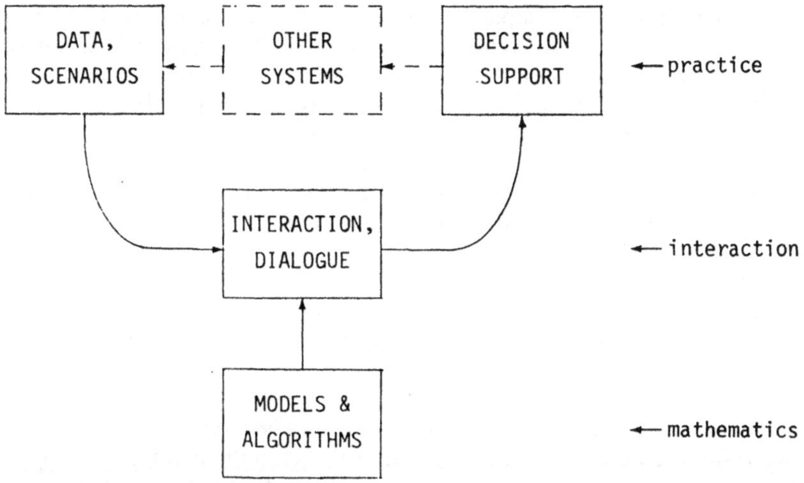

FIGURE 9. Structure of a DSS.

Figure 9 displays the structure of a DSS. The top level represents practice. The system receives data and tentative decisions from the outside world and returns decision support. At the bottom, we have models and algorithms, a collection of black boxes that contain the quantitative tools on which the DSS relies. It is illuminating to distinguish two types of models. *Evaluative* models are designed to answer the question: given a decision, what is its quality? *Generative* models do, in some sense, the reverse: given a desired quality, what is an appropriate decision? This distinction reflects two functions that the system should be able to perform: either it *assists* in representing and evaluating decisions proposed by the user; or it *advises* and generates complete plans by itself. These two roles are the extremes of the spectrum, and there is much in between. An important feature of a DSS is that it is always the user who is in charge, irrespective of the mode of operation.

The middle level is the interface where the interaction between human insight and experience and algorithmic power and precision takes place. It is the core of the system. A few remarks are in order here.

Interaction is possible in a technical sense, but why is it desirable? The brief answer is that practical planning problems tend to be both *hard* and *soft*. Hard on grounds of complexity considerations; in order to obtain solutions of an acceptable quality within an acceptable amount of time, one has to resort to approximation algorithms, and even to heuristic procedures in which man and machine divide the tasks in accordance with their respective capabilities. Soft because notions like feasibility and optimality are not as precise as in mathematics but are carried implicitly in the value judgement of the decision maker; interaction is the obvious way to cope with this. As a result, interaction adds to effectivity, efficiency, and acceptability. Better solutions are obtained faster, and an interactive planning system is more readily adopted than a black box approach.

While the user interface is the most visible part of a DSS, its only purpose is to create the opportunity to manipulate information in a convenient way. Whether information and manipulation make sense depends on the context, which consists of the planning situation on the one hand and the models and methods on the other. One might say that the role of information technology pertains to the form, while practice and mathematics provide the substance.

Much more could be said about the desirable functional properties of a DSS, about the design of a user interface, and about the need to turn the art of representation into a science. But due to limitations of time, I will leave it at this.

Propositions

I would like to finish with a number of propositions. Some of you may disagree with some of them. Please view them as a contribution towards the discussions during this meeting.

Proposition 1. *A heuristic is not necessarily an approximation algorithm.*
An algorithm is an unambiguous recipe and belongs to the normative science of mathematics, whether it is deterministic or randomized, and whether it optimizes, approximates, or performs some other function. A heuristic, in the original meaning of the word, is an empirical approach to practical problem solving by trial and error.

A DSS is a typical form of a heuristic. It may rely on all kinds of things, including approximation algorithms. It would clean up OR terminology if we would distinguish these concepts more carefully.

Proposition 2. *There is no gap between theory and practice.*
I know that some of my colleagues are exclusively interested in theory, while others are completely occupied by practice. For many of them, the gap does exist. What I really want to say is that *the gap is in the mind of the beholder.*

Many of the truly outstanding OR projects over the past fifteen years were carried out by people who knew how to create innovative modelling concepts, develop original solution techniques, and achieve successful implementations. They are generally too busy to realize the distinction between theory and practice, let alone to worry about a gap. What they do is true OR.

PROPOSITION 3. *There is no crisis in OR.*
I am not sure if I agree with this myself. After all, as far as the *systems* is concerned, we see that OR is in danger of becoming dominated by the areas of computer engineering and information systems. And as to the *ideas*, one could defend the point that the easy work has been done and that the hard work is safer in the hands of mathematicians and computer scientists.

However, it is the task of OR to bring ideas and systems together, and there is a bright future ahead. The use of OR models and methods requires computing facilities, and it is only in the last few years that these are becoming widely available at affordable prices. The current revolution in information technology will have a dramatic impact on the demand for OR techniques.

PROPOSITION 4. *Away with applications.*
Both theory and practice are vital to OR, so what is wrong with applications? They are of no concern to either side.

Theoreticians are interested in theorems, not in applications. If they talk about applications, they usually refer to examples. Although these can be very helpful, their purpose is to illustrate theory, without any practical motive.

Practitioners are not interested in applications but in solutions. They are not concerned with problem types or models, but they are faced with a problem situation. They do not ask for the application of some form of abstraction, but for a practical solution - whether you arrive at it by mathematics or by black magic.

Thank you for your attention. I wish you a successful meeting.

BIBLIOGRAPHICAL NOTES
I have used material from [3], [2] and [1] in the sections on complexity, parallelism and interaction, respectively. The quotation from [4] was translated by Joke Sterringa.

REFERENCES

1. J.M. ANTHONISSE, J.K. LENSTRA, M.W.P. SAVELSBERGH (1988). Behind the screen: DSS from an OR point of view. *Decision Support Systems*, to appear.
2. G.A.P. KINDERVATER, J.K. LENSTRA, A.H.G. RINNOOY KAN (1988). Perspectives on parallel computing. *Oper. Res.*, to appear.
3. J.K. LENSTRA (1986). Interfaces between operations research and computer science. Γ.Π. Πραστακος (ed.). *Επιχειρησιακη Ερευνα και Ηλεκτρονικοι Υπολογιστες*, EEEE, Athens, 35-48.
4. D. ZONDERLAND (1969). *De reisavonturen van professor Zegellak*, Het Spectrum, Utrecht.

Numerical Methods for Nonlinear Programming Problems

Klaus Ritter, München

1. Introduction

The purpose of this paper is to describe some basic ideas of algorithms for solving nonlinear programming problems. A short description of optimality conditions in Section 3 is followed by a discussion of superlinearly convergent methods for unconstrained problems in Section 4. An extension of these methods for linearly constrained problems is outlined in Section 5. Nonlinear inequality constraints are discussed in Section 6. Problems of this type are usually solved by constructing and solving a sequence of simpler, i.e., unconstrained or linearly constrained, minimization problems. The final section deals with the application of automatic differentiation in nonlinear programming.

2. Formulation of the problem, definitions and notation

Let $x \in E^n$ and assume that $F(x)$ and $h_1(x),\ldots,h_m(x)$ are given real-valued functions. Set

$$R = \{x \in E^n \mid h_i(x) \leq 0,\ i = 1,\ldots,m\}\ .$$

We consider the problem of determining an $\bar{x} \in R$ such that

$$F(\bar{x}) \leq F(x) \quad \text{for all } x \in R \tag{2.1}$$

The function $F(x)$ is called the objective function and the functions $h_i(x)$ are referred to as constraints. Every $x \in R$ is said to be a feasible solution, whereas every $\bar{x} \in R$ satisfying (2.1) is termed an optimal solution. For any $\hat{x} \in R$, the

Operations Research Proceedings 1987

constraint $h_i(x) \leq 0$ is said to be active at $\hat{x}$ if $h_i(\hat{x}) = 0$. If $h_i(\hat{x}) < 0$, then the i-th constraint is inactive at $\hat{x}$. If $h_i(x)$ is differentiable we denote its gradient by $\nabla h_i(x)$. For the gradient of $F(x)$ at some point x_j we use the notation $\nabla F(x_j)$ or g_j. For any column vector x and any matrix M we denote the transpose by x' and M', respectively. The symbol $\|x\|$ is used for the Euclidean norm of x.

Throughout this paper we will make the following assumption.

Assumption 1.

i) There is $x_o \in R$ such that the set

$$S_o = \{x \in R \mid F(x) \leq F(x_o)\}$$

is bounded.

ii) $F(x)$ and $h_1(x),\ldots,h_m(x)$ are continuously differentiable.

iii) For every $x \in R$ the gradients of the constraints, active at x, are linearly independent.

Clearly, the first two parts of the above assumption imply the existence of an optimal solution to the minimization problem

$$\min\{F(x) \mid x \in R\} \ . \tag{2.2}$$

The last part of the assumption is used to avoid some technical difficulties which arise if gradients of active constraints are linearly dependent.

3. Optimality conditions

Our first task is to present necessary conditions for an optimal solution to (2.2) in the form of the following version of the Kuhn-Tucker-theorem. For a proof of this theorem the reader is referred to MANGASARIAN [8].

Theorem 1.

Let Assumption 1 be satisfied and let $\bar{x}$ be an optimal solution to the problem

$$\min\{F(x) \mid x \in R\} .$$

Then there are numbers $\lambda_1,\dots,\lambda_m$ such that

i) $\nabla F(\bar{x}) = \sum_{i=1}^{m} \lambda_i \nabla h_i(\bar{x})$

ii) $\lambda_i h_i(\bar{x}) = 0 , \qquad i = 1,\dots,m$

iii) $h_i(\bar{x}) \leq 0, \quad \lambda_i \leq 0, \quad i = 1,\dots,m .$

Every $\bar{x}$ satisfying the conditions i) - iii) of Theorem 1 is said to be a stationary point. The numbers $\lambda_1,\dots,\lambda_m$ are called Lagrange multipliers or dual variables.

Condition (ii) is called the complementary slackness condition. It implies that, for every i, at least one of the numbers λ_i and $h_i(\bar{x})$ is zero. Thus the above theorem states that at an optimal solution the gradient of the objective function F(x) is a linear combination of the gradients of the constraints active at $\bar{x}$. The coefficients λ_i in this linear combination are nonpositive. If $h_i(\bar{x}) = 0$ implies $\lambda_i < 0$ for $i = 1,\dots,m$, i.e., if the Lagrange multipliers corresponding to active constraints are strictly negative, then the strict complementary slackness condition is said to be satisfied at $\bar{x}$.

The above theorem tells us that the stationary points are the only candidates for optimal solutions. Unfortunately, not every stationary point is in general an optimal solution. This is only true if we impose strong convexity assumptions.

Theorem 2.

Let F(x) be pseudo-convex and let $h_1(x),\dots,h_m(x)$ be quasi-convex. Then every stationary point is an optimal solution of problem (2.2).

For a proof of this theorem and a definition of pseudo- and quasi-convex functions the reader is again referred to [8].

The above theorems provide us with a characterization of optimal solutions. It is, however, in general not possible to obtain a stationary point directly from the conditions of Theorem 1. In order to calculate optimal solutions we have in general to use iterative methods. These methods generate a sequence of points which under appropriate assumptions converges to an optimal solution. For numerical purposes it is convenient to look first at unconstrained minimization problems.

4. Unconstrained minimization problems

In this section we describe iterative methods for minimizing a function F(x) without constraints. We consider algorithms which generate sequences $\{x_j\}$, $\{s_j\}$, and $\{\sigma_j\}$ such that, for every j,

$$x_{j+1} = x_j - \sigma_j s_j , \qquad F(x_{j+1}) < F(x_j) . \tag{4.1}$$

The vector s_j and the scalar $\sigma_j > 0$ are called the search direction and the step size, respectively.

Since we have no constraints it follows from Theorem 1 that every $\bar{x}$ with $\nabla F(\bar{x}) = 0$ is a stationary point and hence a candidate for an optimal solution. In order to show that, under appropriate assumptions, the sequence $\{x_j\}$ converges to a stationary point we have to impose certain conditions on $\{s_j\}$ and $\{\sigma_j\}$. We assume that s_j is choosen in such a way that, for every j,

$$g_j s_j \geq \epsilon_1 \|g_j\| \|s_j\| \qquad \text{for some } 0 < \epsilon_1 < 1 \tag{4.2}$$

and

$$\epsilon_2 \|g_j\| \leq \|s_j\| \leq \epsilon_3 \|g_j\| \qquad \text{for some } 0 < \epsilon_2 < \epsilon_3 . \tag{4.3}$$

These conditions imply that the angle between search direction and gradient of $F(x)$ at x_j is bounded away from 90 degree and that the search direction and the gradient have the same order of magnitude.

The step size σ_j has to satisfy the condition that, for every j and some $\epsilon_4 > 0$ and $\epsilon_5 > 0$,

$$g_j's_j \geq \epsilon_4 \qquad \text{implies} \qquad F(x_j - \sigma_j s_j) \leq F(x_j) - \epsilon_5 . \qquad (4.4)$$

Since $-g_j's_j$ is the derivative of $F(x_j - \sigma s_j)$ with respect to σ at $\sigma = 0$, the above condition states that whenever this derivative is less than or equal $-\epsilon_4 < 0$, then σ_j must be choosen in such a way that $F(x)$ decreases by at least the fixed amount ϵ_5.

An abvious way to determine a search direction for which conditions (4.2) and (4.3) are satisfied is to set $s_j = g_j$ as in the so-called method of steepest descent. As an example of a step size procedure for which condition (4.4) is satisfied we discuss the Armijo-Goldstein method which is as follows.

Step size procedure (Armijo-Goldstein)

Choose a constant $0 < \delta < 0.5$. Determine the smallest integer $\upsilon_j \geq 0$ such that

$$F(x_j) - F(x_j - (0.5)^{\upsilon_j} s_j) \geq (0.5)^{\upsilon_j} \delta g_j's_j \qquad (4.5)$$

Set $\sigma_j = (0.5)^{\upsilon_j}$.

In this method we first try the step size 1. If it satisfies (4.5) it is accepted. It it does not satisfy (4.5), we try the step size 1/2, then 1/4 and so on.

By Taylor's theorem there is $v_j \in \{v | v = x_j - t\sigma s_j,\ 0 \leq t \leq 1\}$ such that

$$\begin{aligned} F(x_j - \sigma s_j) &= F(x_j) - \sigma \nabla F(v_j)'s_j \\ &= F(x_j) - \sigma g_j's_j - \sigma(\nabla F(v_j) - g_j)'s_j . \end{aligned}$$

Therefore,

$$\frac{F(x_j) - F(x_j - \sigma s_j)}{\sigma g_j s_j} = 1 + \frac{(\nabla F(v_j) - g_j)' s_j}{g_j' s_j} .$$

Since $\|\nabla F(v_j) - g_j\| \to 0$ as $\sigma \to 0$, it follows that the σ_j in the Armijo-Goldstein procedure is well defined and that, for every j with $g_j' s_j \geq \epsilon_4$, we have $\sigma_j \geq \omega$ for some $\omega > 0$, independent of j. Thus $g_j s_j \geq \epsilon_4$ implies

$$F(x_j) - F(x_{j+1}) \geq \delta\sigma_j g_j s_j \geq \delta\omega\epsilon_4 > 0 ,$$

i.e.

$$F(x_{j+1}) \leq F(x_j) - \epsilon_5 ,$$

where $\epsilon_5 = \delta\omega\epsilon_4$.

The main convergence result for a method with the above properties is contained in the following theorem.

Theorem 3

Let Assumption 1 be satisfied and assume that an algorithm generates sequences $\{x_j\}$, $\{s_j\}$, $\{\sigma_j\}$, and $\{g_j\}$ for which the conditions (4.1) - (4.4) are satisfied. Then

i) $g_j \to 0$ as $j \to \infty$

ii) Every cluster point of $\{x_j\}$ is a stationary point

iii) If $\|x_{j+1} - x_j\| \to 0$ as $j \to \infty$ and $\bar{x}$ is an isolated cluster point of $\{x_j\}$, then $x_j \to \bar{x}$ as $j \to \infty$.

Proof:

i) Suppose $\{g_j\}$ does not converge to zero. Then there are $\epsilon > 0$ and an infinite subset $J \subset \{0,1,2,\ldots\}$ such that

$$\|g_j\| \geq \epsilon \quad \text{for} \quad j \in J .$$

By (4.2) and (4.3) we have, for $j \in J$,

$$g_j's_j \geq \epsilon_1 \|g_j\| \|s_j\| \geq \epsilon_1 \epsilon_2 \|g_j\|^2 \geq \epsilon_4 \ .$$

where $\epsilon_4 = \epsilon_1 \epsilon_2 \epsilon^2 > 0$. Hence, it follows from (4.1) and (4.4) that $F(x_j) \to -\infty$ as $j \to \infty$. Since $F(x)$ is bounded from below this contradiction implies that $g_j \to 0$ as $j \to \infty$.

ii) Let $\bar{x}$ be a cluster point of $\{x_j\}$. Then there is an infinite set $J \subset \{0,1,2,\ldots\}$ such that $x_j \to \bar{x}$ as $j \to \infty$, $j \in J$. Thus

$$\nabla F(\bar{x}) = \lim_{\substack{j\to\infty \\ j\in J}} g_j = 0 \ .$$

iii) Assume that $\bar{x}$ is an isolated cluster point of $\{x_j\}$, i.e., that there is some $\epsilon > 0$ such that $\bar{x}$ is the only cluster point of $\{x_j\}$ in the set

$$U = \{x \,|\, \|x - \bar{x}\| \leq 2\epsilon\} \ .$$

If $\{x_j\}$ does not converge to $\bar{x}$, then there is some cluster point $\hat{x} \neq \bar{x}$ of $\{x_j\}$. Since $\hat{x} \neq U$ and $\|x_{j+1} - x_j\| \to 0$ as $j \to \infty$, it follows that the set

$$U_1 = \{x \,|\, \epsilon \leq \|x - \bar{x}\| \leq 2\epsilon\}$$

contains infinitely many members of the sequence $\{x_j\}$. Thus it contains some cluster point z, say. Because $z \in U_1 \subset U$ and $z \neq \bar{x}$ we have a contradiction which proves that $x_j \to \bar{x}$ as $j \to \infty$.

The results of the above theorem apply to all functions $F(x)$ which are continuously differentiable and have bounded level sets. If we make additional assumptions for $F(x)$, we obtain stronger convergence results. If, e.g. we assume that $F(x)$ is convex, it follows from Theorem 2 and Theorem 3 that every cluster point of $\{x_j\}$ is an optimal solution. If $F(x)$ is strictly convex, it has a unique global minimizer $\bar{x}$, say. It follows from the above theorem that $\{x_j\}$ converges to $\bar{x}$.

Numerical experience has shown that the sequence $\{x_j\}$ generated by the method of steepest discent, i.e., the choice $s_j = g_j$, converges in general very slowly and is, therefore, not suitable for solving practical problems. In addition to establish the convergence of $\{x_j\}$ to a stationary point it is important to study the rate of convergence. For practical purposes algorithms with superlinear convergence are particularly interesting.

Definition 1.

A sequence $\{x_j\}$ is said to converge superlinearly to some $\bar{x}$ if

$$\frac{\|x_{j+1} - \bar{x}\|}{\|x_j - \bar{x}\|} \to 0 \quad \text{as} \quad j \to \infty .$$

Unfortunately, it does not seem possible to obtain superlinear convergence without some assumptions concerning second order derivatives of the function $F(x)$.

Assumption 2

The sequence converges to some $\bar{x}$ such that

i) $\nabla F(\bar{x}) = 0$, $F(x_{j+1}) < F(x_j)$ for all j;

ii) $F(x)$ is twice continuously differentiable in some neighborhood of $\bar{x}$ and $G = G(\bar{x})$, the Hessian matrix of $F(x)$ at $\bar{x}$, is positive definite.

Using Taylor's theorem the following result can be shown (see e.g. [13]).

Theorem 4

Let Assumption 2 be satisfied and $x_{j+1} = x_j - \sigma_j s_j$. Then $\{x_j\}$ converges superlinearly to $\bar{x}$ if and only if

$$\left\| \frac{g_j}{\|g_j\|} - G \frac{\sigma_j s_j}{\|g_j\|} \right\| \to 0 \quad \text{as} \quad j \to \infty .$$

If the matrix G were known, it would be easy to satisfy the conditions of the theorem by setting

$$s_j = G^{-1}g_j \quad \text{and} \quad \sigma_j = 1 \ .$$

Since $\bar{x}$ and hence $G = G(\bar{x})$ is not known we have to approximate $G^{-1}g_j$. In most practical methods this is done by setting

$$s_j = M_j g_j \tag{4.6}$$

where M_j is an (n,n)-matrix associated with x_j. The sequence $\{M_j\}$ should be determined such that

$$\frac{1}{\|g_j\|} \|(M_j - G^{-1})g_j\| \to 0 \quad \text{as} \quad j \to \infty \ . \tag{4.7}$$

It can be shown (see e.g. [13]) that the Armijo-Goldstein method gives $\sigma_j = 1$ for j sufficiently large, if s_j is determined by (4.6) and (4.7) holds.

Since it follows from (4.6), (4.7) and $\sigma_j = 1$ for j sufficiently large that the condition of Theorem 3 is satisfied, the crucial point for superlinear convergence is (4.7).

An example of a method for which (4.7) holds is the Newton-method in which we set

$$M_j = G_j^{-1} \ ,$$

where $G_j = G(x_j)$ is the Hessian matrix of F(x) at x_j. Because $x_j \to \bar{x}$ as $j \to \infty$, it follows that $G_j^{-1} \to G^{-1}$ as $j \to \infty$. This implies (4.7). The main disadventage of the Newton-method is the necessity to calculate all second order partial derivatives of F(x).

Variable metric methods intend to construct a sequence $\{M_j\}$ with property (4.7) without calculating any second order derivative of F(x). They can be motivated as follows. If we set

$$d_j = \frac{g_j - g_{j+1}}{\|\sigma_j s_j\|} \qquad \text{and} \qquad p_j = \frac{s_j}{\|s_j\|} \ .$$

it follows from Assumption 2 and Taylor's theorem that

$$d_j = Gp_j + w_j \qquad \text{with} \qquad \|w_j\| \to \infty \quad \text{as} \quad j \to \infty .$$

Therefore,

$$\|G^{-1}d_j - p_j\| \to \infty \quad \text{as} \quad j \to \infty .$$

Because we try to construct a sequence $\{M_j\}$ such that $M_j \to G^{-1}$ as $j \to \infty$ it is reasonable to require M_j to satisfy the quasi-Newton equation

$$M_{j+1}d_j = p_j .$$

Usually, M_{j+1} is obtained from M_j by adding a rank two matrix. A particularly successfull variable metric method is due to BROYDEN [2], FLETCHER [6], GOLDFARB [7], and SHANNO [14]. It uses the update formula

$$M_{j+1} = M_j + \frac{d_j'p_j + d_j'M_jd_j}{(d_j'p_j)^2} p_jp_j' + \frac{p_jd_j'H_j + H_jd_jp_j'}{d_j'p_j} . \tag{4.8}$$

Even though in general the resulting sequence $\{M_j\}$ does not converges to G^{-1}, it can be shown that under appropriate assumption the condition (4.7) is satisfied (see e.g. POWELL [9]).

5. Linearly constrained minimization problems

In this section we describe some basic properties of iterative algorithms which can be used to solve problems of the form

$$\min\{F(x) \,|\, a_i'x \le b_i,\ i = 1,\dots,m\} .$$

Here $a_1,\dots,a_m$ are n-dimensional column vectors and $b_1,\dots,b_m$ are real numbers.

If an initial point $x_o \in R$ is not available it can be obtained by solving the following linear minimization problem

$$\min\{-\lambda \mid a_i'x + \lambda \leq b_i,\ i = 1,\dots,m,\ \lambda \leq 0\}$$

for which $x = 0$ and $\lambda = \min\{0,b_1,\dots,b_m\}$ is a feasible solution. Let $(\bar{x},\bar{\lambda})$ be an optimal solution. If $\bar{\lambda} < 0$, it follows immediately that R is empty. If $\bar{\lambda} = 0$, then $\bar{x} \in R$.

Let $x_j \in R = \{x \mid a_i'x \leq b_i,\ i = 1,\dots,m\}$. To be specific assume that

$$a_i'x_j = b_i\ , \qquad i = 1,\dots,q\ ,$$
$$a_i'x_j < b_i\ , \qquad i = q+1,\dots,m\ .$$

Set

$$T_j = \{x \mid a_i'x = 0,\ i = 1,\dots,q\}$$
$$S_j = \{x \mid x = \lambda_1 a_1 + \dots + \lambda_q a_q\}\ .$$

Then g_j, the gradient of $F(x)$ at x_j, can be written in the form

$$g_j = \tilde{g}_j + \hat{g}_j\ , \quad \tilde{g}_j \in S_j\ , \quad \hat{g}_j \in T_j\ ,$$

where

$$\tilde{g}_j = \lambda_1 a_1 + \dots + \lambda_q a_q\ .$$

Therefore, it follows from Theorem 1 in Section 3 that the iterate x_j is a stationary point if and only if

$$\hat{g}_j = 0 \quad \text{and} \quad \lambda_i \leq 0\ , \quad i = 1,\dots,q\ .$$

If these conditions are satisfied, then the algorithm terminates with the stationary point x_j. Otherwise a search direction s_j has to be determined such that

$$g_j's_j > 0 \quad \text{and} \quad a_i's_j \geq 0\ , \quad i = 1,\dots,q\ .$$

If $\hat{g}_j \neq 0$, then $s_j = \hat{g}_j \in T_j$ is a search direction with the above properties. Indeed,

$$g_j's_j = (\tilde{g}_j + \hat{g}_j)'\hat{g}_j = \hat{g}_j'\hat{g}_j > 0$$

and

$$a_i's_j = a_i'\hat{g}_j = 0\ , \quad i = 1,\dots,q\ . \tag{5.1}$$

On the other hand if $\lambda_q > 0$, say, then it follows that, for every s_j with

$$a_i's_j = 0\ , \quad i = 1,\dots,q-1, \quad a_q's_j = 1\ , \tag{5.2}$$

we have $s_j \notin T_j$ and

$$g_j's_j = (\tilde{g}_j + \hat{g}_j)'s_j = (\lambda_1 a_1 + \dots + \lambda_q a_q)'s_j = \lambda_q > 0\ .$$

Substituting $x_j - \sigma s_j$ into the active constraints we observe from (5.1) and (5.2) that for all $\sigma > 0$

$$a_i'(x_j - \sigma s_j) = a_i'x_j - \sigma a_i's_j = b_i\ , \quad i = 1,\dots,q-1\ .$$

Thus these constraints remain active. If $s_j = \hat{g}_j$, then it follows that the q-th constraint remains active too. If s_j has the property (5.2), then

$$a_q'(x_j - \sigma s_j) = a_q'x_j - \sigma < b_q, \quad \sigma > 0\ .$$

Therefore, this constraint will not be active at the next point.

In order to determine the largest σ for which $x_j - \sigma s_j$ remains feasible we substitute this expression into the inactive constraints. After rearranging we have

$$-\sigma a_i's_j \le b_i - a_i'x_j\ , \quad i = q+1,\dots,m\ .$$

If $-a_i's_j \le 0$ for $i = q+1,\dots,m$, then $x_j - \sigma s_j \in R$ for all $\sigma \ge 0$. Otherwise,

$$\sigma_j^* = \min\left\{\frac{b_i - a_i'x_j}{-a_i's_j}\,\middle|\,\text{all } i \text{ with } -a_i's_j > 0\right\}$$

is the maximal step size, i.e., the largest value of σ such that $x_j - \sigma s_j \in R$.

Now the Armijo-Goldstein method with 1 replaced by $\min\{1,\sigma_j^*\}$ can be used to determine the step size σ_j. The new iterate is then $x_{j+1} = x_j - \sigma_j s_j$. If $\sigma_j = \sigma_j^*$, then a new constraint will be active at x_{j+1}. If $\sigma_j < \sigma_j^*$, no new constraint becomes active at x_{j+1}.

It remains to give a rule which decides whether $s_j \in T_j$ or not. Such a rule can be based on a comparison of $\|\hat{g}_j\|$ and $\max\{\lambda_1,\ldots,\lambda_q\}$. It states that s_j should be in T_j if either

$$\|\hat{g}_j\| \geq \max\{\lambda_1,\ldots,\lambda_q\}$$

or $\hat{g}_j \neq 0$ and a new constraint became active at x_j.

The second part of the above rule is needed to prevent the algorithm from zig-zagging.

Algorithms which use the above ideas for choosing the search direction and the step size can be implemented in many ways. Details of such implementations are beyond the scope of this paper. The reader is referred to [12] and [13].

Let $\{x_j\}$ be a sequence of iterates generated by such an algorithm. It can be shown that every cluster point of such a sequence is a stationary point. Furthermore, we have the following interesting result. Suppose the sequence $\{x_j\}$ converges to $\bar{x}$ and

$$a_i'\bar{x} = b_i \ , \qquad i \in I \subset \{1,\ldots,m\}$$
$$a_i'\bar{x} < b_i \ , \qquad i \notin I \ .$$

If the strict complementary slackness condition is satisfied at $\bar{x}$, then there is some j_o such that, for every $j \geq j_o$,

$$a_i'x_j = b_i \ , \quad i \in I \ , \qquad a_i'x_j < b_i \ , \quad i \notin I \ ,$$

i.e., for j sufficiently large the set of active constraints is equal to the set of constraints which are active at the limit point $\bar{x}$. This implies that, for $j \geq j_o$, we always have

$$s_j \in T_j \quad \text{and} \quad \sigma_j < \sigma_j^* \ .$$

Because of this property the same ideas as in the unconstrained case can be used to construct algorithm which generate superlinearly convergent sequences $\{x_j\}$. For details the reader is again referred to [12] and [13].

6. Nonlinear inequality constraints

In this section we discuss some methods for solving the general nonlinear programming problem

$$\min\{F(x) \mid h_i(x) \leq 0,\ i = 1,\dots,m\} \tag{6.1}$$

In most practical algorithms (6.1) is not solved directly. Instead a sequence of simpler optimization problems is constructed and solved in such a way that the sequence of solutions to these auxiliary problems converges to the optimal solution of (6.1).

As a first example for this approach we will describe a penalty function method, see e.g. FIACCO and McCORMICK [3]. A typical penalty function for (6.1) can be defined as follows.

$$P(x;\mu_1,\dots,\mu_m) = F(x) + \sum_{i=1}^{m} \mu_i (h_i(x)_+)^2 , \tag{6.2}$$

where $\mu_i \geq 1$, $i = 1,\dots,m$, are constants, the so called penalties, and $y_+ = \max\{0,y\}$. Clearly, $P(x;\mu_1,\dots,\mu_m) \geq F(x)$ for all x, and equality holds if and only if $x \in R$, i.e., x is a feasible solution to (6.1).

The penalty function method now proceeds as follows. For an arbitrary initial point $\hat{x}_o$ a global minimizer $\hat{x}_1$ of the function (6.2) is determined. If $\hat{x}_1 \in R$, then $\hat{x}_1$ is an optimal solution to (6.1). If $\hat{x}_1$ is not a feasible solution of (6.1), then the penalties corresponding to constraints with $h_i(\hat{x}_1) > 0$ are multiplied by 10 and the new penalty function is minimized with $\hat{x}_1$ as starting point. In this way a sequence $\{\hat{x}_j\}$ is generated which undet suitable assumptions converges to the global solution $\hat{x}$ of (6.1), (see e.g. [3]). Furthermore, it can be shown that, for $i = 1,\dots,m$,

$$-2\mu_i h_i(\hat{x}_j)_+ \to \hat{\lambda}_i \quad \text{as} \quad j \to \infty , \tag{6.3}$$

where $\hat{\lambda}_i$ are the Lagrange multipliers associated with the optimal solution $\hat{x}$ of (6.1).

From (6.3) we observe that, if $\hat{\lambda}_i < 0$, then the penalty μ_i has to become very large if $\hat{x}_j$ is to be close to $\hat{x}$. It is well known that this fact causes numerical problems with the penalty function methods. They are in general efficient methods to compute a reasonable approximation to $\hat{x}$ in a few iterations. However, if a very good approximation to $\hat{x}$ is desired, then other methods have to be used. In the following we describe such a method which is due to ROBINSON [12]. In this method a sequence of linearly constrained minimization problems is solved.

Robinson's method requires as a starting point an approximation $(x_o,\lambda_{1o},\ldots,\lambda_{mo})$ to a stationary point $\hat{x}$ of (6.1) and its associated Lagrange multipliers $\hat{\lambda}_1,\ldots,\hat{\lambda}_m$. With this information the linear functions

$$Lh_i(x;x_o) = h_i(x_o) + \nabla h_i(x_o)'(x - x_o)\ , \qquad i = 1,\ldots,m$$

are defined and the minimization problem

$$\min\{F(x) - \sum_{i=1}^{m} \lambda_{1o}[h_i(x) - Lh_i(x;x_o)] \mid Lh_i(x;x_o) \leq 0,\ i = 1,\ldots,m\} \qquad (6.4)$$

is solved. Let

$$(x_1,\hat{\lambda}_{11},\ldots,\hat{\lambda}_{m1}) \qquad (6.5)$$

be an optimal solution to (6.4) and its associated Lagrange multipliers, respectively. Using (6.5) we define a new problem of the form (6.4) and solve it. Continuing in this way we obtain a sequence $\{x_j\}$.

Under appropriate assumptions it can be shown (see [12]) that there are positive constants γ_1 and γ_2 with the following properties. If

$$\|(x_o,\lambda_{1o},\ldots,\lambda_{mo}) - (\hat{x},\hat{\lambda}_1,\ldots,\hat{\lambda}_m)\| \leq \gamma_1$$

then the sequence $\{x_j\}$ converges to $\hat{x}$ and

$$\|x_j - \hat{x}\| \leq \gamma_2(0.5)^{2^j}$$

Numerical experience shows that the assumption that the starting point $(x_o, \lambda_{1o}, \ldots, \lambda_{mo})$ is sufficiently close to $(\hat{x}, \hat{\lambda}_1, \ldots, \hat{\lambda}_m)$ is crucial for the success of Robinson's method, i.e., it is in general a typically local method. Since we have seen that the penalty function method is suitable for computing an approximation x_j to $\hat{x}$ and, by (6.3), approximations to $\hat{\lambda}_i$, it is reasonable to try to combine the two methods.

In the combined method a few iteration with the penalty function method are performed to obtain $\hat{x}_j$ and $\hat{\lambda}_{ij} = -2\mu_i h_i(x_j)_+$. Then $(\hat{x}_j, \hat{\lambda}_{1j}, \ldots, \hat{\lambda}_{mj})$ is used as starting point for the Robinson method. The details and the convergence properties of the algorithm are given in [1].

7. Automatic Differentiation

As we have seen in the previous sections gradient and Hessian-matrix play an important role in the construction of efficient algorithm. In the following we will outline how they can be obtained by using automatic differentiation as suggested by RALL [10], [11]. Details of the method are described by FISCHER in [4] and [5].

Assume

$$r: D \subseteq \mathbb{R}^n \to \mathbb{R}$$

is a rational function in explicit form. This means, for $r(x)$ we have a formula which only contains the components $x_1, x_2, \ldots, x_n$ of x, some real numbers, the four arithmetic opterations addition, subtraction, multiplication and division, and parentheses at proper places.

Since we have to deal with gradient and Hessian matrix of various functions, let us indicate the gradient and the Hessian by the subscripts G and H respectively.

Looking at the formula for $r(x)$, we shall distinguish three cases.

<u>case: r is primitive</u>

$r(x) = x_i$ = i-th component of x, for some $i \in \{1,2,\dots,n\}$.

Then $r_G(x)$ = i-th unit-vector, $r_H(x)$ = zero-matrix.

<u>case: r is constant</u>

$r(x)$ = const = c, for some $c \in \mathbb{R}$.

Then $r_G(x)$ = zero-vector, $r_H(x)$ = zero-matrix.

<u>case: r is neither primitive nor constant</u>

The general idea of Automatic Differentiation may be expressed by the rule

split and differentiable!

In splitting the formula for r(x) we obtain one of the four sub-cases

(A) $r(x) = a(x) + b(x)$

(S) $r(x) = a(x) - b(x)$

(M) $r(x) = a(x) \cdot b(x)$

(D) $r(x) = a(x) / b(x)$

where a and b are rational functions. Furthermore, these two functions are available in explicit form and the formulas for a(x) and b(x) are shorter than the formula for r(x).

To follow the rule, we differentiate the function r and get table 1.

type	function	gradient
A	$r = a + b$	$r_G = a_G + b_G$
S	$r = a - b$	$r_G = a_G - b_G$
M	$r = a \cdot b$	$r_G = b \cdot a_G + a \cdot b_G$
D	$r = a / b$	$r_G = (a_G - r \cdot b_G)/b$

Table 1: Formula for the gradient

This table reveals the following fact:

For any $x \in D$, the pair $r(x), r_G(x)$ can be computed from the pairs $a(x), a_G(x)$ and $b(x), b_G(x)$.

We differentiate the function r twice and get table 2.

type	function	Hessian matrix
A	$r = a + b$	$r_H = a_H + b_H$
S	$r = a - b$	$r_H = a_H - b_H$
M	$r = a \cdot b$	$r_H = b \cdot a_H + a_G \cdot b_G^t + a \cdot b_H + b_G \cdot a_G^t$
D	$r = a / b$	$r_H = (a_H - r_G \cdot b_G^t - r \cdot b_H - b_G \cdot r_G^t)/b$

Table 2: Formula for the Hessian

From table 1 and table 2 we conclude:

For any $x \in D$, the triple $r(x), r_G(x), r_H(x)$ can be computed from the triples $a(x), a_G(x), a_H(x)$ and $b(x), b_G(x), b_H(x)$.

Now it is obvious how to deal with an explicitely given rational function $f: D \subseteq \mathbb{R}^n \to \mathbb{R}$. The splitting of $f(x)$ yields a characterizing sequence $f_1, f_2, \ldots, f_s$ of functions $f_i: D \to \mathbb{R}$ such that

(1) for $i \in \{1,2,\ldots,n\}$

$f_i(x) = x_i$ = i-th component of x

(2) for $i \in \{n+1, n+2, \ldots, n+d\}$ with some $d \in \{0,1,\ldots\}$

$f_i(x) = c_i$ = real constant

(3) for $i \in \{n+d+1, n+d+2, \ldots, s\}$

$f_i(x) = a_i(x) * b_i(x)$ with $* \in \{+,-,\cdot,/\}$ and $a_i, b_i \in \{f_1, f_2, \ldots, f_{i-1}\}$

(4) $f_s(x) = f(x)$

The characterizing sequence $f_1, f_2, \ldots, f_s$ has nothing to do with Automatic Differentiation, it merely is a convenient means to describe the function f. To compute the gradient of f we use the

Gradient-Method

(0) choose $x \in D$

(1) for $i = 1,\ldots,n$ set $f_i(x) = x_i$, $f_{iG}(x)$ = i-th unit-vector

(2) for $i = n+1,\ldots,n+d$ set $f_i(x) = c_i$, $f_{iG}(x) = \emptyset$

(3) for $i = n+d+1,\ldots,s$ compute $f_i(x), f_{iG}(x)$, with table 1

(4) then $f_s(x) = f(x)$, $f_{sG}(x) = f_G(x)$ = wanted

To compute the Hessian matrix of f we use the

Hessian-Method

(0) choose $x \in D$

(1) for $i = 1,\ldots,n$ set $f_i(x) = x_i$, $f_{iG}(x)$ = i-th unit-vector, $f_{iH}(x) = \emptyset$

(2) for $i = n+1,\ldots,n+d$ set $f_i(x) = c_i$, $f_{iG}(x) = \emptyset$, $f_{iH}(x) = \emptyset$

(3) for $i = n+d+1,\ldots,s$ compute $f_i(x), f_{iG}(x), f_{iH}(x)$ with table 1, 2

(4) then $f_s(x) = f(x)$, $f_{sG}(x) = f_G(x)$, $f_{sH}(x) = f_H(x)$ = wanted.

References

[1] M.J. Best, J. Bräuninger, K. Ritter and S.M. Robinson, A globally and quadratically convergent algorithm for general nonlinear programming problems, Computing 26 (1981) 141-153.

[2] C.G. Broyden, The convergence of a class of double-rank minimization algorithms 2. The new algorithm, J. Inst. Math. Appl. 6 (1970) 222-231.

[3] A.V. Fiacco and G.P. McCormick, Nonlinear Programming: Sequential Unconstrained Minimization Techniques (Wiley, New York, 1968).

[4] H. Fischer, Automatic Differentiation: How to Compute the Hessian Matrix. Technical Report # 104, Inst. f. Angewandte Mathematik u. Statistik, Tech. Univ. München, 1987.

[5] H. Fischer, Some Aspects to Automatic Differentiation. Technical Report # 107, Inst. f. Angewandte Mathematik u. Statistik, Tech. Univ. München, 1987.

[6] R. Fletcher, A new approach to variable metric algorithms, Comput. J. 13 (1970) 317-322.

[7] D. Goldfarb, A family of variable-metric methods derived by variational means, Math. Comput. 24 (1970) 23-26.

[8] O.L. Mangasarian, Nonlinear Programming (McGraw-Hill, New York, 1969).

[9] M.J.D. Powell, Some global convergence properties of a variable metric algorithm for minimization without exact line searches, in: Nonlinear Programming, SIAM-AMS Proceedings, Vol. 9 (American Mathematical Society, Providence, RI, 1976).

[10] L.B. Rall, Automatic Differentiation: Techniques and Applications. Lecture Notes in Computer Science No. 120, Springer-Verlag, Berlin - Heidelberg - New York, 1981.

[11] L.B. Rall, The Arithmetic of Differentiation. Mathematics Magazine 59, 275-282, 1986.

[12] K. Ritter, Convergence and Superlinear convergence of Algorithms for Linearly constrained Minimization Problems, in: Nonlinear Optimization, L.C.W. Dixon, E. Spedicato, G.P. Szegö, ed., Birkhäuser 1980.

[13] K. Ritter, Numerical methods for nonlinear programming. in: Modern Applied Mathematics, B. Korte, ed., North-Holland 1982.

[14] D.F. Shanno, Conditioning of quasi-Newton methods for function minimization, Math. Comput. 24 (1970) 647-656.

Interval uneffectiveness distribution for parallel redundant reliability systems with repair

Matthieu van der Heijden, Eindhoven

When designing a production system, the performances of a number of alternative system configurations are compared. An important performance measure, to which a lot of attention has been paid in literature, is the system long term uneffectiveness. This is defined as the long run fraction of the system capacity which cannot be used due to failure of the systems components. However, the *stationary interval uneffectiveness distribution* is a subject to which relatively little attention has been paid. This distribution represents the probability that the system uneffectiveness will not exceed some pre-specified level in a time interval of given length, when the system has reached statistical equilibrium. This performance measure is of great importance in practice, e.g. when some pre-specified production level has to be attained in a given time period in order for sales contracts to be met.

In the contribution to the DGOR/NSOR conference 1987, a computationally tractable method to approximate the interval uneffectiveness distribution is presented for the special case of parallel redundant systems with repair. These systems consist of n identical units of which at most k units are operating simultaneously. If less than k units are available, the system operates at a proportionally reduced level. The model under consideration includes ample repair facilities, so all failed units can be under repair simultaneously.

The approximate method for the interval uneffectiveness distribution is discussed and its accuracy is evaluated by comparison to results from Monte Carlo simulation. The method appears to be accurate enough for most practical situations. Also, sensitivity results are given, showing in particular a decreasing sensitivity of the measures of performance to the distributional form of the unit lifetimes and repair times as the size of the system increases. Furthermore, it is found that the effect of the distributional form of the unit lifetimes dominates that of the unit repair times. An extensive method description and numerical results can be found in van der Heijden and Schornagel [1]. For the special case with n=2 and k=1, a more accurate method is possible, which is described in van der Heijden [2].

1. Heijden, M.C. van der, and Schornagel, A., "Interval uneffectiveness distribution for a k-out-of-n multistate reliability system with repair", *European Journal of Operational Research*, 1988 (to appear).

2. Heijden, M.C. van der, "Interval availability distribution for a 1-out-of-2 reliability system with repair", *Probability in the Engineering and Informational Sciences* 1, 1987, 211-224.

Entwicklung von PC-Software zur Linearen Optimierung mit einem Anwendungsbeispiel zur Produktionsplanung im Steinkohlenbergbau

Ralph Ohmann, Freiburg/Br.

Entwickelt wurde ein Programm LP86 mit der Zielsetzung, dem Anwender ein Instrument in die Hand zu geben, mit dem er vor allem das Formulieren von Modellen im MPSX-Standardformat und das Interpretieren der Lösungen (einschließlich der Dualwerte) üben kann. LP86 eignet sich deswegen besonders für betriebswirtschaftliche Ausbildungskonzepte, beispielsweise für Lehrveranstaltungen auf den Gebieten Unternehmensplanung, Produktion oder Logistik und für Grundkurse wie "Mathematik für Wirtschaftswissenschaftler" oder "Quantitative Verfahren der Betriebswirtschaftslehre".

Bei LP86 wird besonderer Wert auf eine benutzerfreundliche Menügestaltung, auf Prüfroutinen und auf Fehlerhinweise (z.B. über fehlende oder unkorrekte Eingabedaten) sowie auf Bedienerhinweise gelegt. Der Benutzer kann unter zwei Genauigkeitsstufen wählen. Über eine Programmschnittstelle ist auch die Weiterverarbeitung der auf dem PC mit LP86 erfaßten Modelle auf Großrechnern mit MPSX möglich. LP86 beruht auf der revidierten Simplex-Methode und ist in der Programmiersprache PASCAL geschrieben. Mit LP86 können Modelle bis zur Größe von 400 Strukturvariablen und 80 Restriktionen zuzüglich 800 Restriktionen in Form von oberen und unteren Schranken bearbeitet werden. Es ist lauffähig auf den Rechnern IBM PC, XT und AT oder auf dazu kompatiblen Geräten mit mindestens 256 KB RAM, Arithmetic-Coprozessor 8087, Farbbildschirm und Betriebssystem PC-DOS ab Version 2.0. Als Demonstrationsbeispiel dient die vereinfachte Fassung eines LP-Modells für die Produktionsplanung im Steinkohlenbergbau /1/.

LP86 wurde mit dem Studentenpreis 1987 der DGOR ausgezeichnet. Eine um ein Unterprogramm zur Sensitivitätsanalyse (Ranging) und um zusätzliche Demonstrationsbeispiele erweiterte Fassung wurde inzwischen unter der Bezeichnung LP90 in der Buchreihe "Anwendungsprogramme für dem IBM PC" (Programmdiskette mit Handbuch und ausführlicher Dokumentation) veröffentlicht /2/.

Literatur

/1/ STAHLKNECHT, P., SEIFERT, O.: Produktionsplanung für ein Steinkohlenbergwerk. In: SPÄTH, H. (Hrsg.): Fallstudien Operations Research, Band 3. München/Wien 1980, S. 71 - 91

/2/ STAHLKNECHT, P., OHMANN, R.: Lineare Programmierung auf dem PC. München/Wien 1987

AN EXTENSION OF KARMARKAR'S ALGORITHM FOR BOUNDED LINEAR PROGRAMMING PROBLEMS

Angelika Steger, Bonn

Abstract: In this paper we describe Karmarkar's algorithm and present an extension that works with problems expressed in standard form. We require no a priori knowledge of the optimal value, but assume that the set of optimal solutions is bounded.

1. Introduction

In 1947 G.D. Dantzig developed the simplex method for solving linear programming problems. Since then the simplex method and its many variations have been used very successfully in practice. In theory, however, one drawback remained: the simplex method is not a polynomially bounded algorithm.

For a long time it was even an open question whether solving linear programming problems is an NP–complete problem or not. In 1979 the answer was given by the Russian mathematician Khachian and his polynomially bounded ellipsoid method [3]. However, superior to the simplex method by the worst case criterion, the ellipsoid method could not compete in practice.

Once again researchers got excited when in July 1984 Karmarkar published his "New Polynomial–Time Algorithm for Linear Programming" [2] and claimed that it works 50 times faster than the simplex method. A lot of research has been done and is still going on to establish or disestablish this claim.

Special effort was made to develop variations of the algorithm which can solve linear programming problems under less restrictive assumptions on the problem than required by the original version. Here we want to mention in particular M.J. Todd and B.P. Burrell's work [6]. Their extension of Karmarkar's algorithm using dual variables was a starting point for our variant. Algorithms essentially equivalent to the one we describe have been derived independently by other authors, see for example Gay [1].

In section 2 we describe Karmarkar's original algorithm. It can solve problems satisfying certain restrictive assumptions: the optimal value of the problem has to be zero and, in addition, to the constraint $Ax = 0$ we need one of the form $\sum_{i=1}^{n} x_i = 1$. Todd developed an extension of the algorithm using dual variables which can solve any linear programming problem in the usual form if we know an upper bound for the sum of the components of optimal solutions. In other words, we have to know a bound M such that adding the constraint $\sum_{i=1}^{n} x_i \leq M$ does not change the solutions of the problem. Another drawback of this algorithm is that we have to add a dense row and two dense columns to a constraint matrix which might have been very sparse.

Operations Research Proceedings 1987

In section 3 we propose a variant which can solve any linear programming problem under the assumption:

there exists a bound M such that $\sum_{i=1}^{n} x_i \leq M$ for any optimal solution.

We call a problem bounded, if it satisfies this assumption. Our algorithm has two main advantages: first of all we don't have to know M explicitly and secondly, we only add two dense columns but no dense row to the constraint matrix.

2. Karmarkar's Algorithm

In this section we will discuss how to solve linear programming problems of the form

$$(P) \qquad \begin{aligned} \min\ & c^T x \\ Ax &= 0 \\ e^T x &= 1 \\ x &\geq 0 \end{aligned}$$

satisfying the assumptions

(A1) A has full row rank,

(A2) $x^0 = \frac{1}{n}e = (\frac{1}{n}, \ldots, \frac{1}{n})$ is an initial feasible solution,

(A3) the minimal value of (P) is zero.

Karmarkar's algorithm generates a sequence of strictly positive feasible solutions x^k such that $0 \leq c^T x^k \leq \gamma^k c^T x^0$ where γ is some constant less than one independent of the problem. If the data are integer this implies as in the ellipsoid method that the algorithm can be terminated after a polynomial number of steps and an exact solution can then be obtained by rounding. The iterates x^k do not necessarily satisfy $c^T x^k \leq \gamma c^T x^{k-1}$ which would clearly imply the above inequalities. Instead Karmarkar considers a potential function

$$f(x, c) = n \ln c^T x - \sum_{j=1}^{n} \ln x_j.$$

A straightforward calculation shows that if all iterates satisfy

$$(*) \qquad f(x^k, c) \leq f(x^{k-1}, c) - \delta$$

for some positive constant δ then

$$c^T x^k \leq (e^{-\frac{\delta}{n}})^k c^T x^0.$$

So it is sufficient to generate iterates that fulfill $(*)$. We now describe how this can be done.

The feasible region of problem (P) is the simplex in $S^n = \{x \in I\!R^n \mid e^T x = 1,\ x \geq 0\}$ intersected with the vectorspace $\{x \mid Ax = 0\}$. The center of the simplex is by assumption (A3) an initial feasible solution. Later we will describe a simple procedure to obtain a new feasible point x^1 such that $f(x^0, c) \leq f(x^1, c) - \delta$. Indeed, taking a step in the steepest descent direction guarantees this.

If we could now apply the same procedure again and again using the new points x^1, x^2 etc. we would obtain the desired sequence. Unfortunately, the procedure depends strictly on having the center of the simplex as input point and does not work otherwise.

Karmarkar's idea now was to use a projective transformation to transform x^1 back into the center of the simplex. This transformation is given by

$$\begin{aligned} T &: S^n \rightarrow S^n \\ x &\rightarrow \frac{D^{-1}x}{e^T D^{-1} x} \qquad \text{where } D = \mathrm{diag}(x_1^1, \ldots, x_n^1). \end{aligned}$$

It is easy to show that T is a one–to–one mapping of S^n onto S^n and that the inverse of T is given by $T^{-1} : \overline{x} \rightarrow \frac{D\overline{x}}{e^T D\overline{x}}$.

Therefore, the image of (P) under T is

$$(T) \qquad \begin{aligned} \min & \frac{(Dc)^T x}{e^T D x} \\ ADx &= 0 \\ e^T x &= 1 \\ x &\geq 0. \end{aligned}$$

Thus our linear function turns out to be a ratio of linear functions in the transformed space. We approximate the ratio by $(Dc)^T x$ and thereby obtain a linear program of the same form as (P) also satisfying (A1)–(A3). In the transformed space we again use the above mentioned procedure to obtain a new feasible point $\overline{x}$ such that

$$f(\overline{x}, Dc) \leq f(\frac{1}{n}e, Dc) - \delta.$$

Transforming back we define x^2 as $x^2 = T^{-1}\overline{x}$. It is a direct consequence of the following lemma, which states an important characteristics of the potential function, that x^2 indeed satisfies

$$f(x^2, c) \leq f(x^1, c) - \delta.$$

Lemma 2.1: *Let f be a potential function as defined above. Then*

$$f(Tx, Dc) = f(x, c) + \ln(\det D).$$

This basically defines the algorithm. The only missing part is the procedure to obtain a feasible point $\overline{x}$ with potential function value

$$f(\overline{x}, c) \leq f(\frac{1}{n}e, c) - \delta.$$

Karmarkar's idea was to pick a suitable direction d and move a fixed fraction of the radius $r = \frac{1}{\sqrt{n(n-1)}}$ of the inscribed sphere in this direction:

$$\overline{x} = \frac{1}{n}e + \alpha\frac{d}{\|d\|} \qquad \text{where } \alpha = \frac{\lambda}{\sqrt{n(n-1)}},\ 0 < \lambda < 1.$$

What happens to the potential function if we move away from the center in the direction of d? Picking the direction correctly the first term, $n \cdot \ln c^T x$, will decrease. Hence, we might want to move as far as possible. However, the second term of the potential function, $-\ln \prod_{j=1}^{n} x_j$, is minimized at the center of the simplex. Therefore we shouldn't move too far. The step size α has to be a compromise. Note, that by definition $\overline{x}$ will stay in the interior of the simplex. We also have to assure somehow that $\overline{x}$ will again be a feasible solution.

The direction of steepest descent of the objective function in problem (P) is in direction of $-c$. In order to stay feasible we project c onto the nullspace of $\binom{A}{e^T}$. Recall that for a $p \times q$ matrix M having full row rank the projection onto the nullspace of M is given by

$$P_M = I - M^T(MM^T)^{-1}M.$$

Note, that since A has full row rank by assumption (A1) and since $Ae = 0$, $\binom{A}{e^T}$ has full row rank. So, let

$$\begin{aligned} d &= -P_{\binom{A}{e^T}}(c) = -P_A \cdot P_{e^T}(c) \\ &= -P_A(c - \frac{1}{n}e^T c \cdot e). \end{aligned}$$

We have to show that

$$f(\frac{1}{n}e + \alpha\frac{d}{\|d\|}, c) \le f(\frac{1}{n}e, c) - \delta$$

where $d = -P_{\binom{A}{e^T}}(c)$.

Proving this inequality is the main work in proving the correctness and polynomiality of Karmarkar's algorithm. We outline a proof which combines elements of Karmarkar's [2], Todd's [6] and Padberg's [4] approaches. The idea is to bound the two ratios

$$\frac{c^T\overline{x}}{c^T x^0} \quad \text{and} \quad \frac{\prod_{j=1}^{n} x_j^0}{\prod_{j=1}^{n} \overline{x}_j}$$

separately, where we set $x^0 = \frac{1}{n}e$ and $\overline{x} = \frac{1}{n}e + \alpha\frac{d}{\|d\|}$. In a second step the two bounds are combined and yield, after some calculation, the desired inequality.

The advantage of this proof is that it can be modified to proof the convergence of the extension we will describe in the next section.

We close this section by stating Karmarkar's algorithm formally:

Algorithm A *(Karmarkar's Algorithm):*

step 0: (initialization) let $x^0 = \frac{1}{n}e$.

step k: let $D = \mathrm{diag}(x_1^{k-1}, \ldots, x_n^{k-1})$;

calculate $d := -P_{\binom{AD}{e^T}}(Dc)$ and $\overline{x} = \frac{1}{n}e + \alpha\frac{d}{\|d\|}$;

set $x^k = \frac{D\overline{x}}{e^T D\overline{x}}$.

If $c^T x^k$ is less than required stop,

otherwise return to step k with "k" replaced by "$k+1$".

3. Extension for bounded problems

In this section we develop an extension of Karmarkar's algorithm that can solve any bounded linear programming problem. Suppose the linear programming problem we want to solve is

$$(P) \qquad \begin{aligned} \min\ & c^T x \\ Ax &= b \\ x &\geq 0. \end{aligned}$$

The only assumptions we make are

(B1) A has full row rank,

(B2) $\min\{c^T x \mid Ax = 0,\ e^T x = 1,\ x \geq 0\} > 0$,

(B3) the optimal value of (P) is l^* : we don't assume that we know l^*, but we assume that we know a lower bound for it, say l^0. (It doesn't have to be a good bound, anyone will do, eg. -10^{20}).

Note, that assumption (B2) implies that the optimal value of (P) is finite, but is a little bit stronger. It is well known that the polyhedron of the feasible solutions of a linear programming problem can be written as the sum of the polytope generated by the extreme points and the recession cone. Assumption (B2) requires that all rays of the cone have positive value; non-negativity would be enough to guarantee that (P) has a finite optimal value. Note also that assumption (B3) implies that the problem is feasible.

As in Karmarkar's algorithm we will need a feasible interior solution as a starting point. If no strictly positive solution for (P) is known, we add an artificial variable x_{n+1} with high cost c_{n+1} :

$$\begin{aligned} \min\ (c^T, c_{n+1}) \begin{pmatrix} x \\ x_{n+1} \end{pmatrix} & \\ (A, -\sum_{i=1}^{n} a_i + b) \begin{pmatrix} x \\ x_{n+1} \end{pmatrix} &= b \\ x, x_{n+1} &\geq 0 \end{aligned}$$

where a_i are the column vectors of A. Thus e, the vector of all ones, is feasible.

To get a right–hand–side of zero, we introduce another variable and fix it at one. Note, that assumption (B3) implies that (P) is feasible, therefore the above linear programming problem will also have optimal value l^*, that is $x_{n+1} = 0$ in any optimal solution. By subtracting l^* from the objective function we obtain an equivalent linear problem with optimal value zero:

(P')

$$\min(c^T, c_{n+1}, -l^*)\begin{pmatrix} x \\ x_{n+1} \\ x_{n+2} \end{pmatrix}$$

$$(A, -\sum_{i=1}^{n} a_i + b, -b)\begin{pmatrix} x \\ x_{n+1} \\ x_{n+2} \end{pmatrix} = 0$$

$$x_{n+2} = 1$$

$$x, x_{n+1}, x_{n+2} \geq 0.$$

Now consider the transformation $\hat{T}$ which maps $R^{n+1} \times \{1\}$ onto the simplex in R^{n+2} :

$$\begin{aligned} \hat{T} : R_+^{n+1} \times \{1\} &\rightarrow S^{n+2} \\ x &\rightarrow y \text{ such that } y_j = \frac{x_j}{1+\sum_{i=1}^{n+1} x_i} \quad (j = 1, \ldots, n+1) \\ y_{n+2} &= 1 - \sum_{i=1}^{n+1} y_i = \frac{1}{1+\sum_{i=1}^{n+1} x_i}. \end{aligned}$$

Note that $\hat{T}x$ is in the interior of S^{n+2} whenever x is a strictly positive point and note that $\hat{T}e$ is the center of the simplex.

One can show that the transformation $\hat{T}$ transforms (P') into the linear program

(P")

$$\min\ (c^T, c_{n+1}, -l^*)y$$

$$(A, -\sum_{i=1}^{n} a_i + b, -b)y = 0$$

$$e^T y = 1$$

$$y \geq 0$$

with the properties

(1) x feasible in (P')$\Leftrightarrow$ $y = \hat{T}x$ feasible in (P") and

(2) Optimizers are preserved under $\hat{T}$, i.e. x optimizes (P')$\Leftrightarrow$ $y = \hat{T}x$ optimizes (P").

In order to develop an algorithm to solve linear programming problems of the form (P) it is therefore sufficient to develop one to solve linear programming problems of the form

($\hat{P}$)

$$\min\ c^T x$$

$$Ax = 0$$

$$e^T x = 1$$

$$x \geq 0$$

with the assumptions:

(1) $x^0 = \frac{1}{n}e$ is a feasible solution,

(2) $\min\{c^T x \mid Ax = 0,\ e^T x = 1,\ x \geq 0,\ x_n = 0\} > 0$,

(3) The optimal value of $(\hat{P})$ is zero for $l = l^*$, a lower bound for l^* is known to be l^0.

An additional difficulty is that only solutions with $x_n \neq 0$ correspond to optimal solutions of (P'). We have to keep this in mind. We can now start to develop the algorithm. We want to generate sequences $\{x^k\}$ and $\{l^k\}$ such that

(i) x^k are feasible solutions of $(\hat{P})$ and converge to an optimal solution of $(\hat{P})$,

(ii) $l^k \leq l^*$ and l^k converge to l^*.

Our starting points are by construction $x^0 = \frac{1}{n}e$ and l^0. Assume we already constructed x^{k-1} and l^{k-1}. How can we find x^k and l^k? As before we transform x^{k-1} back into the center of the simplex via the transformation

$$T : x \to \overline{x} = \frac{Dx}{e^T Dx} \qquad \text{where } D = \text{diag}(x_1^{k-1}, \ldots, x_n^{k-1}).$$

Then

$$\begin{array}{lllll} \hat{P}(l) & \min\ (c - l \cdot e_n)^T x & & \overline{P}(l) & \min\ (D(c - l \cdot e_n))^T x \\ & Ax = 0 & & & ADx = 0 \\ & e^T x = 1 & \text{and} & & e^T x = 1 \\ & x \geq 0 & & & x \geq 0 \end{array}$$

have optimal values of the same sign since $D > 0$. Similarly, their duals are

$$\begin{array}{lllll} \hat{D}(l) & \max\ z & & \overline{D}(l) & \max\ z \\ & A^T y + z \cdot e \leq c - l \cdot e_n & \text{and} & & DA^T y + z \cdot e \leq D(c - l \cdot e_n) \end{array}$$

and we know that $\hat{P}(l)$ and $\overline{P}(l)$ have the same value as $\hat{D}(l)$ and $\overline{D}(l)$, respectively. Also, for $l = l^*$ all four programs have optimal value zero.

Now let $y = (AD^2A^T)^{-1}AD^2(c - l^{k-1}e_n)$ and $z = \min_j(D(c - l^{k-1}e_n) - DA^T y)_j$. (y, z) is feasible for $\overline{D}(l^{k-1})$ because y and z are unrestricted in sign.

We now consider two different cases. If $z \leq 0$ we let $l^k = l^{k-1}$ and continue as in Karmarkar's algorithm. The case $z > 0$ indicates that the current lower bound for l^* is wrong: we found a feasible solution (y, z) for the dual $\overline{D}(l^{k-1})$ with value greater than zero. But we want $(\hat{P})$, and this implies $\overline{D}(l^{k-1})$, to have value equal to zero. We increase the lower bound for l^* in an appropriate way and can then continue as before.

Precisely the algorithm looks like:

Algorithm B:

step 0: (initialization) let $x^0 = \frac{1}{n}e$ and let $l^0 \leq l^*$ be given.

step k: let $D = \mathrm{diag}(x_1^{k-1}, \ldots, x_n^{k-1})$ and calculate $z = \min_j(D(c - l^{k-1}e_n) - DA^T y)_j$;

if $z > 0$ find $l^{k-1} \leq l^k \leq l^*$ such that

$\min_j(D(c - l^k e_n) - DA^T y)_j = 0 \quad \text{for} \quad y = (AD^2A^T)^{-1}AD^2(c - l^k e_n)$;

otherwise let $l^k = l^{k-1}$;

compute $d = -P_{\binom{AD}{e^T}}(D(c - l^k e_n))$ and $\overline{x} = \frac{1}{n}e + \alpha\frac{d}{\|d\|}$;

finally, set $x^k = \frac{D\overline{x}}{e^T D\overline{x}}$.

If $c^T x^k$ is less than required stop,

otherwise return to step k with "k" replaced by "$k+1$".

The following theorem shows that Algorithm B indeed converges.

Theorem 3.1: *Algorithm B generates a sequence of points such that*

(1) $l^* - \frac{c^T x^0 - \frac{1}{n} l^0}{\beta}\gamma^k \leq l^k \leq l^*$ *and*

(2) $l^* \leq \frac{c^T x^k}{x_n^k} \leq l^* + \frac{c^T e - l^0}{\beta}\gamma^k$

where β is a positive constant depending on the problem and $\gamma < 1$ is the constant given by Karmarkar's algorithm.

For a proof and more detailed comments see Steger [5].

References

[1]: Gay, D.M.: Karmarkar's Algorithm for Problems in Standard Form. *Mathematical Programming 37 (1987), 81–90.*

[2]: Karmarkar, N.: A New Polynomial–Time Algorithm for Linear Programming. *Combinatorica 4 (4)(1984), 373–395.*

[3]: Khachiyan, L.G.: A Polynomial Algorithm in Linear Programming. *Doklady Akademii Nauk SSSR 244 (1979), 1093–1096, translated in Soviet Mathematics Doklady 20 (1979), 191–194.*

[4]: Padberg, M.: A Different Convergence Proof of the Projective Method for Linear Programming. *Operations Research Letters 4 (1986), 253–257.*

[5]: Steger, A.: An Extension of Karmarkar's Algorithm for Bounded Linear Programming Problems. *M.S. thesis, State University of New York at Stony Brook (August 1985).*

[6]: Todd, M.J. and B.P. Burrell: An Extension of Karmarkar's Algorithm for Linear Programming Using Dual Variables. *Algorithmica 1 (1986), 409–424.*

Dynamische Bestellmengen- und Losgrößenplanung
Verfahrensübersicht und Vergleich

Andreas Robrade und Klaus Zoller, Hamburg

Zusammenfassung: Die neuere Literatur bietet eine Vielzahl von Heuristiken zur dynamischen Bestellmengen-/Losgrößenplanung. In einem strukturellen Vergleich von Konstruktionsmerkmalen und in numerischen Tests werden diese auf ihre relative Leistungsfähigkeit untersucht. Dabei zeigt sich, daß bei regelmäßiger Bedarfsentwicklung die Abbruchregel von Groff und bei stärker schwankendem Bedarf ein von den Verfassern entwickelter Algorithmus den in kommerziellen Softwaresystemen implementierten Verfahren vorzuziehen ist.

Summary: Recent research on single-item dynamic lot sizing produced a number of different heuristic approaches to the problem. This paper compares their formal structures and reports on the result of extensive numerical tests. It is suggested that (1) the heuristics most frequently implemented in commercial software packages are inefficient by to-day's standards, (2) Groff's stopping rule should be used when demand is fairly regular, and (3) an algorithm proposed by the authors prevails under conditions of erratic demand.

Für die Planung isolierter Fertigungslose/Bestellmengen bieten zahlreiche Hersteller EDV-Unterstützung an. Eine Bestandsaufnahme bei 10 kommerziellen Softwarepaketen zeigt, daß hier mit wenigen und recht betagten Verfahren operiert wird (Tab.1), obwohl die deutsche und angelsächsische Literatur ein breites Spektrum von Verfahren

Tab.1: Bestellmengen-/Losgrößenberechnung in Standard-Softwaresystemen

Hersteller	System	Verfahren: Harris	LUC[1]	PPR[2]
Bull:	IMS-TD/MIACS-TD	x	x	
CTM:	PLANOS-MB	x		
Digital:	VAX-Profi			x
HP:	HP-MPN-MM3000 und PM3000	x		x
IBM:	COPICS	x	x	x
Nixdorf:	COMET-TOP FEROS und LAWI	x		
Philips:	Fertigungsorganis. 4000	x		x
SAP:	System RM		x	x
Siemens:	IS-BS PBV	x	x	
Sperry:	UNIS	x	x	

1) vgl. Abschn. 2.1 2) vgl. Abschn. 2.2

unterschiedlicher Leistungsfähigkeit aufweist (Tab.2). Eine vergleichende Untersuchung bietet sich an.

Tab.2: Heuristische Verfahren der deterministisch-dynamischen Bestellmengen-/Losgrößenplanung

Abbruchregeln	Algorithmen
LUC: Least Unit Cost	• Incremental Order-Algorithmen
PPR: Part Period-Regel (DeMatteis/Mendoza 1968)	IOA-TR: Trux (1972)
	IOA-GA: Gaither (1981)
SMR: Silver/Meal-Regel (1973)	• Part Period-Algorithmen
	PPA-FB: Forward / Backward (DeMatteis/Mendoza 1968)
GRR: Groff-Regel (1979)	PPA-BM: Blackburn/Millen (1979)
IOQ: Incremental Order Quantity (Boe/Yilmax 1983)	PPA-MG: Maximum Gain (Axsäter 1980/Karni 1981)
	• Silver/Meal-Algorithmus
	SMA-AM: Absolutes Minimum (Blackburn/Millen 1980)
	• Kombinationsalgorithmen (Robrade/Zoller 1987)
	K-GRsr: auf Basis GRR
	K-SMsr: auf Basis SMR

Im folgenden werden nach einer Erörterung der Planungssituation (Abschn.1) zunächst die in Abbruchregeln (Abschn.2) und Algorithmen (Abschn.3) gruppierten Verfahren kurz skizziert (Details in der zitierten Literatur und in Robrade/Zoller 1987). Es schließt sich eine Darstellung umfangreicher numerischer Tests an (Abschn.4), die Anlaß für eine Überarbeitung gewerblich angebotener Softwaresysteme geben sollte (Abschn.5).

1. Planungssituation

Dynamische Bestellmengen- und Fertigungslosgrößenrechnung geht von einem zeitlich differenzierten Netto-Bedarf d_t, $t=1,2,...,T$, aus. Verbindlich entschieden wird jeweils nur über den Umfang (x) bzw. die Reichweite ($\tau \leq T$) der anstehenden (t=1) Vorratsergänzung, $x_1 = \Sigma_{h=1}^{\tau} d_h$. Erst nach deren Verbrauch - planmäßig also zu Beginn der Periode $(\tau+1)$ - folgt die Entscheidung über $x_{\tau+1}$; dies jedoch auf der Basis einer aktualisierten Bedarfszeitreihe $d_1^a, d_2^a, ..., d_T^a$, deren Werte mit den ursprünglichen Daten $d_{\tau+t}$ im Überlappungsbereich $t = 1.2,...,T-\tau$ nicht übereinstimmen müssen.

Die Kosten des einzelnen (hier: ersten) Zyklus kann man mit

$$K(\tau) = C1 + C2 \cdot \Sigma_{h=1}^{\tau}(h-1) \cdot d_h$$

angeben, wobei

C1 die fixen (Bestell-, Einrichte-, Umrüst-) Kosten

C2 die von der vermeidbaren Lagerzeit (s.u.) abhängigen Kosten für Lagerung und Kapitalbindung pro ZE und ME angibt und unterstellt wird, daß die Bedarfsmenge d_t jeweils zu Beginn der Periode t verfügbar sein muß, eventuelle Lagerungs- und Kapitalbindungskosten der Bedarfsperiode selbst also als unvermeidlich außer Betracht dieser Planungsrechnung bleiben (vgl. dazu ausführlich Knolmayer 1985, S.413ff).

Unterstellt man weiter, daß die Bedarfe d_t, t=1,2,...,T, zuverlässig bekannt sind und daß am Ende der Periode T das Lager leer (der zyklische Vorrat abgebaut) sein muß, so kann man mit einem allerdings sehr aufwendigen Verfahren (Abschn.4.1.a) die über T Perioden kostenminimale Sequenz von Bestellmengen/Fertigungslosen berechnen (Wagner/Whitin 1958). Damit ist eine Grundlage für Vergleichsrechnungen gegeben.

2. Abbruchregeln

Myopische Heuristiken oder Abbruchregeln erhöhen schrittweise die Reichweite von x_1, τ=1,2,..., bis die Überschreitung eines Grenzwerts signalisiert, daß τ^* erreicht ist. Sie unterscheiden sich hinsichtlich der Konstruktion dieses Grenzwerts:

- Minimierung der Stückkosten: LUC
- Gleichheit von Lager- und Bestellfixkosten pro Zyklus: PPR
- "Minimierung" der Kosten pro Zeiteinheit: SMR
- Ausgleich der (zeitlichen) Grenzkosten für Einrichtung/Bestellung und Lagerkosten: GRR
- Ausgleich von Lagerkosten eines Periodenbedarfs d_t und Bestellfixkosten: IOQ

Im unmittelbaren Vergleich stellen sich die hier erörterten Abbruchregeln folgendermaßen dar (vgl. Robrade/Zoller 1987):

$$K(\tau^*)\begin{cases} \leq C2\cdot\tau^*\cdot\Sigma_{h=1}^{\tau^*}d_h & \text{LUC} \quad (1)\\ < C2\cdot\Sigma_{h=1}^{\tau^*}(h-1)\cdot d_h + C2\cdot\Sigma_{h=1}^{\tau^*+1}(h-1)\cdot d_h & \text{PPR} \quad (2)\\ < C2\cdot(\tau^*)^2\cdot d_{\tau^*+1} & \text{SMR} \quad (3)\\ \leq C2\cdot\Sigma_{h=1}^{\tau^*}(h-1)\cdot d_h + \frac{1}{2}\cdot C2\cdot\tau^*\cdot(\tau^*+1)\cdot d_{\tau^*+1} & \text{GRR} \quad (4)\\ \leq C2\cdot\Sigma_{h=1}^{\tau^*+1}(h-1)\cdot d_h & \text{IOQ} \quad (5)\end{cases}$$

Grundsätzliche methodische Einwände lassen sich nur gegen LUC und IOQ vorbringen:

- LUC ist einseitig rückwärts orientiert und macht die Reichweiten-Entscheidung von dem in diesem Zusammenhang wenig sinnvollen Kriterium "Kosten der Lagerung der Gesamtmenge x_1 für die Gesamtdauer τ" abhängig;
- IOQ liefert i.d.R. erhebliche Überreichweiten (vgl. (2) und (5)) und erreicht in der statischen (Grenz-) Situation im Gegensatz zu den anderen Abbruchregeln keine optimalen Lösungen.

Die Auswahl einer Abbruchregel konzentriert sich daher auf PPR, SMR und GRR.

3. Algorithmen

Abbruchregeln enden nach τ^* bzw. (τ^*+1) Iterationen mit einem definitiven Ergebnis. Für Algorithmen gilt dies nicht: Sie suchen die Entscheidung über τ^* durch Einbeziehung zusätzlicher Informationen (Exploration) und/oder Vergleichsrechnungen zu verbessern bzw. abzusichern.

Obwohl z.T. eigenständig entwickelt, lassen sich diese Verfahren den Kriterien IOQ, PPR und SMR zuordnen. Zu GRR existiert u.W. bisher kein Algorithmus. SMR und GRR lassen sich jedoch in einem Kombinationsalgorithmus vorteilhaft verknüpfen.

IOQ Algorithmen:

- IOA-TR: Für eine zunächst mit der IOQ bestimmte obere Schranke τ^* wird anschließend geprüft, ob die korrespondierende Menge kostensenkend in zwei Teilmengen aufgespalten werden kann (Lagerkostenreduktion durch das zweite Los übersteigt zusätzliche Bestellfixkosten).
- IOA-GA: Mit der IOQ werden zwei aufeinanderfolgende Lose/Bestellmengen, τ_1^* und τ_2^*, bestimmt. Anschließend wird geprüft, ob die Ausgliederung von $d_{\tau_1^*}$ Kosten einspart (Lagerkostenreduktion für $d_{\tau_1^*}$ größer als zusätzliche Lagerkosten für das zweite Los).

PPR Algorithmen:

- PPA-FB: Nach einer vorläufigen Bestimmung von τ^* mit PPR folgt ein Forward/Backward-Test. Die Reichweite wird auf (τ^*-1) gekürzt, falls $2 \cdot d_{\tau^*} \leq d_{\tau^*+1}$, ansonsten auf den höchsten Wert τ erhöht, für den stets noch $(\tau-1) \cdot d_\tau \leq d_{\tau+1}$ gilt.
- PPA-BM: Für das durch die PPR ermittelte τ^* gilt, daß die Lagerkosten gerade noch kleiner sind als C1. PPA-BM wählt die Reichweite, deren betragsmäßige Differenz aus Lagerkosten und C1 am geringsten ist (Differenzvergleich für τ^* und (τ^*+1)).
- PPA-MG: Das PPR-Kriterium wird genutzt, um einen iterativen Prozeß zu steuern. Dessen Ausgangslösung $x_t = d_t, \forall t$, (Lot for Lot), ist schrittweise durch Zusammenlegung zweier aufeinanderfolgender Lose/Bestellmengen zu verbessern. Die Auswahl der beiden Lose/Bestellmengen erfolgt durch Ermittlung des "maximum gain" (Karni 1981) bzw. maximalen "saving" (Axsäter 1980).

SMR-Algorithmus:

- SMA-AM: Das der SMR zugrunde liegende Kriterium "Kosten pro Zeiteinheit" ist nicht konvex. SMA-AM ermittelt das absolute Minimum, indem er über das erste Minimum (SMR) hinaus weiterprüft.

Kombinationsalgorithmen

Die Kombinationsalgorithmen wurden im Rahmen dieser Untersuchung entwickelt und werden daher hier ausführlicher dargestellt. Unter mehreren Varianten (vgl. Robrade/Zoller 1987) erweist sich ein Algorithmus als zweckmäßig, der Reichweiten nach GRR berechnet, vergleichende Bewertungen hingegen auf der Grundlage von Kosten pro ZE (SMR) vornimmt.

Referenzlösung:

Mit GRR Berechnung von τ_1^* und τ_2^*, so daß

$$x_1 = \Sigma_{h=1}^{\tau_1^*} d_h \quad , \quad x_{\tau_1+1} = \Sigma_{h=\tau_1^*+1}^{\tau_1^*+\tau_2^*} d_h$$

und

$$k_0 = (K(\tau_1^*) + K(\tau_2^*))/(\tau_1^* + \tau_2^*)$$

(Definition: $\tau_k' = \tau_k^*(t_{k-1})$ bezeichne die nach GRR kostengünstigste Fortsetzung einer Folge von Reichweiten $\tau_1',\ldots,\tau_{k-1}'$ mit $t_{k-1} = \Sigma_{h=1}^{k-1}\tau_h'$, $k \geq 2$.)

Revision:

Für $i = -2,-1,+1,+2$: $\tau_1' = \tau_1^* + i$ und Berechnung von $\tau_k' = \tau_k^*(t_{k-1})$ für $k = 2,3,\ldots,m$ mit

$$t_{m-1} < t_2^* \leq t_m \quad , \quad t_2^* = \tau_1^* + \tau_2^*$$

und Auswertung

$$k_i = \Sigma_{h=1}^{n} K(\tau_h')/\Sigma_{h=1}^{n}\tau_h'$$

mit

$$n = \begin{cases} m-1, & \text{falls } t_2^* - t_{m-1} \leq |t_2^* - t_m| \\ m & \text{sonst} \end{cases}$$

Dann ist

$$k_{i^*} = \min_{-2\leq i\leq +2} \{k_i\} \quad => \quad \tau^{**} = \tau_1^* + i^*$$

In der hier dargestellten Form

Abbruchregel GRR Stufen der Referenzlösung: $s = 2$ Revisionstiefe: $r = 2$	K-GR22

erweist sich der Kombinationsalgorithmus als besonders leistungsfähig.

4. Numerische Tests

Einer übersichtsweisen Zusammenfassung und Bewertung der Ergebnisse umfangreicher Simulationsrechnungen (Abschn.4.2) stellen wir die Beschreibung der dafür gewählten Versuchsanordnung voran, da diese erheblichen Einfluß auf die Aussagefähigkeit solcher Tests besitzt (s.a. Knolmayer 1985, S.422).

4.1 Versuchsanordnung

Die wesentlichen Determinanten der Versuchsanordnung sind (a) das Konzept einer optimalen Lösung, (b) die Wahl der Teststrecken und (c) die Maßstäbe zur Bewertung der getesteten Heuristiken.

(a) Optimalität

Zuverlässige Aussagen über die Qualität dynamischer Planungsverfahren erfordern eine Betrachtung längerer Zeiträume. Optimalität kann überdies nur im Hinblick auf vollständig definierte, zeitlich exakt abgegrenzte Planungsaufgaben festgestellt werden, angesichts der Planungsunsicherheit längerer Zeiträume also i.d.R. nur rückblickend. Immerhin ermöglichen solche ex post-Betrachtungen statistisch abgesicherte Aussagen über die relative Leistungsfähigkeit unterschiedlicher Planungsverfahren. Den Maßstab hierfür liefert der dynamische Algorithmus von Wagner/Whitin (1958).

Als optimal i.S. der operativen Aufgabenstellung kann dessen Lösung jedoch nur dann gelten, wenn die Planungsaufgabe auch ex ante vollständig definiert und exakt begrenzt ist:

1. Sind d_t Prognosewerte und deshalb mit zunehmender Unsicherheit behaftet, so erwächst daraus bei hinreichend großem Planungshorizont T ein systematisches Problem, das kaum befriedigend zu lösen ist.
2. Ist hingegen T zu knapp bemessen, und reicht der Bedarf über T hinaus, so geht hiervon eine kaum allgemein abschätzbare kostenmehrende Wirkung auf die - formal weiterhin "optimale" - Lösung aus, wie Abb.1 für T = 7 / 10 / 13 / 16 zeigt.

Insoweit erfährt der Optimalitätsbegriff des WW-T bei rollierender Planung eine nicht zu unterschätzende Relativierung, reduziert er sich auf eine Kategorie von eher experimenteller als praktischer Relevanz.

(b) Teststrecken

Als exogene Determinanten der Leistungsfähigkeit von Heuristiken erwiesen sich bereits in früheren Untersuchungen (z.B. Ohse 1969, Knolmayer 1985) die Relation von Bestellfix- und Lagerkosten C1:C2 (part periods) sowie die numerische Struktur der Bedarfsdaten.

Erstere wurden in dieser Untersuchung über

C1:C2 = 1.000 / 2.500 / 5.000 / 7.500 / 10.000 / 15.000

variiert. In Verbindung mit einem mittleren Bedarf von 1.000 ME/ZE führt dies zu mittleren Reichweiten $1 \leq \tau \leq 6$.

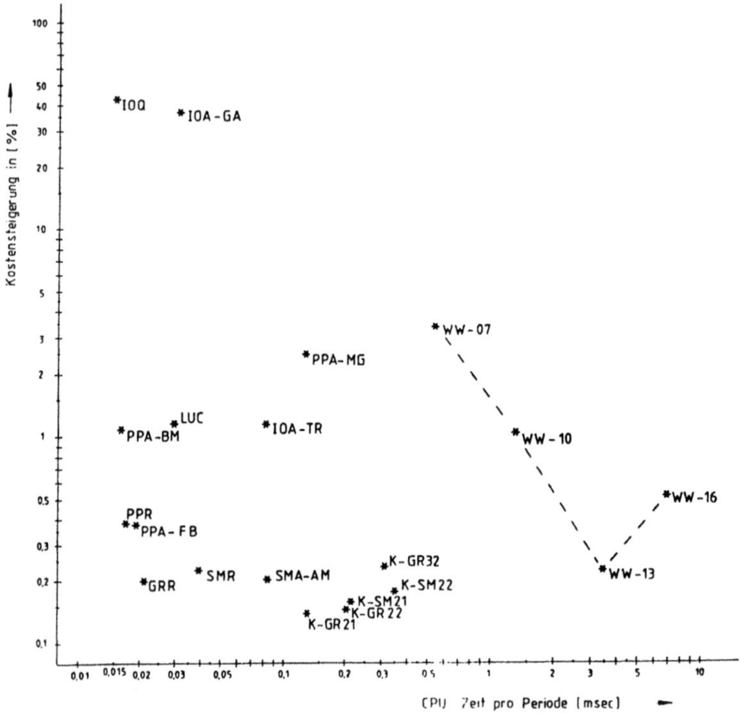

Abb.1: Testergebnisse - erratischer Bedarf (einer von 30 Tests)

Mit Rücksicht auf die Entstehung von Bedarfsdaten in der Praxis - zum einen aus Prognosen mit mathematischen Funktionen (Regressionsrechnung), zum anderen aus solchen über Koeffizienten (z.B. Saisonmodelle) und aus deterministischen Sekundärbedarfsrechnungen - wurde hier unterschieden zwischen

- systematischem Bedarf, der einer mathematischen Gesetzmäßigkeit folgt: linearer, progressiver und degressiver Trend, ohne und mit Überlagerung durch additive und multiplikative Saisoneinflüsse sowie Stufenformen - insgesamt 11 Teststrecken mit je 500 Perioden (Details in Robrade/Zoller 1987),
- erratischem Bedarf, simuliert als stationärer Prozeß mit Erwartungswert 1.000 ME/ZE und gleichverteilter Störung im Intervall ±20%, ±40%, ±60%, ±80% und ±100% - insgesamt 5 Teststrecken mit je 4.000 Perioden, auf denen wahrscheinlichkeitstheoretisch fundierte Aussagen gewonnen werden können.

In beiden Fällen wird jedoch deterministischer Bedarf angenommen. Damit kann - anders als in der Praxis - die eigentliche Leistungsfähigkeit der Verfahren unverfälscht von Prognosefehlern beobachtet werden.

(c) Bewertung

Grundsätzlich interessieren die zu erwartenden Mehrkosten der Heuristiken gegenüber vergleichbaren optimalen Lösungen, das Risiko höherer als der erwarteten Mehrkosten im Einzelfall sowie die benötigten Rechenzeiten. In den hier referierten Tests wurden diese Größen wie folgt gemessen:

Auf Abschnitten von jeweils $T = 50$ Perioden erzeugen die Verfahren Reichweiten-Sequenzen $\tau_1^*, \ldots, \tau_m^*$. Dabei wird m durch

$$\Sigma_{h-1}^{m-1} \tau_h^* < 40 \leq \Sigma_{h-1}^{m} \tau_h^* =: T^* < 50$$

bestimmt und somit sichergestellt, daß die (bei rollierender Planung artifizielle) Endlichkeit des Informationshorizonts T ohne verfälschende Auswirkungen auf das Ergebnis bleibt (anders z.B. Ohse 1969).

Mit den o.e. mittleren Reichweiten $\bar{\tau}$ ergeben sich hieraus jeweils $8 \leq m \leq 40$ Zyklen. Deren gesamte Kosten $C(\tau_1^*, \ldots, \tau_m^*)$ werden mit den korrespondierenden Minimalkosten nach WW-T*, $K_{T^*}^*$, verglichen. Die relativen Differenzen

$$\kappa = \frac{C(\tau_1^*, \ldots, \tau_m^*) - K_{T^*}^*}{K_{T^*}^*} \cdot 100 \qquad [\%]$$

konstituieren jeweils eine Beobachtung.

- Ein Test mit systematischem Bedarf liefert $500/50 = 10$ Beobachtungen, die sich durch systematische Variation der Funktionsparameter voneinander unterscheiden. Da sie somit nicht statistisch i.e.S. sind, dient ihr Mittelwert $\bar{\kappa}$ als Maß für die zu erwartenden Mehrkosten und der Maximalwert κ_{max} als Indikator für das Mehrkosten-Risiko einer Heuristik. Insgesamt wurden mit 6 Kombinationen C1/C2 und 11 Teststrecken 66 Tests mit systematischen Bedarfen durchgeführt.
- Ein Test mit erratischem Bedarf liefert $4.000/50 = 80$ statistische Beobachtungen, deren Mittelwert $\bar{\kappa}$ die erwarteten Mehrkosten angibt und deren Varianz s_κ^2 gemeinsam mit diversen Quantilen der Häufigkeitsverteilung von κ als Risikomaß dient. Insgesamt wurden mit 6 Kombinationen C1/C2 und 5 Teststrecken 30 Tests mit erratischen Bedarfen durchgeführt.

Rechenzeiten wurden als reine CPU-Zeit (UNIVAC 1100-82) in separaten Produktionsläufen über 500 bzw. 4.000 Perioden gemessen und in msec/Periode umgerechnet, um die Einflüsse unterschiedlicher Reichweiten (s.o., in der einzelnen Beobachtung: $\bar{\tau} = T^*/m$) auszuschalten.

4.2 Ergebnisse

Damit ist eine dreiwertige Zielvorstellung erklärt sowie ein Spektrum von 96 erratischen und systematischen Tests, deren Ergebnisse hier nur in zusammengefaßter Form wiedergegeben werden. Eine detaillierte Dokumentation (Robrade/Zoller 1987) kann bei den Verfassern angefordert werden.

Bei systematischem Bedarf sind die Abbruchregeln allen Algorithmen, einschließlich WW-T ($T \leq 16$) überlegen. Dies äußert sich in

- Mittelwerten $\bar{\kappa} \leq 0{,}1\%$ bei leichten Vorteilen für GRR, PPR, SMR,
- Maximalwerten κ_{max}, die mit $\bar{\kappa}$ hochgradig rangkorreliert sind (Spearman's Rangkorrelationskoeffizient $r^S \geq 0{,}98$ in 58 von 66 Tests; zu r^S vgl. Hartung/Elpelt 1986, S.191f) und durchweg im Bereich $\kappa_{max} \leq 0{,}9\%$ liegen,
- Rechenzeiten unter 0,06 msec/Periode (Faktor 10 schneller als die heuristischen Algorithmen, Faktor 100 schneller als WW-T) bei leichten Vorteilen für GRR und PPR.

Damit erscheint es zulässig und sinnvoll, bei systematischem Bedarf

- auf das offensichtlich redundante Kriterium κ_{max} zu verzichten und
- mit anzunehmenden Kosten von nicht mehr als 2.000 DM/CPU-h (≙ 0,0005 DM/msec) die Rechenzeiten als rangniedrigeres zweites Kriterium zur lexikographischen Definition einer vollständigen Ordnung (Bamberg/Coenenberg 1985, S.43ff) der Heuristiken heranzuziehen ("tie breaker").

Tab.3: Rangmatrix systematischer Bedarf[1)]

Verfahren	Ränge														
	1	2	3	4	5	6	7	8	9	10	11	12	13	14	15
LUC		1	2	25		6	4	3	2	15	2		6		
PPR	29	17	3	12		1		1	3						
SMR		3	3	8	35	11	1	1	1	3					
GRR	24	13	21	4	2			1	1						
IOQ	13											2	7	4	40
IOA-TR				3	11	1					15	11	17	7	1
IOA-GA	5	6										5	4	40	6
PPA-FB	8	17	21	3		2			1	1	2			5	6
PPA-BM											19	22	5	8	12
PPA-MG	1				1	1	8	6	2	1	12	12	21		1
SMA-AM	1		5	5	11	25	1	2	1	1	1	7	4	2	
K-SM21			1	2		5	8	10	28	9	3				
K-SM22				1	1		5	8	11	24	9	5	2		
K-GR21	2			6	3	8	28	8	9	2					
K-GR22		1	1		3	4	10	25	7	10	3	2			

1) Eintragungen x_{ij}: Verfahren i erreichte in 66 Tests x-mal den Rang j

Das Ergebnis dieser Ordnung ist in Tab.3 wiedergegeben, aus der hervorgeht, daß GRR und PPR mit insgesamt 62 (61) Nennungen auf den ersten vier Rängen vorzuziehen sind. Die (in $\bar{\kappa}$ und κ_{max}) keineswegs schlechteren Algorithmen sind wegen ihres höheren Rechenzeitbedarfs hier nicht konkurrenzfähig.

Bei erratischem Bedarf wandelt sich dieses Bild. Bei näherer Betrachtung der Test-Ergebnisse treten hier die Stärken der heuristischen Algorithmen deutlich zutage. Dies äußert sich in

- deutlich geringeren Mittelwerten $\bar{\kappa}$, wobei (nur) K-GR22 durchweg $\bar{\kappa} \leq 1\%$ aufweist und damit um 2/3 besser abschneidet als die beste Abbruchregel (GRR), während bei anderen Verfahren bis zu 40% (erwartete) Mehrkosten auftreten.
- deutlich geringeren Standardabweichungen s_κ, wobei auch hier zwischen $\bar{\kappa}$ und verschiedenen Quantilen (50%, 67%, 75%, 95%) jeweils sehr hohe Rangkorrelationen zu beobachten sind.

Einen deutlicheren Eindruck von der relativen Leistungsfähigkeit vermittelt die analog zu Tab.3 gebildete Rangmatrix für erratischen Bedarf (Tab.4), die K-GR22 vor K-GR21, K-SMrs vor den übrigen Algorithmen und diese in ihrer Mehrzahl nochmals klar vor den myopischen Abbruchregeln (außer GRR und SMR) zeigt.

Tab.4: Rangmatrix erratischer Bedarf[1)]

Verfahren	Ränge														
	1	2	3	4	5	6	7	8	9	10	11	12	13	14	15
LUC									1	1	3	3	11	6	5
PPR							1	2	5	4	16	2			
SMR					4	9	6	3	5		1	2			
GRR					9	7	7	2		5					
IOQ								2	3		1		2	1	21
IOA-TR				2	3	1	5	7	2	2	5		3		
IOA-GA						3	2			1	1	4	2	17	
PPA-FB						2	1	5	10	7		2	3		
PPA-BM										3		13	4	6	4
PPA-MG	2	1		1	8	1	1	3	1	4	3	3	2		
SMA-AM					3	4	7	6	3	3		1	3		
K-SM21	1	1	6	16	3	3									
K-SM22	5	10	9	6											
K-GR21	8	4	14	4											
K-GR22	14	14	1	1											

1) Eintragungen x_{ij}: Verfahren i erreichte in 30 Tests x-mal den Rang j

5. Zusammenfassung

Die vorliegende Untersuchung erhebt keinen Anspruch auf Vollständigkeit und bestätigt z.T. (hinsichtlich der hier als Abbruchregeln klassifizierten Verfahren) bekannte Ergebnisse - vgl. insbesondere Knolmayer (1985) und die dort angegebene Literatur. Gleichwohl läßt sie einige u.E. weiterführende Schlußfolgerungen zu:

(a) Da Rechenzeiten mittlerweile eine eher untergeordnete Rolle spielen, ist den durch methodische Schwächen von Heuristiken verursachten Mehrkosten verstärkt Beachtung zu schenken. Hier interessiert jedoch neben den zu erwartenden auch die Gefahr höherer als der erwarteten Mehrkosten, d.h. die Zuverlässigkeit des Verfahrens. Die Untersuchung zeigt, daß Heuristiken mit hoher (geringer) erwarteter Leistungsfähigkeit durchweg auch (un)zuverlässiger sind.

(b) Dies unterstreicht den Imperativ, in kommerziellen (v.a. PPS-) Software-Systemen Verfahren entsprechender Leistungsfähigkeit auch dann vorzuhalten, wenn nicht als gesichert gelten kann, daß das relevante Kostenvolumen (aller betroffenen Artikelpositionen) bei allen Anwendern sehr groß ist. Hieraus ergibt sich - Tab.5 unterstreicht dies in Zusammenfassung der Tests mit den (für die Praxis bedeutsameren) erratischen Bedarfen - , daß die bisher implementierten Verfahren jedenfalls durch GRR, besser jedoch durch K-GR22 ersetzt werden sollten.

Tab.5: Rangmatrix für vier Heuristiken - erratischer Bedarf[1)]

Verfahren	Ränge 1	2	3	4
LUC				30
PPR		1	29	
GRR		29	1	
K-GR22	30			

1) Eintragungen x_{ij}: Verfahren i erreichte in 30 Tests x-mal den Rang j

(c) In methodischer Hinsicht unterscheiden sich die Kombinationsalgorithmen von den anderen hier vorgestellten Verfahren insofern, als sie sich weder am myopischen Kriterium der Kosten anstehender Entscheidungen noch an dem (bei rollierender Planung irreführenden) Merkmal der auf den Informationshorizont T bezogenen Gesamtkosten (Wagner/Whitin 1958, Axsäter 1980, Karni 1981) orientieren: Als Zielgröße dienen vielmehr die mittelfristigen Kosten pro ZE, wobei der Bezugszeitraum selbst (t_s^*, hier: s = 2) intrinsisch bestimmt wird.

Gegenstand dieser Untersuchung waren ausschließlich Heuristiken zur isolierten Bestimmung vermeintlich oder tatsächlich unabhängiger Bestellmengen und Fertigungslose. Daß damit nur ein kleiner Ausschnitt der im übrigen von Interdependenzen (z.B. Kapazitäts- und Koordinationsproblemen, Zoller 1980) gekennzeichneten logistischen Praxis getroffen wird, darf nicht verkannt werden (Axsäter 1980), sollte aber auch nicht dazu verleiten, durch verbesserte Planungsmethoden eröffnete Rationalisierungsmöglichkeiten ungeprüft zu übergehen.

Literatur

1. Axsäter S (1980) Economic lot sizes and vehicle scheduling. EJOR 4: 395-398
2. Bamberg G, Coenenberg A G (1985) Betriebswirtschaftliche Entscheidungslehre. 4. Aufl., Vahlen, München
3. Blackburn J D, Millen R A (1979) Selecting a Lot-Sizing Technique for a Single Level Assembly Process: Part I - Analytical Results. Production and Inventory Management 20, Nr. 3: 42-47
4. Blackburn J D, Millen R A (1980) Heuristic Lot-Sizing Performance in a Rolling Scheduling Environment. DS 11: 691-701
5. DeMatteis J J (1968) An economic lot-sizing Technique. I: The Part-Period Algorithm. IBM Systems Journal 7: 30-38
6. Gaither N G (1981) A Near-Optimal Lot-Sizing Model for Material Requirements Planning Systems. Production and Inventory Management 22, Nr. 4: 75-89
7. Groff G K (1979) A Lot Sizing Rule for Time-phased Component Demand. Production and Inventory Management 20, Nr. 1: 47-53
8. Hartung J, Elpelt B (1986) Multivariate Statistik. 2. Aufl., Oldenbourg, München u.a.
9. Karni R (1981) Maximum Part-Period Gain (MPG) - A Lot Sizing Procedure for unconstrained und constrained Requirements Planning Systems. Production and Inventory Management 22, Nr. 2: 91-98
10. Knolmayer G (1985) Zur Bedeutung des Kostenausgleichsprinzips für die Bedarfsplanung mit PPS-Systemen. ZfbF 37, Nr. 5: 411-427
11. Mendoza A G (1968) An economic lot-sizing Technique. II: Mathematical Analysis of the Part-Period Algorithm. IBM Systems Journal 7: 39-46
12. Ohse D (1969) Lagerhaltungsmodelle für deterministisch schwankenden Absatz. APF 10: 309-322
13. Robrade A, Zoller K (1987) Dynamische Bestellmengen- und Losgrößenplanung. Bericht des Instituts für Betriebliche Logistik und Organisation der Universität der Bundeswehr Hamburg, 87/1
14. Silver E A, Meal H (1973) A Heuristic for selecting Lot-Size Quantities for the Case of a deterministic Time-varying Demand Rate and discrete Opportunities for Replenishment. Production and Inventory Management 14, Nr. 2: 64-74
15. Trux W R (1972) Einkauf und Lagerdisposition mit Datenverarbeitung. 2. Aufl., Moderne Industrie, München
16. Wagner H M, Whitin T M (1958) Dynamic Version of the Economic Lot Size Model, MS 5: 89-96
17. Zoller K (1980) Effiziente Bestellpolitiken in Mehrproduktlägern. In: Proc. Op. Res. 9, hrsg. von J. Schwarze, L. von Dobschütz, B. Fleischmann, Ch. Schneeweiß und H. Steckhan. Würzburg, Wien: 79-85

Hedging and Standard - MRP

K. van Donselaar and J. Wijngaard, Eindhoven

Zusammenfassung: In MRP-Paketen ist es möglich, Normen für Sicherheitsbestände zu verwenden. Jedoch keines der MRP-Pakete bietet Hedging als eine alternative Möglichkeit für Sicherheitsnormen an, obwohl diese Technik aus theoretischer Sicht mehrere Vorteile bietet. An der Universität in Eindhoven hat man versucht, Hedging in ein existierendes MRP-Paket einzubauen. Der funktionale Entwurf zeigte: Was auf den ersten Blick aussah wie eine einfache Ergänzung eines MRP-Pakets, hatte schließlich einen großen Einfluß auf viele MRP-Module.

Summary: In standard-MRP packages it is possible to use safety stocknorms. In none of the standard-MRP packages however, it is possible to use hedging as an alternative technique for safety stocknorms, although this technique has several advantages from a theoretical point of view. At Eindhoven University of Technology an attempt is made to incorporate hedging in an existing MRP-package.
From the functional design it appeared, that what originally seemed to be a simple addition to the options available in an MRP-package, turned out to have a rather large impact on many of the MRP-modules.

1. Introduction.

The goal of this paper is to investigate the flexibility of standard-MRP with respect to building in an extra option called hedging. The logic of MRP is assumed to be known here. Details on MRP can be found in Orlicky's book [1]. Hedging is an alternative for the use of safety stock norms and is described in an article by Miller [2].

A general introduction on hedging is given below together with an argumentation on the merits and demerits of hedging. In section 3 two possible ways to implement hedging in an MRP-package are mentioned. The impacts of hedging on MRP are discussed in Section 4.

Operations Research Proceedings 1987

2. Hedging.

* Introduction on hedging.

Hedging is an alternative for the use of norms for safety stock or safety time. Ultimately it will result in extra physical stock, which can be used to absorb all kind of fluctuations, especially fluctuations in demand. Hedging consists of raising the requirements for a product by adding time-phased extra requirements called hedges. Hedging is usually performed at the Master Production Schedule (MPS)-level.

Example: Assume the following requirements and hedges for a product called C:

period	1	2	3	4	5	6	7
Requirements	10	10	10	10	10	10	10
Hedges	5	-	-	3	-	-	-
Total requirements	15	10	10	13	10	10	10

Figure 1. Requirements planning for product C in period 1.

The actual production for product C is based on total requirements. Instead of 10 units in period 1 (based on the requirements only), 15 products are made. A safety stock of 5 products will be the result.

With a leadtime of 2 periods for product C the hedge in period 4 does not influence the order release for C in the current period. It does however enlarge the planned order release of C in period 2. Consequently the requirements for a component of C, say product D, are enlarged. If D has a leadtime of 6 periods, the order release of D is affected by the hedge in period 4 and in the end there will be created a safety stock of three units of D.

At the beginning of the next period (period 2) the hedge in period 4 will enlarge the order release for C and thus result in safety stock for product C unless action is taken. If for example every period the hedge is put off one period (see figure 2), the hedge in period 4 still leads to safety stock for D, the component of C, but not to safety stock for product C itself. In effect hedging distributes the safety stock over the production chain.

period	2	3	4	5	6	7
Requirements	10	10	10	10	10	10
Hedges	5	-	-	3	-	-
Total requirements	15	10	10	13	10	10

Figure 2. Requirements planning for product C in period 2.

The benefit of having a hedge in period 4 is, that the requirements for C three periods from now on may be raised with 3 products without endangering the material availability of the components. With no safety stock and a leadtime of 6 periods for D this would not have been possible. Note that the safety stock for D is controlled by the hedges of C.

* Advantages and disadvantages of hedging.

Advantages of hedging are:

1. The safety stocknorm is implemented on the MPS-level. Due to the per definition limited number of items at this level the control of the stocknorms is relatively simple.
2. The hedge is implemented as a time-phased extra requirement. This results in a reduction of the number of reschedule-in messages compared to a safety stock norm in standard MRP-systems. In standard MRP systems the safety stock norm is implemented as an immediate requirement, equivalent to one hedge in period 0. In case the safety norm is implemented as a time-phased extra requirement which is put off every period, the total extra requirements until a fixed point in the future will decrease as time goes by.

   Assume in the previous example that the leadtime of C equals 4 periods. According to figure 1 the hedges up to period 4 were 5+3=8 in the current period: period 1. One period later the hedges up to period 4 were 5 (fig. 2). This matches reality: As time goes by it becomes more evident what the real demand will be and so less safety is required.

   In case the safety norm is implemented as an immediate requirement, a constant amount of safety is required all the time. As a consequence hedging will yield more reschedule-out and thus less reschedule-in messages compared to the systems currently available. The increase in reschedule-out messages can be dampened.
3. The stocknorm for a component is called integral if the norm may be covered by inventory of the component as well as by inventory of any of its parent-parts. The hedges work as integral stocknorms. This prevents that components are ordered unnecessary in situations where enough inventory is available of the parent-parts of these components.

   In case of hedging the stocknorm for the MPS-product as well as for the components is implemented as an extra requirement on the MPS-level. This implies that if the MPS-product has sufficient inventory available, the stocknorm for the component is netted by the inventory of the MPS-product. Thus no components are ordered if there are no gross requirements for the component. So in hedging the safety stocknorm is used as an integral norm.

Safety stocknorms for the absorption of demand forecast errors should indeed be integral norms. This is a consequence of the fact, that if demand forecast errors can be absorbed by inventory at the component level, these can also be absorbed by inventory at the parent-part level.

Recall that standard-MRP does generate an order for the safety norm of the components regardless of how many inventory of the MPS-product is available. So standard MRP interpretes the safety stocknorm as a local norm. That means, that the safety stock norm for the components always leads to physical stock at the component level.

4. In assembly environments the safety stocks of components from one parent-part are controlled in a coördinated way: At the MPS-level it is decided how many safety stock should be available and by MRP-explosion this is translated to all components. In this way it is impossible, that one component strives for a safety stock of 100, while another component with the same parent-part strives for no safety stock.
5. The use of safety stocks is better supported with hedging. In standard MRP the master scheduler does not know whether a change in the MPS is realistic. With hedging he may raise the level of the MPS up to the level of the original MPS plus the hedges and still be sure, that the material to allow such a raise is available. The MPS no longer needs to be constant. The only constraint is that the MPS plus the hedges remain constant.

Disadvantages of hedging are:

1. The MRP-scheme for periods beyond the leadtime is garbled because its input is no longer the expected requirements, but rather the maximal requirements (the expected requirements plus hedges). Particularly the Available Balance, Planned Orders and Planned Order Release do not resemble the expected values anymore. As a result the Capacity Requirements Planning, which is based on the Planned Orders, overestimates the future capacity requirements and thus sees problems where there are none.
2. By hedging an amount of safety is implemented at the MPS-level. This safety is then exploded to all components and consequently all components receive an equivalent amount of safety. If some components are common while others are not, the required safety is not the same for each component.

   Likewise, if the supply and/or production processes for each of the components have very different characteristics, the uncertainties and the corresponding safety norms per component may differ. To deal with these differences safety stocknorms still have to be implemented for each component. It appears that in most situations hedging cannot fully replace the safety stocknorms.

3. The implementation of hedging in MRP-software.

There are two possible ways to implement hedging in standard-MRP-software. The first one stems from the observation, that hedges increase the requirements or forecasts of the MPS-items. From this observation the hedges are interpreted as "additional forecasts" and treated exactly the same way as the real forecasts. So the records, screens and programs, which were meant to deal with the forecasts only, are then used for the hedges too. The second way is to make separate records, screens and programs for the hedges.

The first way requires less adaptation of the software, but has three severe disadvantages:

1) The forecast is connected with one specific need date. It is characteristic for the hedges that their need-date changes every review-period. If the need-date of the "extra forecast" does not change, it will ultimately result in physical stock at the MPS-level: An expensive solution.
2) The planner cannot distinguish the real forecasts from the additional forecasts or hedges.
3) The Rough Cut Capacity Planning (RCCP) is no longer valid because the total forecast does no longer represent the average requirement.

Designing separate records, screens and programs for the hedges seems to be necessary.

4. Major impacts of hedging on the MRP-philosophy.

Although at first glance hedging seemed to be a fairly simple addition to an MRP-package, detailed elaboration of the concept revealed, that hedging does not neatly fit within standard-MRP-logic. The main cause for this is that in case of hedging all requirements are maximal requirements, whereas MRP is based on expected requirements. As a consequence many of the concepts and functions in MRP get a different interpretation or become garbled. (see disadvantage 1 of hedging in section 2).

The RCCP is a good example to illustrate the effort, which is needed to restore the effects caused by the implementation of hedging. Performing the RCCP according to standard MRP when the requirements are raised with hedges leads to overestimating the future requirements. To avoid this, the RCCP should be performed on an MPS without hedges, i.e. before the hedges are added. The hedges are added to the requirements of the MPS-item before these requirements are netted and lot-sized. (otherwise the hedges in the first periods will not be netted even though there is plenty of inventory available). These two facts lead to the conclusion, that the MPS should be the requirements of the MPS-item before netting and lot-sizing.

So it appears, that due to hedging the MPS should be seen as an agreement between Sales and Production expressed in terms of requirements rather than in terms of production orders as it is in standard-MRP. That implies, that the MPS-level has to be moved towards the point after the stockpoint for the MPS-product.

In order to avoid radical changes in the MRP-software-architecture the following solution is suggested to allow hedging in MRP-software: The user must enter a twin-MPS-item for every MPS-item. This twin-item is an MPS-controlled item; it is being used to determine the MPS without lot-sizing and netting. If this MPS is determined by the master scheduler it makes up together with the hedges the dependent demand for the original MPS-item. The original item is now interpreted as an MRP-controlled item and for this item the planned orders are calculated from the dependent demand, inventory on hand and the open orders, taking lot-sizing into account. The new productstructure

is:

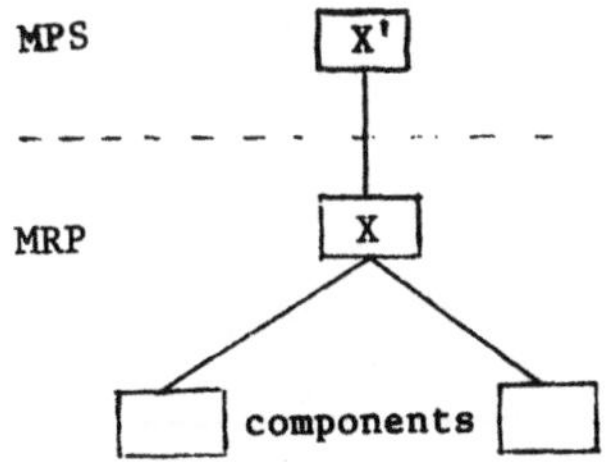

and was:

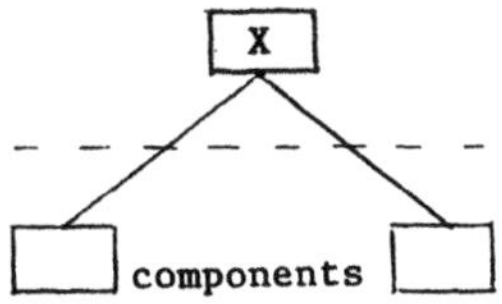

It should be noted here, that in this option the Available to promise is based on the MPS without lot-sizing and netting plus the hedges. That implies that e.g. stock due to lot-sizing is not incorporated. Compared to the current situation more safety stock is incorporated, since all the hedges for the components are now available to promise.

The fact that the MPS takes no lot-sizing into account implies, that the input for the RCCP is changed and the RCCP-calculations should be adapted to this. The variable resource requirements should be raised with an additional set-up-charge and the fixed resource requirements should be set equal to zero. The additional set-up-charge is equal to the set-up-time divided by the average production.

Possible capacity-inventories are not distinguished by most MRP-software-packages and the RCCP can therefore not take account of them if the inventory on hand is left out of the RCCP-calculations.

The above illustrates, that the implementation of hedging in standard-MRP-packages requires artificial methods of adaptions.

Acknowledgements.

The authors would like to thank Dr. S.D. Flapper from Eindhoven University of Technology for his contribution in the research on hedging.

Literature.

[1] Orlicky, J., Material Requirements Planning, Mc Graw-Hill, 1975.

[2] Miller, J.G., Hedging the Master Schedule, appeared in "Disaggregation Problems in Manufacturing and Service Organizations", Martinus Nijhoff Publishing, 1979, pp.237-256.

ECONOMIC ENERGY PROCESSING IN A SMALL-SCALE POWER-STATION

Günter Fandel and Joachim Reese, Hagen

The handling of constant power-heat-relations is a well known problem in the field of a firm's energy generation for own purposes. Nevertheless, there exist none but a few vague approaches, which can deal with an exact evaluation of the solutions generally practiced. In a case study for a small-scale power-station it is shown, that the structure of the problem can be expressed by a linear programming formulation, to which standard optimization procedures are applicable.

The fundamental description of the observed case is as follows: Steam is generated alternatively in three boilers and, subsequently, transformed in power as well as in heat by turbines. The constant relations of the respective output quantities depend obviously on the type of turbine involved in the energy generation process. Surplus production of heat is not feasible, as the quantities, which exceed the periodical demand, cannot be emitted due to air pollution acts. This is why the planning staff must be engaged in solving this problem of joint production by optimum machine employment. The influence of the choice done on the resulting energy costs is demonstrated. Furthermore, additional market transactions with regard to the generated quantity of electric power often are inevitable. Some concluding remarks refer to the robustness of the optimal solution, as far as the development of the processing costs is concerned.

NUMERICAL EVALUATION OF HEURISTICS FOR THE MULTI-ITEM SINGLE-LEVEL CAPACITATED LOT-SIZE PROBLEM

Hans-Otto Günther, Mannheim

The problem considered is the determination of lot sizes for multiple products to be produced on a single production facility with limited capacity. Demand is assumed to be deterministic and time-varying and must be met without backordering. The objective is to minimize the sum of setup and inventory holding costs.

An optimal solution to this lot-sizing problem within reasonable computer time is not available so far. However, various heuristic solution procedures have been proposed in the literature (e.g. [1]). Heuristics of the period-by-period type are easy to implement and quite effective as well. It is common to these heuristics that they develop production quantities for the current period using some basic lot size criterion, then proceed to the next period and repeat the process until the end of the horizon is reached. In addition, so called improvement steps can be used in order to improve the final solution obtained by the lot-sizing heuristic.

In a simulation study several well known lot-sizing heuristics of the period-by-period type are compared and the effectiveness of additional improvement steps is investigated. A large set of test problems was randomly generated reflecting different demand patterns, cost structures and tightness of capacity constraints. Numerical results show that the lot-sizing heuristics considerably differ in their cost performance especially when capacity constraints are tight and the average time between orders is high. The relative performance of the lot sizing heuristics differ by up to 40% under certain test conditions.

(The complete version of the paper is available upon request from the author.)

Reference

Günther, H.-O., Planning lot sizes and capacity requirements in a single stage production system, European Journal of Operational Research, 31(1987), 223-231.

FLEXIBILIZATION OF SEQUENCING PRIORITY RULES

Reinhard Haupt, Köln

Priority rules can be defined as sequencing decision criteria according to which a job among those waiting for being processed by a machine is selected at the time a machine becomes free. Such scheduling procedures (e.g. "shortest-processing-time" or "minimum-slack-time" rules) are extensively discussed in the heuristic scheduling literature.
One predominant characteristic of commonly used priority rules consists of the limitation of the sequencing information input to the actual set of jobs in the queue before the machine in question. Such a local information basis for example may exclude an urgent job that is going to be completed on its predecessor facility in the very near future from the actual scheduling decision, whereas an implementation of a "look-ahead" rationale into the priority rule would take into account this urgent job to arrive soon.
In general, flexibilization of priority rules requires the extension of the sequencing decision horizon to a global (instead of a local) information basis. This may be accomplished e.g. by anticipating jobs arriving from other queues, or by anticipating queues the jobs in the queue in question will proceed to next. Further elements of flexibilization may be involved by a change of the priority rule over time, or by basing the priority assignment on the evaluation of the performance of alternate hypothetical job selections.
The efficiency of such flexibilization components is investigated by means of a job shop scheduling simulation. Various experimental conditions, such as the capacity load level, the tightness of job due-dates, and the extent to which dates can be set internally by the production control system point out the following results: When used per se, most of those "flexibilizers" exhibit modest performance measures (in terms of mean flow time, and of mean tardiness, respectively); but integrated into classical basic priority rules they yield (under well-defined shop load and job due-date conditions) remarquable sequencing improvements. This is especially true for a flexibilization approach, the priority assignment of which looks for an evaluation of alternate hypothetical job selections.

INTERACTIVITY IN PRODUCTION CONTROL SIMULATION

Jiří Herman, Prague

Summary: Intermittent production systems work practically always under conditions of uncertainty. In the whole range of the system behaviour (from the ideal to the catastrophic one), the behaviour of a typical production system can be specified as a mutable or disturbant.

One of the basic features of such a behaviour is that the system achieves another than expected or desirable goal state, without any possibility of some corrective action in the real time. Moreover, it is not only uncertainty, but also the intermittent structure of the production process itself that leads to this type of the behaviour.

A production process and control simulation based on a heuristic model respecting the real, initial state of the production system can reveal those situations, in which the system tends to leave its presumed goal behaviour. Thus, on the one hand, the reality affects the model, which is natural and necessary. On the other hand, the simulation model with its dynamic features and prediction ability can affect the real system directly and in advance, in order to minimize the possible deviation from its desirable function in the future.

Various measures and actions modifying the system structure or functions usually require not only some lead-time, but also some human supported information.

Therefore, a computer-oriented or initiated interactive approach to the process simulation can help to solve this problem of mutual relationship between the reality and the model in the field of computer aided production control systems. The interaction is based on a forward, into the future moving time distance between the decision maker and the computer model and on some combination of the actual and simulated decisions.

Using an LP model for flow planning

P.C.T. van der Hoeven, IJmuiden

At Hoogovens IJmuiden, the Dutch steel company, an LP model has been developed in order to support the flow planning in the order-driven part of the production process. The context is seen figure 1.

```
orderpressure - -> capacityplan
    :       |          |
    :       v          v
    :      FLOW PLANNING
    :            :
    v            v
  orderplanning
       :
       v
   factories
```

figure 1

```
                      K1 - CA
                     /  \ /  \
SF - WB - BT          X       NW - EV
                     \  / \  /
                      K2 - BA
```

figure 2

The objective of the order planning is to distribute weekly the available capacities over the booked orders according to the order priorities. This planning is made in two stages: first a flow plan is made and then the flow capacities are distributed over the orders. In figure 2 the 4 different flows are sketched, which lead to the production of tin plate, the raw material for cans. The abbreviations in this figure all stand for an installation. As should be noticed the flows share part of the installations. Other flows lead to the raw material for tubes or for cars and all these flows use shared plants.

For the flow planning a large LP model is used (approximately 6.000 constraints, 10.000 variables, solved after 5.000 iterations). The decisions variables in this model are the quantities of throughput for each flow (circa 70), each treatment (between 2 and 7), each priority class (5) and each week (between 2 and 7). The main constraints are the capacities; the objective function is given by penalties on not processing material and in some cases on underloading certain plants (many plants are made a bottleneck and hence the capacity plan is leading).

The model is run weekly in a rather stochastic environment. The randomness in the planning environment is due on the one hand to really unpredictable events and on the other hand to the fact that not all details are modeled. The resulting desire to describe more details is also inspired by the wishes of the final user.

In the model described above some details are brought into account, e.g. the different processing times, the efficiencies and the events occuring between the registration of the stocks and the beginning of the planned period. On the other hand, the description of too many details leads to modeling errors, because the model and its outcomes become too complex to be understood. Each week other special problems arise, which we deliberately did not model.

Many modelers consider it a challenge to model all kind of details but we advise them to be reticent towards this challenge because in the end the user is better off with a mean that does the bulk of the planning and that leaves the details to be solved by the human mind. In this manner OR techniques really become decision supporting tools.

The model is operational since summer 1986. As a follow-up research a simulation model is now being built, which should form a testing environment for ideas about future logistics.

UNABHÄNGIGE PLANUNGSZEITRÄUME BEI MEHRPERIODIGER CHARGENPRODUKTION

Heinz D. Mathes, Frankfurt/Main

Zusammenfassung : Bei mehrperiodigen Planungsproblemen scheitert die Ableitung einer vollständigen optimalen Politik häufig daran, daß nicht für den gesamten vorgegebenen Planungszeitraum sondern höchstens für ein Teilintervall die relevanten Daten mit hinreichender Genauigkeit prognostiziert werden können. Für ein einfaches Modell der mehrperiodigen Chargenproduktion wird deshalb der Frage nachgegangen, ob und unter welchen Bedingungen die auf der Grundlage der verfügbaren Informationen ermittelten optimalen Teilpolitiken ihre Optimalität bei einer Erweiterung der Datenbasis im Zeitablauf beibehalten.

Summary : In multi-period planning problems the numerical derivation of a complete optimal policy is frequently impossible, because the relevant data cannot be forecasted with sufficient precision for the whole given planning interval but only for some subinterval. In the context of a simple model of multi-period batch production the question is therefore investigated if and under which conditions optimal subpolicies which have been derived on the basis of the limited attainable information remain optimal when the data basis is extended in course of time.

1. Problemstellung

Betrachtet wird das Problem, die chargenweise Produktion eines ungeformten Gutes über einen festgelegten Zeitbereich $Z := \{1,...,T\}$ (den Planungszeitraum) kostenminimal zu planen. Aufgabe der Planung sei es, die Anzahl der aufzulegenden Chargen - die Chargenserie - für alle $t \in Z$ so zu bestimmen, daß die Nachfrage eines jeden Zeitpunkts vollständig befriedigt werden kann, und die Summe der Produktions- und Lagerkosten über Z ein Minimum annimmt. Neben die technisch bedingte Forderung nach Ganzzahligkeit der Produktionsmengen können dabei noch zusätzliche Nebenbedingungen wie etwa Kapazitätsrestriktionen treten.

Entscheidungsprobleme mit dieser Struktur lassen sich als endlichstufige dynamische Optimierungsprobleme mit beschränkten Steuerbereichen formulieren, wobei die Steuervariablen überdies nur ganzzahlige Werte annehmen dürfen. Die numerische Ermittlung einer optimalen Lösung, (Entscheidungsfolge, Politik) die für den gesamten Zeitbereich Z eine kostenminimale Produktionssteuerung angibt, setzt voraus, daß für alle

Operations Research Proceedings 1987

Zeitpunkte $t \in Z$ die erforderlichen Planungsdaten bereits im Planungszeitpunkt vollständig vorliegen, d.h., der Prognosehorizont T_p mindestens so groß ist wie der Planungshorizont T ($T_p \geqq T$). In vielen Fällen sind die Zeitreihen der relevanten Daten - etwa der Preise der Einsatzgüter oder der Nachfrageverlauf - aber höchstens für ein Teilintervall $Z' \subset Z$ mit hinreichender Genauigkeit (oder zu vertretbaren Kosten) prognostizierbar, so daß $T' = T_p < T$ gilt. Wird jedoch wegen der beschränkt verfügbaren Informationen nur für Z' eine optimale Produktionssteuerung bestimmt, so muß als Folge der Existenz zeitlicher Abhängigkeiten zwischen den einzelnen Produktionszeitpunkten davon ausgegangen werden, daß diese Politik im allgemeinen nurmehr suboptimal ist, wenn der Prognosehorizont (die Datenbasis) im Zeitablauf erweitert wird.

Man kann sich in diesem Zusammenhang die Frage stellen, ob und unter welchen Bedingungen der Zeitbereich Z in disjunkte Teilintervalle zerlegt werden kann, welche die Eigenschaft aufweisen, daß die ihnen zugeordneten optimalen Politiken nicht von den Daten der jeweils nachfolgenden Zeitpunkte bzw. Intervalle abhängen. Ein Teilintervall mit dieser Eigenschaft heißt dann unabhängiger Planungzeitraum Z* /5/. Läßt sich nun innerhalb eines gegebenen Prognosehorizonts $T_p < T$ ein solcher unabhängiger Planungszeitraum $Z^* := \{1,...,T^* \leqq T_p\}$ angegeben, dann kann dessen optimale Politik ohne Berücksichtigung der im Planungszeitpunkt noch unbekannten zukünftigen Datenentwicklung abgeleitet werden. Daraus aber folgt weiterhin, daß diese auf der Basis unvollständiger Informationen ermittelte Politik immer auch eine Teilpolitik einer optimalen Produktionspolitik für jedes Zeitintervall Z' mit $T' > T^*$ bildet.

Nachfolgend werden für das einfache Chargenserienproblem die Voraussetzungen für die Existenz unabhängiger Planungszeiträume untersucht. Ausgehend von entsprechenden Ergebnissen für dynamische Losgrößenmodelle wird gezeigt, wie diese durch geeignete Modifikationen auf den Fall von ganzzahligen Steuervariablen übertragen und hinreichende Bedingungen für die Existenz solcher Zeiträume formuliert werden können.

2. Das Produktionsmodell

Über einen Planungszeitraum Z, der die Zeitpunkte $\{1,...,T\}$ umfaßt, soll ein ungeformtes Gut (z.B. Farbe, Tapeten, Schokoladenrohmasse) so erzeugt werden, daß dessen Nachfrage $d_t \in \mathbb{R}_+$ zu jedem Zeitpunkt $t \in Z$ vollständig befriedigt werden kann. Die Produktionsmenge u_t

eines jeden Zeitpunkts kann dabei entweder unmittelbar zur Befriedigung der Nachfrage dieses Zeitpunkts verwendet oder auf Lager genommen und zur Bedarfsdeckung der Folgezeitpunkte eingesetzt werden. Die Produktion werde chargenweise durchgeführt. Durch die technischen Eigenschaften des verwendeten Produktionsmittels (Kessel, Druckbehälter etc.) insbesondere durch dessen Fassungsvermögen, sei eine einheitliche Chargengröße $m \in \mathbb{N}$, $m > 1$, vorgegeben. Weiterhin werde unterstellt, daß der Bestand an Produktionsmitteln so dimensioniert sei, daß mit Kapazitätsengpässen zu keinem Zeitpunkt $t \in Z$ gerechnet werden müsse und $d_t > m$ gelte . Der Lagerbestand , der zur Deckung der Nachfrage in t verwendet werden kann, sei mit z_t bezeichnet. Da Fehlmengen unzulässig sind, muß gelten $z_t \in \mathbb{R}_+$. Die Produktionskosten eines jeden Zeitpunktes setzen sich aus einem von der Seriengröße unabhängigen Term (chargenfixe Kosten) und den chargenproportionalen Kosten zusammen. Die chargenfixen Kosten $K_t(u_t)$ umfassen etwa die Kosten des Einrichtens und Überwachens der Chargenserie, während die chargenvariablen Kosten im wesentlichen von den Materialkosten und der Chargengröße abhängen, so daß mit $p_t \cdot m$ die proportionalen Kosten einer Einheitscharge m bezeichnet werden. Die chargenfixen Kosten seien wie folgt definiert :

$$K_t(u_t) = \begin{cases} 0 \text{ für } u_t = o \\ K_t > 0 \text{ für } u_t > 0 \end{cases} \qquad \text{für alle } t \in Z$$

Neben den Produktionskosten fallen Lagerkosten i_t je Einheit und Periode an, die der Periode der Entstehung des Lagerbestands zugerechnet werden.

Das Entscheidungsmodell zur Bestimmung der kostenminimalen Produktionspolitik läßt sich damit als diskret ganzzahliges dynamisches Optimierungsproblem (P) formulieren :

$$Z\ (u^*,z^*) = \min_{(u,z)} \left\{ \sum_{t=1}^{T} (K_t(u_t) + p_t \cdot u_t + i_t \cdot z_{t+1}) \right\}$$

$$\left.\begin{array}{ll} (1) & z_{t+1} = z_t + u_t - d_t, \quad z_{t+1} \geqq 0 \\ (2) & u_t \in U_t := \{0, m, 2m, \ldots\} \end{array}\right\} \text{ für alle } t \in Z$$

$$(3) \quad z_1 = 0\ , \ d_t \in \mathbb{R}_+\ , \ m \in \mathbb{N},\ m > 1$$

Eine zulässige Lösung (Produktionspolitik, -steuerung) dieses Problems besteht aus einer Folge von Steuervariablen u_t und Zustandsvariablen z_{t+1}, welche für alle t die Bedingungen (1) und (2) erfüllen : $(u,z) = (u_t, z_{t+1})_1^T = (u_1, z_2, u_2, z_3, \ldots, u_T, z_{T+1})$: Die Menge der zulässigen Lösungen werde mit L(P) symbolisiert, die Menge der optimalen Lösungen mit L*(P) und eine optimale Lösung entsprechend $(u^*, z^*) = (u_t^*, z_{t+1}^*)_1^T$. Zur Vereinfachung soll hier auch von einer "T-optimalen Lösung (Politik, Steuerung)" gesprochen werden.
Ein unabhängiger Planungszeitraum Z* läßt sich dann wie folgt definieren : Sei (u*,z*) eine T*-optimale Politik. $Z^* = \{1, \ldots, T^*\}$ heißt genau dann unabhängiger Planungszeitraum, wenn für alle T > T* gilt

$$(u_t^*, z_{t+1}^*) = (u_t', z_{t+1}') , \quad t = 1, \ldots, T^* ,$$

und

$$(u_t', z_{t+1}')_1^T \text{ eine T-optimale Politik ist.}$$

3. Das transformierte Problem

Ersetzt man die Bedingung (2) durch die schwächere Forderung $u_t \in [0, +\infty)$ für $t \in Z$, erhält man die relaxierte Form des Problems (P). Diese weist die typische Struktur des dynamischen Losgrößenproblems ohne Zustandsrestriktionen auf (sieht man von der Nichtnegativitätsbedingung einmal ab). Für diese Problemstrukturen sind bereits in /1,2,6,7 / mit unterschiedlichen Annahmen über die jeweiligen Kostenverläufe Existenzbedingungen für unabhängige Planungszeiträume diskutiert worden. Diese Ergebnisse beruhen u.a. auf der Prämisse, daß die verwendete Zielfunktion über der nach unten beschränkten konvexen Lösungsmenge konkav verläuft und ihr Minimum annimmt. Optimale Lösungen liegen dann stets in einem oder mehreren Extremalpunkten der Lösungsmenge /2/. Da diese Bedingungen bei ökonomischen Anwendungen für das relaxierte Problem im allgemeinen erfüllt sind und trivialerweise gilt, daß L(P) in dessen Lösungsmenge enthalten ist, stellt sich die Frage, ob diese Existenzaussagen auch auf den Fall ganzzahliger Steuervariablen übertragen werden können. Den Grundgedanken bildet dabei die Überlegung, die Lösungsmenge des relaxierten Problems so zu reduzieren, daß die entstehenden Extremalpunkte bezüglich u_t Koordinaten ganzzahlig modulo (m) aufweisen.

Derartige Schnittebenen lassen sich durch die Ungleichungen $z_t \geqq z_t^o$ für alle $t \in Z$ einführen, wobei z_t^o durch

$$z_t^o = \min\left\{z : z \geqq 0,\ \sum_{i=1}^{t-1} d_i + z \equiv 0 \bmod (m)\right\}$$ definiert wird.

In ökonomischer Hinsicht stellt z_t^o einen von den Parametern d_t und m abhängigen Mindestlagerbestand dar, der durch die Wahl der Produktionspolitik nicht beeinflußt werden kann. Für z_1^o wird gesetzt: $z_1^o = 0$. Führt man durch die Transformation $v_t = z_t - z_t^o$ für alle t neue Zustandsvariablen v_t ein, läßt sich ein Optimierungsproblem (Q) formulieren, das die angestrebten Eigenschaften besitzt:

$$K(u^*,v^*) = \min_{(u,v)} \left\{ \sum_{t=1}^{T} (K_t(u_t) + p_t \cdot u_t + i_t \cdot v_{t+1}) \right\}$$

$$\left.\begin{array}{ll} (1) & v_{t+1} = v_t + u_t - d_t^o, \quad v_{t+1} \geqq 0 \\ (2) & u_t \in [0, +\infty) \\ (3) & d_t^o = d_t - z_t^o + z_{t+1}^o \end{array}\right\} \quad \text{für alle } t \in Z$$

$$(4) \quad v_1 = 0$$

Zur Notation wird vereinbart: $(u,v) = (u,z-z^o) = (u_t, z_{t+1} - z_{t+1}^o)_1^T$.
Den Zusammenhang zwischen (P) und (Q) beschreiben folgende Eigenschaften:

(P1) $(u,z-z^o) \in L(Q)$ für $(u,z) \in L(P)$

(P2) $(u,z) \in L(P)$. Dann gilt: $Z(u,z) - K(u,z-z^o) = \sum_{t=1}^{T} i_t \cdot z_{t+1}^o = \text{const.}$

(P3) Das Minimum von $K(u,v)$ wird in einem oder mehreren Extremalpunkten von L(Q) angenommen /2/.

Für L(Q) gilt zudem

<u>Satz 1</u>: Sei (u',v') Extremalpunkt von L(Q). Dann folgt $u_t' \equiv 0 \bmod (m)$ und $v_{t+1}' \equiv 0 \bmod (m)$ für alle $t \in Z$.

Zum Beweis zeigt man zunächst, daß aus der Definition von z_t^o unmittelbar folgt $d_t^o \equiv 0 \bmod (m)$. Nach einem Theorem von Tompkins-Heller-Gale /3/ läßt sich zeigen, daß die Koeffizientenmatrix des durch (1) definierten Gleichungssystems vollständig unimodular ist. Zusammen mit d_t^o ganzzahlig mod (m) ergibt sich damit aus dem Theorem von Hoffman-Kruskal zur ganzzahligen Optimierung /4/ sofort die Behauptung.
Unter Verwendung von (P1) - (P3) und Satz 1 folgt als zweites wichtiges Ergebnis dieses Abschnitts

Satz 2: $(u^*,z^*) \in L^*(P)$ genau dann, wenn $(u^*,z^*-z^o) = (u^*,v^*) \in L^*(Q)$ und (u^*,v^*) Extremalpunkt von $L(Q)$.

Beweis: Sei $(u^*,z^*) \in L^*(P)$. Dann ist wegen (P1) $(u^*,z^*-z^o) \in L(Q)$. Es existiere ein $(u',v') \in L^*(Q)$ mit $K(u',v') < K(u^*,v^*)$. Aus (P3) folgt, daß (u',v') Extremalpunkt von $L(Q)$ ist, dessen Koordinaten nach Satz 1 ganzzahlig mod (m) sind. Damit aber gilt $(u',v'+z^o) \in L(P)$ und aus (P2) folgt im Widerspruch zur Annahme
$Z(u',v'+z^o) = Z(u',z') < Z(u^*,z^*)$.
Sei $(u^*,z^*-z^o) = (u^*,v^*) \in L^*(Q)$. Aus (P3) und Satz 1 ergibt sich $(u^*,z^*) \in L(P)$. $(u^*,z^*) \in L^*(P)$ folgt unmittelbar aus (P2).

Satz 2 gewährleistet damit, daß jede extremale Optimallösung von (Q) in eine optimale Lösung von (P) transformiert werden kann. In der gewählten Formulierung weist (Q) die typische Struktur des dynamischen Losgrößenmodells auf, für das die in /1,2,6,7 / angegebenen Aussagen über die Eigenschaften optimaler Politiken in vollem Umfang gelten. Satz 2 sichert nun, daß für das Chargenproduktionsproblem (P) unter Beachtung der Transformation optimale Politiken mit eben diesen Eigenschaften existieren und somit auch hinreichende Bedingungen für unabhängige Planungszeiträume angegeben werden können.

4. Unabhängige Planungszeiträume bei ganzzahligen Steuervariablen

Die wichtigste Voraussetzung für die Existenz unabhängiger Planungszeiträume - neben dem Vorliegen bestimmter Kostenverläufe - stellt die Existenz optimaler Politiken mit der Regenerationseigenschaft dar. Bei Fehlen der Ganzzahligkeitsbedingung wird dadurch gefordert, daß kostenminimale Produktionssteuerungen auftreten, bei denen nur dann produziert wird, wenn der Lagerbestand auf Null abgesunken ist. Unter Beachtung der Transformation läßt sich diese Eigenschaft für das Chargenproduktionsproblem entsprechend formulieren:

(P4): Für das Problem (P) existieren T-optimale Politiken (u^*,z^*) mit der Eigenschaft

$$u_t^* \cdot (z_t^* - z_t^o) = 0 \text{ für alle } t \in Z$$

(Regenerationseigenschaft)

(P4) fordert damit, daß optimale Politiken gefunden werden können, für die nur dann $u_t \neq 0$ gilt, wenn der Lagermindestbestand z_t^o erreicht worden ist.
Beachtet man, daß $d_t^o \equiv 0 \bmod (m)$, so folgt, da Fehlmengen nicht zugelassen sind, für optimale Politiken mit der Regenerationseigenschaft

(P5): Für T-optimale Politiken (u*,z*) mit der Regenerationseigenschaft gilt

$$u_t^* \in \left\{ 0 \, , \sum_{i=t}^{t'} d_i^o \quad \text{mit } t \leqq t' \leqq T \right\} \quad \text{für alle } t \in Z$$

Die Existenz optimaler ganzzahliger Politiken mit den Eigenschaften (P4) und (P5) eröffnet nunmehr - zusammen mit einer wenig einschränkenden Annahme über den Verlauf der variablen Kosten - die Möglichkeit, hinreichende Bedingungen für unabhängige Planungszeiträume bei Chargenproduktion zu formulieren. Ein Beweis des nachfolgenden Satzes kann weitgehend analog / 1 / geführt werden:

<u>Satz 3</u>: Sei durch (u*,z*) eine (T*+1)-optimale Politik mit der Regenerationseigenschaft gegeben. $(T^*+1) \leqq T$.

Gelten die Bedingungen

(a) $$p_{T^*+1} \leqq \min_{1 \leqq t \leqq T^*} \left\{ p_t + \sum_{j=t}^{T^*} i_j \right\}$$

(b) $$z^*_{T^*+1} = z^o_{T^*+1} \quad ,$$

dann ist $Z^* := \{ 1, \ldots, T^* \} \subset Z$ ein <u>unabhängiger Planungszeitraum</u>.

Satz 3 liefert nicht nur hinreichende Bedingungen, sondern ist auch konstruktiv in dem Sinne, daß durch die Bedingungen (a) und (b) eine für alle (T*+1), $1 < (T^*+1) \leqq T = T_p$, durchzuführende Folge von Prüfschritten definiert wird, mit der sichergestellt werden kann, ob innerhalb des Prognosehorizonts T_p ein unabhängiger Planungszeitraum existiert.

/1/ Eppen,G.D., Gould,F.J., Pashigian,B.P., Extensions of the Planning Horizon Theorem in the Dynamic Lot Size Model, in: MS, Vol.15, 5, 1969, S.268-277.

/2/ Hadley,G., Nichtlineare und dynamische Programmierung, Würzburg 1969.

/3/ Heller,I., Tompkins,C.B., An Extension of a Theorem of Dantzig's, in: Linear Inequalities and Related Systems, hrsg. von Kuhn,H.W., Tucker,A.W., Princeton 1956, S.247-254.

/4/ Hoffman,A.J., Kruskal,J.B., Integral Boundary Points of Convex Polyhedra, in: Linear Inequalities and Related Systems, hrsg. von Kuhn,H.W., Tucker,A.W.,Princeton 1956, S.223-246.

/5/ Müller-Merbach,H., Operations Research, München 1971, S.355.

/6/ ter Haseborg,F., Optimale Lagerhaltungspolitiken für Ein- und Mehrproduktläger, Göttingen 1979.

/7/ Wagner,H.M., Whitin,T.M., Dynamic Version of the Economic Lot Size Model, in: MS, Vol.5, 1958, S.89-96.

VERGLEICH VON VERSCHNITTSOFTWARE

Ute Milautzki-Finke, Harald Dyckhoff, Hermann-Josef Kruse

Hagen

Zuschneideprozesse werfen für Unternehmen eine Reihe von Problemen auf, bei deren Lösung eine Computerunterstützung von großer Bedeutung, wenn nicht sogar unabdingbar ist. Im Vordergrund stehen Fragen der Materialwirtschaft, speziell die Frage, wieviel Rohmaterial durch Optimierung des Zuschneideprozesses und Verringerung des Verschnitts eingespart werden kann. Aber auch andere Aspekte einer computergestützten Produktionsplanung und -steuerung bis hin zur Einbettung der Zuschneideprozesse in ein umfassendes CIM-System werden berührt. Wegen eines starken Wettbewerbsdrucks und damit verbundener Rationalisierungs- und Modernisierungszwänge sehen sich mit Verschnittproblemen konfrontierte Unternehmen immer mehr vor die Entscheidung gestellt, geeignete DV-Lösungen zu finden. Der daraus resultierenden Nachfrage nach Standardsoftware für Verschnittprobleme steht allerdings nur ein unübersichtliches Angebot gegenüber.

Ziel des Beitrags ist es, für mehr Transparenz hinsichtlich des Angebotes an Verschnittsoftware zu sorgen. In einer empirischen Untersuchung wurden 36 Programmpakete ermittelt, die von 21 verschiedenen Firmen als Standardsoftware zur DV-mäßigen Unterstützung und Optimierung industrieller Zuschneideprozesse auf dem deutschen Markt angeboten werden. Anhand auswertbarer Angaben der Anbieter werden 28 dieser Programme - differenziert nach verschiedenen, für einen potentiellen Anwender relevanten Kriterien - vergleichend gegenübergestellt.

Literatur

Dyckhoff, H., Kruse, H.-J., Milautzki, U.: Standardsoftware für Zuschneideprobleme, in: Zeitschrift für wirtschaftliche Fertigung und Automatisierung, 82. Jg., H. 8, 1987, S. 472 ff.

Integrierte hierarchische Ersatzteilbemessung und Fertigungssteuerung in Reparatursystemen

Ch. Schneeweiß, Mannheim
H. Schröder, Frankfurt

Die Planung in Reparatursystemen reicht von langfristigen Investitionsentscheidungen bis hin zu ganz kurzfristigen Fertigungssteuerungsentscheidungen. Alle Teilplanungen hängen miteinander zusammen und üben Wechselwirkungen aufeinander aus. Dadurch ist man gezwungen, den gesamten Problemkomplex hierarchisch zu strukturieren und die einzelnen Planungsmodelle auf jeder Hierarchieebene zu integrieren.
Entwickelt wurde ein zweistufiges Entscheidungssystem, in dem die mittelfristige Ersatzteilbemessung (ETB) mit der kurzfristigen Fertigungssteuerung (FST) abzustimmen war.
Die ETB wird unter einer ökonomischen Zielsetzung durchgeführt, während die FST mit einem technischen Zielkriterium erfolgt. Die entscheidende Größe bei der ETB ist die Anzahl Teile eines Typs, die sich zu einem beliebigen Zeitpunkt im Reparaturprozeß befindet. Die Verteilungen dieser stochastischen Größen werden mit Hilfe eines analytischen Prioritätswarteschlangensystems ermittelt. Dabei ist eine wertebezogene Prioritätssteuerung angesetzt, durch die der im Mittel in der Warteschlange gebundene Materialbestandswert minimiert wird.
Das Prinzip der konzipierten FST besteht darin, jeweils dem Teil die höchste Bearbeitungspriorität zu geben, bei dem die potentiellen negativen Folgen durch eine mögliche Verschiebung des Reparaturbeginns am größten sind. Dabei werden die negativen Folgen für die einzelnen Teile mit den Risikozunahmen für das Auftreten von Fehlmengen und Fehlzeiten bewertet.
Die Abstimmung der ETB mit der FST geschieht sowohl durch eine Abwärts- als auch durch eine Feedback- und Feedforward-Aufwärtsintegration. Das Zusammenwirken der analytisch entwickelten Heuristiken für die ETB und FST wurde anhand von Simulationsuntersuchungen kritisch überprüft. Dabei fanden echte Datensätze aus den Elektronik- und Triebwerkswerkstätten der Deutschen Lufthansa Verwendung.

Schröder, H.: Integration von Planung und Steuerung in Reparatursystemen, Minerva Publikation 1988

Ein Vergleich zweier Lösungsansätze zur eindimensionalen Verschnittoptimierung

Hartmut Stadtler, Lüneburg

Zur Bestimmung einer kontinuierlichen Lösung eindimensionaler Verschnittprobleme werden in der Literatur das 'spaltenerzeugende Verfahren' von Gilmore und Gomory (1961) sowie das 'Einzelschnittmodell' von Dyckhoff (1981) vorgeschlagen. Der Vortrag zielt darauf ab, Gemeinsamkeiten und Unterschiede der beiden Lösungsansätze herauszuarbeiten.

Es wird gezeigt, daß das 'spaltenerzeugende Verfahren' zur Bestimmung eines vollständigen Schnittmusters - bei einer geeigneten DP-Formulierung des Rucksackmodells - ebenfalls auf Einzelschnitten beruht. Die Mengen der Einzelschnitte sind für die beiden Lösungsansätze deckungsgleich.

Dies läßt u.a. den Schluß zu, daß alle Verschnittprobleme, die sich mit dem 'Einzelschnittmodell' abbilden lassen, grundsätzlich auch mit dem 'spaltenerzeugenden Verfahren' gelöst werden können (aber nicht umgekehrt). Als weitere Konsequenz der gefundenen Übereinstimmung ist es nunmehr möglich, die für das 'Einzelschnittmodell' entwickelten Reduktionsregeln auf das DP-Rucksackmodell zur Verminderung des Rechenaufwandes zu übertragen.

Bei der Implementation der beiden Verfahren auf einem Personal Computer wurde für das 'spaltenerzeugende Verfahren' das Softwarepaket MPCODE (Land und Powell, 1973) um ein DP-Verfahren zur Rucksackoptimierung erweitert und für das 'Einzelschnittmodell' die Software Hyper Lindo (Schrage, 1986) eingesetzt. Der Rechenaufwand zur Lösung von drei aus der Literatur bekannten Verschnittproblemen war für das 'Einzelschnittmodell' erheblich höher (> Faktor 10) als für das 'spaltenerzeugende Verfahren'.

Literatur:

1) Dyckhoff H (1981) A new linear programming approach to the cutting stock problem. Operations Research 29: 1092 - 1104
2) Gilmore PC, Gomory RE (1961) A linear programming approach to the cutting stock problem. Operations Research 9: 849 - 859
3) Land AH, Powell S (1973) Fortran codes for mathematical programming, John Wiley & Sons, London New York Sydney Toronto
4) Schrage L (1986) Linear,integer,and quadratic programming with LINDO, The Scientific Press, Redwood city, USA

DAS AGGREGATIONSPROBLEM IN DER HIERARCHISCHEN PRODUKTIONSPLANUNG

Marion Switalski, Bielefeld

Die hierarchische Produktionsplanung (HPP) ist eine Weiterentwicklung des Sukzessivplanungskonzepts, bei der die Zerlegung des Gesamtproblems und die Abstimmung der Teilprobleme aufgrund der vorgegebenen hierarchischen Struktur der Planungsaufgabe erfolgt. Um für jede Planungsebene die angemessene zeitliche Reichweite und sachliche Komplexität des Problems zu erhalten, ist die Aggregation von detaillierten Daten erforderlich.

Aggregation bedeutet die Zusammenfassung von ähnlichen Objekten zu einem gemeinsamen Oberbegriff, dem Aggregat. Durch die Disaggregation erfolgt der Rückschluß von den Ergebnissen des aggregierten Modells auf konkrete Handlungsanweisungen. Da eine perfekte bzw. konsistente Aggregation, wie sie in der Volkswirtschaftslehre diskutiert wird, in der Regel nicht möglich ist, werden für die HPP andere Zugänge zum Aggregationsproblem notwendig:

- approximative Aggregation: Da die zugrundeliegenden Daten häufig Zufallsgrößen sind, wird eine weitgehende Übereinstimmung der Ergebnisse des aggregierten und des detaillierten Modells verlangt, z.B. durch Minimierung der euklidischen Abweichungen. Eine Anwendung in der HPP wird von Axsäter et al. beschrieben.
- LP–Aggregation: Die oberste Planungsstufe, d.h. die aggregierte Planung im engeren Sinne, ist in den Ansätzen zur HPP häufig als lineares oder gemischt–ganzzahliges Programm formuliert, das durch eine geeignete lineare Abbildung aggregiert werden kann.
- heuristische Aggregation: Entsprechend der empirischen Struktur der Planungsaufgabe werden die detaillierten Informationen von unten nach oben immer weiter verdichtet. Dieses Vorgehen wird in den meisten bekannten HPP–Modellen angewendet.

Diese drei Ansätze geben unterschiedliche Hilfen bei der Aggregation für die HPP. Da sie bislang in der Literatur fast vollständig separat diskutiert wurden, liegt es nahe, ihre Vorteile im Rahmen einer Aggregationsstrategie zu verbinden.

Literatur

Axsäter [1981]: Axsäter, S., Aggregation of Product Data for Hierarchical Production Planning, Operations Research, Vol. 29, No. 4, 1981, S. 744 – 756

Bartels [1986]: Bartels, S., Entwicklung und Beurteilung von Aggregationsstrategien für lineare Entscheidungsmodelle, Verlag Peter Lang GmbH, Frankfurt am Main – Bern – New York 1986

Bitran / Haas / Hax [1981]: Bitran, G. R., Haas, E. A., Hax, A. C., Hierarchical Production Planning: A Single Stage System, Operations Research, Vol. 29, Nr. 4, 1981, S. 717 – 743

Hax / Meal [1975]: Hax, A. C., Meal, H. C., Hierarchical Integration of Production Planning and Scheduling, in: Geisler, M. A. (Ed.): Logistics, TIMS Studies in the Management Sciences, North Holland, Amsterdam 1975, S. 53 – 69

Liesegang [1980]: Liesegang, G., Aggregation bei linearen Optimierungsmodellen – Beiträge zur Konzipierung, Formalisierung und Operationalisierung, Habilitationsschrift Köln 1980

A METHODOLOGY FOR CUSTOMER ORDERS PLANNING AND DELIVERY DATES ASSIGNMENT

Ilías P. Tatsiópoulos, Athens

Summary: An interactive decision framework is presented for negotiating delivery dates and prices in small job-shops. The system consists of the following procedures:

a) QUOTCLAS: Classification of customer enquiries according to the elasticity of prices and delivery dates.

b) LEADTEST: Delivery lead times are checked against norms set by the management and shortening options are examined.

c) TOTBA: The impact of new orders on existing work backlogs is examined against norms set by the management.

d) QUOTCALC: Costs are calculated in cases of disruptions of lead time and backlog norms. Price discounts are premiums are considered for alternative delivery lead times.

A. Introduction

Engineering industry includes a great number of small jobbing component manufacturing firms producing entirely according to customers' specifications, so that no product or production process standardisation is possible. Most of the jobs are for one piece components released in small batches in a functionally organised shop.

These firms sell to their customers not only an item made to desired specifications at an agreed price but also an item delivered at a specified time. The Customers Orders Planning and Delivery Dates Assignment function has to solve the following problem: "To negotiate the timing of deliveries with customers on a realistic basis reflecting the presence of other orders, the capabilities of the shop, and the "costs" of achieving a certain timing, as well as the value of the timing to the customer". Quotations to customer enquiries should reflect a combined price/delivery date policy aiming at the synchronization of demand and capacity by attracting or deterring jobs in particular time periods. Customers who accepted long delivery times or placed their orders in underloaded periods would benefit substantially in quoted prices, while customers pressing for particularly short delivery times in overloaded periods would be charged increased prices. Of course such a policy should take into account the elasticity of

Operations Research Proceedings 1987

prices and delivery dates and the importance of the particular customer. There is a scarcity of models in the delivery dates assignment literature (Baker and Bertrand, 1981; Eilon and Chowdhury, 1976; Seidman and Smith, 1981; Weeks and Fryer, 1977; Weeks, 1979) trying to integrate in this way the pricing and delivery dates assignment decisions.

The purpose of this study was to develop an interactive decision framework for the above problem to suit the practical needs of real world situations. The proposed model seeks no "optimal solution". Instead, the manager is able to try and evaluate various alternatives under the guidance of the computerised system.

The problem nature assumes a close cooperation of the production and marketing functions in the firm. The task of integration is considerably easier in small firms where there are no organisational and behaviourial difficulties due to clearly set boundaries between the marketing and production control departments as can happen in larger firms. A small computer system easily accessed by a small number of people responsible for both actions may be enough.

B. Lead time as a control problem

Lead time in practically any company is a function of backlog (Plossl and Wight, 1973). Based on this truth, the technique called Input/Output Control (IOC) has been developed in order to manage backlogs of work-in-process by controlling the input of new orders against output of completed work, i.e. against capacity. The approach was meant to be used at the shop orders release decision level. However, the total manufacturing lead time, defined as the time elapsed between the arrival of raw material and the delivery date, includes not only the time spent on the shop-floor (throughput time) but also the time spent in the pool of unreleased jobs (pool delay). Reducing throughput times and balancing the work on the shop-floor using IOC methods, leads almost inevitably to increases in the waiting time in the pool of unreleased jobs. The total manufacturing lead time, which actually determines the delivery lead time being of primary interest to the customer, remains equal or even longer as in the case where jobs are released as soon as they become available.

In this study we take the view that both throughput time and total manufacturing lead time have to be controlled to desired norms by using IOC methods (Tatsiópoulos, 1983). This requires a close cooperation of the production and marketing functions of the firm with the help of a hierarchical backlog control system as displayed in Exhibit 1. There are two decision levels. At the higher level where decisions on delivery dates and prices are taken, the purpose is to control the pool of unreleased jobs and the total backlog (which is equal to the sum of the released backlog and the pool). The total backlog length and the pool length are used as state variables. The goal variables are the norms for the total backlog length and the pool delay. The controllable variables are the input rate of customer orders from the market and the capacity (output rate). The input rate is influenced by offering discounts to customer orders that contribute to the norms and charging premiums to orders that disturb those norms. The output rate is influenced by determining mean production rates for every work centre and the shop as a whole.

At the lower level, jobs are chosen from the pool to be released to the shop. The released backlog length is used as a state variable. The goal variables are the norms for the released backlog length. The controllable variable is the input rate from the pool to the released backlog.

C. Norms of the system

Setting of performance norms by higher management for the purpose of controlling workload in a job-shop depends on the configuration of the shop and the competitive market environment. The norms to be set are:

— Lead time norms
  - Average total manufacturing lead time
  - Average pool delay
  - Average shop-floor throughput time
  - Average slack per operation

— Backlog length norms
  - Maximum and minimum total backlog length
  - Maximum and minimum released backlog length
  - Maximum and minimum pool length
  - Maximum and minimum queue length

— Capacity utilisation norm

— Priority norms

  Maximum quota of pool priority jobs. A pool priority job will be released at once without having to wait in the pool of

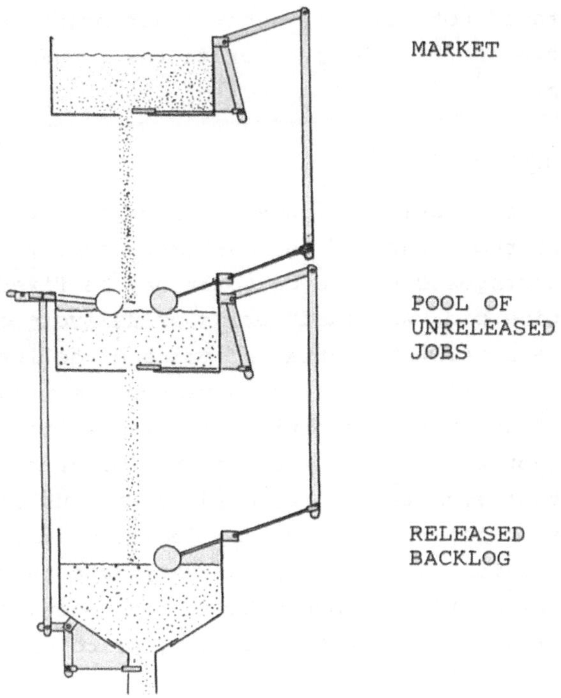

Exhibit 1: Hierarchical backlog and lead time control

unreleased jobs.
Maximum quota of hot priority jobs. A hot priority job will not have to wait either in the pool or in the queues of individual work centres.

D. Procedure QUOTCLAS

The number of delivery dates assignment and price setting decision options depends on the nature of the customer enquiry. The price and the delivery date proposed by the customer may be fixed, negotiable or open for competitive bid. A "fixed" price or delivery date is open only to a yes/no answer. A "negotiable" price or delivery date means that the customer is willing to discuss reasonable proposals. A price or delivery date "open for competitive bid" means that the firm has to make a one-shot quotation, along with other similar firms in the area, which will be either accepted or rejected by the customer. Procedure QUOTCLAS sets the "rules of the game" by deciding, according to the appropriate combination of price and delivery date characteristics, whether the quotation will be competitive or non-competitive and whether the forwards or backwards scheduling principle will be used by the system.

E. Procedure LEADTEST

Decisions of the customer orders planning function are taken every time a customer enquiry arrives in the shop. First the delivery lead time imposed by the customer is checked against the "should be" delivery time as calculated using lead time norms set by the management. If the customer's delivery lead time is short, the manager has to decide whether he will assign a later delivery date to the customer or he will undertake the cost of various alternatives for cutting down the manufacturing lead time (expediting, splitting, alternative routing).

Assigning expediting costs to individual jobs is an extremely difficult, if not impossible, task. In order to overcome this difficulty, the system provides for norms of permissible percentages of pool and hot priority jobs over the total number of jobs in the shop. If these maximum percentages are exceeded, expediting is not considered and lead times are shortened using either job splitting or alternative routing.

The extra costs associated with job splitting are direct costs due to the additional set-up hours. Therefore, they can be easily calculated and assigned to a particular job.

An alternative processing plan (e.g. on a CNC machine) may result in shorter lead time at the expense of higher hourly manufacturing costs. This extra cost can be calculated and assigned to individual jobs.

F. Procedure TOTBA

Exceeding of the maximum total backlog length norms at any work centre by the addition of the new customer order to the existing backlog leads to the following alternative actions in order to maintain the backlog lengths according to the norms: later delivery, extra capacity and reallocation of machine operators (capacity reallocation).

The capacity reallocation procedure is only possible if the maximum backlog length of the shop as a whole is not exceeded and a reallocation of multi-task operators is possible.

If the problem cannot be solved by the reallocation procedure, then extra capacity must be provided, usually in the form of overtime. The extra capacity is used to adjuct the excessively long backlog lengths back to their normal level.

In the case of shortfall of work, a new customer order helps to keep the shop within the norms if: a) The custoner's delivery lead time is long enough to allow a large time buffer in order to move the order over time and cover underloaded periods. b) The order arrives in underloaded period which is characterized by backlog lengths lower than the minimum norms.

Cost evaluations of disruptions or contributions to the norms imposed by the above alternative actions and the consequent alterations of the price/delivery dates proposals in the negotiating environment are processed by the following procedure QUOTCALC.

G. Procedure QUOTCALC

The general methodology of Cosmetatos and Eilon (1981) will be used in this study in order to calculate cost changes. According to the above methodology, absolute changes in costs are replaced by relative ones. This implies that the change in the unit production cost relative to some initial value (rather than the absolute value of the unit production cost itself) will be considered. Following a suitable decomposition of the unit production cost a model can be developed which relates changes in unit production cost components with variables than can be directly affected by management decisions concerning inputs and outputs expressed in either physical or financial terms.

The standard hour rate which is usually incorporated in job cost estimating procedures is an average value taken from the costs of past operations in the shop. However, this unit production cost may vary heavily in the short-term, depending on the level of output and the effort required to maintain this level (capacity changes). If the input flow of customer orders is to be smoothed over time by establishing cause and effect relationships between inputs and performance norms, then the short-term variation of the unit production cost has to be monitored and a model has to be built in order to predict those changes as a result of changing the input and output flow of work in an effort to maintain the backlog length and lead time norms.

The replacement of absolute changes in costs by relative ones enables the manager to avoid relying on unreliable historical mean values of the standard hour rate (unit production cost) and translate changes of input and output flow decisions into percentage price changes (discounts or premiums).

In the present study, the standard hour rate (SHR) which is considered to be the unit production cost, is decomposed as follows:

$$SHR = W/V + E/V$$

where: W = direct wages
E = other production costs such as depreciation charges, indirect labour costs, etc.
V = volume of output measured in standard hours of manufacturing over a period of time.

This breakdown of the unit production cost conforms with the one usually adopted by accountants; therefore, the numerical values of W and E can be assumed to be known.

For the purpose of this study, it is considered that wages paid to the employees consist of two elements only, basic wages and overtime pay:

$$W/V = W_b/V + W_o/V$$

The unit basic wage cost W_b/V and the unit overtime cost W_o/V depend on the total hours of normal work T_b and the total hours of overtime work T_o during the planning time period, as well as the level of output.

$$W_b/V = w_b T_b/V$$

$$W_o/V = w_o T_o/V$$

where: w_b = the average basic rate
w_o = the average overtime rate

The number of hours of normal work per planning period may increase by increasing the number of workers or decrease by cutting down the workforce or the working days per week. The number of overtime hours depends on decisions about the overtime production rate which are taken in the context of assigning total production rates to work centres.

The overhead cost E is considered to be fixed in the short and medium term, i.e. no major changes of equipment or organisational structure occur. Therefore, the unit overhead cost E/V depends only on the level of output.

Relative changes of the standard hour rate SHR* occur as a result of management action concerning relative changes of the number of hours of normal work T_b^* per period, overtime work T_o^* and the output V* in standard manufacturing hours. T_b depends on the size of the workforce and the working days per week. The output V is considered equal to the production rate set by the management.

According to the decomposition of the standard hour rate as the unit production cost, presented in the previous sections, the following relations will hold:

$$SHR + \delta SHR = (W_b(T_b+\delta T_b) + w_o(T_o+\delta T_o)+E)/(V+\delta V)$$

$$SHR^* = \delta SHR/SHR = (w_b T_b + w_o T_o + E)/(V+\delta V)SHR +$$

$$+ (w_b \delta T_b + w_o \delta T_o)/(V+\delta V)SHR - SHR/SHR =$$

$$= -\delta V(w_b T_b + w_o T_o + E)/V(V+\delta V)SHR + w_b \delta T_b + w_o \delta T_o/(V+\delta V)SHR =$$

$$= 1/(V+\delta V)SHR\, [\, w_b(\delta T_b - T_b \delta V/V) + w_o(\delta T_o - T_o \delta V/V) - E\delta V/V\,] =$$

$$= 1/(V+\delta V)SHR\, [\, T_b w_b(\delta T_b/T_b - \delta V/V) + T_o w_o(\delta T_o/T_o - \delta V/V) - E\delta V/V)\,] =$$

$$\doteq (T_b w_b/(1+\delta V/V)VSHR)(\delta T_b/T_b - \delta V/V) +$$

$$+ (T_o w_o/(1+\delta V/V)VSHR)(\delta T_o/T_o - \delta V/V) - E\delta V/V(1/(1+\delta V/V)VSHR)$$

If we put, $T_b w_b/V \cdot SHR = l_b f_w$, $T_o w_o/V \cdot SHR = l_o f_w$, $E/V \cdot SHR = f_\varepsilon$ and,

$T_b^* = \delta T_b/T_b$, $T_o^* = \delta T_o/T_o$, $V^* = \delta V/V$ we get,

$$SHR^* = (1/(1+V^*))\ l_b f_w (T_b^* - V^*) + (1/(1+V^*))\ l_o f_w (T_o^* - V^*) - f_e V^*/(1+V^*)$$

The above formula permits the calculation of relative changes of the standard hour rate induced by changes of the workforce size, overtime work and the production rate.

The customer orders schedule is updated every time that a new customer order enters the shop and the resulting total backlog lengths are calculated. The distance of the backlog lenghts representing the new schedule from the maximum and minimum norms as compared to the distance from the norms of the old schedule's backlog lengths define the relative impact of the new customer order on the norms, i.e. whether the new order helps maintaining the norms or causes disruptions.

The contribution in the case of shortfall of work can be measured in cost terms by calculating the relative change of the standard hour rate as a result of the increased output due to the contributed load in underloaded periods. The standard hour rate is relatively decreased during these periods because of a decrease of the unit overhead cost. Therefore, the relative cost decrease can be transformed into a relative price discount which is offered to the customer in order to attract jobs into such periods.

In case of overloads, the release of overtime work cannot correspond exactly to the amount and timing required for a specific customer order. Instead, a decision is made which concerns the average production rate and the average number of overtime hours per time period and for the next few periods. In this sense it is impossible (and unfair) to assign the extra capacity to one particular order. What can be done is to calculate the relative change of the standard hour rate in the short-term and then use this increased rate to calculate the relative cost increases of the new job and all the future jobs that may arrive during the time periods with increased overtime work. The relative cost increase is transformed to a relative premium on the price.

References

BAKER, K. and BERTRAND, J., 1981, An investigation of Due-date assignment rules with constrained tightness, Journal of Operations Management, 1 (3), pp.109-120.

COSMETATOS, G. and EILON, S., 1981, Analysis of unit cost-A modelling approach, in: Falster, P. and Rolstadas, A., Production Management Systems, IFIP, North Holland, pp.167-178.

EILON, S. and CHOWDHURY, I., 1976, Due dates in job-shop scheduling, International Journal of Production Research, 14 (2), pp.223-237.

PLOSSL, G. and WIGHT, O., 1973, Capacity planning and control, Production and Inventory Management, 3rd. Qtr., pp.31-67.

SEIDMANN, A. and SMITH, M., 1981, Due Date assignment for production systems, Management Science, 27 (5), pp.571-581.

TATSIOPOULOS, I. and KINGSMAN, B., 1983,Lead time management. A review, European Journal of Operational Research, 14 (4), pp.351-358.

WEEKS, J. and FRYER, J., 1977, A methodology for assigning minimum cost due dates, Management Science, 23 (8), pp.872-881.

WEEKS, J., 1979, A simulation study of predictable due dates, Management Science, 25 (4), pp.363-373.

Flexible Manufacturing Systems: background examples and models

W.H.M. Zijm, Eindhoven

Abstract:

In this paper, we discuss recent innovations in manufacturing technology and their implications on the design and control of manufacturing systems. Recognizing the need to respond properly to rapidly changing market demands, we discuss several types of flexibility that can be incorporated in our production organisation to achieve this goal. We show how the concept of a Flexible Manufacturing System (FMS) naturally arises as an attempt to combine the advantages of traditional Job Shops and dedicated production lines.

The main body of the paper is devoted to a classification of FMS problem areas and a review of models developed to understand and solve these problems. For each problem area, a number of important contributions in the literature is indicated. The reader, interested in the applications of Operations Research models but not familiar with the technical background of FMS's, will find the descriptions of some essential FMS elements useful. Some final remarks and directions for future research conclude the paper.

Zusammenfassung:

In dieser Arbeit diskutieren wir neuere Entwicklungen in der Fertigungstechnologie und ihren Einfluß auf den Entwurf und die Steuerung von Fertigungssystemen. In Anbetracht der Notwendigkeit, schnell und in der richtigen Weise auf sich ändernde Marktsituationen zu reagieren, besprechen wir unterschiedliche Typen von Flexibilität, die in unsere Produktionsorganisation eingebaut werden können. Wir zeigen, wie das Konzept eines flexiblen Fertigungssystems (FFS) auf natürliche Weise aus dem Versuch entsteht, die Vorteile von herkömmlichen Job Shops und Fertigungslinien zu kombinieren.

Der Schwerpunkt dieser Arbeit liegt auf einer Klassifikation der FFS Problemgebiete und einer Übersicht über Modelle, die zum Verständnis und zur Lösung dieser Probleme entwickelt wurden. Zu jedem Problemgebiet werden wichtige Beiträge aus der Literatur angegeben. Der an der Anwendung von Operations Research-Modellen interessierte Leser, der jedoch nicht mit dem technischen Hintergrundwissen von FFS vertraut ist, wird die Beschreibung wichtiger FFS-Elemente als nützlich anerkennen. Zum Abschluß dieser Arbeit werden Anregungen für weitere Forschungen gegeben.

Operations Research Proceedings 1987

1. Introduction

The need for a more flexible response to rapidly changing market demands has caused a rather dramatic change in the philosophy on the design and the layout of production systems, as well as on production organisation and even product development. Order leadtimes and production throughputtimes have to be shortened, final stock and work-in-process inventory levels have to be reduced and at the same time service levels are to be increased. The implications of these objectives can be observed at many different levels in the industrial organisation.

In this paper, we focus exclusively on the implications at the shopfloor level. This paper is based on Zijm[1987] where a large number of models for the design and control of Flexible Manufacturing Systems are treated in extenso. For a discussion of the relationship between the need for an increased flexibility and methods for integral goodsflow control, we refer to Zijm[1988]. It should be understood that the two subjects are closely related to each other; it makes no sense to invest in the installation of flexible technology, without adapting at the same time the entire logistics organisation to the possibilities, induced by the increased flexibility. In other words: moving towards flexible manufacturing means more than only installing automated production systems.

The main objective of this paper is to review a number of important contributions in the literature on Operations Research models developed to evaluate FMS investment decisions as well as FMS performance. We present a classification of FMS models, covering justification, design and operational control problems.

The structure of the paper is as follows. In the next section, we highlight some background issues of flexible manufacturing. In particular, different types of flexibility are characterized. We dwell on the advantages and disadvantages of Job Shops and dedicated production lines and show how the development of the FMS-concept arises naturally.

Section 3 provides a reference framework for those readers not familiar with the technical background of an automized manufacturing system. A number of basic concepts and elements of Flexible Manufacturing Systems are briefly described.

In section 4, we present a classification of FMS problems and their mutual relationships. This classification is neither entirely new nor unique; earlier work has been published by e.g. Kusiak[1986], Stecke[1985], Van Looveren et. al.[1986] and Kalkunte et. al.[1986]. Future developments on the path towards the "Factory of the Future" are nicely described in Bullinger et. al.[1986] and Meredith[1987]. With respect to each problem area, a number of important contributions in the literature is indicated. This literature review is not exhaustive; despite of the fact that the area is relatively new, the number of papers, conferences and journals devoted to FMS's is growing rapidly.

In section 5, finally, some conclusions are formulated and directions for future research are indicated.

In this paper, we deal almost exclusively with flexible machining systems, not with flexible assembly systems. The nature of flexible assembly systems demands for a different approach in some important aspects, both in design and with respect to operational control. In Zijm[1987], flexible assembly systems are discussed separately for this reason.

A number of models, mentioned only briefly now but discussed more thoroughly in Zijm[1987], differ in important aspects from previous ones published. With respect to justification, it should be said that we do not believe in one Operations Research model to represent all possible trade-offs; however, in highlighting the benefits of a FMS, models can play a significant role.

Finally, we remark that the classification presented here reflects our idea to decompose problems as much as possible, in order to provide insights and to facilitate quick and, if necessary, repeated solution of these problems. All these issues are discussed more extensively in Zijm[1987], which forms the basis of this paper.

2. Flexibility and manufacturing

Flexibility in manufacturing means: a quick response to the market. To develop, manufacture and distribute, with short leadtimes, products of high quality, if possible on an order based planning rather than on a forecast based planning.

Hence, flexibility means more that just installing automized production systems; it has farreaching consequences also for product development, as well as for materials management and physical distribution management. In general, we distinguish between three different, but interrelated types of flexibility.

- Volume flexibility is related to the possibility to react to volume fluctuations in sales. The basic idea is that leadtimes should be so short that production will be able to follow these fluctuations. This in turn requires flexibility in capacity but the alternative is to keep large stocks of final products, a situation which is generally recognized to be higly undesirable by most industrial companies. An appropriate logistics organisation is a key factor in attaining volume flexibility.

- Mix flexibility refers to the rate at which changes with respect to the specifications of a product can be reflected on the shop floor. Throughputtimes and set-up times in the manufacturing process are the key parameters here.

- Innovation flexibility finally deals with the ability to get new products, or new versions of a product, on the market in time. Product life cycles become shorter and shorter; we are forced to reduce development leadtimes drastically. Also, when products are designed, care should be taken to make them "manufacturable". In fact, product and process development should go hand in hand.

Although all three types of flexibility have their impact on the design of production systems, Flexible Manufacturing Systems particularly seem to be the key element towards achieving mix flexibility.

- A Flexible Manufacturing System is a production system capable of manufacturing a large variety of different parts or products through each other, while maintaining high overall production rates and short throughputtimes. Often, these systems are characterized by a high degree of automation, such as an automated Material Handling System (MHS), Computer Numerically Controlled (CNC) machines, while also the complete shopfloor process is under rigid computer control. The absence of large set-up times is a key feature of these systems. When operational, a FMS can be (re)planned with a high frequency (e.g. every four hours).

Other views on flexibility in manufacturing are described by various authors, see e.g. Buzacott[1982] and Slack[1983].

It is interesting to compare a FMS with more traditional manufacturing systems such as the classical Job Shop or, on the other hand, the dedicated production line, as found in traditional automobile factories. In the latter ones, a high productivity and short throughputtimes were achieved indeed, but at the cost of an extreme inflexibility with respect to variations in the output. On the other hand, the traditional Job Shop made it possible to supply a fairly varied range of products, but, due to large changeover times, at the cost of high intermediate stocks and extremely long throughputtimes.

Flexible Manufacturing Systems are ultimately an attempt to combine the advantages of both the Job Shop and the dedicated production line, i.e. a high degree of diversity and, at the same time, a high level of efficiency with short throughputtimes, and to avoid the disadvantages. In passing, we mention another advantage of short throughputtimes, with respect to quality: faults are detected earlier and information about these can therefore be fed back more quickly. High intermediate stocks, on the other hand, only help to conceal faults (Schonberger[1982]).

In the next section, we describe some key elements of a metal cutting FMS, to provide the reader with some basic understanding of the (computer controlled) hardware components, that are the key towards more flexibility.

3. Elements of Flexible Manufacturing Systems

This section is devoted to a description of some of the basic hardware components that can be found in a metal cutting Flexible Manufacturing System. Examples of these kind of systems are described in several papers (e.g. Stecke and Solberg[1981]). Examples of Flexible Assembly Systems can be found in Wittrock[1985] and Zijm[1987].

Let us start with the heart of a metal cutting FMS, a CNC-machine. Fig. 3.1 shows an integrated milling and drilling machine. The main difference, compared with conventional machinery, is the fact that this CNC-machine is capable to perform a, usually large, set of different operations, without human intervention. Each operation is described by a NC-program, specifying among others which tools (and in what sequence) are needed to perform that operation. On completion, the part is automatically unloaded and replaced by another. After identification, a new program is loaded and the next operation may start. It is emphasized that one operation may require the subsequent use of several different tools.

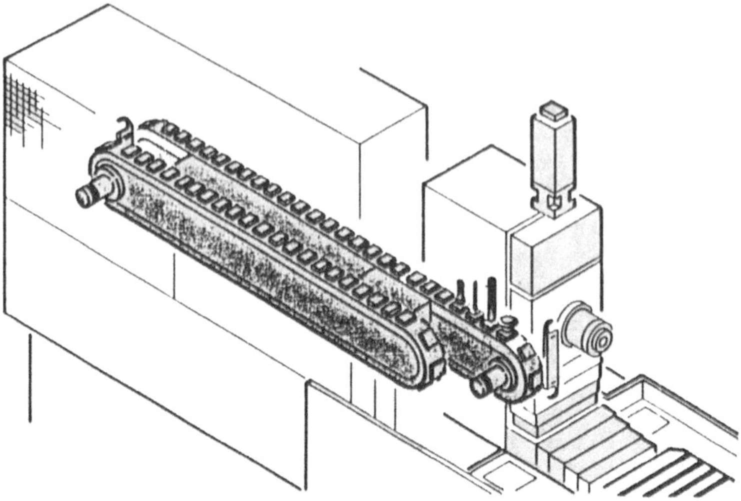

Fig. 3.1. An integrated milling and drilling CNC-machine.

Coupled with the machine is a tool magazine, holding a large variety of different milling cutters and drills. An automatic tool changer interchanges tools between the drive mechanism and the tool magazine when needed, in only a few seconds. The tool magazine is loaded with all tools, needed to perform operations in a fixed period of four hours, say. It is this tool magazine, together with the automatic tool changing mechanism, which gives the machine its inherent flexibility.

Modern machines are usually provided with an inspection mechanism, to check the state of all tools on operation. It is not uncommon that a tool is worn off after an exploitation of totally 30 minutes (using it intermittently).

Associated with the machine (or with a group of similar machines) is a (usually small) set of specially designed pallets that can hold one or more parts or workpieces. The correct positioning of the parts on the pallets is achieved by means of special fixtures or clamping devices. The pallet with the workpiece is placed and fixtured on a worktable which can be rotated on two or three axis.

The machine may be equipped with a number of special features such as (vision) systems for identifying the parts and checking their correct positions, while also the performance of each operation is supervised continuously.

Fig. 3.2 shows a Flexible Manufacturing System, consisting of a number of the above described CNC-machines. The machines are basically similar, however, due to the fact that each machine may be loaded with a different set of tools, the set of operations assigned to each machine may be different (the assignment of operations to machines is one of our major control problems, compare the next section). The machines are in general connected by an automated Material Handling System (MHS), in our example an Automated Guided Vehicle System (AGVS). Other Material Handling Systems may be a railcart system, an automatic conveyor system

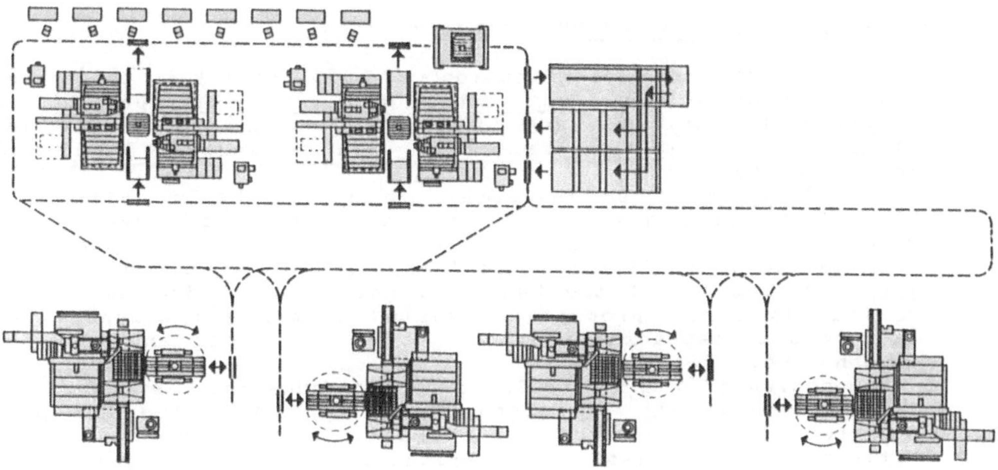

Fig.3.2. A Flexible Manufacturing System with AGV transport.

or, when machines are physically placed together in a <u>workcell</u>, a <u>handling robot</u>. If buffer space is provided, it is usually integrated with the MHS. Pallets, fixtures, grippers and tools are shared by the machines.

The machines, including the tool magazines, the tool changers and the pallet positioning mechanisms, are usually controlled by local micro-processors. The interaction between different machines in the system and the operation of the MHS is supervised by a central computer. The latter one also communicates with the computers of the production planning department.

Stecke and Solberg[1981] describe a Flexible Manufacturing System at Caterpillar, consisting of three drilling machines, four milling machines, two vertical turret lathes and an inspection machine, connected by a railcart system. Other examples of FMS's in metal cutting are given by e.g. Groover and Zimmers[1984], Ranky[1983] and by Kearney and Tracker[1980], a large manufacturer of automated production systems. These systems are typically meant to produce an almost unlimited variety of parts in very small batches, on order. Special designs can be implemented by specifying a new set of programs. Typical products of such a system are e.g. gear boxes, compressor houses, etc.

In the next section, we will define a classification of FMS problem areas and discuss a number of relevant models, suggested in the literature. In order to get a clear picture of these problems and their mutual relationships, the background material presented above may serve as a reference to the reader.

4. A classification of FMS problems

The hierarchical classification, we will present in this paper, distinguishes between justification problems, design problems and operational problems. Each of these three groups can be subdivided again. Another important area is what we shall call interface problems.

Justification of an investment in a FMS should be based on clearly defined objectives such as a desired market share, reduction of stocks and leadtimes and the like, in particular when replacing traditional machinery. Quantitative models should yield estimations of the benefits, and of the reductions in the various types of production related costs, and balance these against implementation costs in order to select an acceptable alternative.

The design of a FMS is related to such questions as: central or local buffers, the operation of the MHS but also the parts spectrum to be produced on the system. Design conflicts should ultimately been solved by studying their impact on performance criteria, in view of the desired objectives.

Operational problems primarily deal with detailed shopfloor planning and scheduling decisions, on a daily basis. It is the responsibility of the shopfloor controller to operate the system in an optimal way, within the constraints imposed by the design and the layout of the system.

Interface problems finally consider the interaction with the environment of the FMS. Flexibility implies that one is able to produce a large variety of products in a short time. Material availability and modular product structures are then crucial factors. Another important factor relates to the adaptability of the organisation, in particular in the field of production planning and distribution, to an increased flexibility in the factories.

In the next subsections, justification problems are briefly discussed, after which we pay more detailed attention to design and operational problems. Interface problems are not discussed at all, for an elaborate treatment of this subject we refer to Zijm[1987].

4.1. Justification problems

A frequently raised argument to invest in flexible technology is the loss of market share in the long run if one doesn't. This idea is usually based on the observation that diversity in demand has increased, and will further increase, rapidly, leading to either an explosion of the set-up's (and hence a loss of capacity) in traditional machining centres, or long leadtimes and high work-in-process levels and, consequently, high final inventory levels, all in line with the usual economies-of-scale considerations. At the same time, decreasing product life cycles make it risky to keep large amounts of final products in stocks, apart from the fact that high interest rates make stock reductions inescapable. Therefore, we should focus on economies of scope, rather than on economies of scale (Goldhar and Jelinek[1983]).

Despite all this, the range of OR-models that address in particular an investment analysis of FMS's, in view of these operational cost characteristics, is very limited. One reason may be the difficulty in determining future effective demand, by the interaction between manufacturing technology and market characteristics (Burstein and Talbi[1985]). Also, sound measurements of market competitiveness, by means of for example portfolio analysis methods, are usually based on doubtful assumptions, if developed at all.

Whether or not classical capital budgetting techniques should be used to evaluate investment decisions in flexible technology, is the subject of an ongoing debate in the literature. Burstein and Talbi[1985], Michael and Millen[1985] and Primrose and Leonard[1986] express the opinion that these traditional approaches fail to capture adequately the essential features (and, in particular, the benefits) of a FMS; some alternative approaches are suggested. On the other side of the spectrum we find for instance Kulatilaka[1985] who exploits a risk-adjusted discounted cash flow model in calculating net present values. Strategic issues with respect to FMS-investments are furthermore discussed by Gaimon[1985],[1986] and Srinivasan and Millen[1986]. An interesting scoring model for flexible manufacturing systems project selection is presented by Nelson[1986]. Fine and Freund[1986] develop a convex quadratic programming method to decide upon a portfolio of flexible and nonflexible capacity.

Papers that address the reduction of specific operational costs are those of Porteus[1985], [1986] and of Van Beek and Van Putten[1987]. These authors do not try to develop an overall investment analysis method but merely try to quantify integral effects of for instance reductions in set-up costs or overall leadtimes. Unfortunately, they all consider the single product situation whereas the benefits of a FMS particularly arise from the possibility to produce a large variety of different items with a negligible loss because of set-up times, hence in small batches and with short throughputtimes. Zijm[1987] advocates the use of a slightly extended version of a model, developed originally by Silver[1975] (see also Silver and Peterson[1985]) to quantify the effects of reductions in changeover costs in a Group Technology environment (cf. Burbidge[1975]). In particular, a distinction is made between changeover costs from one part family to another one and changeover costs within one family. Modern Group Technology approaches, combined with sophisticated machine loading techniques, enable planners to define large part families, where the inter-family changeover costs drop to zero. The advantages of these achievements are demonstrated in Zijm[1987].

4.2. Design problems

Once a decision has been made with respect to a possible investment in a Flexible Manufacturing System, the question arises what is the best system layout and what basic operation rules should be implemented in order to exploit the installed equipment in an optimal way. For example, choices have to be made about sizes and locations of buffers, release rules for dispatching products to the system should be developed, etc. Performance indices may be a desired throughput and/or throughputtimes, along with the desired work-in-process levels. Note that we are still in a situation where actual production needs are not specified; at the best, we have a global indication about the long term production ratio's for families of items. A careful long-term part family selection, based on similar operating characteristics as well as on similar geometrical and technological attributes (Group Technology) should be included in the design phase.

With respect to system layout and system performance, queueing network models have provided valuable insights into a number of design issues. In particular, closed queueing networks or restricted open queueing networks yield an acequate description of a FMS, in principle at least. The ratio behind this is that in most FMS's the number of parts in the system at any moment is limited, due to the limited number of available pallets or because of a strict workload control.

Recently, we have seen a tremendous increase in papers dealing with queueing network models applied to a FMS-environment. Let us describe the basics of the approach. A FMS is described as a network of single or multi-server stations where each multi-server station denotes a workcell of identical machines. The material handling system (MHS) is usually modelled as a pure delay, i.e. as an infinite server (e.g. in the case of an automatic conveyor belt) but sometimes also as a multi-server station (e.g. if only a limited number of AGV's is available). A basic result in queueing network theory states that under certain assumptions the steady state probabilies of finding n_i jobs at workcell i have a product form, i.e. they separate into a normalization constant and factors that depend exclusively on the characteristics of one workcell (Jackson[1963], Gordon and Newell[1967], Baskett et. al.[1975]). In workcells with a FCFS (First Come First Serve) discipline, exponentially distributed service times have to be assumed in order to make the model exact, while furthermore buffer capacities are unlimited in principle. However, several approximations have been developed for more complicated networks (e.g. Whitt[1983 a], [1983 b], Yao and Buzacott[1985 b], [1986 a]).

Closed queueing networks depict the situation where each finished job is immediately replaced by a new one. This situation can be modelled explicitely by defining a special single server load/unload station. The throughputtime, calculated by Little's formula, is then the time between two successive visits of a job to the load/unload station in the closed queueing network. Special algorithms have been developed to calculate these different performance measures in an effective way, the two most well-known being the convolution method (Reiser and Kobayashi[1975], Reiser[1977]) and the Mean Value Analysis (MVA) algorithm (Reiser and Lavenberg[1980], Reiser[1981]).

Solberg[1977], [1981] was the first author who applied closed queueing network techniques to model and analyze Flexible Manufacturing Systems. Approximate MVA algorithms for FMS's are described in e.g. Suri and Hildebrant[1984] and Shalev-Oren, Seidman and Schweitzer[1985].

The infinite buffer assumption appeared to be a serious obstacle to model realistic FMS's (with usually small buffers). However, in one special case, particularly relevant to FMS's, a nice solution exists. A central server model (CSM) describes the situation where, between any two stations, access is needed to a central server. In most FMS's, the MHS does play the role of such a central server. Under the assumption of exponential service time distributions, these models appear to be time reversible (Keilson[1979], Kelly[1979], even when buffers are assumed to be finite. Note that, due to the limited buffer assumption, blocking may occur when a job attempts entry from the MHS to a work station. The capacity of the MHS on the other hand is assumed large enough to accomodate all jobs in the system. These models have been analyzed by Yao and Buzacott[1985 a], [1986 b]; even in certain state-dependent routing cases, product form solutions could be obtained.

In more general limited capacity models (without the central server assumption), blocking phenomena prevent an exact analysis of problems of realistic sizes. Approximation methods have been developed by Yao and Buzacott[1985 b], Altiok and Perros[1986], Dallery and Yao[1986], Van Dijk[1987] and, using a fluid model approach, by De Koster[1986] and Wessels et. al.[1986]. An interesting alternative method, based on a perturbation analysis of one long simulation experiment, is presented by Ho et. al.[1979]. A detailed analysis of a buffer control model with priorities in the routing of the jobs is provided by Repkes and Zijm[1988].

Furthermore, we advocate the use of aggregation methods (see e.g. Chandy et. al.[1975]), for the situation where some part of a larger network behaves like a closed queueing network (because of a workload dependent admittance rule), and of renewal approximations (Whitt[1983 a], [1983 b]) or an exponentialization method (Yao and Buzacott[1986]) for the case of general service time distributions. Buzacott and Yao[1986] provide a detailed overview of queueing network approaches to model FMS's.

The long term part family selection problem is usually solved by clustering techniques, based on operating characteristics as well as on technological considerations. Since basically the same approach is used for short term FMS planning, in particular when grouping operations for a certain period, we postpone its discussion to section 4.3.

4.3. Operational problems

Once the detailed design and, subsequently, the installation of the FMS have been completed, one should develop a framework for the solution of a number of frequently returning planning and scheduling problems. In this section, we define four, hierarchically coupled, levels of decision making in an operational environment (compare also Kusiak[1985]).

- Planning

Prepare a list of production orders, based on internal or external demand, for instance on a weekly basis.

- Operations Grouping

Each job involves a sequence of operations, each one described by a NC-program. Different operations may require different fixtures or even pallets. Divide the first planning period into a number of time windows. For each time window, varying for example from four to eight hours, select a set of appropriate operations for processing in that window, taking into account eventual precedence constraints, such that refixturing within such a time window is not necessary.

- Tool Loading

Given the outcome of an Operations Grouping procedure for a time window, assign operations and (hence) tools to the tool magazines of the CNC-machines.

- Sequencing

Within one time window, develop dispatching rules for the set of operations, given the outcome of the Tool Loading procedure.

Below, we discuss some approaches to these problems.

4.3.1. Planning

Suppose a long term family structure has been determined for the set of all part types, based particularly on operating characteristics such as required tools, fixtures, etc. In a Flexible Manufacturing System this means that in particular set-up costs only play a role at the level of a complete family, i.e. when changing production from one family to another one. Within one family, set-up costs are not relevant any more, moreover, inventory holding costs do not depend on the particular item within a family structure.

Production planning in the case of a machine shop as the one, discribed in section 3, is usually based on an explosion from a Master Production Schedule (MPS) on end-item level. This is a very common approach, known as Materials Requirement Planning (MRP), see e.g. Orlicky[1975], Silver and Peterson[1985] or Vollman, Berry and Whybark[1984]. In our case, the explosion can be on a family level basis, due to the flexibility of the FMS, and details with respect to part types can be specified at the last moment. This reflects the concepts of Hierarchical Production Planning (HPP), developed by Hax and Meal[1975] and Bitran, Haas and Hax[1981], [1982], see also Hax and Candea[1984], ch. 6.

The reader should recognize that, due to the MRP-type approach, the machine shop is confronted with an essentially deterministic "demand" pattern. However, limited capacity and economic considerations based on set-up and inventory holding costs force the planning department to smooth this pattern and to determine batch sizes and the like. To convert the production requirements to a detailed production plan, we propose the use of a capacitated lotsize models at a family level. Heuristics to solve these models have been proposed by a number of authors, e.g. Dixon and Silver[1981], Dogramaci et. al.[1981] and Lambrecht and Vanderveken[1979]. Numerical comparisons have been presented by Maes and Van Wassenhove[1986 a],[1986 b], a more extensive treatment can be found in Maes[1987].

4.3.2. Operations Grouping

Suppose that for the first week, we have an adequate production plan on a family level basis. In general, the total number of tools needed to execute such a plan, exceeds the total capacity of the tool magazine. Furthermore, each job can in general be split up into a number of operations, where each operation needs special fixtures and sometimes even pallets; such an operation is in general completely specified by a NC program. Recall that one operation should not be identified with one single tool. In general, one operation requires a set op appropriate tools, the cardinality of this set however never exceeds the capacity of one tool magazine. Hence, one operation can (and will) be completely performed by one machine.

Due to the fact that subsequent operations of the same job require in general different fixtures or clamping devices and sometimes different pallets, it seems reasonable to split up the planning period into a number of time windows such that in each time window only operations are planned that do not require some intermediate refixturing. These time windows may vary, dependent on the nature of the underlying machining process, from four to eight hours (for the unmanned night shift), unless automatic refixturing robots are available.

It follows that for each time window, we have to specify a set of operations for immediate and simultaneous processing, taking into account precedence constraints and the total capacity of the tool magazines. Hence, reloading of tools in the magazines, coupled with the machines, only takes place at the end of a time window. In order to minimize the number of time windows (or to maximize the average size of the windows), it is appropriate to select "schedulable" operations such that the corresponding sets of required tools have maximal intersections, where the union to these sets should not exceed the total available capacity. This approach leads in a natural way to a clustering formulation.

King[1980] and King and Nakornchai[1986] have developed a Rank Order Clustering (ROC) algorithm in order to solve these types of problems in a more traditional Group Technology (GT) environment (for an introduction into GT concepts, see Burbidge[1975]). Although certainly not the most efficient one, the matrix formulation proposed by these authors gives a nice insight into the way tool requirements influence the operations grouping. Extended versions of these type of algorithms (MODROC) have been presented by Chandrasekharan and Rajagopalan[1986].

Kusiak[1987] and Kumar, Kusiak and Vannelli[1986] present several algorithms for clustering parts according to the tool types needed. These methods are based on similarity measures between different operations (the p-median method) or on graph-theoretical approaches (in particular by trying to find so-called k-decompositions).

From the resulting set of clusters (where each cluster corresponds to a set of operations and a set of tools), we select as many as total tool magazine capacity allows. If the intersections of different clusters with respect to tool requirements is non-void, double or triple loading of similar tool types may be needed. Another reason for multiple loading of tool types may be to provide extra flexibility in the ultimate scheduling problem.

4.3.3. Tool loading

Once a set of operations, and a corresponding set of tools, is specified by the Operations Grouping procedure for execution in the next time window, the question arises how to assign these tools to the tool magazines coupled with the CNC machines. In particular, one may wonder what objectives should be persued when assigning these tools.

Stecke[1983] presents a number of integer programming formulations and develops heuristics to solve these tool loading problems. Central in her approach is the concept of pooling. A set of machines is said to be pooled, if they can perform exactly the same set of operations. Note that this does not exclude the situation where an operation is assigned to more than one pooled group of machines. However, in general one can say it is better to have a minimum of groups of pooled machines (compare also Stecke and Morin[1985], Stecke and Solberg[1985].

What kind of criterion function should be taken depends on the problem situation. Several criteria have been investigated in Stecke[1983]. For example, one may wish to balance the machine workloads, or to minimize the number of movements from machine to machine, or to fill the tool magazines as densely as possible.

Quite a different formulation is given in Zijm[1987], based on work of Zeestraten[1987]. The objective is here to maintain a maximum of flexibility in the underlying sequencing problem. The idea is that, for each machine, one should

- maximize the amount of time, needed for operations that may be performed by that machine, and
- minimize the amount of time, needed for operations that have to be performed by that machine exclusively.

Another formulation has been presented by Rajagopalan[1986].

4.3.4. Scheduling

As a result of the tool loading procedure, we know exactly which operations can be performed on which machines. The final task left is now to schedule the operations such that for example the total makespan is minimized. The resulting scheduling problem can be formulated as a mixed integer program but is in general extremely difficult to solve exactly (Florian et. al.[1980]). Also, the fact that there may be a number of stochastic interrupts, due to for instance breakdown of tools or other unexpected failures, leads to the use of simple priority rules such as SPT (Shortest Processing Time First) or SPT/TOT (compare e.g. Stecke and Solberg[1984]). However, recent results in job shop scheduling theory justify the exploitation of more advanced algorithms in our FMS scheduling problem.

To illustrate the difficulties of our scheduling problems, the reader should recall that an operation can be performed on more than one machine in general. This introduces parallelism but the parallel machine structure is highly operation dependent. For example, operation A may be performed on either machine 1 or machine 2, operation B on either machine 2 or 3, operation C on either machine 1 or 3, etc. To further complicate matters, we have precedence constraints between operations that actually belong to the same job and are performed in the same time window (in the case that refixturing is not necessary). The resulting problem is a strong generalization of the classical job shop problem.

In the case where machines are pooled together (compare section 4.3.3.) and no operation is assigned to more than one group of pooled machines, we still have a, more natural, generalization of the job shop problem. Only when each pool contains only one machine, the classical job shop formulation arises.

If, on the other hand, pools contain more machines, but no consecutive operations of one job are scheduled in the same time window (hence no precedence constraints exist), we are left with a number of parallel machine problems for which good heuristics are developed, e.g. the LPT (Longest Processing Time First) rule or the MULTIFIT method (cf. Coffman, Garey and Johnson[1978]).

Promising new methods for the job shop scheduling problem have been presented by Adams, Balas and Zawack[1986]. They use an iterative procedure with Carlier's algorithm for the one machine scheduling problem with release and due dates as a building block (compare Carlier[1982]).

The literature related to scheduling problems in FMS's is rapidly growing although most problem formulations are highly simplified versions of the one discussed here. We mention in particular Morton and Smunt[1986] who present a more general framework for planning and scheduling and Shaw and Whinston[1986] and Subramanyam and Askin[1986] who exploit results from Artificial Intelligence. Finally, although meant for Flexible Assembly Systems (which differ in many important aspects from Flexible Manufacturing Systems, see also Zijm[1987]), we mention the work of Wittrock[1985], who develops a periodic scheduling algorithm for a printed circuit board assembly line, actually a heuristic procedure to minimize cycle time. Pinedo et. al.[1986] also discuss a periodic scheduling algorithm for a flexible assembly line, but include a finite buffer constraint.

This concludes our subsection on operational problems and therebye also the complete hierarchical classification. As the reader may have noticed, we believe that stochastic network methods are particularly suitable for design problems, whereas operational problems lend themselves more to a mathematical programming formulation. We will come back to this point in the final section.

5. Conclusions and directions for future research

There is certainly a lack of Operations Research models that address in particular the justification of investments in flexible technology. As mentioned before, most of the papers use capital budgetting techniques such as Discounted Cash Flow (DCF) analysis (see e.g. Lutz[1982]). However, most of the supposed benefits of a FMS are treated as given parameters in these models. We believe that exactly this quantification of possible benefits is a major problem, for which more research is needed. Also, the cost of installing a FMS are in general not easy to estimate, being not only dependent on hard- and software, but also on for example education and training. Total installation costs are therefore in general not linear in the required capacity. Some kind of break-even analysis (see e.g. Thuesen [1982]) may be necessary to make a total cost-benefit analysis.

As mentioned before, we believe that queueing network techniques are suitable to analyze in particular design problems whereas operational problems can be better formulated by using mathematical programming models. Solutions to design problems have to be robust with respect to a large and varying set of operational conditions, which cannot even be specified at this design phase. This leads in a natural way to a stochastic formulation. On the other hand, the short term planning environment, although highly diverse and complex, is in general deterministic in nature.

It is in general not very realistic to model a Flexible Manufacturing System as a product form network, due to finite buffer constraints, state-dependent routing and job priorities, together with the fact that "aggregate" service times are in general not exponential. More research on approximation methods which capture effectively these and other limitations, inherently present in FMS's, is needed. Also, the need for good traffic control rules for the MHS justifies further research.

With respect to operational problems, we feel that the impact of different grouping strategies on the loading problem, and of different loading strategies on the final scheduling problem, is not well understood. Furthermore, as sketched in the preceding section, the loading and scheduling problems itselves are extremely difficult; there is still a lot of room for better heuristics.

Several problems we did not mention at all. Machine control problems belong to the field of CAM (Computer Aided Manufacturing). For example, if operations are very short, the sequence in which the tools are loaded in the tool magazines becomes important. This leads to a formulation, known as the Travelling Salesman Problem. On the other side of the spectrum we have what we called FMS interface problems. Material procurement becomes a very critical factor when exploiting a Flexible Manufacturing System. Also, due to the flexibility of these production systems, production planning can be much more of a hierarchical nature, where details are filled in literally at the last moment. The ultimate success of a transfer to flexible technology is highly dependent on a good appreciation of these factors. The need for Flexible Management Systems to fully exploit the benefits of Flexible Manufacturing Systems is recognized by now, but research on these interface problems has hardly begun.

References

Adams, J., E. Balas and D. Zawack[1986], The shifting bottleneck procedure for job shop scheduling, Management Science Research Report No. MSRR-525, Carnegie Mellon University, Pittsburgh,

Altiok, T. and H.G. Perros[1986], Open Networks of queues with blocking: Split and Merge Configurations, IIE Transactions, Vol. 18 (5), pp. 251-261,

Baskett, F., K.M. Chandy, R.R. Muntz and F.G. Palacios[1975], Open, closed and mixed networks of queues with different classes of customers, Journal of the Association of Computing Machinery 22 (2), pp. 248-260,

Bitran, G.R., E.A. Haas and A.C. Hax[1981], Hierarchical Production Planning: A Single Stage System, Operations Research, Vol. 29 (4), pp. 717-743,

Bitran, G.R., E.A. Haas and A.C. Hax[1982], Hierarchical Production Planning: A Two Stage System, Operations Research, Vol. 30 (2),

Bullinger, H.-J., H.J. Warnecke and H.-P. Lentes[1986], Towards the Factory of the Future, International Journal of Production Research, Vol. 24 (4), pp. 697-741,

Burbidge, J.L.[1975], The Introduction of Group Technology, Wiley, New York,

Burstein, M.C. and M. Talbi[1985], Economic evaluation for the optimal introduction of Flexible Manufacturing Technology under rivalry, Annals of Operations Research, Vol. 3, pp. 81-112,

Buzacott, J.A.[1982], The fundamental principles of flexibility in manufacturing systems, Proc. 1st Int. Conf. on FMS's, Brighton, U.K., pp. 13,

Buzacott, J.A. and D.D. Yao[1986], On Queueing Network Models of Flexible Manufacturing Systems, Queueing Syst. 1, pp. 5-27,

Carlier, J.[1982], The one-machine sequencing problem, European Journal of Operational Research 11, pp. 42-47,

Chandrasekharan, M.P. and R.J. Rajagopalan[1986], MODROC: an extension of rank order clustering for group technology, International Journal of Production Research, Vol. 24 (5), pp. 1221-1233,

Chandy, K.M., U. Herzog and L.S. Woo[1975], Approximate analysis of general queueing networks, IBM Journal of Research and Development 19, pp. 43-49,

Charles Stark Draper Research Laboratory[1982], The Flexible Manufacturing Systems Handbook, Cambridge, Massachusetts,

Coffman, E.G., M.R. Garey and D.S. Johnson[1978], An application of Bin-Packing to Multi-Processor Scheduling, SIAM Journal of Computing, Vol. 7, pp. 1-16,

Dallery, Y. and D.D. Yao[1986], Modeling a System of Flexible Manufacturing Cells, in Kusiak[1986 b], pp. 289-300,

Dixon, P.S. and E.A. Silver[1981], A Heuristic Solution Procedure for the Multi Item, Single level, Limited Capacity, Lotsizing Problem, Journal of Operations Management, Vol. 2 (1), pp. 23-39,

Dogramaci, A. J.E. Panayiotopoulos and N.R. Adam[1981], The Dynamic Lotsizing Problem for multiple items under limited capacity, AIIE-Transactions, Vol. 13 (4), pp. 294-303,

Fine, Ch.H. and R.M. Freund[1986], Economic Analysis of Product-Flexible Manufacturing Systems Investment Decisions, in Stecke and Suri[1986], pp. 55-68,

Florian, M., J.K. Lenstra and A.H.G. Rinnooy Kan[1980], Deterministic Production Planning: Algorithms and Complexity, Management Science, Vol. 26 (7), pp. 12-20,

Gaimon, C.[1985], The dynamical optimal acquisition of aùtomation, Annals of Operations Research, Vol. 3, pp. 59-79,

Gaimon, C.[1986], The continuous acquisition of automation to reduce production and in-process inventory costs, in Kusiak[1986 a], pp. 29-43,

Goldhar, J.D. and M. Jelinek[1983], Plan for Economies of Scope, Harvard Business Review, Vol. 61 (6),

Gordon, W.J. and G.F. Newell[1967], Closed queueing systems with exponential servers, Operations research 15, pp. 254-265,

Groover, M.P. and E.W. Zimmers, Jr.[1984], CAD/CAM: Computer Aided Design and Manufacturing, Prentice-Hall, Inc., Englewood Cliffs, NJ,

Hax, A.C. and D. Candea[1984], Production and Inventory Management, Prentice-Hall, Inc., Englewood Cliffs, NJ,

Hax, A.C. and H.C. Meal[1975], Hierarchical Integration of Production Planning and Scheduling, in M. Geisler (ed.) [1975], TIMS Studies in Management Science, Vol. 1, Logistics, North Holland/American Elsevier, New York,

Ho, Y.C., M.A. Eyler and T.T. Chien, A gradient technique for general buffer storage design in a serial production line, International Journal of Production Research 17 (6), pp.557-580,

Jackson, J.R.[1963], Jobshop-like Queueing systems, Management Science 10 (1), pp. 131-142,

Kalkunte, M.V., S.C. Sarin and W.E. Wilhelm[1986], Flexible Manufacturing Systems: A Review of Modeling Approaches for Design, Justification and Operation, in Kusiak[1986 a], pp. 3-25

Kearney and Tracker Corporation[1980], Understanding Manufacturing systems, Vol. 1, Milwaukee, WI,

Keilson, J.[1979], Markov Chain models: Rarity and Exponentiality, Springer Verlag, New York,

Kelly, F.P.[1979], Reversibility and Stochastic Networks, Wiley, New York,

King, J.R.[1980], Machine-component grouping in production flow analysis: An approach using a rank order clustering algorithm, International Journal of Production Research 18, pp. 213,

King, J.R. and V. Nakornchai[1986], An interactive data-clustering algorithm, in A. Kusiak[1986], pp. 285-300,

Koster, M.B.M. de[1986], Approximate Analysis of Production Systems, Report BDK/ORS/86/09, Department of Industrial Engineering, Eindhoven University of Technology, The Netherlands,

Kulatilaka, N.[1985], Capital Budgetting and Optimal Timing of Investments in Flexible Manufacturing Systems, Annals of Operations Research, Vol. 3, pp. 35-57,

Kumar, K.R., A. Kusiak and A. Vannelli[1986], Grouping of parts and components in Flexible Manufacturing Systems, European Journal of Operations Research 24, pp. 387-397,

Kusiak, A.[1985], Flexible Manufacturing Systems: a structural approach, International Journal of Production Research, Vol. 23 (6), pp. 1057-1073,

Kusiak, A. (ed.) [1986 a], Flexible Manufacturing Systems, Methods and Studies, North Holland, Amsterdam,

Kusiak, A. (ed.) [1986 b], Modelling and Design of Flexible Manufacturing Systems, Elsevier, Amsterdam,

Kusiak, A.[1986 c], Application of Operational Research models and techniques in Flexible Manufacturing Systems, European Journal of Operational Research 24, pp. 336-345,

Kusiak, A.[1987], The generalized group technology concept, International Journal of Production Research, Vol. 25 (4), pp. 561-569,

Lambrecht, M.R. and H. Vanderveken[1979], Heuristic Procedures for the Single Operation, Multi Item Loading problem, AIIE-Transaction, Vol. 11 (4), pp. 319-326,

Lutz, R.P.[1982], Discounted Cash Flow Techniques, in G. Salvendy (ed.) [1982], Handbook of Industrial Engineering, Wiley, New York, pp. 9.3.1-9.3.25,

Maes, J.[1987], Capacitated Lotsizing Techniques in Manufacturing Resources Planning, Dissertation, University of Leuven, Belgium,

Maes, J. and L.N. van Wassenhove[1986 a], Multi Item Single Level Capacitated Lotsizing Heuristics: A Computational Comparison. Part 1: Static Case, IIE Transactions, Vol. 18 (2), pp. 114-123,

Maes, J. and L.N. van Wassenhove[1986 b], Multi Item Single Level Capacitated Lotsizing Heuristics: A Computational Comparison. Part 2: Rolling Horizon, IIE Transactions, Vol. 18 (2), pp. 124-129,

Meredith, J.R.[1987], Implementing the Automated Factory, Journal of Manufacturing Systems, Vol. 6 (1), pp. 1-13,

Michael, G.J. and R.A. Millen[1985], Economic Justification of modern computer-based Factory Automation Equipment: a status report, Annals of Operations Research, Vol. 3, pp. 25-34,

Morton, T.E. and T.L. Smunt[1986], A Planning and Scheduling System for Flexible Manufacturing, in Kusiak[1986 a], pp. 151-164,

Nelson, G.A.[1986], A scoring model for Flexible Manufacturing Systems project selection, European Journal of Operational Research 24, pp. 346-359,

Orlicky, J.A.[1975], Materials Requirements Planning, McGraw-Hill, New York,

Pinedo, M.L., B. Wolf and S.Th. McCormick[1986], Sequencing in a Flexible Assembly Line with Blocking to Minimize Cycle Time, in Stecke and Suri[1986], pp. 499-508,

Porteus, E.L.[1985], Investing in reduced set-up in the EOQ model, Management Science 31 (8), pp.998-1010,

Porteus, E.L.[1986], Optimal lot sizing, process quality improvement and setup cost reduction, Operations Research 34 (1), pp. 137-144,

Primrose, P.L. and R. Leonard[1986], Conditions under which Flexible Manufacturing is Financially Viable, in Kusiak[1986 a], pp.355-366,

Rajagopalan, S.[1986], Formulation and Heuristic solutions for Parts Grouping and Tool Loading in flexible Manufacturing Systems, in Stecke and Suri[1986], pp.311-320,

Ranky, P.G.[1983], The Design and Operation of Flexible Manufacturing Systems, IFS Publications, Kempston, Bedford, U.K.,

Reiser, M.[1977], Numerical methods in separable queueing networks, Studies in the Management Sciences 7, pp.113-142,

Reiser, M.[1981], Mean-Value Analysis and Convolution Method for Queue-Dependent Servers in Closed Queueing Networks, Performance Evaluation 1, pp. 7-18,

Reiser, M. and H. Kobayashi[1975], Queueing networks with multiple closed chains: theory and computational algorithms, IBM Journal of Research and Development 19, pp.283-294,

Reiser, M. and S.S. Lavenberg[1980], Mean Value Analysis of closed multichain queueing networks, Journal of the Association of Computing Machinery 22, pp. 313-322,

Repkes, H. and W.H.M. Zijm[1988], A Preemptive Repeat Priority Analysis of a buffer model in a Flexible Manufacturing System, in preparation,

Schonberger, R.J.[1982], Japanese Manufacturing Techniques, The Free Press (MacMillan), New York,

Shalev-Oren, S., A. Seidmann and P.J. Schweitzer[1985], Analysis of Flexible Manufacturing Systems with priority scheduling: PMVA, Annals of Operations Research 3, pp. 115-139,

Shaw, M.J.P. and A.B. Whinston, Applications of Artificial Intelligence to Planning and Scheduling in Flexible Manufacturing, in Kusiak[1986 a], pp. 223-242,

Silver, E.A.[1975], Modifying the Economic Order Quantity (EOQ) to Handle Coordinated Replenishments of two or more items, Production and Inventory Management, Vol. 16 (3), pp. 26-38,

Silver, E.A. and R. Peterson[1985], Decision Systems for Inventory Management and Production Planning, 2nd. ed., Wiley, New York,

Slack, N.[1983], Flexibility as a Manufacturing Objective, International Journal of Operations and Production Management 3, pp. 4-13,

Solberg, J.J.[1977], A Mathematical Model of Computerized Manufacturing Systems, Proceedings 4th Int. Conf. on Production Research, Tokyo, Japan,

Solberg, J.J.[1981], Capacity Planning with a Stochastic Workflow Model, AIIE Transactions, Vol. 13 (2), pp. 116-122,

Srinivasan, V. and R.A. Millen, Evaluating Flexible Manufacturing Systems as a strategic investment, in Stecke and Suri[1986], pp. 83-93,

Stecke, K.[1983], Formulation and solution of nonlinear integer production planning problems for Flexible Manufacturing Systems, Management Science, Vol. 29 (3), pp. 273-286,

Stecke, K.[1985], Design, Planning, Scheduling and Control problems of Flexible Manufacturing Systems, Annals of Operations Research, Vol. 3, pp. 3-12,

Stecke, K. and Th. L. Morin[1985], Optimality of balancing workloads in certain types of Flexible Manufacturing Systems, European Journal of Operational Research 20 (1), pp. 68-82,

Stecke, K. and J.J. Solberg[1981], Loading and Control policies for a Flexible Manufacturing System, Int. Journal of Production Research, Vol. 19 (5), pp. 481-490,

Stecke, K. and J.J. Solberg[1985], The optimality of unbalancing both workloads and machine group sizes in closed queueing networks of multiserver queues, Operations Research 33 (4), pp. 822-910,

Stecke, K. and R. Suri (ed.) [1986], Flexible Manufacturing Systems, Operations Research models and applications, Elsevier, Amsterdam,

Subramanyam, S. and R.G. Askin[1986], An Expert System Approach to Scheduling in Flexible Manufacturing Systems, in Kusiak[1986 a], pp. 243-256,

Suri, R. and R.R. Hildebrant[1984], Modelling Flexible Manufacturing Systems Using Mean-Value Analysis, Journal of Manufacturing Systems, Vol. 3 (1), pp. 27-38,

Thuesen, G.J.[1982], Project Selecion and Analysis, in G. Salvendy (ed.)[1982], Handbook of Industrial Engineering, Wiley, New York, pp. 9.4.1-9.4.27,

Van Beek, P. and C. van Putten[1987], OR contributions to flexibility improvement in production/inventory systems, European Journal of Operational Research 31, pp. 52-60,

Van Dijk, N.M.[1987], Simple product form bounds for flexible manufacturing systems, presented at the 16th DGOR-meeting, Veldhoven, The Netherlands,

Van Looveren, A.J., L.F. Gelders and L.N. van Wassenhove[1986], A review of FMS Planning models, in Kusiak[1986 b], pp. 3-31,

Vollman, Th. E., W.L. Berry and D.C. Whybark[1984], Manufacturing Planning and Control Systems, Dow Jones - Irwin, Homewood, Illinois,

Wessels, J., J. Hontelez and W.H.M. Zijm[1986], Eine Produktionslinie mit unzuverlässigen Komponenten (in German), presented at the 15th DGOR-meeting, Ulm, FRG,

Wittrock, R.J.[1985], Scheduling algorithms for flexible flow lines, IBM Journal of Res. and Develop., Vol. 29 (4), pp. 401-412,

Whitt, W.[1983], The Queueing Network Analyzer, Bell System Technical Journal, Vol. 62 (9), pp. 2779-2816,

Whitt, W.[1983], Performance of the Queueing Network Analyzer, Bell System Technical Journal, Vol 62 (9), pp. 2817-2843,

Yao, D.D. and J.A. Buzacott[1985 a], Modeling a class of state-dependent routing in flexible manufacturing systems, Annals of Operations Research 3, pp. 153-167,

Yao, D.D. and J.A. Buzacott[1985 b], Modeling the performance of flexible manufacturing systems, International Journal of Production Research 23, pp. 945-959,

Yao, D.D. and J.A. Buzacott[1986 a], The exponentialization approach to flexible manufacturing system models with general processing times, European Journal of Operational Research 24, pp. 410-416,

Yao, D.D. and J.A. Buzacott[1986 b], Models of flexible manufacturing systems with limited local buffers, International Journal of Production Research 24, pp. 107-118,

Zeestraten, M.[1987], De Loading Functie (in Dutch), Internal Report, Delft University of Technology, The Netherlands,

Zijm, W.H.M.[1987], Flexible Manufacturing Systems: an Operational Research approach, CQM-report, Centre for Quantitative Methods, Eindhoven,

Zijm, W.H.M.[1988], Integral Goodsflow Control: an Operational Research approach, in preparation.

EVOLUTIONARY TRANSFORMATION OF PRODUCTION SYSTEMS TOWARDS COMPUTER INTEGRATED MANUFACTURING SYSTEMS

Janez Dekleva, Emil Zavadlav, Ljubljana

The production system structure encountered in metal-cutting industries in Yugoslavia is classical work-shop. It offers the greatest flexibility of production yet at the cost of huge work in process and long manufacturing lead-times.

Therefore, flexible automation is a must. The question remains, though, how to achieve it, as industries in developing countries have neither skills nor financial resources necessary.

We propose gradual, evolutionary approach. The first step is transformation into cellular manufacturing systems and then automation of individual production cells and subsystems like transportation or storage/retrieval.

Methods used belong to OR. The most helpful among them is discrete simulation.

EXPERT SYSTEM IN PRODUCTION FLOW OPTIMIZATION

Janez Dekleva, Janez Kusar, Darko Menart

SUMMARY - The new manufacturing philosophy characterized by flexible manufacturing systems and production cells with their cost and benefits are being applied in industry, above all in metal cutting industry.
Production Flow Analysis (PFA) /1/, a method for identification of families of parts and corresponding groups of machines is shortly described.
The Extended version of PFA /2/,/3/ is added to solve the problems appearing due to the lack of optimal data sets. Special emphasis were given to the Group Analysis (GA, i.e. the second phase of PFA). In the extended version of GA the modified incidence matrix is used for selecting the initial groups of machines. The building blocks of the extended GA are added to the original scheme.
Decision rules for GA and the knowledge base to the intelligent problem solving are discussed. It is our belief that within the domain sphere of the proposed expert system in addition to the already mentioned identification procedure there exist certain diagnostics problems, e.g. the evaluation of the applied routings and the estimation of organizational aspects of manufacturing systems. New facts and rules are added into proposed knowledge base with the necessary inference engine.

REFERENCES

/1/ Burbidge J.L.: Production Planning, Heinemann, London, 1971.

/2/ Dekleva J., Menart D.: Advantages and Disadvantages of Different Production Cell Identification Methods, Advances in Production Management Systems 85, North-Holland, Amsterdam, 1986, pp. 73-82.

/3/ Dekleva J., Menart D.: Extensions of Production Flow Analysis, Journal of Manufacturing Systems, Vol. 6, No. 2, 1987, pp. 93-105

Simple product form bounds for flexible manufacturing systems

Nico M. van Dijk , Enschede

At present it is commonly known that flexible manufacturing components such as assembly lines, conveyer belts, material handling systems and automated guided transport systems can be looked upon as a network of interconnected service stations. The classical theory of queueing networks might therefore be applicable for estimating system performance measures such as throughputs, blocking probabilities and sojourn times. This paper aims to provide the insight in results that can be adopted from networks of queues for FMS-purposes. These results are twofold: (i) explicit formulae (ii) simple bounds.

In queueing networks an essential distinction can be made between product-form and non-product form systems. For the first category relatively simple explicit formulae are available while for the second there are not. In a first part it will be argued that this classification can be based upon a so-called notion of "job-local-balance". By means of various examples it will be illustrated up to what extend this notion can be satisfied for FMS.

It so turns out that FMS mostly pertains to the non-product form category due to typical phenomena as blocking, dynamic routing, dynamic server allocation, job priorities or machine breakdowns. A variety of approximative algorithms are currently available for non-product form systems such as finite production lines. Generally these require a large amount of computational effort. For flexible manufacturing purposes, however, one may rather be interest in robust but simple bounds as to get a fast impression of the system performance at hardly computational expense.

In a second part, therefore, it will be outlined and illustrated that this same notion of job-local-balance can also be utilized to obtain simple and insensitive bounds for non-product form situations. Typical FMS-components as overflow stations, facilities with breakdowns and finite production lines will be included. Numerical support indicates a practical usefulness of these bounds as fast indicators. This is illustrated also by an application to optimal design.

In a final part the bounding method will be briefly evaluated and its application to larger structures by means of aggregation or decomposition will be addressed.

APPROXIMATE MEAN VALUE ANALYSIS AND FLEXIBLE MANUFACTURING SYSTEMS

Jan B.M. van Doremalen, Eindhoven

In the last decade the use of queueing network models as tool for the (rough-cut) analysis of (flexible) manufacturing systems has drawn much attention, cf.[1].

The analysis of large scale queueing network models has benefitted very much from the development of the Mean Value Analysis (MVA). This technique provides a means to evaluate the behavioral characteristics of product-form or separable queueing networks in a recursive manner, cf.[2].

One of the most promising features of MVA is that it provides a basis for the development of intuitive approximation methods for those queueing network models that are to complex or that are not separable. In [3] we have discussed MVA as basis for the development of approximation methods at length.

References.

[1] J.A. Buzacott and D.D. Yao, On queueing network models of flexible manufacturing systems, Queueing Systems 1(1986)5-27.

[2] M. Reiser and S.S. Lavenberg, Mean Value Analysis of closed multichain queueing networks, Journal of the A.C.M. 27(1980)313-322.

[3] J.B.M. van Doremalen, Approximate analysis of queueing network models, Thesis, Eindhoven University of Technology, Department of Mathematics and Computing Science, 1986.

Workload control in FMS environments

Paul Giesberts
René de Koster
Jacob Wijngaard

Eindhoven

A study is being made of the design of a highly automated production line for the assembly and test of certain electro mechanical products. It is prespecified that the line must be able to produce 240 different product types in a random sequence. A high performance is required with respect to throughput and throuphputtimes. The flexibility to meet these requirements can be obtained by using robots and programmable product identifiers. The desired throughput can be obtained by controlling the line with workload control. The advantage of this control system is that it can be implemented in a hierarchical control system.

To evaluate the effect of the proposed control rules on the performance of the line a mathematical model of the line was built and analysed both with simulation and numerical techniques.

UNTERSUCHUNG ZYKLISCHER FERTIGUNGSPROZESSE MIT HILFE VON ZEITEREIGNISGRAPHEN

Herve HILLION Klaus MEIER Jean-Marie PROTH

Vandoeuvre les Nancy Cedex

Betrachtet werden Werkstückstypen, die mit einer gebenen Produktionsratio zyklisch zu fertigen sind. Der Fertigungsprozeβ typengleicher Werkstücke sei durch eine Maschinenfolge sowie den zugehörigen Maschinenbelegungszeiten festgelegt. Gegeben sei weiter ein Systembeschickungszyklus sowie für jede Maschine ein Arbeitszyklus. Der Produktionsratio entsprechend legt der Systembeschickungszyklus fest, wie das Fertigungssystem mit Werkstücken zu beschicken ist. Der Arbeitszyklus einer Maschine legt fest, in welchem Zyklus die Maschine Arbeitsgänge durchführen soll.

Der Fertigungsprozeβ kann durch einen stark-zusammenhängenden Zeitereignisgraphen modelliert werden. Als Zeitereignisgraphen werden spezielle Zeit-Petri-Netze bezeichnet, bei denen zu jeder Stelle genau ein Pfeil hin- und von jeder Stelle genau ein Pfeil wegführt. Stark-zusammenhängende Zeitereignisgraphen haben folgende Eigenschaften : (E1) die Anzahl der in einem elementaren Kreis zirkulierenden Marken ist konstant, (E2) eine Schaltsequenz führt genau dann zu einer totalen Verklemmung, wenn es einen elementaren Kreis gibt, der keine markierte Stelle enthällt, (E3) die Periodendauer einer Schaltsequenz ist gleich der maximalen Zykluszeit der elementaren Kreise, wobei die Zykluszeit eines elementaren Kreises als Quotient (Summe der Schaltzeiten aller Transitionen im Kreis / Anzahl der im Kreis zirkulierenden Marken) definiert ist.

Im Modell wird der Fertigungsprozeβ jedes Werstücks des Systembeschickungszyklus durch einen Prozeβkreis dargestellt. Eine Transition wird als Arbeitsgang und eine Stelle als Zwischenlager betrachtet. Die in einem Prozeβkreis zirkulierenden Marken stellen Werkstücke dar. Arbeitsgänge, welche durch die gleiche Maschine zu realisieren sind, werden durch einen dem Arbeitszyklus der Maschine entsprechenden Steuerkreis verbunden. Hierbei wird zwischen zwei aufeinanderfolgende Transitionen eine Stelle eingeführt. Eine in einem Steuerkreis zirkulierende Marke bestimmt, welcher Arbeitsgang als nächstes realisiert wird. Führt man nun noch in jedem Prozeβkreis eine zusätzliche Transistion ein und verbindet diese Transitionen durch einen Steuerkreis, der den Systembeschickungszyklus darstellt, liegt ein stark zusammenhängender Zeitereignisgraph vor.

Aus den Eigenschaften stark-zusammenhängender Zeitereignisgraphen kann man folgern, daβ eine Ausgangsverteilung von Werkstücken, bei der zwischen zwei aufeinanderfolgenden Arbeitsgängen mindestens ein Werkstück zwischengelagert ist, im stationären Fall immer zur vollen Auslastung der Bottleneckmaschine führt und somit maximale Produktion erreicht wird. Von diesem Ergebnis ausgehend ist es interessant, das Prozeβverhalten so einzustellen, daβ bei minimaler Zwischenlagerung maximale Produktion erreicht wird. Das Optimierungsproblem kann als lineares Programm mit (0,1) Variablen formuliert werden. Auf die detaillierte Beschreibung des Programms sowie einer effizienten Heuristik, die für diese Problemstellung entwickelt worden ist, kann in dieser Kurzfassung nicht eingegangen werden.

Operational rules for the design of straight production lines

A.G. de Kok
Eindhoven

World-wide in Philips plants production systems are installed consisting of a number of work stations and an automatic transport system connecting these work stations. An important problem is the design of these production systems in such a way that targets set by plant management with respect to production capacity and throughput time are met. Both the Department of Mechanization Principles of the Centre for Manufacturing Technology (CFT) and the Centre for Quantitative Methods (CQM) operate as internal consultants within Philips to support management on this subject and both groups do research in the field of performance evaluation of production systems. Together CFT and CQM have started a project to analyze basic production systems often met in practice and to find operational rules (rules of thumb) that can help to do global design of production systems.

This paper considers production systems consisting of identical machines (i.e. identical cycle times and break-down behaviour) with finite intermediate buffers, the so-called transfer lines or straight production lines. Even these relatively simple production systems are hard to handle analytically. Therefore we use a hybrid approach applying both accurate analytical approximations for transfer lines with perfect machines and exponential cycle times, and inter- and extrapolation rules. The combination of these methods yields very accurate approximations for the throughput and the throughput time of the transfer line, which can be calculated using a pocket calculator. The table below gives some idea of the accuracy of the approximations. We consider machines with fixed cycle times equal to 1. The number of machines is denoted by N, the size of the intermediate buffers is denoted by K. μ denotes the stand-alone efficiency of a machine. Let E[S] and c_s^2 denote the mean and squared coefficient of the break-down time, respectively. The throughput of the transfer line is denoted by τ and the throughput time by DLT. The indices a and s refer to the approximations and simulations, respectively.

Table. Comparison of approximations with simulation for fixed cycle times.

				$c_s^2 = 1/2$				$c_s^2 = 2$			
N	K	μ	E[S]	τ_a	τ_s	DLT_a	DLT_s	τ_a	τ_s	DLT_a	DLT_s
5	4	0.99	5	0.96	0.97	13.3	13.4	0.95	0.97	13.4	13.5
5	4	0.99	20	0.95	0.96	13.4	13.2	0.95	0.96	13.4	13.6
9	4	0.99	5	0.96	0.96	25.7	26.4	0.93	0.96	26.5	25.9
9	4	0.99	20	0.92	0.94	26.6	27.8	0.92	0.93	26.6	26.4

Reference De Kok, A.G., 1987, Een rekenmodel voor het bepalen van de performance van gebalanceerde produktielijnen met eindige buffers, CQM-proj. rep. 174-014-02.

CAPN - AN OPEN NETWORK OF PRODUCTION CONTROL - CONCEPT AND DISCUSSION

Winfried Matthes, Cologne

As a potential kernel of a controlling system of computer-integrated manufacturing (CIM) or general computer-integrated product processing (CIP) an open network model of distributed and connected production contol modules is designed and discussed with regard of the following aspects and premises:

- networks of production processes with variable programs/product qualities and quantities, methods, lot sizes, schedules, ressourses, and financing are primary objects of planning and control;
- hierarchical and heterarchical institutions of planning and control develop, change and purse conflicting goals in a polycentric and in-acting systems;
- knowledge-and-computer-based methods like heuristics, simlators, sub-optimizers, and interactive adaptive control systems are applied to support single and group decision processes to structurize and control product processing in coordination with financing;
- computer-based communication techniques (local area networks, connected data and program systems) help to coordinate partial and general production plans in an open way;
- the computer aided planning network (CAPN) interacts with central accounting systems and related superior institutions within the management system of the firm.

Modelling the planning network contains the phases:

- construction of modules (coordination of fields, goals, and instruments of process control) according to practically accepted or acceptable heuristic rules related to
- determination of controlling cuts (active and passive cutting parameters, internal and external planning variables resp.) and to
- planning integration (structurization of coordinating processes and strategies) of modules and superior controlling institutions with a local area network.

Modelling and testing the CAPN as a tool of collectively generating, simulating, selecting, controlling, and adapting processing structures will stimulate analyzing and synthesizing practicable process control systems within operative and strategic fields of the firm.

VERFAHREN ZUR OPERATIONENREIHENFOLGEBILDUNG IN FERTIGUNGSSYSTEMEN

Reiner Pietrzak, Hannover und Josef Kurpiers, Hannover

1 Einleitung

"Losgröße 1", "rüstzeitfreier Auftragswechsel" u. ä. Schlagworte kennzeichnen Zielsetzungen für den Betrieb flexibel automatisierter Fertigungssysteme /GR/. Realisiert ist dies zur Zeit allenfalls für Fertigungen mit hohem Wiederholanteil am Werkstückspektrum, also für die Fertigung kleiner bis mittlerer Serien. In der Einzel- und Kleinserienfertigung, gekennzeichnet durch eine größere Vielfalt an benötigten Werkzeugen und Spannmittelvarianten, ist der "rüstzeitfreie Auftragswechsel" häufig nicht zu realisieren /BA/.

Vor allem eine steigende Werkzeugvielfalt würde zu großen Werkzeugmagazinen mit einem zunehmenden Anteil an selten genutzten aber teuren Werkzeugen führen. Hier setzen Überlegungen zur werkzeugbezogenen Reihenfolgeoptimierung von abzuarbeitenden Aufspannungen (Operationen) an.

2 Struktur des Betriebsmittelbedarfs

Betrachtet wird hier z. B. ein kurzfristiger Auftragsbestand an einem flexiblen Bearbeitungszentrum für den Zeitraum von einer Woche. Bild 1 zeigt für einen Ausschnitt aus diesem Auftragsbestand beispielhaft die Struktur des Betriebsmittelbedarfs für Engpaßspannmittel und Werkzeuge.

Erste Erfahrungen wurden mit dem Verfahren der Clusteranalyse gewonnen /FR/. Die Operationen wurden zunächst nach ihrem Spannmittelbedarf vorgruppiert, da der Umbau von Engpaßspannmitteln mit erheblichem Aufwand verbunden ist und nach Möglichkeit vermieden werden sollte. Es wurden dann Gruppen von in ihrem Werkzeugbedarf her ähnlichen Operatio-

Operations Research Proceedings 1987

nen bei gleichem Spannmittelbedarf gebildet. Für die Reihenfolge der Operationen mit ähnlichem Werkzeugbedarf bietet es sich an, sogenannte Differenzwerkzeugsätze zu bilden. In Bild 2 ist das Verfahren dargestellt.

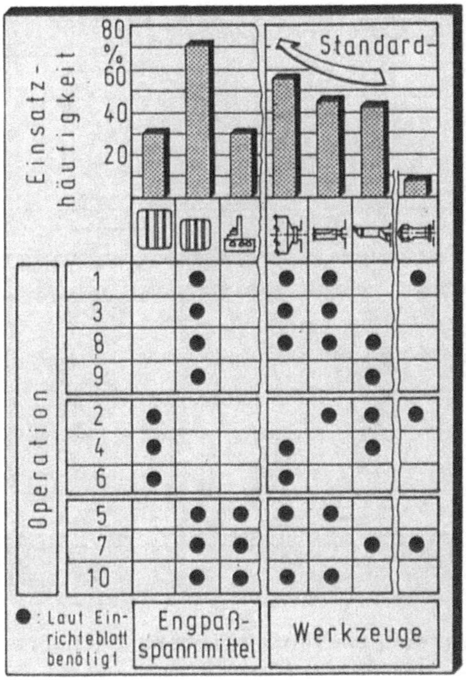

Bild 1: Struktur des Betriebsmittelbedarfs für einen kurzfristigen Auftragsbestand

Der in einem Differenzeinrichteblatt enthaltene Differenzwerkzeugsatz kennzeichnet diejenigen Werkzeuge, die beim Übergang von einer Operation zur Nachfolgeoperation gegen Werkzeuge ausgetauscht werden müssen, die für die Nachfolgeoperation nicht mehr benötigt werden. Nicht der komplette Werkzeugsatz wird auftragsbezogen ausgetauscht, sondern nur die Werkzeuge, die tatsächlich neu benötigt werden oder aus Verschleißgründen ausgetauscht werden müssen.

Eine in dieser Weise starr festgelegte Reihenfolge kann die Flexibilität des Fertigungsablaufs stark einschränken. Überlegungen zur Durchlaufzeitverkürzung zielen beispielsweise auch darauf ab, Auftragsbestände an der Maschine zu verringern. Diese Überlegungen dürfen durch werkzeugbezogene Optimierungen nicht beeinträchtigt werden. Es kann deshalb sinnvoll sein, in einem Auftragsbestand häufig vorkommende Werkzeuge zu einem Standardwerkzeugsatz zusammenzufassen (Bild 1).

Dadurch können reihenfolgeunabhängig Rüstzeiteinsparungen bezogen auf Werkzeugwechsel erzielt werden.

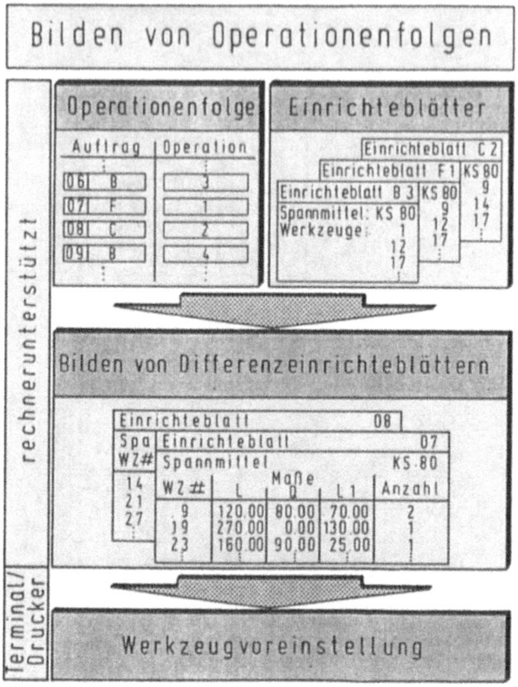

Bild 2: Bilden von Differenzeinrichteblättern nach Operationenfolgen

3 Eingesetzte Verfahren zur werkzeugbezogenen Rüstzeitoptimierung von Operationenfolgen

Ausgehend von Verfahren der Clusteranalyse wurden mathematische Verfahren untersucht, die Reihenfolgeprobleme effizienter behandeln können. Auch die Bildung von Standardwerkzeugsätzen wurde weiter verfolgt.

3.1 Verfahren zur Behandlung von Reihenfolgeproblemen

Alle betrachteten Verfahren für die Reihenfolgebildung erfordern eine Datenaufbereitung in der in Bild 3 dargestellten Form. Zunächst wird eine Matrix mit den möglichen Übergängen von einer Operation zu einer anderen gebildet. Als Kenngröße für einen Übergang wird der Überdeckungsgrad zweier aufeinander folgender Operationen bezüglich ihres Werkzeugbedarfs gebildet. Können alle Werkzeuge weiter verwendet werden, ist der Überdeckungsgrad "1". Die Diagonale der Matrix kennzeichnet für die einzelnen Operationen, welches Engpaßspannmittel benötigt wird und wie lange es durch Rüsten, Bearbeiten und Abrüsten für die jeweilige Operation belegt ist. Werkzeugbedarf, Bearbeitungszeit, werkzeugbezogene Rüstzeiten und Spannmittelbedarf werden aus den Einrichteblättern zu den Operationen gewonnen. Aus einer Kennzeichnung der Engpaßspannmittel und aus dem Auftragsbestand ergeben sich Abfolgebindungen, die bestimmte Übergänge in der Matrix ausschließen. Diese Übergänge werden mit dem überdeckungsgrad "0" gesperrt.

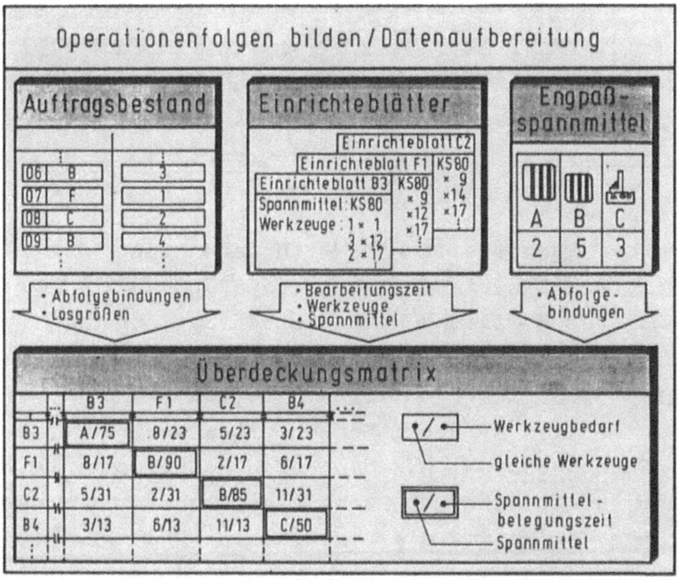

Bild 3: Datenaufbereitung für die Operationenfolgebildung

Als prinzipiell einsetzbare Verfahren zur Behandlung dieses Reihenfolgeproblems wurden die vollständige Enumeration, Entscheidungsbaumverfahren und heuristische Verfahren betrachtet /MÜ/. Die für das Verfahren benötigte Rechenzeit wurde gegenüber der erforderlichen Speicherkapazität und der tatsächlichen Auffindung eines Optimums als entscheidendes Kriterium für eine Auswahl angesehen. Diese Überlegungen führten zum Einsatz heuristischer Verfahren, bestehend aus einem Eröffnungsverfahren und einer nachfolgenden iterativen Optimierung, die bei geringer Erfolgsaussicht auf weitere Optimierung abgebrochen wird. Als Eröffnungsverfahren wurde der Greedy-Algorithmus /MÜ/ verwendet. Die Anwendung dieses Algorithmus kann an Bild 4 nachvollzogen werden.

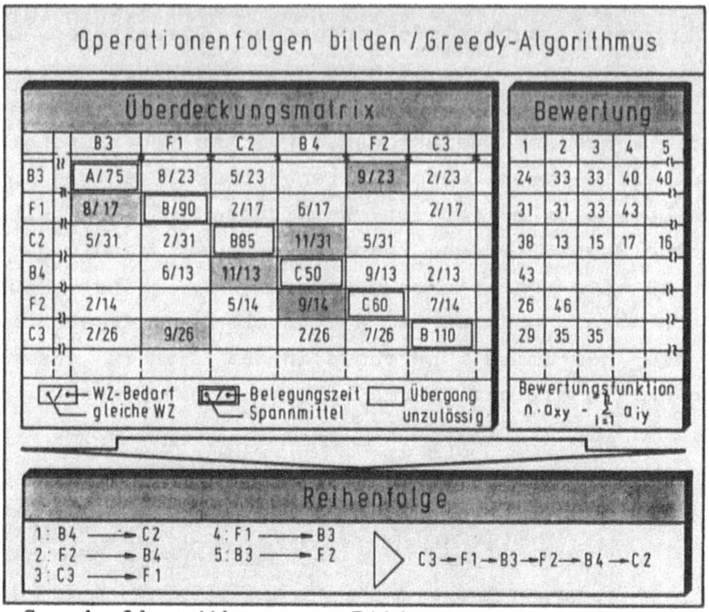

Bild 4: Der Greedy-Algorithmus zur Bildung von Operationenfolgen

Mit Hilfe der dargestellten Bewertungsfunktion werden die einzelnen Zeilen der Matrix bewertet. Der höchste überdeckungsgrad in der Zeile mit der höchsten Bewertung kennzeichnet den für den jeweiligen Rechengang günstigsten Übergang. Beim ersten Rechengang ist dies der Übergang B4->C2. Spalte und Zeile zu diesem Übergang werden gestrichen. Leere Plätze in der Matrix kennzeichnen unzulässige Übergänge. Bei Durchführung der Rechengänge 1 bis 5 - die Rasterunterlegung kennzeichnet jeweils den günstigsten Übergang - ergibt sich für den dargestellten Ausschnitt aus der Matrix die beschriebene Reihenfolge. Bei sehr kleinen Auftragsbeständen kann es vorkommen, daß eine vollständige Reihenfolge aufgrund zu vieler unzulässiger Übergänge nicht gebildet werden kann. In einem solchen Fall kann nur die Bildung eines Standardwerkzeugsatzes zu Rüstzeiteinsparungen führen.

3.2 Verfahren zur Standardwerkzeugsatzbildung in einem kurzfristigen Auftragsbestand

Aus dem Auftragsbestand und den zur Bearbeitung dieses Auftragsbestandes benötigten NC-Programmen und Einrichteblättern für eine betrachtete Maschine können entsprechend Bild 5 Häufigkeitsverteilungen des Werkzeugbedarfs ermittelt werden. Die Magazinkapazität der Fertigungseinrichtung begrenzt die Anzahl gleichzeitig für den Einsatz verfügbarer Werkzeuge. Je nach Höhe des Maschinenstundensatzes und der Höhe der Kosten für die Werkzeugbereitstellung können unterschiedliche Grenzhäufigkeiten ermittelt werden. Die Grenzhäufigkeit kennzeichnet die Einsatzhäufigkeit, bei der sich die Kosten für auftragsbezogenen und verbrauchsbezogenen Werkzeugwechsel gerade die Waage halten. Unterhalb dieser Grenzhäufigkeit ist der auftragsbezogene, oberhalb der Grenzhäufigkeit der verbrauchsbezogene Werkzeugwechsel günstiger.

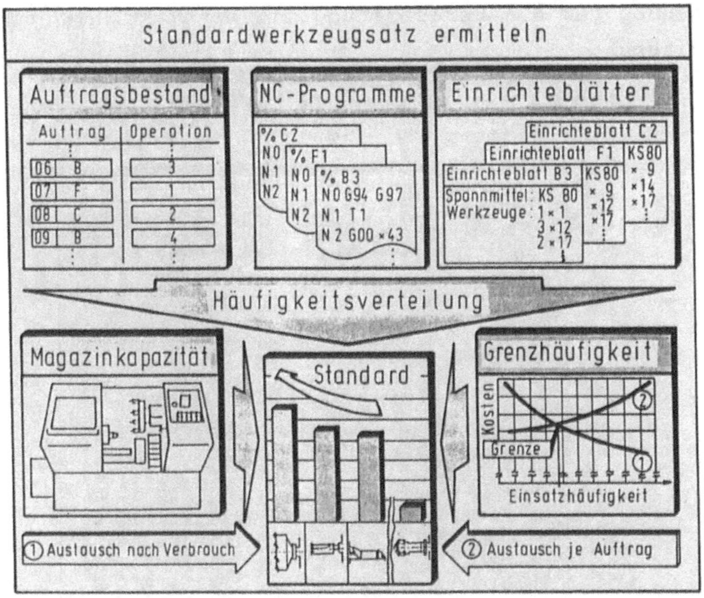

Bild 5: Ermittlung eines Standardwerkzeugsatzes nach Auftragsbestand

Für eine Fertigungseinrichtung in einem Unternehmen des Großwerkzeugmaschinenbaus wurde auf der Grundlage von auswertbaren NC-Programmen eine vergleichende Betrachtung dieser Möglichkeiten zur Rüstzeiteinsparung durchgeführt. Bei der Fertigungseinrichtung handelt es sich um ein Bearbeitungszentrum, auf dem Spindelkästen, Bett- und Supportteile u. ä. Großteile bearbeitet werden. Für die Auswertung wurden aus dem gesamten Spektrum von 35 NC-Programmen jeweils 15 Programme als kurzfristiger Auftragsbestand (eine Woche) zusammengestellt. Es wurde je eine Versuchsreihe mit 20 Auftragsbeständen bei starrer Reihenfolgebildung und mit lediglich Standardwerkzeugsatzbildung durchgeführt. Im Ergebnis konnten durch Standardwerkzeugsatzbildung immerhin die Hälfte der durch starre Reihenfolgebildung erzielbaren Rüstzeiteinsparungen erreicht werden. Der Aufwand für Reihenfolgebildung ist nur dann zu rechtfertigen, wenn der Kapazitätsgewinn in der Fertigung auch genutzt werden kann, es sich also um eine Engpaßfertigungseinrichtung handelt.

4 Integration in die kurzfristige Werkstattsteuerung

Die bisherigen Erfahrungen zeigen, daß beide Verfahren geeignet sind, bei kurzfristig auftretenden Kapazitätsengpässen Hilfestellung zur störungsfreien Bewältigung zu geben. Dabei werden die Datenbestände für die Werkstattsteuerung in einem relationalen Datenbanksystem gehalten. Diese Daten werden zur Laufzeit der Einplanung eines kurzfristigen Auftragsbestandes für die Verarbeitung mit einem PETRI-Netz-Interpreter /TÖ/ aufbereitet.

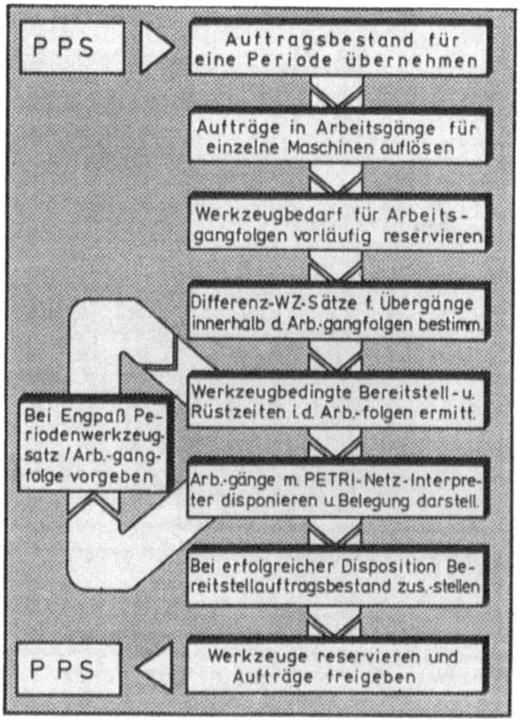

Bild 6: Ablauf der werkzeugbezogenen Auftragsdisposition

Zur Datenaufbereitung wird auf die bei der Reihenfolgeplanung und Standardwerkzeugsatzbildung verwendeten Algorithmen zurückgegriffen. Bild 6 zeigt den Ablauf der werkzeugbezogenen Auftragsdisposition. Engpässe an einzelnen Fertigungseinrichtungen werden an einer grafischen Darstellung der Kapazitätsbelegung erkennbar, die auf dem Bildschirm ausgegeben wird. Die Standardwerkzeugsatz- und Operationenreihenfolgebildung kann genutzt werden, diese Engpässe auszuregeln.

5 Schlußfolgerungen und Ausblick

Zur Datenaufbereitung für die Standardwerkzeugsatz- bzw. Operationenreihenfolgebildung verwendete Algorithmen wurden eingesetzt, um Auftrags- und zugehörige Werkzeugdaten aus einer Datenbank zu verdichten. Die Verdichtung der Daten ermöglicht eine Vereinfachung der Struktur des PETRI-Netzes zur Abbildung einer Modellwerkstatt. Anstelle einzelner Werkzeuge, die den Werkzeugkreislauf von der Kommissionierung über Montage, Voreinstellung, Transport, Einsatz und Aufbereitung durchlaufen, wurden Werkzeugsätze und die zugehörigen Bereitstell- und Rüstzeiten betrachtet. Auf diese Weise konnten für einen Modellauftragsbestand kurze Laufzeiten für den PETRI-Netz-Interpreter erzielt werden. Für die Zukunft muß nun geprüft werden, in welchem Maße die Ausweitung der Komplexität des PETRI-Netzes bei der Betrachtung von Einzelwerkzeugen zu Laufzeitveränderungen gegenüber der eingesetzten Mischung mit algorithmischen Verfahren führt.

6 Literatur

/MÜ/ Müller-Merbach, H.; Optimale Reihenfolgen; Springer-Verlag, Berlin-Heidelberg 1970

/FR/ Freist, Chr.; Einsatzmöglichkeiten statistischer Verfahren in CAD/CAM-Systemen; Dr.-Ing.-Diss. Universität Hannover 1984

/GR/ Granow, R.; R. Schmiedeskamp; J. Balbach; Werkzeugverwaltung - Einstieg in die rechnerunterstützte Werkstattkommunikation; VDI-Z 126 (1984), 17, S. 617-623

/BA/ Balbach, J. u.a.; Reorganisation des Betriebsmittelwesens in einem Unternehmen des Schwermaschinenbaus; VDI-Z 126 (1984), 19, S. 733-739

/TÖ/ Tönshoff, H.K. und A. Horns; Petri-Netze als spezielle Inferenzsysteme und ihre Anwendung in der Werkstattsteuerung und -simulation; Fortschr.-Ber. d. IPA-Arbeitstagung 18 (1986), Springer-Verlag, Berlin 1986

VERFAHREN ZUR KAPAZITÄTSPLANUNG FÜR FLEXIBLE FERTIGUNGSSYSTEME

Horst Tempelmeier, Darmstadt

Bei der Konfiguration eines flexiblen Fertigungssystems (FFS) sind Entscheidungen über die Art und Anzahl der mit dem System zu bearbeitenden Produktarten, die Art und Anzahl der dafür einzusetzenden Maschinen, die Anzahl von Werkstückträgern (Paletten) sowie über die Kapazität des Materialflußsystems und die Anzahl und Dimension der Werkstückpuffer zu treffen. In den letzten Jahren sind zahlreiche analytische Verfahren zur Abschätzung der Auswirkungen dieser Konfigurationsvariablen eines FFS auf die relevanten Zielgrößen (z.B. Produktionsraten, Auslastungen, Durchlaufzeiten, Lagerbestände) entwickelt worden, die auf der Darstellung des FFS als geschlossenem Warteschlangennetzwerk basieren. Die Anwendbarkeit dieser Verfahren - zu nennen sind hier insb. CANQ (Computer Analysis of Networks of Queues) und MVA (Mean-Value Analysis) - wurde anhand von Simulationsuntersuchungen nachgewiesen, die sich vor allem auf hypothetische FFS beziehen.

In der vorliegenden Untersuchung wurden für in der betrieblichen Praxis implementierte FFS mit unterschiedlichen Systemstrukturen die ökonomisch relevanten Auswirkungen vorgegebener Konfigurationen mithilfe der analytischen Verfahren CANQ und MVA ermittelt. Zusätzlich wurde als einfache Alternative zu den analytischen Verfahren jeweils ein SIMAN-Grobsimulationsmodell erstellt. Zur Überprüfung der Qualität dieser "Grobmodelle" wurde für jedes FFS ein detailliertes SIMAN-Feinsimulationsmodell entwickelt. Der Ergebnisvergleich zeigt, daß sowohl die analytischen Verfahren als auch die Grobsimulation eine gute Abschätzung des Systemverhaltens ermöglichen, wobei die Abweichungen von den Ergebnissen der Feinsimulation i.a. nur wenige Prozent betragen.

Literatur:

Shalev-Oren, S., Seidmann, A. und P.J. Schweitzer, Analysis of of Flexible Manufacturing Systems with Priority Scheduling: PMVA, in: Annals of Operations Research 3(1985), S. 115-139

Solberg, J.J., A mathematical model of computerized manufacturing systems, in: Proceedings of the Fourth International Conference on Production Research, Tokyo 1977

Suri, R. und R.R. Hildebrant, Modelling Flexible Manufacturing Systems Using Mean-Value-Analysis, in: Journal of Manufacturing Systems 3(1984)1, S. 27-38

Tempelmeier, H., Simulation mit Siman, Arbeitspapier, TH Darmstadt, Institut für BWL, Fachgebiet Fertigungs- und Materialwirtschaft, Darmstadt 1987

SCHEDULING TECHNIQUES WITH LIMITED INTERMEDIATE STORAGE

Auke P. Woerlee, Rotterdam

Given a set of n jobs J_j, j=1,...,n, and a set of m machines M_i, i=1,...,m. A machine environment is called a flow shop, if each job J_j consists of a chain $O_{1j},...,O_{mj}$. O_{ij} has to be processed on M_i during pij time units. These machine environments often occur in a more complex form in assembly lines.
Short-run scheduling techniques can be important controlling these assembly lines. Therefore it is interesting to investigate scheduling algorithms for flow shop models with limited intermediate storage.
An exact algorithm will be shown to solve m-machine flow shop problems with limited intermediate storage. In the algorithm we will use the critical path of a schedule to generate successive nodes in the branch-and-bound tree. Using the critical path we avoid the investigation of nodes in the search tree which are not of any interest. The lowerbounds are obtained from 2-machine problems with limited intermediate storage, which are NP-hard problems, but can be solved reasonable fast.
The difficulty of branch-and-bound algorithms is the amount of time required to solve problems of any practical size. Therefore a truncated branch-and-bound method is proposed to solve these problems.

ZUR STRUKTURIERUNG HIERARCHISCHER PLANUNGSSYSTEME

Werner Delfmann, Frankfurt

Zusammenfassung: Die interdependente Festlegung der Anzahl der hierarchischen Ebenen und der Leitungsspannen innerhalb eines Systems koordinierter Planungen bildet den Gegenstand der Untersuchung. Für dieses Problem wird ein verallgemeinerter Lagrange-Ansatz vorgestellt, der es erlaubt, für unterschiedliche Zielsetzungen und Rahmenbedingungen Aussagen über die vorteilhafte Gestaltung der Planungshierarchie abzuleiten.

1. Einführung

Der hierarchische Aufbau von Unternehmensbereichen wird wesentlich durch die Ausprägung zweier Variablen determiniert:
Die Zahl der hierarchischen Ebenen (Tiefengliederung) und die Leitungs- oder Kontrollspannen (Breitengliederung), d.h. die Anzahl der Untergebenen, welche einem Vorgesetzten innerhalb der Hierarchie unmittelbar unterstellt sind.
Zwischen den beiden Größen besteht ein enger Zusammenhang. Denn - bei gegebener Größe des Ausführungsbereichs der Unternehmung - sinkt die Zahl der hierarchischen Ebenen tendenziell mit zunehmenden Leitungsspannen und umgekehrt.
Erst seit Ende der 60er Jahre wird verstärkt versucht, Aussagen über die Leitungsspannen durch modelltheoretische Analysen zu fundieren. In diesen frühen Arbeiten wird das eigentliche Kernproblem jedoch umgangen. Die Interdependenzen zwischen den Leitungsspannen und der Anzahl der hierarchischen Ebenen werden in unzulässiger Weise zerschnitten. Erst Arbeiten der neueren amerikanischen Organisationstheorie etwa von WILLIAMSON, MACKENZIE, KEREN und LEVHARI oder auch Martin J. BECKMANN gehen auf dieses Kernproblem ein. Sie bilden die Grundlage der folgenden Überlegungen.
Und zwar soll der hierarchische Aufbau der Unternehmung nicht generell, sondern unter einem ganz speziellen Blickwinkel analysiert werden. Nämlich im Hinblick auf den Aufbau des Planungssystems der Unternehmung.
Dabei wird von der Grundüberlegung ausgegangen, daß die Unternehmensplanung in Form

Operations Research Proceedings 1987

eines hierarchisch gegliederten Systems von koordinierten Teilplanungen unterschiedlichen Umfangs und Detailliertheitsgrades zu konzipieren ist.
Die folgende Analyse beschäftigt sich mit der Frage, wie viele hierarchische Ebenen und welche Leitungsspannen - oder in diesem Zusammenhang vielleicht besser Planungsspannen - unter jeweils gegebenen Rahmenbedingungen im Planungssystem der Unternehmung zu realisieren sind.

2. Modellkonstruktion

a) Vorüberlegungen

1. Betrachtet wird eine Unternehmung, in der der erforderliche Output der Planung nach Umfang und Detailliertheit festgelegt ist. Das sind praktisch jene Planungsgrössen der untersten Planungsebene, die unmittelbar in Ausführungshandlungen umgesetzt werden.
Diese Einzelpläne müssen koordiniert sein. Natürlich muß diese Koordination nicht die gesamte Unternehmung umfassen. Sie kann via Entscheidungsdelegation durchaus auf Teilbereiche begrenzt sein. Für die Koordination sorgt letztlich der für diesen Teilbereich zuständige Planungschef an der Spitze der Hierarchie.

2. Unterstellt man einen hierarchischen Aufbau des Planungssystems in Form einer Linienorganisation, so läuft der Planungsprozeß etwa folgendermaßen ab.
Die Stellen der untersten Ebene sammeln die relevanten Informationen über die ihnen jeweils zugeordneten Bereiche, also z.B. Marktentwicklungen, Finanzierungsbedingungen, Beschaffungspreise usw. Diese Informationen werden in verdichteter Form an die nächst höhere Planungsebene weitergegeben, bis die Gesamtheit aller Informationen der Spitze der Planungshierarchie vorliegt. Natürlich in besonders stark aggregierter Form.
Auf dieser Grundlage wird hier ein sehr globaler Plan für den gesamten Unternehmensbereich erstellt. Er wird der zweiten Planungsebene als Rahmen vorgegeben. Dort werden hieraus mehrere detailliertere Pläne für kleinere Teilbereiche entwickelt, die wiederum der nächsten Ebene als Rahmenpläne vorgegeben werden. An der Basis der Planungspyramide liegen schließlich als Endergebnis dieses stufenweisen Vorgehens die unmittelbar umsetzbaren Größen in dem erforderlichen Umfang und Detailliertheitsgrad vor.

b) Prämissen

1. Für die Modellkonstruktion wird nun von folgenden Prämissen ausgegangen: Mit dem zu konzipierenden Planungssystem sollen Pläne vorgegebenen Umfangs und vorgegebener Detailliertheit für zuvor festgelegte Bereiche der Unternehmung entwickelt werden. Und zwar seien genau P Pläne mit jeweils D Plandaten zu erstellen. Der Ausführungsbereich der Unternehmung ist also quasi in homogene Teileinheiten zerlegt worden,

für die Teilpläne gleicher Komplexität zu entwickeln sind.
Die Planungshierarchie soll - wie eben betrachtet - als reine Linienorganisation ausgelegt sein, in der keine nicht-hierarchischen Informations- und Planungskanäle existieren. Die Ebenen der Hierarchie sind von 1 - der untersten Ebene, die die unmittelbar umsetzbaren Pläne erarbeitet - bis N - der Hierarchiespitze - durchnumeriert.

Die Anzahl der Planer auf jeder Ebene sei mit x_n bezeichnet, ihre Planungsspannen sind s_n. Beide Größen sind Variable des Problems. Zur Besetzung der einzelnen Ebenen des Planungssystems stehen hinreichend jeweils gleich qualifizierte Planer zur Verfügung (Homogenitätsannahme).
Jedes Mitglied der Ebene n ist einem Mitglied der nächst höheren Ebene n+1 unterstellt, an das es die von der nächst niedrigeren Ebene erhaltenen Informationen verdichtet weitergibt und von dem es seinerseits den Rahmenplan vorgegeben bekommt. Diese eigenen Planungen sind detaillierter. Ihre Ergebnisse werden den direkt unterstellten Mitgliedern der nächst niedrigeren Ebene wiederum als Rahmen vorgegeben.

Jeder Planer muß also seine Arbeitszeit auf verschiedene Aktivitäten aufteilen: Entgegennahme und Weitergabe von Planungen über- und Informationen untergeordneter Ebenen und eigene Planungstätigkeit.
Als Kriterium für die Leistungsfähigkeit des Planungssystems dient der Zeitbedarf, der von der Erfassung der Daten bis zur Erstellung der unmittelbar umsetzbaren Pläne der letzten Ebene anfällt.
Dieser Zeitbedarf hängt aber von der Struktur des Planungssystems ab. Also davon, wieviele Mitarbeiter insgesamt in der Planung eingesetzt sind und wie ihre Planungszuständigkeiten organisatorisch geregelt sind.
Grundlegend ist nun die folgende Annahme: Der Zeitbedarf z_n eines Planers hängt von der Komplexität der Planungsaufgabe und der Anzahl der von ihm direkt zu koordinierenden Teilplanungen ab. Er setzt sich aus zwei Teilen zusammen.

$$z_n = a_n(D) + b_n(D) \cdot s_n = a_n + b_n \cdot s_n \tag{1}$$

Die erste Komponente kennzeichnet den Zeitbedarf für die eigentliche Planungstätigkeit. Er hängt nur von der Komplexität der Planungsaufgabe ab - hier durch die Anzahl der pro Teilplan vorzugebenden Plandaten D ausgedrückt. Die zweite Komponente beschreibt hingegen den koordinationsbedingten Zeitbedarf. Er hängt neben der Komplexität auch von der Planungsspanne s_n ab. Sie entspricht der Anzahl der pro Planer direkt zu koordinierenden Teilpläne der nächsten Ebene. Die beiden Komponenten werden im folgenden vereinfacht als reine Planungszeit und als Koordinationszeit bezeichnet.

2. Wenn man davon ausgeht, daß die Planer einer Ebene gleichzeitig arbeiten, dann kann man aufgrund der Homogenitätsannahme über die Qualifikation der Planer folgendes sagen: Der Zeitbedarf wird von dem Planer der Ebene mit dem größten Koordinationsbe-

darf, d.h. der größten Planungsspanne bestimmt. Nun kann die Koordination auf den nächst höheren Ebenen jeweils erst dann einsetzen, wenn alle Informationen der vorgelagerten Ebenen vorliegen. Um eine harmonische Auslastung der einzelnen Planungsebenen sicherzustellen, ist es deshalb sinnvoll, die Planungsspanne auf einer Ebene jeweils einheitlich zu bemessen. Unter dieser Voraussetzung entspricht der Zeitbedarf einer Planungsebene gerade dem Zeitbedarf jedes einzelnen Planers dieser Ebene, wie ihn Gleichung (1) beschreibt. Und die Gesamtplanungszeit ergibt sich unmittelbar als Summe der Planungszeiten aller einzelnen Ebenen.

$$z_N = \sum_{n=1}^{N} z_n = \sum_{n=1}^{N} (a_n + b_n \cdot s_n) \qquad (2)$$

Hierbei steht s_n für die Planungsspanne der n-ten Ebene, die man jetzt, da einheitlich, auch als Verhältnis der Mitgliederzahlen der nächst niedrigeren Ebene n-1 zu jener Ebene n ausdrücken kann.

$$s_n = \frac{x_{n-1}}{x_n} \qquad (3)$$

Damit ist der Zusammenhang zwischen der Struktur des hierarchischen Planungssystems und dem gesamten Planungszeitbedarf hergestellt.

c) Zielfunktion

Unterschiedliche Strukturen von Planungssystemen zeichnen sich also auf der einen Seite durch einen unterschiedlichen Planungszeitbedarf, auf der anderen Seite durch einen unterschiedlichen personellen Einsatz auf den einzelnen hierarchischen Ebenen aus. Diese beiden Merkmale gilt es auf ihre Kostenwirkung hin zu bewerten. Die direkten Kosten des Planungssystems mögen in erster Linie die Gehaltskosten der Planer sein.

Von der Gesamtplanungszeit hängt es ab, inwieweit die jeweils ausgeführten Pläne auf die aktuellen Rahmenbedingungen abgestimmt sind, in welchem Umfang demzufolge die Gewinnchancen ausgenutzt werden. Hier erscheint die Annahme plausibel, daß mit steigender Planungszeit die Gewinnchancen immer weniger ausgenutzt werden können. Steigende Planungszeit bedeutet demnach zunehmenden Gewinnentgang.

Damit stellt sich insgesamt das Problem, das hierarchische Planungssystem so zu strukturieren, daß ein Ausgleich zwischen den Gehaltskosten als direkten Planungskosten und dem Gewinnentgang bei steigenden Planungszeiten hergestellt wird.

Als Zielfunktion des Modells wird also die Summe aus 2 Komponenten minimiert. Erstens die Gehaltskosten aller Ebenen der Planungshierarchie. Sie ergeben sich für jede Ebene n als Produkt aus dem Gehalt pro Planer g_n und deren Anzahl x_n. Beide Größen können im allgemeinen von Ebene zu Ebene variieren. Die zweite Komponente bilden die Opportunitätskosten der Gesamtplanungszeit. Hierbei wird von der nicht sehr restriktiven Annahme ausgegangen, daß die Opportunitätskostenfunktion linear oder progressiv steigend verläuft, formal also konvex ist (OK(0) = 0, OK'>0, OK''> 0).

$$K_N = \sum_{n=1}^{N} g_n \cdot x_n + OK(Z_N) = Min! \qquad (4)$$

Bei der Minimierung der Zielfunktion sind zwei Bedingungen zu berücksichtigen. Einmal der Zusammenhang zwischen der Anzahl der Planer auf benachbarten Ebenen und den Planungsspannen, der in Gleichung (3) beschrieben ist. Und zum zweiten der Zusammenhang zwischen dem Gesamtplanungszeitbedarf und den Planungsspannen, der in der Funktion (2) beschrieben ist. Dies führt zu dem Lagrange-Ansatz

$$L = OK\,(Z_N) + \sum_{n=1}^{N} g_n \cdot x_n - \lambda \cdot \left[Z_N - \sum_{n=1}^{N} a_n + b_n \frac{x_{n-1}}{x_n} \right] = Min! \qquad (5)$$

Variablen der Zielfunktion sind die gesamte Planungszeit (Z_N) und die Anzahl der Planer Ebene n (x_n), wobei $x_N = 1$ und $x_o = P$ vorgegeben sind.

3.Modellanalyse

a) Mathematische Lösung

Die Lösung des Lagrange-Ansatzes führt auf ein System rekursiver Bestimmungsgleichungen für die Planungsspannen und damit gleichzeitig für die Anzahl der Mitarbeiter der einzelnen hierarchischen Ebenen. Dabei sind drei Einflußgrößen bestimmend:

- Der Koordinationszeitbedarf auf den einzelnen Ebenen (b_n)
- die marginalen Gehaltskosten auf den Ebenen (g_n) und
- die marginalen Opportunitätskosten der Planungszeit (OK').

$$s_n^* = s_{n+1}^* \cdot \left(\frac{b_{n+1}}{b_n} + \frac{g_n}{b_n \; OK'} \cdot x_{n+1} \right) \quad \text{für alle } n=1,..N-1 \qquad (6)$$

Dieses Gleichungssystem löst den Lagrange-Ansatz zunächst nur implizit. Um explizite Optimumaussagen abzuleiten, kann man sich u.U. eines Ansatzes der mathematischen Kontrolltheorie bedienen. Und zwar wird das stetige Äquivalent der skizzierten Minimierungsaufgabe formuliert und unter Anwendung des Pontrjagin'schen Maximumprinzips in Verbindung mit einem iterativen Approximationsverfahren gelöst (ähnlich Keren/ Levhari, S.1170). Für die ökonomische Interpretation der Modellanalyse reichen die Lösungsgleichungen (6) jedoch vollkommen aus.

b) Ökonomische Interpretation

1. Die Gleichung (6) läßt erkennen, daß im allgemeinen die Planungsspannen von Ebene zu Ebene größer werden, wenn man von der Spitze der Hierarchie zur Basis fortschreitet. Denn: Alle Größen der Klammer sind nicht negativ, und man kann davon ausgehen,

daß der Koordinationszeitbedarf zur Spitze hin nicht abnimmt, b_{n+1}/b_n ist also mindestens gleich 1. Der Klammerausdruck ist damit insgesamt > 1.
Um die grundlegenden Zusammenhänge erkennen zu können, sei im folgenden von einem gleichbleibenden Koordinationszeitbedarf (6) pro Teilplan auf allen Ebenen ausgegangen und außerdem seien linear steigende Opportunitätskosten unterstellt, d.h. die marginalen Opportunitätskosten sind positiv und konstant. Dann vereinfacht sich die Lösungsgleichung zu

$$s_n^* = s_{n+1}^* \cdot \left(1 + \frac{g_n}{c} \cdot x_{n+1} \right), \text{ mit } c = b \cdot OK' = \text{const.} \tag{7}$$

2. Es wird jetzt deutlich, wie die Lösung von der Gehaltsstruktur abhängt.
a) Solange die Gehaltskosten im Verhältnis zu den marginalen Opportunitätskosten sehr gering sind, ist der zweite Summand in Gleichung (7) zu vernachlässigen. Dann unterscheiden sich die Planungsspannen benachbarter hierarchischer Ebenen und damit aller Ebenen nur unwesentlich. Dieses Ergebnis deckt sich mit jenem, das man bei der Minimierung des Planungszeitbedarfs erhalten würde. Je weniger die Gehaltskosten eine Rolle spielen, desto mehr entspricht die Kostenminimierung einer Minimierung der Opportunitätskosten, und da diese mit dem Planungszeitbedarf steigen, einer Minimierung der Planungszeit. Hieraus kann man folgern, daß z.B. in hoch dynamischen Märkten, wo eine verzögerte Anpassung an geänderte Marktbedingungen besonders nachteilig ist und damit die marginalen Opportunitätskosten besonders hoch sind, in der Tendenz eher einheitliche Planungsspannen zu realisieren wären als unter wenig veränderlichen Marktbedingungen. Eine höhere Planungshierarchie und insgesamt kleinere Planungsspannen wären die Folge. Man kann dieses Ergebnis auch als analytische Begründung für eine verstärkte Spezialisierung der einzelnen Planungsbereiche unter dynamischen Marktbedingungen verstehen.

b) Ein deutlich anderes Ergebnis zeigt sich, wenn die Gehälter im Vergleich zu den marginalen Opportunitätskosten von nennenswerter Bedeutung sind. Dann nehmen nämlich die Planungsspannen von der Spitze der Hierarchie zur Basis hin kontinuierlich zu. Nehmen wir zunächst an, es werde auf allen Ebenen der Planungshierarchie ein einheitliches Gehalt gezahlt. Dann ist der Faktor vor x_{n+1} in Gleichung (7) konstant. Die Zunahme der Planungsspanne hängt daneben allein von der Anzahl der Mitglieder der nächst höheren hierarchischen Ebene ab.

$$s_n^* = s_{n+1}^* \left(1 + \frac{g}{c} \cdot x_{n+1}\right) \tag{8}$$

Diese Zahl steigt, je weiter man sich von der Spitze der Planungshierarchie entfernt. Deshalb ist die Zunahme der Planungsspannen im Vergleich zur nächst höheren Hierarchieebene um so größer, je weiter die Ebene von der Hierarchiespitze entfert ist. Die Planungspyramide wird flacher als bei einheitlichen Planungsspannen.

c) Die Annahme gleicher Gehälter auf allen Ebenen ist sicherlich wenig wirklichkeitsnah. Was läßt sich für den Fall sagen, daß die Gehälter auf höheren Ebenen auch höher sind?

Gehen wir zur Verdeutlichung vereinfacht davon aus, daß das Gehaltsgefüge gerade umgekehrt proportional zur Anzahl der Mitglieder auf jeder Ebene der Hierarchie ist.

$$g_n = g \cdot \frac{1}{x_n} = g_n \cdot x_n = g = \text{const.} \qquad (9)$$

Dann ist die Gehaltssumme auf jeder Ebene konstant. Da höhere Ebenen weniger Mitglieder haben, erhält jedes Mitglied hier mehr Gehalt als auf den niedrigeren Ebenen.

$$s_n = s_{n+1} \left(1 + \frac{W}{C}\right) \qquad (10)$$

Der rechte Teil der Klammer hängt dann nicht mehr von der Anzahl der Mitglieder der Ebene ab. Die Planungsspannen nehmen nach oben hin gleichbleibend ab, also stärker als im Fall des Einheitsgehalts. Die Folge ist eine sich nach oben hin noch schneller verjüngende, besonders flache Planungspyramide.

3. Insgesamt zeigt sich:
a) Je stärker die Gehaltsstruktur mit den hierarchischen Ebenen variiert, desto flacher wird die Pyramide. Je flacher die Pyramide, desto niedriger ist auch die Personenzahl, die in der Planung eingesetzt wird.

b) Je stärker das Gewicht der Kosten der Planungshierarchie im Vergleich zu den marginalen Opportunitätskosten der Planungszeit und je stärker differenziert die Gehaltsstruktur ist, desto flacher ist die Pyramide, desto geringer die optimale Gesamtzahl der einzusetzenden Planer und desto höher ist die insgesamt benötigte Planungszeit.

c) Bei zunehmendem Gewicht der Gehaltskosten im Verhältnis zu den Opportunitätskosten der Planungszeit werden also Lösungen vorteilhaft, die sich immer stärker von jener Lösung unterscheiden, die bei der Zielsetzung Minimierung der Planungszeit realisiert werden. Die Vorteile niedriger Opportunitätskosten bei kurzer Planungszeit werden bei zunehmenden Gehaltskosten immer stärker aufgezehrt.

4. Abschließend sei noch auf die Abhängigkeit der Planungsspannen von der Anzahl zu erstellender Einzelpläne (P) eingegangen. Dies ist die Frage nach der optimalen Betriebsgrößenvariation im Planungssystem, d.h. der Anpassung des Planungssystems in wachsenden Unternehmungen.
Eine multiple Betriebsgrößenvariation des Planungssystems (unveränderte Planungsspannen) ist angebracht bei vernachlässigbaren Gehaltskosten und wenn einheitlichen Gehältern konstante marginale Opportunitätskosten gegenüberstehen. Hier ändert sich die Gestalt der Planungshierarchie nicht, wenn die Basis wächst.

Eine mutative Betriebsgrößenvariation des Planungssystems hingegen ist vorteilhaft bei wachsenden marginalen Opportunitätskosten und einheitlichem Gehalt. Hier sinken die optimalen Planungsspannen mit wachsender Zahl der Einzelpläne. Die Planungspyramide wird schlanker und höher.
Bei konstanten marginalen Opportunitätskosten und mit der Höhe der hierarchischen Ebenen zunehmenden Gehälter steigen die Planungsspannen hingegen an. Die Planungspyramide ist gedrungener und flacher.

Literatur:

(1) Beckmann, M.J.: Rank in organizations, Heidelberg 1978.

(2) Keren, M./Levhari, D.: The optimum span of control in a pure hierarchy, Management Science, Vol.25, No.11, Nov.1979, S.1162 - 1172.

(3) McKenzie, K.D.: Measuring a person's capacity for interaction in a problem solving group, Organizational Behavior and Human Performance, Vol.12, No.2, Oct.1974, S.149 - 169.

(4) Williamson, O.E.: Hierarchical control and optimum firm size, J.Political Economy, Vol.75, No.2, April 1967, S.123 - 138.

AXIOMATISCHE FUNDIERUNG INTERTEMPORALER NUTZENFUNKTIONEN

Harald Dyckhoff, Hagen

Aus der Investitionstheorie ist bekannt, daß die üblichen finanzmathematischen Vorteilhaftigkeitskriterien für Investitionsvorhaben, wie z.B. der Kapitalwert, unter der Voraussetzung der Fiktion eines vollkommenen Kapitalmarktes ökonomisch sinnvoll sind ("Fisher-Separation"). Im allgemeinen können Investitions-(und Finanzierungs-)Entscheidungen jedoch nicht losgelöst von den persönlichen Konsumpräferenzen des Entscheidungsträgers bzw. derjenigen Personen, über deren Gelder disponiert wird, gefällt werden. Bei als begrenzt unterstelltem Planungszeitraum kennt die herrschende Theorie drei Fälle unterschiedlicher Konsumpräferenzen, genannt Vermögensstreben, Einkommensstreben und Wohlstandsstreben. Die beiden ersten Fälle werden konkretisiert als Maximierung des Vermögens am Planungshorizont bei vorgegebenem zwischenzeitlichen Konsum bzw. Maximierung der Breite des Konsumstroms bei vorgegebener Struktur des Konsumstroms und vorgegebenem Endvermögen. Das Wohlstandsstreben als allgemeiner Fall intertemporaler Nutzenmaximierung wird in der Regel weder weiter konkretisiert noch eingehender diskutiert. In diesem Beitrag sollen verschiedene Formen intertemporaler Nutzenfunktionen aus Axiomen abgeleitet und im Hinblick auf die von ihnen implizierte Konsum- und Zeitpräferenz diskutiert werden.

Zur Wahl der Vertragsdauer bei Kreditarrangements unter asymmetrischer Information

Jan Pieter Krahnen, Berlin

Summary: The european literature on financial management has always emphasized the importance of matching the term structure of corporate assets and liabilities. In this paper an attempt is made to analyse the optimal credit contract with respect to its duration by means of a simple example. Using an agency-theoretic framework the choice between short term and long term contract is shown to depend primarily on the degree of reversibility (or the share of sunk costs in total costs) of the investment project.

Kurzfassung: Die Vielfalt vertraglicher Gestaltungsformen auf Kapitalmärkten ist seit jeher ein Untersuchungsgegenstand der vornehmlich beschreibenden Finanzlehre. Sowohl auf dem Markt für Eigen- wie auch für Fremdkapital werden unterschiedliche Kombinationen typischer Vertragselemente beobachtet. In diesem Beitrag soll unter Verwendung eines einfachen Beispiels ein bisher wenig beachteter Aspekt vertraglicher Gestaltungsmöglichkeiten untersucht werden. Dabei handelt es sich um die Laufzeit eines Darlehnvertrags, der zwischen einem Kreditgeber und einem Kreditnehmer vereinbart wird. Die Ausgangsfrage lautet: Kann eine plausible Begründung für die Verwendung kurz- oder langfristiger Kreditverträge gegeben werden? Während traditionellerweise Unternehmensaktiva in der BWL nach dem bilanzorientierten Kriterium der Kapitalbindungsdauer unterschieden werden, wird in diesem Beitrag das kostenorientierte Kriterium der Irreversibilität von Investitionsentscheidungen verwendet. Nach diesem Kriterium unterscheiden sich Aktiva im Hinblick auf den Anteil versunkener Kosten an den gesamten Investitionsauszahlungen. Anhand eines Zwei-Perioden (Zahlen-)Beispiels wird gefragt, ob ein unmittelbarer Zusammenhang zwischen dem Ausmaß der Irreversibilität von Investitionsprojekten einerseits und der gewählten Vertragszeit andererseits bestehen kann. Dabei werden Anreizprobleme zwischen Kreditgeber und -nehmer durch das Risiko eines unbeobachtbaren Projektwechsels nach Vertragsabschluß berücksichtigt. Im Beispiel verhält sich die Laufzeit von Kreditverträgen proportional zur Höhe der versunkenen Kosten. Bei hohen (niedrigen) "sunk costs" kann der zweiperiodische (einperiodische) Kreditvertrag als 'optimal' identifiziert werden. Die im Mehrperiodenfall erkennbare Relevanz versunkener Kosten eröffnet einen neuen Zugang, nun aus entscheidungstheoretischer Sicht, zu der seit Ende der sechziger Jahre ruhenden Diskussion über die Verwobenheit finanz- und produktionswirtschaftlicher Strukturen.

ZUR BEDEUTUNG VEREINFACHENDER VERFAHREN DER WIRTSCHAFTLICHKEITSRECHNUNG BEI DELEGATION VON INVESTITIONSENTSCHEIDUNGEN

Felix Liermann, Frankfurt am Main

Zusammenfassung: Vereinfachende Investitionsrechenverfahren werden als Entscheidungskriterium für einen einzelnen Investor überwiegend kritisch beurteilt. Demgegenüber sind sie bei Delegation von Investitionsentscheidungen als Verhaltensnorm für einen nachgeordneten Mitarbeiter ebenso zu erwägen wie komplexere Entscheidungskriterien. Als Beispiel hierzu wird gezeigt, daß unter bestimmten Bedingungen das Kriterium der (dynamischen) Amortisationsrechnung dem Kapitalwertkriterium als Verhaltensnorm überlegen sein kann.

Summary: Simplified investment decision rules e.g. the pay back method are often criticized for suppressing information. As it is shown this argument does not hold in a principal-agent-context. If information and incentive problems arise simplified decision rules could prove more useful than complex ones for controlling the agents activities.

(1) Die Investitionstheorie beschäftigt sich mit der Frage, wie die Vorteilhaftigkeit von Kapitalanlagemöglichkeiten, die sich hinsichtlich der Höhe und der zeitlichen Struktur der erwarteten Ein- und Auszahlungen unterscheiden, beurteilt werden und wie eine Anlageentscheidung getroffen werden kann, die der Zielsetzung des Investors bestmöglich entspricht. Das Ergebnis sind Entscheidungskriterien für unterschiedliche Problemsituationen, an denen sich ein Investor im Einzelfall orientieren kann.

Die entwickelten Entscheidungskriterien lassen sich u.a. nach dem Umfang der im Kalkül berücksichtigten Informationen über die zu beurteilenden Projekte klassifizieren:

- Die klassischen dynamischen Verfahren (Kapitalwertmethode, Methode des internen Zinsfußes, Annuitätenmethode) berücksichtigen jeweils sämtliche mit dem Projekt erwarteten künftigen Ein- und Auszahlungen sowie die zeitliche Struktur der Zahlungsreihe.

- Vereinfachende Verfahren beschränken sich demgegenüber auf einen Teil der finanziellen Konsequenzen (z.B. Gewinn-, Kostenvergleichsrechnung) bzw. auf einen begrenzten Planungshorizont (z.B. Amortisationsrechnung)

Operations Research Proceedings 1987

und vernachlässigen in der statischen Version die zeitliche Struktur der Projektzahlungsreihe bzw. verwenden anstelle der periodenabhängigen Ein- und Auszahlungen Durchschnittswerte.

Unter Informationsgesichtspunkten sind die komplexeren Verfahren den vereinfachenden Kriterien überlegen; die vereinfachenden Kriterien werden in der investitionstheoretischen Literatur dementsprechend kritisch beurteilt. Dagegen ist in der Unternehmenspraxis die Vorgabe dieser Kriterien als Richtlinie für nachgeordnete Investitionsentscheidungen weit verbreitet. Empirische Untersuchungen (z.B. GRABBE 1976) ermitteln, in welchem Umfang welche Entscheidungskriterien für Investitionsentscheidungen in der Praxis verwendet werden. Daß im Zeitablauf eine zunehmende Akzeptanz der klassischen dynamischen Verfahren, z.B. der Kapitalwertmethode, beobachtet werden kann, wird darin positiv bewertet; die noch verbliebene "Kluft" zwischen Theorie und Praxis wird mit einer vermeintlichen Zurückhaltung der Praktiker gegenüber komplexeren Rechenverfahren erklärt.

In solchen Überlegungen bleibt jedoch unberücksichtigt, daß sich Investitionsentscheidungen in der Praxis als multipersonale Entscheidungsprozesse darstellen, während die Investitionstheorie die Entscheidungssituation eines einzelnen Investors betrachtet. Zwar gehört die Investitionspolitik einer Unternehmung zu den wesentlichen Aufgaben der Unternehmensleitung (Instanz); diese verfügt jedoch nur über eine begrenzte Kapazität zur Informationsbeschaffung und -verarbeitung, so daß es für sie notwendig bzw. sinnvoll ist, den Investitionsentscheidungsprozeß (ganz oder teilweise) an nachgeordnete Mitarbeiter (Entscheidungsträger) zu delegieren. Je nach dem Ausmaß der Delegation wird dann die Entscheidung mehr oder weniger zentral bzw. dezentral getroffen. Im Falle einer vollständigen Delegation der Investitionsentscheidung an einen nachgeordneten Mitarbeiter hat dieser die Aufgabe, selbständig nach erfolgversprechenden Investitionsmöglichkeiten zu suchen, deren im Zeitablauf zu erwartenden Konsequenzen zu prognostizieren und zu bewerten und schließlich die zu realisierenden Projekte auszuwählen.

Anstelle des ursprünglichen (und vergleichsweise einfachen) Investitionsentscheidungsproblems hat die Instanz jetzt ein Investitionsorganisationsproblem zu lösen: Sie trifft die Investitionsentscheidung nur mittelbar, indem sie die Aktivitäten des nachgeordneten Entscheidungsträgers durch organisatorische Maßnahmen in ihrem Sinne steuert. In einer Verhaltensnorm kann sie ihm hierzu ein Ziel vorgeben und dadurch seinen Handlungspielraum beschränken. Die Eignung einer Verhaltensnorm ist daran zu messen, inwieweit sie - eingebunden in das Bündel sonstiger organisatorischer Regelungen - den Entscheidungsträger zu einem Verhalten veranlaßt, das den (Investitions-)Zielen der Instanz möglichst gut entspricht.

Wenn ausschließlich die Verbreitung investitionstheoretischer Entscheidungskriterien in der Praxis beurteilt wird, liegt dabei stillschweigend die Annahme zugrunde, es sei sinnvoll, einem Entscheidungsträger als Verhaltensnorm gerade jenes Entscheidungskriterium vorzugeben, an dem sich die Instanz bei der Beurteilung der Vorteilhaftigkeit von Investitionsmöglichkeiten orientiert. Im folgenden soll gezeigt werden, daß es z.B. aufgrund einer Informationsasymmetrie zwischen Instanz und Entscheidungsträger (vgl. FRANKE, 1975) unter bestimmten Bedingungen gerade sinnvoll sein kann, dem Entscheidungsträger ein von dem der Instanz abweichendes Entscheidungskriterium als Verhaltensnorm vorzugeben (vgl. hierzu auch DEMSKI/FELTHAM, 1976; WEINGARTNER, 1977).

(2) In einem einfachen Beispiel wird das Kapitalwertkriterium hinsichtlich seiner Eignung als Verhaltensnorm mit dem Kriterium der (dynamischen) Amortisationsrechnung verglichen (zu ähnlichen Überlegungen im Einpersonenkontext vgl. WEINGARTNER, 1969; HAEGERT/WITTMANN, 1977):

1. Wir nehmen an, eine Instanz habe die routinemäßige Entscheidung über die Ersatzbeschaffung von Aggregaten an den Leiter einer Produktionsabteilung delegiert. Sie selbst orientiert sich bei ihren Investitionsentscheidungen an der Kapitalwertmethode. Danach erweisen sich von einer Menge zur Wahl stehender Investitionsmöglichkeiten jene Investitionsprojekte als vorteilhaft, bei denen die Summe der mit dem gewählten Kalkulationszinsfuß p abgezinsten künftigen Einzahlungsüberschüsse aus dem jeweiligen Projekt die Anschaffungsauszahlungen übersteigen, d.h. für deren Kapitalwert KW gilt:

$$KW = \sum_{t=0}^{T} e_t \, (1+p)^{-t} \geq 0.$$

Von zwei sich ausschließenden Projekten ist jenes mit dem höheren positiven Kapitalwert vorzuziehen. Dabei bezeichnen e_t, t=0, ..., T die Einzahlungsüberschüsse in den Zeitpunkten t und T den Zeitpunkt der letzten Zahlung.

Zwischen Instanz und Entscheidungsträger bestehe folgende Informationsasymmetrie: Der Entscheidungsträger hat einen guten Überblick über Investitionsalternativen in seinem Bereich und kann die Projektkonsequenzen - innerhalb der Rahmenbedingungen, die von der Instanz als Plandaten vorgegeben werden - mit Sicherheit prognostizieren. Mangels eigener Informationen über Gesamtzusammenhänge rechnet er mit dem Eintreten der vorgegebenen Plandaten. Die Instanz ihrerseits verfügt über Informationen, die ihr einen Rückschluß auf die begrenzte Dauerhaftigkeit der von ihr durch die vorgegebenen Planwerte gesetzten Rahmenbedingungen ermöglicht; über das Entscheidungsfeld des Entscheidungsträgers ist sie hingegen nur unzureichend informiert.

Eine Veränderung der Plandaten kann in dem betrachteten Entscheidungsbereich zu einer Revision der zuvor getroffenen Investitionsentscheidung noch vor Ablauf der geplanten Projektdauer führen, weil sich unter den geänderten Bedingungen nun eine andere Investitionsalternative als optimal erweist.

Für das Beispiel genügt es, einen Zeitraum von zwei Perioden zu betrachten; der Kalkulationszinsfuß betrage p=10%. Anreizprobleme treten nicht auf.

2. Zu Beginn der Periode 1 (im Zeitpunkt t=0) stehen dem Entscheidungsträger die Investitionsalternativen I_1 und I_2 zur Auswahl, für die er unter den gegebenen Bedingungen des Unternehmensgesamtplans folgende Zahlungsreihen ermittelt hat:

t	0	1	2	KW_2		KW_1	
I_1	-100	100	100	73,55	D_2=18,18	-9,09	D_1=-18,18
I_2	-100	120	56	55,37		9,09	

KW_2 bezeichnet die Summe der abgezinsten Einzahlungsüberschüsse der Zeitpunkte 0, 1 und 2; KW_1 berücksichtigt dagegen nur die Einzahlungsüberschüsse der Zeitpunkte 0 und 1. Orientiert sich der Entscheidungsträger bei seiner Entscheidung ebenso wie die Instanz am Kapitalwertkriterium, so wählt er das Projekt mit dem höheren Kapitalwert, also I_1, dessen Kapitalwert $KW_2(I_1)$ um D_2 = 18,18 höher ist als der von I_2, das allerdings auch die längere Amortisationsdauer aufweist. Im Gegensatz zu Projekt I_2 reicht bei I_1 der Einzahlungsüberschuß des Zeitpunkts 1 noch nicht aus, um die Anschaffungsauszahlung zuzüglich Zinsen zu decken: Bei Projekt I_1 ist die Summe $KW_1(I_1)$ der abgezinsten Einzahlungsüberschüsse der Zeitpunkte 0 und 1 um D_1 = -18,18 geringer als der entsprechende Wert $KW_1(I_2)$ bei I_2.

3. Die Instanz rechnet damit, daß mit einer Wahrscheinlichkeit w die Gesamtunternehmensplanung im Zeitpunkt t=1 Korrekturen erfährt, die die Fortführung des vom Entscheidungsträger begonnenen Investitionsprojekts im Vergleich zu einer Revision der Investitionsentscheidung (Realisation eines Projekts I_3) nachteilig macht. Der Restwert von I_1 bzw. I_2 sei jeweils null und das dann vorteilhaftere Projekt I_3 sei von I_1 bzw. I_2 unabhängig. Wäre es der Instanz möglich, in t=0 die Investitionsentscheidung im Produktionsbereich selbst zu treffen, so würde sie die Möglichkeit einer Revision bereits bei der Ausgangsentscheidung berücksichtigen und die Konsequenzen abschätzen, mit denen im Fall eines vorzeitigen Abbruchs in t=1 zu rechnen ist. Hierzu sind die Kapitalwerte $KW_1(I_i)$ der erwogenen Projekte miteinander zu vergleichen.

Die Vorteilhaftigkeit des Projekts I_1 gegenüber I_2 ergibt sich (unabhängig von den Einzahlungsüberschüssen des Folgeprojekts I_3) aus der Gegenüberstellung der im Fall des Abbruchs bzw. der Fortführung jeweils erzielten Kapitalwerte:

$$D = KW(I_1) - KW(I_2) = w \cdot D_1 + (1-w) \cdot D_2$$

mit

$$D_1 = KW_1(I_1) - KW_1(I_2),$$
$$D_2 = KW_2(I_1) - KW_2(I_2).$$

Gilt D>0, so ist in t=0 aus Sicht der Instanz die Wahl von I_1 vorteilhafter; im Fall D<0 ist I_2 vorzuziehen.

4. Der Entscheidungsträger wählt in t=0 gerade dann das Projekt I_2, wenn ihm als Verhaltensnorm anstelle des Kapitalwertkriteriums das Kriterium der (dynamischen) Amortisationsrechnung vorgegeben wird. Danach ist dasjenige Projekt vorteilhaft, dessen Amortisationsdauer eine vorgegebene maximale Amortisationsdauer T_a nicht überschreitet bzw. für das (bei Normalinvestitionen) gilt:

$$KW_{Ta} = \sum_{t=0}^{T_a} e_t \, (1+p)^{-t} \geq 0.$$

Im Beispiel gilt T_a=1: Wegen $KW_1(I_2) \geq 0$ und $KW_1(I_1) < 0$ wählt der Entscheidungsträger das Projekt I_2.

5. Die Vorteilhaftigkeit der erwogenen Verhaltensnormen hängt im Bei-

spiel von den Kapitalwertunterschieden D_2 - bezogen auf den gesamten Planungshorizont - und D_1 - bei Betrachtung nur der Periode 1 - sowie von der Wahrscheinlichkeit w einer Planrevision ab. Liegt (wie im Beispiel) eine Situation vor, in der die nachteiligen Konsequenzen der Projektwahl bei vorzeitigem Abbruch des Projekts die vorteilhaften Konsequenzen bei Fortführung gerade aufwiegen ($D_2 = -D_1 = 18{,}18$), so hängt die Entscheidung für die eine oder andere Verhaltensnorm nur von der Wahrscheinlichkeit w einer erforderlichen Revision ab: Gilt $w<0{,}5$, so ist als Verhaltensnorm das Kapitalwertkriterium vorteilhafter; falls $w>0{,}5$, ist das Kriterium der Amortisationsrechnung vorteilhafter. Im Fall $w=0{,}5$ sind beide Verhaltensnormen äquivalent.

(3) Aufgrund ihrer beschränkten Kapazität zur Informationsbeschaffung und -verarbeitung kann die Instanz bei ihrer Entscheidung über die vorzugebende Verhaltensnorm den daraus resultierenden Planungsprozeß des Entscheidungsträgers und dessen Konsequenzen jedoch nicht exakt antizipieren. Würde sie sich nämlich umfassend über das Entscheidungsfeld des Entscheidungsträgers informieren, so könnte sie die Entscheidung auch selbst treffen; die Delegation würde die Instanz nicht entlasten (Dilemma der Organisationstheorie). Die Instanz muß sich vielmehr darauf beschränken, die erwarteten Konsequenzen der erwogenen Verhaltensnormen nur mehr oder weniger grob abzuschätzen. Bei der Entscheidung über die vorzugebende Verhaltensnorm ist die Instanz auf vereinfachende Beurteilungskriterien angewiesen. Die folgende Überlegung führt zu Effizienzkriterien, an denen sich die Instanz bei der Beurteilung alternativer Verhaltensnormen orientieren kann (LAUX/LIERMANN, 1987):

Die Befolgung einer Verhaltensnorm im Sinne der Instanz stellt an den Entscheidungsträger gewisse Anforderungen. Inwieweit diese Verhaltensnorm die gewünschte Steuerungswirkung zeigt, hängt (auch) davon ab, ob der Entscheidungsträger aufgrund seiner Fähigkeiten und Handlungsmöglichkeiten in der Lage ist, den damit an ihn gestellten Anforderungen zu genügen, und ob er gewillt ist, die Verhaltensnorm zu befolgen. Im einzelnen ist zu beurteilen,

- ob und in welchem Ausmaß der Entscheidungsträger die für eine angemessene Befolgung der Verhaltensnorm erforderlichen Informationen bereits besitzt oder sich beschaffen kann (Informationskompatibilität),

- ob und in welchem Ausmaß der Entscheidungsträger in der Lage ist, Informationen angemessen zu verarbeiten, insbesondere aus den vorliegenden Informationen gute Prognosen der künftigen Einzahlungsüberschüsse zu erstellen (Kalkülkompatibilität),

- ob und in welchem Ausmaß die Befolgung der Verhaltensnorm mit der Zielsetzung des Entscheidungsträgers übereinstimmt, oder ob mit der Befolgung eigener (abweichender) Ziele zu rechnen ist (Anreizkompatibilität).

Im Lichte dieser Effizienzkriterien ist die Eignung einer Verhaltensnorm danach zu beurteilen, welche Kompatibilitätseigenschaften sie aufweist. Auch anhand dieser Kriterien kann verdeutlicht werden, daß es in unterschiedlichen Situationen erwägenswert ist, dem Entscheidungsträger für dessen Investitionsentscheidung anstelle des Kapitalwertkriteriums das Kriterium der (dynamischen) Amortisationsrechnung vorzugeben und damit den Planungshorizont zu begrenzen:

1. Der Entscheidungsträger verfügt hinsichtlich der Bestimmungsfaktoren der Einzahlungsüberschüsse früher Planperioden über einen guten Informationsstand; Indikatoren für die Projektentwicklung in späteren Perioden kann er hingegen nicht beobachten. Dies ist z.B. dann der Fall, wenn zwischen Entscheidungsbereichen Interdependenzen derart bestehen, daß die Erfolgsdeterminanten des einen Bereichs Entscheidungsvariablen eines anderen Bereiches sind, oder wenn - wie im vorhergehenden Beispiel - der Zeitraum der strategischen Unternehmensplanung geringer ist als die zu erwartende Projektlebensdauer. Der Entscheidungsträger wird dann von einem Zeitpunkt an die Daten der letzten Periode für alle folgenden Perioden fortschreiben: Die Informationskompatibilität nimmt mit zunehmendem Planungshorizont ab.

2. Der Entscheidungsträger verfügt über eine gute Prognosefähigkeit, die jedoch mit der Entfernung vom Planungszeitpunkt abnimmt. Von einem bestimmten Zeitpunkt an ist die Zuverlässigkeit der Prognose so gering, daß die hierfür ermittelten Einzahlungsüberschüsse für die Projektbeurteilung nicht mehr aussagekräftig sind: Die Kalkülkompatibilität nimmt mit zunehmendem Planungshorizont ab.

3. Der persönliche Planungshorizont des Entscheidungsträgers als Organisationsmitglied ist kürzer als die Projektlebensdauer. Die Gefahr von Fehlentscheidungen hängt dann davon ab, inwieweit der Entscheidungsträger durch ein geeignetes Anreizsystem schon frühzeitig an die Projektkonsequenzen späterer Perioden gebunden werden kann. Je weniger dies möglich ist, um so bedeutsamer werden Kontroll- und Sanktionsmechanis-

men, die den Entscheidungsträger noch vor seinem Ausscheiden treffen. Bei unsicheren Erwartungen ist allerdings der Rückschluß auf die Vorteilhaftigkeit der Entscheidung aus frühzeitigen Kontrollen um so schwieriger, je länger die Restlebensdauer des Projekts ist. Entsprechend größer ist dann auch der Manipulationsspielraum, der dem Entscheidungsträger zur Begründung der "Vorteilhaftigkeit" seiner Entscheidung verbleibt. Ist der Entscheidungsträger nach seinem Ausscheiden nicht mehr sanktionsfähig, so nimmt die Anreizkompatibilität mit der Länge des vorgegebenen Planungshorizonts ab.

(4) Die Beurteilung von Verhaltensnormen anhand der Kompatibilitätseigenschaften legt die Frage nahe, ob es dann nicht auch vorteilhaft sein kann, einem nachgeordneten Entscheidungsträger als Verhaltensnorm ein komplexeres Entscheidungskriterium vorzugeben, als es die Instanz bei eigener Entscheidung verwenden würde, wenn dieser über die entsprechenden Informationen und Kalkülfähigkeiten verfügt bzw. sie sich leicht erwerben kann. Daß eine Instanz von qualifizierten Stabsmitgliedern zwar häufig differenzierte und anspruchsvolle Analysen erwartet, jedoch die daraus abgeleiteten Empfehlungen nicht als Handlungsempfehlungen, sondern als zusätzliche Informationen zur Bestätigung oder Korrektur ihrer aufgrund eines einfacheren Entscheidungskriteriums eher intuitiv getroffenen Entscheidung verwendet, weist darauf hin, daß dann die Anreizkompatibilität gering sein kann. Die entsprechenden Anreizprobleme werden in der organisationstheoretischen Literatur als "Stabsproblematik" (unverantwortete Expertenmacht und Frustrationsgefahr der Stabsmitglieder) diskutiert.

<u>Literatur:</u>

DEMSKI, J.S.; FELTHAM, G.A. (1976): Cost Determination: A Conceptual Approach. Ames, Iowa 1976.

FRANKE, G. (1975): Investitionspolitik, betriebliche. In: HWB. GROCHLA, E.; WITTMANN, W. (Hrsg.), 4. Aufl., Bd. I/2, Sp.1996-2004. Stuttgart 1975.

GRABBE, H.-W. (1976): Investitionsrechnung in der Praxis - Ergebnisse einer Unternehmensbefragung. Köln 1976.

HAEGERT, L.; WITTMANN, F. (1977): Zur Eignung der Amortisationsrechnung als Kriterium für Investitionsentscheidungen bei unsicheren Erwartungen. ZfbF 29, S. 475-489 (1977).

LAUX, H.; LIERMANN, F. (1987): Grundlagen der Organisation. Berlin 1987.

WEINGARTNER, H.M. (1969): Some New Views on the Payback Period and Capital Budgeting Decisions. MS 15, S. B-594-B-607 (1969).

WEINGARTNER, H.M. (1977): Capital Rationing: n Authors in Search of a Plot. JoF 32, p. 1403-1431 (1977).

DIE BEWERTUNG DES INFORMATIONSSYSTEMS EINER UNTERNEHMUNG

Joachim Reese, Hagen

Zusammenfassung: Die Bewertung des Informationssystems konzentriert sich in dem hier erörterten theoretischen Ansatz auf den innerbetrieblichen Informationsfluß zwischen den einzelnen Aufgabenstellen sowie die stellenbezogene Informationsverarbeitung zur Vereinfachung dieses Flusses. Es wird vorausgesetzt, daß die verschiedenen Informationsarten voneinander abgrenzbar sind und Informationsnachfrage sowie Informationsangebot der Stellen als bekannt vorliegen. Für die Beschreibung des Informationssystems bedient sich der Ansatz eines algebraischen Instrumentariums, insbesondere von Matrizen zur Abbildung von Informationsfluß, -verarbeitung, -nachfrage und -angebot. Die Systembewertung erfolgt anhand von Effizienz- und Flexibilitätskriterien.

A. Einleitung

Die in einer Unternehmung für die Abwicklung von Informationsprozessen geschaffenen Voraussetzungen und Einrichtungen lassen sich auch als das Informationssystem der Unternehmung begreifen. Das Informationssystem ist zugleich Teil der gesamten Unternehmungsorganisation, zu der beispielsweise auch Befehls-, Kontroll- oder Anreizsysteme gehören. Es kann sowohl formale als auch informelle Komponenten enthalten. Im Hinblick auf die Herbeiführung von Sachentscheidungen fällt dem Informationssystem die Aufgabe zu, alle benötigten Daten zu sammeln, gegebenenfalls zu speichern und an die einzelnen Entscheidungsträger weiterzuleiten (vgl. auch McFARLAND 1979). Für eine Systembeurteilung ist es deshalb zunächst von zentraler Bedeutung, daß die individuellen Informationsnachfragen durch die zu bewältigende Entscheidungsproblematik von vornherein bekannt sind. Das Informationsangebot ist wiederum durch die Na(frage zum größten Teil vorbestimmt; sein Umfang spiegelt sich in den Aufgabenbeschreibungen wider. Das Angebot ist jedoch durch unmittelba)

Operations Research Proceedings 1987

Aufbereitung bzw. Verarbeitung der Ursprungsdaten modifizierbar. Sind dann die Informationsflüsse grundsätzlich an ein bestehendes Befehlssystem gekoppelt, so gründet sich eine vergleichende Bewertung von Informationssystemen vor allem auf die Beurteilung der Verarbeitungsaktivitäten hinsichtlich der Rohdaten. Unter diesen Umständen resultiert ein günstiges Systemurteil daraus, daß es gelingt, die Menge der zu übermittelnden Informationen zu reduzieren, ohne deren repräsentativen Charakter zu gefährden. Umfassender ist das Problem der Bewertung von nicht gebundenen Informationsflüssen. Sofern nämlich das hierarchische Befehlssystem nicht maßgeblich die Einrichtung von Informationskanälen determiniert, bietet sich eine Vielzahl von Gestaltungsmöglichkeiten bzw. direkten Wegen der Informationsübertragung, die gleichzeitig den Bedarf an stelleninterner Informationsaufbereitung in erheblichem Maße senken und dadurch die Beurteilung der Effizienz und Flexibilität eines Informationssystems wesentlich beeinflussen können.

B. Modellierung des Informationssystems

Der Systemansatz erfordert es, bei der Erfassung von Informationsaktivitäten zwischen den betreffenden Elementen und Beziehungen zu unterscheiden. Die Systemelemente lassen sich darüber hinaus in personale Elemente und Datenelemente trennen. Personale Elemente sind, je nach dem Detaillierungsgrad der Betrachtung, zum Beispiel Stellen oder Abteilungen der Unternehmung, soweit sie entscheidungsrelevante Informationen anbieten oder nachfragen. Als Datenelemente gelten demgegenüber die Datenarten, die als Informationen nachgefragt bzw. angeboten werden. Aus der Zuordnung von Datenelementen zu personalen Elementen ergeben sich dann

- einerseits das Informationsangebot, d.h. die Menge der Datenarten, die von einer Stelle entweder selbst erhoben oder von außerhalb der Unternehmung beschafft worden sind,
- und andererseits die Informationsnachfrage, d.h. die Menge der von einer Stelle nachgefragten Datenarten.

Bezeichnen $i = 1,\ldots,I$ die personalen Elemente und $j = 1,\ldots,J$ die Datenelemente des Systems, so können Elementzuordnungen als Binärmatrizen dargestellt werden, die nur Nullen und Einsen enthalten. Jede Eins veranschaulicht dann, daß die Zuordnung erfolgt, während eine Null besagt, daß keine Zuordnung stattfindet. Dementsprechend können sowohl das Informationsangebot als auch die Informationsnachfrage durch I-J-Matrizen abgebildet werden, wobei die Zeilen die personalen Elemente und die Spalten die Datenelemente kennzeichnen. Die Angebotsmatrix wird durch A, die Nachfragematrix durch N symbolisiert. Je nachdem, ob das Matrixelement (i,j) aus A nun 1 oder 0 ist, wird die Datenart j also von Stelle i angeboten oder nicht. Analoges gilt für N und nachgefragte Informationen (vgl. hierzu auch HWA 1976).

Bei den Systembeziehungen ist danach zu differenzieren, ob sie sich zwischen den personalen Elementen entwickeln oder aber auf die Datenelemente erstrecken. Die Beziehungen zwischen den personalen Elementen spiegeln den Informationsfluß wider; sie sind das eigentliche Bewertungsobjekt, nachdem Informationsangebot und Informationsnachfrage bereits vorab festliegen. Daneben betreffen die Beziehungen zwischen den Datenelementen die Informationsverarbeitung. Auch sie können Gegenstand der Systembeurteilung sein, wobei Verarbeitungsaktivitäten auch als eine Form der Aufgabenbeschreibung interpretierbar sind und mit ihrer Bewertung somit zugleich ein anderes Organisationsproblem, nämlich die zweckgerechte Aufgabendelegation, mit berührt wird.

Informationsfluß und Informationsverarbeitung sind als quadratische Matrizen darstellbar. Das Element (i^1,i^2) der I-I-Flußmatrix F beschreibt, ob ein Informationsfluß von i^1 nach i^2 stattfindet, während das Element (j^1,j^2) der J-J-Verarbeitsmatrix V anzeigt, ob die Datenart j^1 unmittelbar zu j^2 verarbeitet werden kann.

C. Bewertungskriterien

Von grundlegender Bedeutung für die Qualität eines innerbetrieblichen Informationssystems ist zunächst die Tatsache, daß die gegebene Informationsnachfrage innerhalb eines angemessenen Zeitrahmens $\bar{T}$ erfüllt wird. Formal gilt hierfür die Bedingung

$$A' \cdot (\sum_{t=1}^{\bar{T}} F^t) \geq N'. \qquad (1)$$

A' und N' bezeichnen die jeweils transponierten Matrizen. F^t entsteht aus F durch (t-1)-fache Multiplikation mit selbst und zeigt an, welche Informationskontakte nach genau t Schritten durch Zwischenschaltung anderer Abteilungen zustande kommen. Dementsprechend gibt die Matrix $\sum_{t=1}^{\bar{T}} F^t$ alle Flüsse wieder, die nach spätestens $\bar{T}$ Schritten ihren Adressaten erreichen und somit noch eine rechtzeitige Nutzung der übermittelten Informationen gewährleisten. Informationssysteme, die Bedingung (1) nicht erfüllen, sind in Anbetracht der vorliegenden Entscheidungssituation unzulässig, da sie eine wesentliche Eigenschaft vermissen lassen.

Unter den zulässigen Informationssystemen kann jedoch durch Zugrundelegung weiterer Kriterien noch feiner differenziert werden. So ergibt sich die Informationsdauer T eines Systems unmittelbar aus den vorangegangenen Überlegungen dadurch, daß neben

$$A' \cdot (\sum_{t=1}^{T} F^t) \geq N' \qquad (1a)$$

auch

$$A' \cdot (\sum_{t=1}^{T-1} F^t) \ngeq N' \qquad (1b)$$

gelten muß, d.h. in T-1 Schritten wird die Informationsnachfrage noch nicht vollständig befriedigt. Dies geschieht erstmals nach T Schritten. Die Informationsdauer ist als Bewertungskriterium besonders dann von erheblicher Relevanz, wenn ein Zeitvorsprung die Entscheidungen selbst oder deren Auswirkungen auf die Erfüllung von Unternehmungszielen beeinflußt.

Was die übergeordnete Effizienz eines Informationssystems anbetrifft, so ist auf jeden Fall sicherzustellen, daß nicht ein anderes System existiert, welches bei einer verfügbaren Zeit $\bar{T}$ mit geringerem Aufwand auskommt. In bezug auf die Einrichtung von Informationsflüssen ist deshalb zu fordern, daß gegenüber der betrachteten Flußmatrix F keine andere Matrix F^* existiert, die Bedingung (1) sowie

$$F^* < F \qquad (3)$$

erfüllt. Ist darüber hinaus jede Zeiteinsparung entscheidungsrelevant, so muß die Effizienzbedingung (3) entsprechend erweitert werden. Es darf kein anderes Informationssystem existieren, für welches

$$(F^*, T^*) < (F, T) \qquad (3a)$$

gilt. D.h. es bleiben auch solche Systeme effizient, die zwar aufwendiger sind und für die Bedingung (3) zutrifft, die zugleich aber Zeitèrsparnisse bewirken.

Die Effizienzüberlegungen lassen sich verschärfen, indem bei gleichermaßen effizienten Systemen unterstützend deren zumeist unterschiedliche Flexibilität mit berücksichtigt wird. Diese Systemflexibilität drückt sich darin aus, daß Informationsflüsse bestehen, die bei der vorherrschenden Angebots- und Nachfragesituation gar nicht beansprucht werden. Ändert sich jedoch beispielsweise die Informationsnachfrage, so können die entsprechenden Flüsse schnellstmöglich aktiviert werden. Die Flexibilität eines Informationssystems läßt sich grundsätzlich durch

$$\text{flex}\ (F,T) = |A' \cdot (\sum_{t=1}^{T} F^t) - N'| \qquad (4)$$

messen, wobei der Operator | | die entsprechende Matrix auf ihre Wertesumme abbildet. Es wird also eine Form von organisatorischem Überschuß ermittelt, der so lange positiv zu bewerten ist, wie die bestehenden Verbindungen nicht dazu genutzt werden, um unnötige Daten zu übertragen. Dieser Fall der Erzeugung von Systemredundanz wird hier nicht weiter betrachtet; er läßt sich durch eine nachfrageorientierte Informationsübermittlung von vornherein ausschließen.

Am exaktesten, aber auch am schwierigsten ist es, die Kosten K(F,T) des Informationssystems zu bestimmen. Sie schlagen sich zum einen als Transaktionskosten in der Einrichtung und Unterhaltung der Informationsflüsse nieder; gleichzeitig können sie aber auch von der beanspruchten Informationsdauer abhängen, ausgedrückt etwa durch Opportunitätskosten infolge von Entscheidungsverzögerungen. Welche Kosten im einzelnen tatsächlich zu berücksichtigen sind und in welcher Höhe sie anfallen, hängt stets maßgeblich von der konkreten Entscheidungssituation ab. Vielfach werden sie sich nur grob schätzen lassen, da für die meisten intern erbrachten Leistungen kaum Marktpreise existieren. Dies gilt insbesondere dann, wenn auch die Verarbeitungsprozesse für die einzelnen Informationsarten in die Bewertung einbezogen werden sollen.

D. Die Bestimmung eines optimalen Informationssystems

Aus den verschiedenen Bewertungsansätzen ergeben sich zwangsläufig auch Erkenntnisse hinsichtlich der bestmöglichen Gestaltung von Informationssystemen. Zugleich wird durch solche Überlegungen erst ersichtlich, wie weit ein zu beurteilendes System vom Optimum noch entfernt ist. Ein optimales Informationssystem kann dadurch bestimmt werden, daß die systembezogenen Kosten unter Einhaltung von notwendigen Nebenbedingungen minimiert werden. Dies führt zu einem ganzzahligen, nicht-linearen Optimierungsproblem, das sich folgendermaßen beschreiben läßt:

Minimiere $K(F,T)$ (5)

unter den Nebenbedingungen

$$A' \cdot \left(\sum_{t=1}^{T} F^t\right) \geq N' \qquad (6)$$

$$0 < T \leq \bar{T} \qquad (7)$$

F Binärmatrix . (8)

Restriktion (6) entspricht Bedingung (1) und garantiert die Zulässigkeit des Systems, während Restriktion (7) sicherstellt, daß die verfügbare Zeit eingehalten wird. In (8) sind die 0-1-Bedingungen für die Flußvariablen formuliert. Für diese Problemdarstellung existiert auf alle

Fälle eine zulässige Lösung, solange jedenfalls die nachgefragten Informationen überhaupt angeboten werden. Sie besteht in einem nach allen Richtungen offenen System, in dem alle Informationen schnellstmöglich - d.h. $T = 1$ - weitertransportiert werden können. Die Flußmatrix F enthält hierbei lediglich Einselemente. Ist sogar $\bar{T} \geq 2$ gegeben, so steht ein weiteres Standardsystem zur Verfügung, das ebenfalls stets zulässig ist. Es zeichnet sich dadurch aus, daß alle Daten zunächst an eine zentrale Stelle übermittelt und von dort dann an die Nachfrager weitergeleitet werden, so daß jede vorhandene Information in zwei Schritten erreichbar ist. Dieses System läßt sich als Datenbanksystem charakterisieren, die zentrale Informationsstelle als Datenbank. Der Vorteil eines Datenbanksystems gegenüber dem offenen System besteht darin, daß die Flußmatrix relativ dünn mit Einselementen besetzt ist. Im übrigen muß keines dieser beiden Systeme für gegebene Informationssituationen optimal sein. Zur Bestimmung des Optimums sind vielmehr geeignete Verfahren auszuwählen.

Die Zielfunktion (5) läßt sich ohne weiteres auf einen Satisfizierungsanspruch reduzieren, was vernünftig ist, wenn die Lösungssuche dadurch wesentlich beschleunigt wird. Auch ist es denkbar, ein anderes Ziel zu verfolgen. Soll beispielsweise eine größtmögliche Flexibilität gemäß (4) angestrebt werden, so müßte zugleich aber auch die Systemeffizienz nach (3) bzw. (3a) im Restriktionensystem zusätzlich Beachtung finden.

Die Problemformulierung kann recht einfach auf Informationsverarbeitungsprozesse ausgedehnt werden, die sich im Gegensatz zu den Übertragungsaktivitäten innerhalb einzelner Stellen abwickeln. Das bisher fest vorgegebene Informationsangebot der einzelnen Stellen wird dadurch variabel; durch interne Aufbereitung der Daten entsteht eine neue Angebotssituation, die eventuell auch dazu beitragen kann, den Datentransport zu erleichtern. Die erweiterte Problemformulierung lautet nun:

Minimiere $K(F,V,T)$ (5a)

unter den Nebenbedingungen

$$(A \cdot \sum_{t=1}^{T^*} V^t)' \cdot (\sum_{t=1}^{T-T^*} F^t) \geq N' \qquad (6a)$$

$$0 < T^* \leq T \leq \bar{T} \qquad (7a)$$

F und V Binärmatrizen. (8a)

Abgesehen von dem veränderten Informationsangebot bleibt zu beachten, daß bei einem gegebenem Zeitrahmen $\bar{T}$ für den Informationsfluß lediglich noch $\bar{T}-T^*$ Schritte zur Verfügung stehen, da die übrige Zeit von Verarbeitungsaktivitäten beansprucht wird. Die Problemerweiterung hat unübersehbare Konsequenzen für die Bewertung von Informationssystemen. In aller Regel werden neue Flußmatrizen zulässig; die Optimallösung paßt sich dementsprechend an. Lediglich wenn jegliche Informationsverarbeitung als unmöglich oder unwirtschaftlich erachtet wird, V also die Einheitsmatrix darstellt, geht der erweiterte Ansatz in die ursprüngliche Formulierung über.

E. Allgemeine Ergebnisse

Liegen die Absichten dieses Bewertungsansatzes auch gerade in der Berücksichtigung unternehmensbezogener Tatbestände begründet, so gestatten sich dennoch einige generelle Bemerkungen zu grundlegenden Typen von Informationssystemen, die unmittelbar verifizierbar sind:

- Ein streng kompetenzgebundenes Informationssystem, das auf horizontale Informationsflüsse verzichtet, birgt häufig den Nachteil langer Informationsdauern und somit die Gefahr der Nichterfüllung von Informationsnachfrage. Außerdem werden die zentralen Instanzen vielfach mit redundanten Daten überlastet, die sie nicht als Nachfrager, sondern lediglich zur Weiterübermittlung erhalten.
- Demgegenüber zeichnet sich das Datenbanksystem durch generell kurze Informationsdauern aus. Zusätzlich kostenwirksam ist allerdings die Einrichtung einer Datenbank einzig zum Zweck der Datenspeicherung und des Datentransports. In einigen Fällen erzeugt dieses System außerdem einen erheblichen organisatorischen Überschuß, der selbst bei veränderter Informationssituation nur zu einem geringem Teil wieder aktiviert wird.
- Das offene System, in dem grundsätzlich jeder mit jedem kommuniziert, bietet theoretisch die Möglichkeit der schnellsten Informationsübertragung. Allerdings werden die einzelnen Informationsflüsse sehr unübersichtlich, so daß beispielsweise Mehrfachinformationen ein hohes Maß an Redundanz verursachen können. Die Kosten der Systemeinrichtung sind zudem beträchtlich.
- Das Umlaufsystem, bei dem Informationen im Umlaufverfahren weitergegeben werden, weist dagegen extrem lange Informationsdauern auf, so daß es höchstens für weniger dringende Informationen in Frage kommt, bei denen das Zeitmoment eine untergeordnete Rolle spielt. Darüber hinaus ist das System vielmals von großer Redundanz, weil die umlaufenden Daten nur vereinzelt auch nachgefragt werden. Der Einrichtungsaufwand ist dagegen minimal; die zugehörige Flußmatrix enthält sehr wenig Einselemente.

Literatur

HWA, C. S., Ein Matrix-Ansatz zur Gestaltung von Management-Informationssystemen, in: E. Grochla (Hrsg.), Organisationstheorie, 2. Teilband, Stuttgart 1976, S. 460 - 467

McFARLAND, D. E., Management, New York und London 1979

DER INFORMATIONSWERT IN ENTSCHEIDUNGSMODELLEN MIT FUZZY-NUTZEN

Heinrich Rommelfanger, Frankfurt am Main

Zusammenfassung: In dieser Arbeit wird untersucht, wie zusätzliche Informationen in Entscheidungsmodellen zu bewerten sind, bei denen die Konsequenzen nur vage in Form unscharfer Mengen vom Entscheidungsträger angegeben werden können. In solchen Fuzzy-Entscheidungsmodellen können Informationen sowohl dazu dienen, die a priori-Verteilung über dem Zustandsraum zu verbessern, als auch, um genauere Nutzenbewertungen zu erhalten. In Anlehnung an die klassische MARSCHAKsche Definition werden für beide Informationsarten Informationswerte formuliert.

Summary: In this paper we discuss the value of additional information in decision models in which consequences are not known exactly but the utilities are allowed to be fuzzy. In fuzzy decision systems of this kind information may improve the a priori-probabilities as well as the utility values. In either case we define values of information which are orientated at the classical concept of MARSCHAK.

1. Einleitung

Die klassische Entscheidungstheorie basiert auf der Annahme, daß Entscheidungsprobleme in Form von Entscheidungsbäumen abgebildet werden können und daß der Entscheidungsträger alle erforderlichen Nutzen- und Wahrscheinlichkeitswerte präzise festlegen kann. Diese grundlegenden Voraussetzungen liegen aber bei realen Entscheidungssituationen nur selten vor, da der Entscheider zumeist nur über unvollständige, ungenaue und unzuverlässige Informationen verfügt. Es ist daher vernünftig, auf die Annahme eindeutiger Dateninputs zu verzichten und zu gestatten, daß er die Nutzen und/oder die Wahrscheinlichkeiten nur näherungsweise in Form unscharfer Mengen angibt.

In Fuzzy-Entscheidungsmodellen können zusätzliche Informationen sowohl dazu dienen, die a priori-Verteilung über dem Zustandsraum zu verbessern, als auch, um genauere Nutzenbewertungen zu erhalten. In dieser Arbeit soll nun untersucht werden, wie diese Informationen zu bewerten sind.

Um die Seitenvorgabe einzuhalten wird die Untersuchung beschränkt auf den Risikofall und den Lösungsansatz auf der Basis unscharfer Erwartungsnutzen. Auch auf die Darstellung eines anschaulichen numerischen Beispiels muß verzichtet werden.

Operations Research Proceedings 1987

2. Der Informationswert im klassischen Entscheidungsmodell

Ausgangspunkt und Vergleichsgegenstand unserer Untersuchung ist das klassische Entscheidungsmodell $< A, S, u(a_i,s_j), p(s_j) >$ mit den Komponenten

- A Menge aller vom Entscheidungsträger als relevant angesehenen Handlungsalternativen a_i, $i = 1,..,m$,
- S Menge der Umweltzustände s_j, $j = 1,...,n$,
- $u(a_i,s_j)$ Nutzen des Entscheidungsträgers, wenn die Alternative a_i realisiert wird und der Umweltzustand s_j eintritt,
- $p(s_j)$ a priori-Wahrscheinlichkeitsverteilung über S.

Nach dem BERNOULLI-Prinzip ist dann die Alternative $a^* \in A$ die optimale, die den Erwartungsnutzen

$$E(a_i) = \sum_{j=1}^{n} u(a_i,s_j)\, p(s_j) \tag{1}$$

maximiert, d.h. $E(a^*) = \underset{a_i \in A}{\text{Max}}\, E(a_i)$.

Existiert zusätzlich eine Informationsmenge $X = \{x_1,...,x_r,..,x_R\}$ und sind die bedingten Wahrscheinlichkeiten $p(x_r|s_j)$ bekannt, so kann bei Beobachtung der Information $x_r \in X$ anstelle der a priori-Verteilung $p(s_j)$ die a posteriori-Verteilung

$$p(s_j|x_r) = \frac{p(x_r|s_j)p(s_j)}{\sum_j p(x_r|s_j)p(s_j)} = \frac{p(x_r|s_j)p(s_j)}{p(x_r)} \tag{2}$$

zur Bewertung des Erwartungsnutzens benutzt werden:

$$E(a_i|x_r) = \sum_{j=1}^{n} u(a_i,s_j)p(s_j|x_r)\,. \tag{3}$$

Die optimale Alternative $a^*(x_r)$ berechnet sich dann aus

$$E(a^*(x_r)) = E(a^*(x_r)|x_r) = \underset{a_i \in A}{\text{Max}}\, E(a_i|x_r)\,. \tag{4}$$

Vor Abzug der Informationskosten ist dann der ex ante-Erwartungsnutzen mit Information gleich

$$E(X) = \sum_{r=1}^{R} E(a^*(x_r))\cdot p(x_r) \tag{5}$$

und der ex ante-Wert der zusätzlichen Information X

$$W(X) = E(X) - E(a^*). \tag{6}$$

3. Fuzzy-Erwartungswerte

Wir wollen nun den Fall betrachten, daß der Entscheidungsträger nicht in der Lage ist, den Alternativen eindeutige zustandsabhängige Nutzenwerte zuzuordnen. Es sei ihm aber möglich, eine größenordnungsmäßige Bewertung in Form unscharfer Nutzenmengen

$$\tilde{U}(a_i,s_j) = \{(u,f_{ij}(u))\,|\,u \in U \subset \mathbb{R}\} \tag{7}$$

anzugeben, wobei U die Menge aller bei diesem Problem möglichen Nutzenwerte ist und $f_{ij} : U \longrightarrow [0,1]$ die Zugehörigkeit eines Nutzenwertes $u \in U$ zur Menge der bei der Konstellation (a_i,s_j) wahren Nutzenwerte widerspiegelt.

Ein Weg zur Bestimmung der optimalen Alternative dieses Fuzzy-Entscheidungsproblems $<A,S,\tilde{U}(a_i,s_j),p(s_j)>$ ist, vgl. WATSON; WEISS; DONELL /8/ die Berechnung der Fuzzy-Erwartungswerte

$$\tilde{E}(a_i) = \{(\bar{u},f_i^E(\bar{u}))\,|\,\bar{u} \in \mathbb{R}\} \tag{8}$$

mit Hilfe des Erweiterungsprinzips von ZADEH, so daß die Zugehörigkeitswerte $f_i^E(\bar{u})$ ermittelt werden gemäß der Formel

$$f_i^E(\bar{u}) = \max_{(u_1,\ldots,u_n) \in U^n} \operatorname{Min}(f_{i1}(u_1),\ldots,f_{in}(u_n)) \tag{9}$$

$$\text{so daß } \bar{u} = \sum_{j=1}^{n} u_j \cdot p(s_j).$$

Die Berechnung des unscharfen Erwartungsnutzens $\tilde{E}(a_i)$ ist i.a. sehr aufwendig, sie wird aber einfach, wenn alle unscharfen Nutzenbewertungen Fuzzy-Zahlen bzw. Fuzzy-Intervalle vom L-R-Typ sind, vgl. DUBOIS/PRADE /1/, und sich daher abgekürzt darstellen lassen als

$\tilde{U}(a_i,s_j) = (u_{ij};\underline{v}_{ij};\bar{v}_{ij})_{LR}$ L-R-*Fuzzy-Zahl* oder

$\tilde{U}(a_i,s_j) = (\underline{u}_{ij};\bar{u}_{ij};\underline{v}_{ij};\bar{v}_{ij})_{LR}$ L-R-*Fuzzy-Intervall* .

Zur Vereinfachung der Darstellung werden wir im weiteren Verlauf dieses Beitrages bei allgemeinen Formulierungen nur die L-R-Darstellung von Fuzzy-Zahlen aufschreiben. Die entsprechende Schreibweise für Fuzzy-Intervalle erhält man daraus, indem man anstelle des eindeutigen Gipfelpunktes jeweils die beiden Endpunkte des Gipfelplateaus berücksichtigt. Der Fuzzy-Nutzenerwartungswert $\tilde{E}(a_i)$ läßt sich mittels der erweiterten Addition darstellen als

$$\tilde{E}(a_i) = \tilde{U}(a_i,s_1)p(s_1) \oplus \cdots \oplus \tilde{U}(a_i,s_n)p(s_n) \tag{10}$$

und unter der Annahme, daß nur Nutzenbewertungen des gleichen L-R-Typs gewählt werden, berechnen als

$$\tilde{E}(a_i) = \left(\sum_{j=1}^{n} u_{ij}\cdot p(s_j);\ \sum_{j=1}^{n} \underline{v}_{ij}\cdot p(s_j);\ \sum_{j=1}^{n} \bar{v}_{ij}\cdot p(s_j)\right)_{LR} . \tag{11}$$

4. Ordnungsrelationen zwischen unscharfen Mengen

Die Frage, welche Alternative im Lichte der Gewinnerwartungswerte $\tilde{E}(a_i)$ die optimale ist, läßt sich bei unscharfer Nutzenbewertung nicht so leicht beantworten wie im klassischen Modell , da hier keine eindimensionale Bewertung vorliegt.

Von ZADEH stammt die nachfolgende Definition einer Ordnungsrelation zwischen unscharfen Mengen, die u.a. auch von ENTA /2/ benutzt wird:

Definition 1: Eine Menge $\tilde{A} \in \tilde{\mathcal{P}}(Y)$ wird einer Menge $\tilde{B} \in \tilde{\mathcal{P}}(Y)$ auf dem Niveau $\rho \in [0,1]$ vorgezogen, und man schreibt $\tilde{A} \succ_\rho \tilde{B}$, wenn gilt

i. $\inf A_\alpha \geq \sup B_\alpha$ für alle $\alpha \in [\rho,1]$ und (12)

ii. $\inf A_\alpha > \sup B_\alpha$ für wenigstens ein $\alpha \in [\rho,1]$

Dabei sind $A_\alpha = \{y \in Y | f_A(y) \geq \alpha\}$ und $B_\alpha = \{y \in Y | f_B(y) \geq \alpha\}$ die α-Niveau-Mengen von $\tilde{A}$ bzw. $\tilde{B}$.

Offensichtlich wird gemäß der Definition 1 die Menge $\tilde{A}$ der Menge $\tilde{B}$ um so "sicherer" vorgezogen, je kleiner ρ ist.

Da die Definition 1 Ausdruck einer recht pessimistischen Grundhaltung ist, wird hier eine weitere Ordnungsrelation vorgeschlagen, die beschränkt bleibt auf den Vergleich von Fuzzy-Zahlen und Fuzzy-Intervallen und in der Extremform $\delta = 0$ auf RAMÍK; RIMANÉK /3/ zurückgeht.

Definition 2: Ein Fuzzy-Intervall $\tilde{A}$ wird einem Fuzzy-Intervall $\tilde{B}$ auf dem Niveau $\delta \in [0,1]$ vorgezogen, geschrieben $\tilde{A} \succ_\delta \tilde{B}$, wenn gilt

i. $\sup A_\alpha \geq \sup B_\alpha$ und $\inf A_\alpha \geq \inf B_\alpha$ für alle $\alpha \in [\delta,1]$ (13)

ii. eine der vorstehenden Ungleichungen für wenigstens ein $\alpha \in [\delta,1]$ im strengen Sinne erfüllt ist.

Es ist leicht einzusehen, daß für Fuzzy-Intervalle des gleichen L-R-Typs die Bedingung (13) der Definition 2 äquivalent ist zu dem Ungleichungssystem

$$\begin{aligned} \inf A_\delta &\geq \inf B_\delta \\ \inf A_1 &\geq \inf B_1 \\ \sup A_1 &\geq \sup B_1 \\ \sup A_\delta &\geq \sup B_\delta \;. \end{aligned} \tag{14}$$

Für Fuzzy-Zahlen stimmen die beiden mittleren Ungleichungen von (14) überein, da definitionsgemäß die Niveaumengen A_1 und B_1 nur einelementig sind.

Ob der Entscheidungsträger ein Sicherheitsniveau δ für ausreichend hält, hängt von seiner subjektiven Risikoeinstellung ab. Ein praktikabler Weg ist die Vorgabe eines Mindestniveaus δ_{Min} und die Wahl der Ordnungsrelation

$$a_i \succ a_k \Longleftrightarrow \tilde{E}(a_i) \succ_\delta \tilde{E}(a_k) \quad \text{mit} \quad \delta \leq \delta_{Min} \;. \tag{15}$$

In der Literatur findet sich noch eine Vielzahl weiterer Methoden, um eine Rangfolge unscharfer Mengen aufzustellen, vgl. z.B. den Übersichtsartikel von ROMMELFANGER /5/. In dieser Arbeit soll nur das Niveau-Ebenen-Verfahren berücksichtigt werden, das bei einem Vergleich mit empirisch ermittelten Rangordnungen unscharfer Menge mit am besten abschnitt und auf beliebige unscharfe Mengen über $\mathbb{R}$ anwendbar ist.

<u>Definition 3:</u> Eine unscharfe Menge $\tilde{A} \in \tilde{\mathcal{P}}(Y)$ wird einer unscharfen Menge $\tilde{B} \in \tilde{\mathcal{P}}(Y)$ vorgezogen, wenn für die mittels der Niveau-Ebenen-Verfahren berechnete Mittelwerte gilt

$$\hat{E}(\tilde{A}) > \hat{E}(\tilde{B}) \quad . \tag{16}$$

Ob ein Abstand $\hat{E}(a_1) - \hat{E}(a_2)$ ausreicht, um daraus auf $a_1 \succ a_2$ zu schließen, hängt wiederum von den subjektiven Vorstellungen des Entscheidungsträgers ab. Zur Vereinfachung der Entscheidung könnte er eine kritische Größe $\gamma > 0$ vorgeben, und nach der Regel

$$a_i \succ a_k \iff \frac{\hat{E}(a_i) - \hat{E}(a_k)}{\mathrm{Max}(\hat{E}(a_1), \ldots, \hat{E}(a_m))} > \gamma \tag{17}$$

verfahren. Alternativen, die nach der Regel (17) nicht unterschieden werden können, sind als äquivalent anzusehen. Um beim Vorliegen mehrerer optimaler Alternativen die zu realisierende auszuwählen, kann der Entscheidungsträger dann zusätzliche Kriterien heranziehen.

Die mittels der a priori-Wahrscheinlichkeitsverteilung ermittelte optimale Alternative wollen wir auch im Falle unscharfer Nutzenbewertung allgemein mit a^* symbolisieren. Der zugehörige Erwartungsnutzen werde bei Vorliegen des L-R-Typs mit $\tilde{E}(a^*) = (u^*, \underline{v}^*, \bar{v}^*)_{LR}$ abgekürzt.

5. Der Informationswert

Liegt eine Informationsmenge $X = \{x_r\}$ vor und sind die bedingten Wahrscheinlichkeiten $p(x_r|s_j)$ bekannt, so kann analog dem Vorgehen im klassischen Entscheidungsmodell die a posteriori-Wahrscheinlichkeitsverteilung berechnet werden.

Bei Beobachtung der Information x_r ergibt sich dann der unscharfe Erwartungswert

$$\tilde{E}(a_i|x_r) = \tilde{U}(a_i, s_1)\ p(s_j|x_r) \oplus \cdots \oplus \tilde{U}(a_i, s_n)\ p(s_n|x_r) \quad . \tag{18}$$

Sind alle Nutzen $\tilde{U}(a_i, s_j)$ Fuzzy-Zahlen des gleichen L-R-Typs, so vereinfacht sich die Berechnung von $\tilde{E}(a_i|x_r)$ analog zu (11) zu

$$\begin{aligned} \tilde{E}(a_i|x_r) &= \left(\sum_{j=1}^{n} u_{ij}\ p(s_j|x_r);\ \sum_{j=1}^{n} \underline{v}_{ij}\ p(s_j|x_r);\ \sum_{j=1}^{n} \bar{v}_{ij}\ p(s_j|x_r)\right)_{LR} \\ &= (u_i(x_r); \underline{v}_i(x_r); \bar{v}_i(x_r))_{LR} \quad . \end{aligned} \tag{19}$$

Zur Vereinheitlichung der Schreibweise wollen wir auch hier die optimale Alternative in Abhängigkeit der beobachteten Information x_r mit $a^*(x_r)$ symbolisieren. Ist der zugehörige Nutzenerwartungswert $\tilde{E}(a^*(x_r))$ vom L-R-Typ, so wird er geschrieben in der Form

$$\tilde{E}(a^*(x_r)) = (u^*(x_r); \underline{v}^*(x_r); \bar{v}^*(x_r))_{LR}.$$

Der ex ante-Erwartungsnutzen mit Information ist dann (vor Abzug der Informationskosten) gleich

$$\tilde{E}(X) = \tilde{E}(a^*(x_1)\ p(x_1) \oplus \cdots \oplus \tilde{E}(a^*(x_R)\ p(x_R), \qquad (20)$$

und hat bei Vorliegen des L-R-Typs die Form

$$\begin{aligned}\tilde{E}(X) &= (u^*(X), \underline{v}^*(X), \bar{v}^*(X))_{LR} \\ &= \left(\sum_{r=1}^{R} u^*(x_r)\ p(x_r), \sum_{r=1}^{R} \underline{v}^*(x_r)\ p(x_r), \sum_{r=1}^{R} \bar{v}^*(x_r)\ p(x_r)\right)_{LR} \qquad (21)\end{aligned}$$

Der Wert der Information X wird nun sichtbar in dem Unterschied zwischen den unscharfen Erwartungsnutzen $\tilde{E}(X)$ und $\tilde{E}(a^*)$.
Aus dem Satz der klassischen Entscheidungstheorie, daß (ohne Berücksichtigung der Informationskosten) der Wert der Information nicht negativ sein kann, folgt, daß für die Gipfelpunkte dieser Erwartungsnutzen gilt

$$u^*(X) > u^* \qquad (22)$$

Daraus folgt unmittelbar, daß weder nach Definition 1 noch nach Definition 2 folgen kann

$$\tilde{E}(a^*) \succ \tilde{E}(X) ,$$

d.h. die Information X kann nicht zu einem schlechteren Erwartungsnutzen führen.

Berücksichtigen wir die Definition 1, so folgt, daß

$$\tilde{E}(X) \succ_{\rho^*} \tilde{E}(a^*) \text{ auf dem Niveau } \rho^* = \mathrm{hgt}(\tilde{E}(X) \cap \tilde{E}(a^*)).$$

Diesem Ansatz äquivalent ist die Definition des *ex ante-Wertes der Information* X (vor Abzug der Informationskosten) als

$$\tilde{W}(X) = \tilde{E}(X) \ominus \tilde{E}(a^*). \qquad (23)$$

Gilt für die Referenzfunktion $L(x) = R(x)$ für alle $x \in \mathbb{R}$, so läßt sich $\tilde{W}(X)$ schreiben als

$$\tilde{W}(X) = (u^*(X)-u^*; \underline{v}^*(X)+\bar{v}^*; \bar{v}^*(X)+\underline{v}^*)_{LR} \quad .$$

Nach dem Possibility-Ansatz von TANAKA;ASAI /6/ ist $\tilde{W}(X)$ "fast immer positiv" auf dem Niveau $h^* = 1-\rho^*$. In der Tatsache, daß bei dieser Bewertung von $\tilde{W}(X)$ die rechte Abweichung vom Gipfelpunkt völlig unberücksichtigt bleibt, wird wiederum die pessimistische Haltung des Ansatzes deutlich.

Ist der Informationswert $\tilde{W}(X)$ nur auf einem Sicherheitsniveau $h^* < 1$ bzw. $\rho^* > 0$ "fast immer positiv", so kann die Frage nach der Höhe akzeptierbarer Informationskosten ebenfalls nur auf einem Sicherheitsniveau h, das dann geringer als dieses h^* sein muß, beantwortet werden. Für ein vorgegebenes Sicherheitsniveau h, lassen sich die *maximal akzeptierbaren Kosten* $K_1(X|h)$ berechnen aus

$$0 = u^*(X)-u^*-K_1(X|h) + (\underline{v}^*(X)+\bar{v}^*)L^{-1}(1-h). \tag{24}$$

als

$$K_1(X|h) = u^*(X)-u^* - (\underline{v}^*(X)+\bar{v}^*)L^{-1}(1-h) . \tag{25}$$

Gemäß der Definition 2 sind Informationskosten $K_2(\delta)$ in Abhängigkeit des Sicherheitsniveaus δ akzeptabel, wenn sie dem Ungleichungssystem

$$\begin{aligned} u^*(X) - K_2(X|\delta)-\underline{v}^*(X)\cdot L^{-1}(\delta) &\geq u^* - \underline{v}^*\cdot L^{-1}(\delta) \\ u^*(X) - K_2(X|\delta) &\geq u^* \\ u^*(X) - K_2(X|\delta)+\bar{v}^*(X)\cdot R^{-1}(\delta) &\geq u^*+ \bar{v}^*R^{-1}(\delta) \end{aligned} \tag{26}$$

genügen.

Die maximale Höhe der akzeptierbaren Informationskosten ist daher

$$\begin{aligned} K_2(X|\delta) = \text{Min}[&u^*(X)-u^*;u^*(X)-u^* - L^{-1}(\delta)\cdot(\underline{v}^*(X)-\bar{v}^*) ; \\ &u^*(X)-u^*+R^{-1}(\delta)\cdot(\bar{v}^*(X)-\bar{v}^*)] . \end{aligned} \tag{27}$$

Bei Verwendung der Definition 3 liegt es nahe, den Wert der Information X zu definieren als

$$K_3(X) = \hat{E}(X) - \hat{E}(a^*). \tag{28}$$

6. Information und Nutzenbewertung

Während in der klassischen Entscheidungstheorie Informationen nur dazu dienen, das Wissen über die Wahrscheinlichkeitsverteilung der Umweltzustände zu verbessern, können bei unscharfen Alternativbewertungen zusätzliche Informationen dazu führen, daß die zuststandsspezifischen Bewertungen "genauer" gegeben werden können. Wir betrachten hier den Fall, daß die linke Ausdehnung $\underline{v}_{ij}$ und die rechte Ausdehnung $\bar{v}_{ij}$ der Fuzzy-Zahl $\tilde{U}(a_i,s_j)$ um die Hälfte kleiner werden.

Informationen dieser Art haben zur Folge, daß die Auswahl der optimalen Alternative im allgemeinen auf einem besseren Sicherheitsniveau erfolgt. (Eine Verschlechterung des Sicherheitsniveaus ist auf jeden Fall ausgeschlossen!)

Auch bei dieser Art von Information könnte man ihren Wert festlegen durch den Preis, den man maximal für diese Information zahlen würde.

Bei Anwendung der Definition 1 könnte man den *Wert der Information* I symbolisiert mit $K_1(I)$, definieren als die größte nicht-negative Zahl, für die gilt

$$\tilde{E}(a^*|I) \ominus K_1(I) \succ_{\rho_i} \tilde{E}(a_i|I) \text{ mit } \rho_i \leq \rho^* \quad \text{für alle } i=1,\ldots,n\;, \quad (29)$$

wobei ρ^* das Niveau ist, auf dem a^* vor Kenntnis der Information I gesichert war.

Bei Berücksichtigung der Definition 2 wollen wir analog vorgehen und *den Wert der Information* I, hier abgekürzt mit $K_2(I)$, definieren als die größte nicht-negative Zahl, für die gilt

$$\tilde{E}(a^*|I) \ominus K_2(I) \succ_{\delta_i} \tilde{E}(a_i|I) \quad \text{mit} \quad \delta_i \leq \delta^* \quad \text{für alle } i=1,\ldots,n\;, \quad (30)$$

wobei δ^* das Niveau ist, auf dem a^* vor Kenntnis der Information I gesichert ist.

Da das Niveau-Ebenen-Verfahren einen Mittelwert liefert, hat eine Information des Typs I keinen Einfluß auf die Werte $\hat{E}(a_i)$ und damit auch nicht auf die hiermit induzierte Rangfolge. Sie könnte aber eine Rolle spielen für den Fall, daß das Niveau-Ebenen-Verfahren nicht zu einer eindeutigen optimalen Alternative führt.

Werden sowohl Informationen des Typs X als auch des Typs I zur Entscheidungsfindung herangezogen, so kann, wie Beispiele zeigen, der Wert der Gesamtinformation höher sein als die Summe der einzelnen Informationswerte.

7. Literatur

/1/ DUBOIS D.; PRADE H.: FUZZY SETS AND SYSTEMS: THEORY AND APPLICATIONS. NEW YORK LONDON TORONTO (1980)

/2/ ENTA Y.: FUZZY DECISION THEORY. YAGER R.R. (ED): FUZZY SETS AND POSSIBILITY THEORY. NEW YORK OXFORD TORONTO (1982)

/3/ RAMÍK J.; ŘIMANÉK J.: INEQAULITY RELATION BETWEEN FUZZY NUMBERS AND ITS USE IN FUZZY OPTIMIZATION. FUZZY SETS AND SYSTEMS 16, 123-138 (1985)

/4/ ROMMELFANGER H.: FUZZY-ENTSCHEIDUNGSMODELLE. DISKUSSIONSPAPIER DES INSTITUTS FÜR STATISTIK UND MATHEMATIK DER UNIVERSITÄT FRANKFURT AM MAIN (1986)

/5/ ROMMELFANGER H.: RANGORDNUNGSVERFAHREN FÜR UNSCHARFE MENGEN. OR-SPEKTRUM 8, 219-228 (1986)

/6/ TANAKA H.; ASAI K.: FUZZY LINEAR PROGRAMMING PROBLEMS WITH FUZZY NUMBERS. FUZZY SETS AND SYSTEMS 13, 1-10 (1984)

/7/ WATSON S.R.; WEISS J.J.; DONELL M.L.: FUZZY DECISION ANALYSIS. IEEE TRANS SYST MAN CYBERN 9, 1-9 (1979)

STRUKTURELLE ANALYSEN DER INTERNEN ORGANISATION

Bernd Schauenberg, Hannover

Empirische Beiträge zur betriebswirtschaftlichen Organisationstheorie haben in den letzten zwei Jahrzehnten u.a. zu folgendem bemerkenswerten Ergebnis geführt: Organisationen scheinen unter "ähnlichen" Randbedingungen und bei "vergleichbarer" Effizienz sehr unterschiedliche Organisationsstrukturen aufzuweisen. Dieser Befund kann kaum auf Mängel der Untersuchungsmethoden zurückgeführt werden. Er dürfte auch nicht durch einen Rückgriff auf subjektive Merkmale des Entscheidungsverhaltens von Organisationsgestaltern (wie z.B. Erwartungen, Risikoneigungen usw.) erklärbar sein. Zudem muß er im Lichte ansonsten bewährter mikroökonomischer Hypothesen überraschen. In dem Beitrag soll der Versuch unternommen werden, dieser Herausforderung dadurch zu begegnen, daß die beobachtbare Vielfalt von Organisationsstrukturen als rationale Reaktion der Organisationsgestalter auf strukturelle Merkmale der internen Organisation rekonstruiert wird. Methodischer Ausgangspunkt sollen spieltheoretische Ansätze (insbes. Koordinationsspiele, spezielle Verhandlungsspiele und Gefangenendilemmata) sein, die die Modellierung von Interaktionsbedingungen mit mehrdeutigen Lösungen erlauben. Das Ziel des Beitrags ist es, die beschriebene Problemstellung zu präzisieren, die entsprechenden sog. institutionalistischen Ansätze zu erläutern sowie erste Ergebnisse vorzustellen und einzuordnen. Die zentrale (und zu belegende) Behauptung ist, daß die angesprochene Vielfalt von Organisationsstrukturen durchaus als Folge von rationalen Entscheidungen der Organisationsgestalter verstanden werden kann.

Literatur:

KLIEMT, H. und SCHAUENBERG, B. (1984): Coalitions and Hierarchies. Some Observations on the Fundamentals of Human Cooperation. In: HOLLER, M. (Hrsg.): Coalitions and Collective Action. Würzburg.

Zur Analyse von Organisationskosten

Wolfgang Schüler, Bielefeld

Kosten werden üblicherweise verstanden als periodenbezogener, bewerteter Verzehr von Produktionsfaktoren zur Erstellung von Gütern und Leistungen. Ihre theoretische Bestimmung auf der Basis von Produktionsfunktionen unterstellt zudem die Realisation effizienter Faktoreinsatzmengenkombinationen und verleiht ihnen damit den Charakter objektiver Unabänderlichkeit.

Werden jedoch organisatorische Sachverhalte in die Analyse einbezogen, so sind subjektive Aspekte, insbesondere solche asymmetrischer Information und divergierender Interessen der Beteiligten, auch in der theoretischen Konzeption zu berücksichtigen.

Die hier gewählte Konzeption versteht das Kernproblem der Organisation auf operativer Ebene, d.h. nach vollzogener Unternehmensgliederung in (Teil-)Verantwortungsbereiche, allgemein als das der Kapazitätsregulierung. Konkretisiert wird es hier so, daß zwischen der Zentrale und dem einzelnen Verantwortungsbereich auf der Basis eines nach bestimmten, den Informationsaustausch betreffenden Regeln ablaufenden Dialogs festzulegen ist, welche Ressourcen diesem Bereich zugewiesen werden und welche Leistungen er damit im Gegenzug zu erbringen sich verpflichtet.

Eine solche Situation läßt sich als Agency-Problem modellieren, in dem der "agent" eine über das zur Erstellung des Leistungsziels unbedingt notwendige Maß hinausgehende Mittelzuweisung erreichen und der "principal" sie gerade verhindern möchte. Mit diesem Modell können nun bestimmte Organisationskosten am konkreten Beispiel der Lebensversicherungsbranche untersucht werden. Die theoretische Erklärung führt zu Aussagen über die Gestalt von Kostenfunktionen und die Interpretation ihrer Parameter. Dem modelltheoretischen Ergebnis wird eine empirische Analyse gegenübergestellt, in der Organisationskosten statistisch geschätzt wurden. Es zeigt sich, daß der empirische Befund in der theoretischen Analyse eine weitgehende Erklärung findet.

DIE KONTROLLE VON ENTSCHEIDUNGSTRÄGERN AUF DER BASIS DES ISTERFOLGES: DAS PROBLEM DER SANKTION BEI UNSICHEREM RÜCKSCHLUSS AUF DIE QUALITÄT DER ENTSCHEIDUNG

Eva Terberger
Frankfurt am Main

Einen der bedeutendsten Problemkomplexe in der Organisationstheorie bilden Fragen im Zusammenhang mit der Delegation von Entscheidungen. Die Delegation von Aufgaben birgt immer die Gefahr in sich, daß Entscheidungen nicht im Sinne der delegierenden Instanz getroffen werden. So kann der beauftragte Entscheidungsträger, den wir der Literatur folgend als Agenten bezeichnen, über mangelnde Fähigkeiten oder unzureichende Informationen verfügen, oder aber persönliche Zielsetzungen können die Aufgabenerfüllung negativ beeinflussen. Mit der Delegation ist demnach immer ein Kontroll- und Steuerungsproblem verknüpft.

Weitverbreitet ist die Praxis, die Qualität einer Aufgabenerfüllung am eingetretenen Erfolg, am sogenannten Isterfolg zu messen. Eine derartige Kontrollmaßnahme bietet sich insofern an, als Informationen über den Isterfolg in der Regel relativ kostengünstig zu beschaffen und intersubjektiv überprüfbar sind. Allerdings läßt die Kontrolle des Isterfolges nur unter sehr restriktiven Bedingungen einen sicheren Rückschluß darauf zu, ob der Agent willens und in der Lage gewesen ist, die ihm gestellte Aufgabe zu erfüllen: Der realisierte Erfolg ist nicht nur von den Handlungen des Entscheidungsträgers abhängig, sondern auch von Umweltbedingungen, die der Agent nicht beeinflussen kann. In diesem Vortrag soll der Frage nachgegangen werden, inwieweit es trotz dieser unvollkommenen Kontrollinformation, die der Isterfolg liefert, sinnvoll sein kann, den realisierten Erfolg als Grundlage für Sanktionsmaßnahmen, wie z.B. die Auflösung der Delegationsbeziehung in der Folgeperiode, heranzuziehen. Diese Frage erscheint besonders deshalb von Bedeutung, weil vollkommenere Kontrollinformationen in vielen Fällen aufgrund von Beobachtbarkeitsproblemen oder prohibitiv hohen Kosten nicht zu beschaffen sind. Es zeigt sich, daß gerade "fehlerhafte" Kontrollinformationen eine positive Anreizwirkung auf das unbeobachtbare Verhalten des Agenten ausüben, so daß die negativen Folgen einer ungerechtfertigten Sanktion überkompensiert werden können.

UNTERSTÜTZUNG VON GRUPPENENTSCHEIDUNGEN DURCH MINIMALE PRÄFERENZMODIFIKATIONEN

Rudolf Vetschera, Wien

1. Problemstellung

Gruppenentscheidungen sind eine verbreitete, aber auch aufwendige Form der Entscheidungsfindung. Durch das Zusammentreffen dieser beiden Faktoren hat sich das Konzept entscheidungsunterstützender Systeme für Gruppen, sogenannter "Group Decision Support Systems" (GDSS) /3/, /5/, /6/, /8/, /10/, /11/, /13/ zu einem bedeutenden Forschungsgebiet entwickelt. GDSS im hier betrachteten Sinn dienen der Unterstützung kooperativer Entscheidungsprozesse im Gegensatz zu Verhandlungssituationen zwischen antagonistischen Parteien. Diese Arbeit behandelt einen Teilbereich der Entwicklung von GDSS, die Interaktionen zwischen Gruppen- und individuellen Entscheidungsprozessen.

In der GDSS-Literatur wird zwischen dem individuellen Meinungsbildungsprozeß und der Aggregation auf Gruppenebene unterschieden. Das Ergebnis der individuellen Meinungsbildung wird im folgenden vereinfachend als individuelle Präferenzordnung bezeichnet. Diese muß jedoch nicht den Präferenzen des Gruppenmitgliedes in individuellen Wahlhandlungen entsprechen, sondern repräsentiert die Ansicht des Gruppenmitgliedes darüber, wie die Gruppe als Ganzes Wahlhandlungen treffen sollte. In einem Prozeß kooperativer Problemlösung kann sich diese Ansicht durch den Austausch von Informationen über Eigenschaften der betrachteten Handlungsalternativen oder über die Bedeutung von Beurteilungskriterien ändern. Derartige Änderungen wurden auch in Experimenten empirisch festgestellt /4/.

In der GDSS-Literatur wurde bisher als Übergang von der Individual- zur Gruppenebene vor allem der Aggregationsaspekt betrachtet und durch Entwicklung spezifischer Algorithmen (z.B. /2/) unterstützt. Der umge-

Operations Research Proceedings 1987

kehrte Effekt der Änderung individueller Meinungen wird zwar im Rahmen eines Verhandlungsprozesses erwartet /9/, wurde aber bisher nicht in formale Ansätze einbezogen.

2. Ein formaler Ansatz zur Ermittlung von Präferenzmodifikationen

2.1. Entscheidungssituation

Das betrachtete Entscheidungsproblem wird durch eine Menge diskreter Handlungsalternativen beschrieben. Die N Alternativen, die mit n indiziert werden, seien durch M Attribute m charakterisiert. Die Gruppe soll eine ordinale Präferenzordnung über die gegebene Alternativenmenge erstellen, die Meinungen der Gruppenmitglieder seien durch eine multiattributive Nutzenfunktion (1) kardinal darstellbar.

$$U(n) = \sum_{m} w_m u_{n,m} \qquad (1)$$

Ohne Beschränkung der Allgemeinheit gelten ferner die Skalierungen $\Sigma w_m = 1$ und $0 \leq u_{n,m} \leq 1$. Ferner liege ein Vorschlag für eine Gruppen-Präferenzordnung vor. Dieser Vorschlag kann z.B. von einem Gruppenmitglied stammen oder von einem GDSS erstellt worden sein. Ohne Beschränkung der Allgemeinheit sei dieser Vorschlag so beschaffen, daß die Alternativen in der Reihenfolge ihrer Indices (1,...,n,...,N) gereiht werden sollen.

Die Nutzenfunktion (1) impliziert eine (vom Vorschlag abweichende) Präferenzordnung. Rückwirkungen vom Gruppenvorschlag auf die individuelle Bewertung führen dazu, daß die individuelle Ordnung an den Vorschlag angenähert wird.

Eine derartige Annäherung kann so weit gehen, daß die individuelle Ordnung völlig dem Vorschlag entspricht. Als schwächere Form der Annäherung definieren wir eine k-Anpassung als eine Modifikation der Nutzenfunktion (1), durch die die individuelle Reihung über die ersten k Alternativen an die Ziel-Präferenzordnung angepaßt wird. Eine k-Anpassung kann durch eine Modifikation der Zielgewichte, der partiellen Nutzenwerte oder beider Komponenten erfolgen.

2.2. Modifikation der Zielgewichte

Die k-Anpassung mit geringster Änderung der Zielgewichte kann durch Lösung des folgenden linearen Programms ermittelt werden:

$$\text{minimiere } \sum_m \delta_m^+ + \sum_m \delta_m^- \qquad (2)$$

u.d.B.

$$\sum_m (w_m+\delta_m^+-\delta_m^-)u_{1,m} > \sum_m (w_m+\delta_m^+-\delta_m^-)u_{2,m} \qquad (Ia)$$

$$\cdots\cdots\cdots\cdots\cdots$$

$$\sum_m (w_m+\delta_m^+-\delta_m^-)u_{k-1,m} > \sum_m (w_m+\delta_m^+-\delta_m^-)u_{k,m}$$

$$\sum_m (w_m+\delta_m^+-\delta_m^-)u_{k,m} > \sum_m (w_m+\delta_m^+-\delta_m^-)u_{k+1,m} \qquad (Ib)$$

$$\cdots\cdots\cdots\cdots\cdots$$

$$\sum_m (w_m+\delta_m^+-\delta_m^-)u_{k,m} > \sum_m (w_m+\delta_m^+-\delta_m^-)u_{N,m}$$

$$\sum_m (\delta_m^+ - \delta_m^-) = 0 \qquad (II)$$

$$w_m - \delta_m^- \geq 0 \text{ für alle } m \qquad (III)$$

Die Abweichungsvariablen δ_m^+ und δ_m^- repräsentieren die in Summe zu minimierenden Änderungen der Zielgewichte. Die vorgegebene Ziel-Präferenzordnung wird durch die Nebenbedingungen (I) erreicht, wobei (Ia) die Reihung der ersten k Alternativen und (Ib) die der anderen Alternativen nach Alternative k bewirken. Nebenbedingung (II) hält die Skalierung der Zielgewichte aufrecht, durch (III) werden negative Gewichte vermieden.

Die Lösung von Modell (2) liefert dem Gruppenmitglied Informationen darüber, wie die im Gruppenvorschlag implizit enthaltenen Zielgewichte von seinen eigenen Gewichten abweichen. Dies kann ein Mitglied dazu veranlassen, die Bedeutung der Kriterien selbst neu zu überdenken und seine individuelle Beurteilung zu ändern. Insbesondere in Situationen, in denen mehrere Ziele als Substitute für ein nicht operationales Oberziel dienen (etwa Umsatz und Marktanteil als Substitute für den langfristigen Erfolg), kann die Berücksichtigung von Urteilen der anderen Gruppenmitglieder eine Verbesserung des Einzelurteils bewirken.

2.3. Modifikation der partiellen Nutzenwerte

Die partiellen Nutzenwerte $u_{n,m}$ können durch ein ähnliches lineares Programm modifiziert werden. Geht man davon aus, daß die Nutzenwerte monotone Transformationen objektiv meßbarer Daten darstellen, so sollte die Reihung der Alternativen in den einzelnen Kriterien erhalten bleiben (II). Durch (III) und (IV) bleiben die Nutzenwerte in jedem Kriterium auf den Bereich von 0 bis 1 skaliert. Mit den Abweichungsvariablen $\Phi_{n,m}^+$ und $\Phi_{n,m}^-$ ergibt sich für die Modifikation der partiellen Nutzenwerte das folgende lineare Programm:

$$\text{minimiere } \sum_{n}\sum_{m} \Phi_{n,m}^{+} + \sum_{n}\sum_{m} \Phi_{n,m}^{-} \qquad (3)$$
u.d.B.

$$\sum_m w_m(u_{1,m}+\Phi_{1,m}^{+}-\Phi_{1,m}^{-}) > \sum_m w_m(u_{2,m}+\Phi_{2,m}^{+}-\Phi_{2,m}^{-}) \qquad (Ia)$$
$$\cdots\cdots\cdots$$
$$\sum_m w_m(u_{k-1,m}+\Phi_{k-1,m}^{+}-\Phi_{k-1,m}^{-}) > \sum_m w_m(u_{k,m}+\Phi_{k,m}^{+}-\Phi_{k,m}^{-})$$

$$\sum_m w_m(u_{k,m}+\Phi_{k,m}^{+}-\Phi_{k,m}^{-}) > \sum_m w_m(u_{k+1,m}+\Phi_{k+1,m}^{+}-\Phi_{k+1m}^{-}) \qquad (Ib)$$
$$\cdots\cdots\cdots$$
$$\sum_m w_m(u_{k,m}+\Phi_{k,m}^{+}-\Phi_{k,m}^{-}) > \sum_m w_m(u_{N,m}+\Phi_{N,m}^{+}-\Phi_{N,m}^{-})$$

$$u_{i,m}+\Phi_{i,m}^{+}-\Phi_{i,m}^{-} > u_{j,m}+\Phi_{n,m}^{+}-\Phi_{n,m}^{-} \qquad (II)$$
$$\text{für } i,j,m\text{: } u_{i,m} > u_{j,m}$$

$$\Phi_{n,m}^{+} = 0,\quad \Phi_{n,m}^{-} = 0 \qquad \text{für } n,m\text{: } u_{n,m} = 0 \qquad (III)$$

$$\Phi_{n,m}^{+} = 0,\quad \Phi_{n,m}^{-} = 0 \qquad \text{für } n,m\text{: } u_{n,m} = 1 \qquad (IV)$$

Durch die Modifikation der partiellen Nutzenwerte übernimmt ein Gruppenmitglied implizit Informationen anderer Mitglieder über die Beurteilung von Alternativen in den einzelnen Kriterien. Dies kann dann von Vorteil sein, wenn der Informationsstand anderer Gruppenmitglieder für besser als der eigene angesehen wird.

Aufgrund der höheren Komplexität von Modell (3) wird in dieser Arbeit, auch in Hinblick auf die Einsatzmöglichkeit in einem interaktiven entscheidungsunterstützenden System, nur Modell (2) weiter betrachtet. Auch die simultane Veränderung von Zielgewichten und partiellen Nutzenwerten, die zu einem nichtlinearen Modell führt, wird hier nicht weiter behandelt.

3. Ermittlung alternativer Modifikationsvorschläge

Es widerspricht dem Konzept interaktiver Entscheidungsunterstützung, dem Benutzer nur eine einzige "optimale" Lösung vorzulegen. Die Lösung von Modell (2) kann dazu führen, daß dem Benutzer die Reduktion der Gewichte von Zielen vorgeschlagen wird, deren Bedeutung für ihn außer Zweifel steht. Möglicherweise existieren aber andere Modifikationen mit geringfügig größerer Abweichungssumme, die dem Benutzer plausibler scheinen.

Zur Ermittlung alternativer Modifikationsvorschläge wird eine Weiterentwicklung der HSJ-Methode /1/ benutzt. Dieses Verfahren generiert alternative Lösungsvarianten mit möglichst unterschiedlicher Struktur, deren Zielfunktionswerte in der Nähe des Optimums liegen. Dazu wird die eigentliche Zielfunktion als Nebenbedingung mit dem um einen Relaxationsfaktor α modifizierten Optimum als oberer Schranke formuliert. Strukturelle Verschiedenheit wird durch Minimierung der Summe der Basisvariablen vorhergehender Lösungen erreicht. Für die beschriebene Anwendung wurde dieser Ansatz insofern modifiziert, als die Variablen mit der Anzahl der bisherigen Lösungen, in denen sie Basisvariable waren, gewichtet werden. Für Modell (2) wird in einem modifizierten HSJ-Schritt daher das folgende lineare Programm gelöst:

$$\text{minimiere } \sum_m n_m^+\delta_m^+ + \sum_m n_m^-\delta_m^- \qquad (4)$$

u.d.B.

$$\sum_m \delta_m^+ + \Sigma\delta_m^- \le \alpha F_{min} \qquad (I)$$

$$\sum_m (w_m+\delta_m^+-\delta_m^-)u_{1,m} > \sum_m (w_m+\delta_m^+-\delta_m^-)u_{2,m} \qquad (IIa)$$

$$\cdots\cdots\cdots\cdots\cdots$$

$$\sum_m (w_m+\delta_m^+-\delta_m^-)u_{k-1,m} > \sum_m (w_m+\delta_m^+-\delta_m^-)u_{k,m}$$

$$\sum_m (w_m+\delta_m^+-\delta_m^-)u_{k,m} > \sum_m (w_m+\delta_m^+-\delta_m^-)u_{k+1,m} \qquad (IIb)$$

$$\cdots\cdots\cdots\cdots\cdots$$

$$\sum_m (w_m+\delta_m^+-\delta_m^-)u_{k,m} > \sum_m (w_m+\delta_m^+-\delta_m^-)u_{N,m}$$

$$\sum_m (\delta_m^+ - \delta_m^-) = 0 \qquad (III)$$

$$w_m - \delta_m^- \ge 0 \qquad \text{für alle } m \qquad (IV)$$

wobei n_m^+ bzw. n_m^- die Anzahl der bisherigen Lösungen darstellen, in denen δ_m^+ bzw. δ_m^- Basisvariable waren. F_{min} ist der durch Lösung von (2) ermittelte Minimalwert, α (>1) ein vom Benutzer festgelegter Relaxationsfaktor.

4. Lösbarkeit

Die Existenz einer Lösung von Modell (2) ist für beliebige Ziel-Präferenzordnungen und beliebige partielle Nutzenfunktionen nicht sichergestellt. Aussagen über die Lösbarkeit können anhand von Simulationsexperimenten getroffen werden.

In praktischen Anwendungen kann davon ausgegangen werden, daß zwischen der von (1) implizierten Präferenzordnung und der Ziel-Präferenzordnung strukturelle Ähnlichkeiten bestehen. Für die Simulationsexperimente wurde unterstellt, daß sich die Ziel-Präferenzordnung ebenfalls aus einer multiattributiven Nutzenfunktion ergibt. Beide Funktionen sollen unterschiedliche Zielgewichte aufweisen und die partiellen Nutzenwerte durch eine monotone Transformation ineinander überführbar sein. Diese Annahmen entsprechen z.B. einer Situation, in der die Gruppenmitglieder zwar bezüglich objektiv meßbarer Attributwerte übereinstimmen, den Attributwerten aber unterschiedliche Nutzengrößen und den Attributen unterschiedliche Gewichte zuordnen.

Wie sich aus Modell (2) zeigen läßt, folgt aus der Lösbarkeit des Modells für ein beliebiges k auch die Lösbarkeit für alle $k' \leq k$. Für gegebene Ziel-Präferenzordnung und partielle Nutzenwerte existiert daher ein k_{max} ($0 \leq k_{max} \leq N$), so daß Modell (2) für alle $k \leq k_{max}$ lösbar ist. Abbildung 1 zeigt die Verteilung der in 500 Simulationsexperimenten mit 10 Alternativen und 4 Kriterien ermittelten Werte von k_{max}.

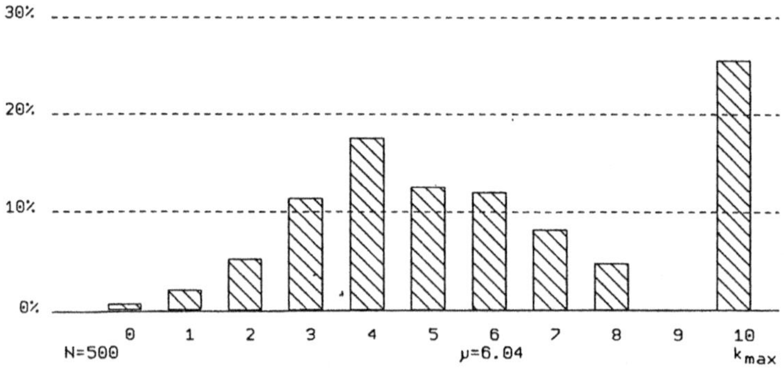

Abbildung 1: Verteilung von k_{max}

In etwa 25% der untersuchten Fälle war eine vollständige Anpassung an die Ziel-Präferenzordnung durch Modifikation der Zielgewichte möglich. Im Durchschnitt konnte die Präferenzordnung bezüglich der ersten 6 der 10 Alternativen angepaßt werden.

5. Anwendungen

Das in dieser Arbeit entwickelte Konzept der minimalen Präferenzmodifikationen kann in mehrfacher Weise zur Unterstützung von Gruppenentscheidungen herangezogen werden.

Die Lösungen der Modelle geben den Gruppenmitgliedern in kompakter Form Informationen über die in den Kompromißvorschlag eingehenden Ansichten der anderen Mitglieder. Sie ermöglichen es den Mitgliedern damit, festzustellen, in welchen Bereichen ihre Ansichten besonders weit von der aggregierten Gruppenmeinung entfernt sind. Diese Bereiche können dann einer erneuten Beurteilung unterzogen werden.

Werden die ermittelten Änderungsvorschläge (zumindest teilweise) realisiert, so kann der Ansatz unmittelbar zur Unterstützung der Kompromißbildung dienen.

Der Ansatz kann aber auch in nicht kooperativen Verhandlungssituationen zur Unterstützung einzelner Parteien dienen. Dazu werden in Modell (2) Annahmen über die Zielgewichte und partiellen Nutzenwerte anderer Verhandlungsparteien eingesetzt. Die Lösung des Modells gibt dann Hinweise, welche Kriterien in der Argumentation besonders betont werden sollen, um eine Annäherung an den eigenen Standpunkt zu erreichen.

6. Zusammenfassung und weitere Entwicklungen

Aushandlungsprozesse in Entscheidungsgremien beinhalten vielfach eine Modifikation der Vorstellungen einzelner Gruppenmitglieder. Ansätze zur Unterstützung von Gruppenentscheidungen können sich daher nicht darauf beschränken, aus den individuellen Präferenzvorstellungen ein Gruppenergebnis zu synthetisieren, sondern müssen auch diese Rückwirkungen berücksichtigen. In der vorliegenden Arbeit wurde ein formaler Ansatz zur Ermittlung minimaler Präferenzmodifikationen für die Anpassung an eine gegebene Ziel-Präferenzordnung vorgestellt. Im Sinne einer flexiblen Entscheidungsunterstützung können dabei alternative Modifikationsvorschläge ermittelt werden.

Die Konzeption des Ansatzes wurde aufbauend auf einer multiattributiven Nutzenfunktion entwickelt. Der Ansatz kann in ähnlicher Form jedoch auch auf andere individuelle Entscheidungsverfahren angewandt werden. So ist z.B. die direkte Übertragung von (2) auf die Prioritätszahlen des Analytic Hierarchy Process von Saaty /12/ und über den Ansatz von Harker /7/ auf die diesen zugrunde liegenden paarweisen Vergleiche möglich.

Die bisher durchgeführten Simulationsexperimente haben gezeigt, daß das formulierte Modell mit vertretbarem Aufwand lösbar ist. Um die praktische Anwendbarkeit zu erproben, wird das Verfahren gegenwärtig in Form eines individuellen Softwaresystems für einzelne Gruppenmitglieder auf einem Personal Computer implementiert. Für spätere Experimente ist vorgesehen, den Ansatz in ein integriertes Group Decision Support System auf einem PC-Netzwerk einzubeziehen.

7. Literatur

/1/ BRILL, E., CHANG, S., HOPKINS, L.: Modeling to Generate Alternatives: The HSJ Approach and an Illustration Using a Problem in Land Use Planning. Management Science 28, 221-235 (1982).

/2/ BUI, T.: N.A.I.: A Consensus Seeking Algorithm for Group Decision Support Systems. Proceedings, 1985 International Conference on Cybernetics and Society. Tucson, Arizona (1985).

/3/ BUI, T., JARKE, M.: Communications Design for Co-oP: A Group Decision Support System. ACM Transactions on Office Information Systems 4, 81-103 (1986).

/4/ CASTORE, C., MURNIGHAN, J.: Determinants of Support for Group Decisions. Organizational Behavior and Human Performance 22, 75-92 (1978).

/5/ DAGNINO, A., HIPEL, K., FRASER, N.: A Decision Support System for Mediation. Working Paper, University of Waterloo (1987).

/6/ DESANCTIS, G., GALLUPE, R.: A Foundation for the Study of Group Decision Support Systems. Management Science 33, 589-609 (1987).

/7/ HARKER, P.: Incomplete Pairwise Comparisons in the Analytic Hierarchy Process. Working Paper 85-03-01, Department of Decision Sciences, The Wharton School, University of Pennsylvania (1986).

/8/ JARKE, M.: Knowledge Sharing and Negotiation Support in Multiperson Decision Support Systems. Decision Support Systems 2, 93-102 (1986).

/9/ JARKE, M., JELASSI, T., SHAKUN, M.: MEDIATOR: Towards a Negotiation Support System. In: Keeney, R. et al. (Hrsg.): Proceedings, International Conference on Multi-Attribute Decision Making via O.R.-Based Expert Systems. Universität Passau (1986).

/10/ LEWANDOWSKI, A., JOHNSON, S., WIERZBICKI, A.: A Prototype Selection Committee Decision Analysis and Support System. SCIDAS: Theoretical Background and Computer Implementation. IIASA Working Paper WP-86-27. Laxenburg (1986).

/11/ NUNAMAKER, J., VOGEL, D.: Negotiations Support Systems - Software and Facilities for Public Sector Issues. Proceedings, 31st Annual Meeting of ISGSR, Budapest (1987).

/12/ SAATY, T.: The Analytic Hierarchy Process. McGraw-Hill, New York (1980).

/13/ STOHR, E.: DSS for Cooperative Decision Making. Working Paper CRIS 19, Center for Research on Information Systems, New York University (1981).

Features of the architecture of decision support systems

Kees M. van Hee Eindhoven

1. Introduction

Decision support systems and expert systems are information systems that assist people by giving advise or solving problems. If this role would be played by human beings we would classify their work as intelligent behaviour. Therefore we may call such systems: intelligent information systems, in contrast with more classical information systems that monitor some process and perform routine functions.

There is a big market for intelligent information systems and therefore it is important to have a good design methodology for these systems. For development approaches for classical information systems seem to be inadequate.

In this paper we study the architecture of intelligent information systems with the purpose to use the architectural characteristics for a designmethodology.

The term "decision support sytem" is used in the area of operations research, while the term "expert system" is coming from the artificial intelligence world. The application area of expert systems seems to be larger than the area of decision support systems [cf. Sowa (1984)].

The following definition of an expert system is due to [Jackson 1986]:

"Computing system capable of representing and reasoning about some knowledge rich domain with a view to solving problems and giving advise".

In the next section we define the concept of a decision support system and in section 3 we consider the model-oriented approach to decision support systems.

In section 4 we consider the process of the construction of a decision as a graph search problem and we indicate how a shell could be constructed for decision situations that can be characterized as "combinatorial planning situations". For a more extended treatment we refer to [Van Hee, Lapinski (1987)].

Operations Research Proceedings 1987

2. Concept of a decision support system

We consider a decision support system (dss) as a subsystem of a control system of some other system, called object or target system. The control system has two tasks: monitoring and steering of the target system. The monitoring task is usually performed partly by a classical information system, also called a database system. The monitoring of the target system is essential for the steering task. The steering task is sometimes completely fulfilled by computers, but in many cases human beings steer the target system. Their control actions are called plans or decisions.

Computer systems that assist decisionmakers by offering facilities for:

i computing effects of plans that are made by the decisionmakers,

ii generating plans within some region of allowable plans,

iii selecting plans from a given finite set of plans and their effects

are called decision support systems.

The first facility may be called 'evaluator' and requires some mathematical model that describes all relevant characteristics of the target system. With this model the effects of plans can be computed. The second facility needs also a model of the target system, however it may be a less precise description, for instance a simplification of the model used to compute the effects. A classical characterisation of an allowable region is the set of optimal plans, where optimality is defined by some criterion. In general this characterisation is a difficult one because the complexity of the computation needed to verify if a given plan is allowable, is the same as generating an optimal one. The last facility performs multicriteria analysis. It is usually studied in combination with linear programming models. The facility is usefull in situations with conflicting evaluation functions, where the vector of evaluation functions does not provide a total ordering on the set of plans. With this facility the decisionmaker may find his utility or weight function over the criteria.

A selection facility does not require any knowledge of the structure of the decision situation or target system, a generating facility requires detailed knowledge of the set of allowable decisions and also knowledge, in the form of a mathematical model, of the target system. An evaluation facility needs the most detailed knowledge of the target system and therefore these subsystems are often developed for a specific decision situation, for instance in the form of a simulationmodel.

Decision support systems that really interact with de decisionmaker, i.e. without a human intermediair, require a high level userinterface. Decision support systems that have a very frequent interaction with decisionmakers, are found on the level of operational management.

3. Model-oriented approach

The model-oriented approach is the approach of operations research. Here the designer or a dss chooses or develops mathematical models that describe relevant aspects of the target system. This is often his starting point. Sometimes the emphasise is on the evaluation of effects of a plan. Examples of this type occur in situations where a new organisation of production resources has to be chosen. In situations where the decision concerns the allocation or assignment of resources the generation of a plan is the main task and there optimization models are often used. It seldom occurs that a target system can adequately be described by only one model. Then we need a set of models, each describing only aspects of the target system. Often these models are interrelated by parameters that are input for one model and output to another. An example of such a situation occurs when for the generation of decisions a simplified model of the target system is used, while for the evaluation a complex simulation model is used. With the simulation model it is practically impossible to optimize, however it computes essential parameters for the simplified optimization model. On the other hand the simulation model can only evaluate plans generated by the optimization model.

In general we may have a network where the models form the nodes and the (directed) arcs are labelled by parameters that are transferred from the source node to the sink node of the arc. Given a value to all parameters in a network and considering them as inputparameters for the models in the network we get for each parameter an output value computed by a model. Hence the network defines a network-function that transforms the set of parameter vectors into itself.

A consistency condition for a model network is that a used parametervector is a fixpoint of the network-function. In general there is very little known about the relationship between input and output parameters of a model-network and therefore very little is known of the network-function. A usual computational procedure is iteration of the network-function until in- and output vectors are close. However in practice there is often no theoretical justification that such a method of successive approximations is converging. Most fixpoint theorems require structural properties of the underlying transformation that are unknown.

In the model-oriented approach the dss is build around the model or the modelnetwork. It means that a datastructure is designed to describe the parameters of the model network and a database system is constructed to manipulate the datastructure. Then each model is translated into a program or module. In case of a network there is a module needed that takes care of the above described successive approximation of a fixpoint. Finally a user-interface is designed which enables the decisionmaker to control the activation of the other modules and with which the database can be questioned. It turns out that decisionmakers usually want to keep more than one parameterset, each one representing a scenario or world. Therefore a database system in a decision support systems needs a facility for version or configuration management.

The model-oriented approach has a disadvantage, namely that the used models are simplifying the decision situation too much. The modelling of the target system is the central goal and not the representation of space of allowable plans. Therefore in many dss's designed according to

the model-oriented approach is the inflexibility for adaptations, in case new constraints arise or the preference pattern over the planning space changes. In many cases these adaptations require fundamental changes of the models and therefore of the algorithms. Such changes cannot be performed by the end-user, the decisionmaker, but they require the involvement of the dss-designer. This fact is probably the cause that there are so little dss's in operation while there are so many decision situations.
In the next section we consider another approach to dss, one that may overcome the problems sketched above.

4. Towards dss-shells

In this section we sketch a model of a dss that can be used to construct a dss-shell. With a dss-shell we mean a software system that accepts a description in a high-level language of a structure of a decision situation. Given this data the shell behaves like a dss for that particular decision situation. This means that it accepts factual data of the decision situation, i.e. data of the actual state of the decision situation, and it interacts with the decisionmaker to construct a plan that satisfies some user-defined requirements.
Such a shell will be very flexible in the sense of section 3 since a decisionmaker or an expert only has to update the high-level description of the decision situation to adapt the dss to changes in the target system. A trade off of the use of such a dss-shell will be a less-efficient system in the sense that there might be model-oriented dss's that give better plans according to criteria suited for the models in such a dss.
A dss-shell may be compared to a database management system and an expert system shell. A database management system gets a description of the datastructure in the form of a scheme and if some database instance is created the query-language offers the user the facility to retrieve data for this particular data structure. If an expert system shell is augmented by a knowledge or rule base then it behaves as a specific expert system.
Now we sketch a model of a dss that is applicable in many decision situations where a plan has to be constructed. The model is described by a tuple
$<D,P,E,T,Q,G,M,SE,SM,U>$
where

D: the set of domains, i.e. an element of D represents the state of the decision situation at some moment;

P: set of plans, i.e. each element is a plan for the decision situation;

E: set of plan-extensions;

T: a plan transformation function, i.e. $T:P \times E \rightarrow P$, so given a plan p and an extension a $T(p,a)$ will be a new plan;

Q: a function that defines the set of allowable plan-extensions of a plan, depending on the domain, i.e. $Q: P \times D \rightarrow 2^E$;

G: a function that determines whether a plan could be accepted or not, i.e. $G: P \times D \rightarrow$ {true, false};

M: a memory structure, i.e. an element m of M contains the set of plans constructed so far and considered relevant. In fact m is a tree, the nodes of which are pairs composed of a plan and a set of extensions. The edges of the tree are labelled by plan-extensions, and if $<p,e>$ is a node and a is an edge starting in this node then the edge ends in a node where the first component equals $T(p,a)$. Further $e \in Q(p,d)$ and $a \in Q(p,d) \backslash e$. Hence the plan-extensions e in a node $<p,e>$ are the free extensions that might be considered in the future. A tree $m \in M$ represents the state of the plan construction process at some stage;

SE: is a selection function that selects an extension from a set of extensions, i.e. $SE: 2^E \rightarrow E$ such that $SE(e) \in e$;

SM: is a selection function that selects a node to extend from a memory state $m \in M$, i.e. $SM: M \rightarrow P \times 2^E$, such that for $m \in M$ $SM(m)$ is a node in the tree m, with a non-empty set of free- extensions;

U: is a memory-update function, i.e. $U: M \times P \times E \rightarrow M$ such that $U(m,p,e)$ is a tree that can be formed out of m by extension of a node $<T(p,e),Q(T(p,e),d)>$ or by transforming the node where p is the first component by deleting e from the set of free extensions, or by deleting leave nodes;

Now we will describe how the shell operates, given these components in some representation. It first asks the user for an initial state for the search process, in fact this will be the root of all memory states during the planning process.

Then the shell will perform the following iteration:

1. select a node from the memory state m according to $SM(m)$, say $<p,e>$,
2. select an extension from $SM(m)$ according to $SE(e)$, say $a(\in e)$,
3. compute $T(p,a)$,
4. if $G(T(p,a),d) =$ true then stop else compute $U(m,p,a)$ and return to step 1.

The functions SE, SM and U may be defined by some formulas in the shell or by the user in an interactive way. The last option means that the shell is asking the user for a selection or update in each step of the process. This may be called hand-control while the first option is full-automatic control of the planning process.

Note that only Q and G depend on domain-data and are defined in fact by the end-user. The other components are domain-state independent and form the so called structural data. It depends on the level of the language in which the components are described and of the expertise of the end-user whether he is able to modify the structural data. In case the language will be a first order logic, it seems that it requires some expert to modify the structural data.

However such an expert can do this very quickly. (We have made some experiments where Prolog was used as description language). Such a shell needs in fact two user-interfaces: one for decisionmakers and one for an expert. Note however that Q and G can be adapted by the decisionmakers.

In [Van Hee, Lapinski (1987)] we showed that it is easy to express all kind of search strategies like depth-first, breadth first, hill climbing etc. by specifying the functions SE, SM and U. There we also considered a jobshop as decision situation as an example.

The representation of plans and extensions is important. If plans are represented by sets of terms (or tuples) and if plan-extension means adding terms (or tuples) then a memory state m may be represented by representing each node only by its difference with its predecessor node. This gives in many cases an interesting reduction of the representation of the state.

We conclude with some remarks.

The approach sketched in this section seems to be suitable for a large class of operational planning problems, specially where a combinatorial model is used in the case of a model-oriented approach.

It will take a lot of experimentations with prototypes before such dss-shells will be able to operate in practice. Two major problems are: the language of the shell to define the structural data and the efficiency of computations if the system is operating under automatic control. The language needs to have great expressive power but should be easy to learn, because a planning expert should use it in stead of a programmer. The computational complexity might be large, however the search graph approach is suitable to apply parallel processing techniques.

5. Literature

Van Hee, K.M., Lapinski, A.
OR and AI approaches to decision support systems.
To appear in Decision Support Systems (1988).

Jackson, P.
Introduction to expert systems.
Addison-Wesley, 1986.

Sowa, J.F.
Conceptual structures: information processing in mind and machine.
Addison-Wesley, 1984.

MESSINA - A Marketing Expert System For The Screening Of Ideas For New-Product-Alternatives

Sigurd Bünte, Lüneburg

In order to achieve a good position in the market place or to maintain a healthy organization, nearly all firms have to develop new products. This involves high risk, because over 50 % of all the products introduced into the market have resulted in flops. Therefore it is necessary to establish a powerful screening instrument that identifies unsuccessful projects in an early stage and thereby reduces costs. Despite this requirement many firms don't possess a powerful screening procedure. On the other side, many results from empirical studies have been published from which it is possible to deduce the knowledge base for such a procedure. As screening is similar to diagnosing, an expert system is appropriate to provide a user-friendly screening procedure.

A comparison of the different techniques for the design and realisation of expert-systems like programming languages of AI, e.g. LISP, programming in logic, e.g. PROLOG, expert-system-shells, e.g. KEE, and procedure programming languages, e.g. PASCAL or COBOL, has shown that PASCAL is the preferred language for our purpose because most firms have installed a personal computer on which a PASCAL compiler may run with only little extra costs. Furthermore, it is possible to choose an inference mechanism that differs from those implemented in PROLOG. Of course, it is easier and less time consuming to develop an expert-system with shells, but often those shells need special hardware, which makes a system more expensive than a PASCAL-based system.

In literature, rules are cited that ease the screening of a product. It is shown how these rules are transformed into production rules which represent a knowledge-representation method. A hierarchy among the rules is established such that the system-performance is tuned.

Finally, a certainty function with a measure of belief and of disbelief has been developed in order to distinguish between those products that possibly become a flop and products that may result in a success.

Decision Support- und Expertensysteme in der Informationsorganisation, oder 'back to the roots of Operations Research'

H.J. Grünwald, Eindhoven

L. Fortuin, Eindhoven

Mit der Verbreitung des Computers nimmt die Informatisierung unserer Welt schnell zu, sie verändert unsere Gesellschaft und auch unsere Organisationen. Decision Support Systemen (DSS) und Expertensystemen (ES) sind Exponenten dieser Entwicklung.

Die Operations Research (OR) Gemeinschaft beansprucht DSS und ES als eigene Teildisziplinen, oder sieht doch zumindest deutliche Zusammenhänge mit OR, was zum Beispiel aus der Zahl der diesen zwei Themen gewidmeten Vorträgen dieser OR Konferenz hervorgeht.

Wir fragen uns, wieweit dieser Anspruch zu Recht erhoben wird. Erweist er sich rechtens, dann wären DSS und ES grossartige "opportunities" für OR. Kann OR dagegen zur Entwicklung und Anwendung von DSS und ES wenig oder gar nichts beitragen, dann sehen sich die viele Kritiker von OR bestätigt, die diesem Fach nur eine kleine ökologische Nische zuweisen in der Evolution von Wissenschaft und Technik innerhalb unserer Gesellschaft.

Die Beantwortung dieser Frage hat also eine strategische Bedeutung für OR. Und umgekehrt könnte richtig eingesetztes, intellektuelles OR Potential von strategischem Wert für Informationsorganisationen sein.

Die Zielgruppen, die Problemstellungen und die Methodiek von DSS und ES entsprechen ursprünglichen Ideen von OR. Es scheint allerdings, dass das heutzutage unter der OR Gemeinschaft herrschende Rationalitätsparadigma das Fach in dem kleinen Teil des "opportunity space" gefangengesetzt hat, der mathematisch zugänglich ist.

Eine Vertiefung des Verständnisses für Rationalität und daraus abgeleitet eine Erweiterung des OR Paradigmas könnte OR freier machen. Und für die sich stürmisch entwickelnde und jeden Intellekt herausfordernde Informationsorganisation wäre zusätzliches, und speziell sogar Or, Talent sehr nützlich. OR nähme dann nicht mehr einen nur marginalen sondern einen zentralen Platz in der Informationsorganisation ein.

Entwurf und Realisierung von PC-gestützten Decision Support-Systemen

Klaus Huckert, Saarbrücken

Decision Support-Systeme sind als Reaktion auf die gescheiterte MIS-Idee entstanden. Zentrale Forderungen an solche Softwareprodukte sind adaptiver Systemaufbau, Benutzerfreundlichkeit, Modellierbarkeit durch den Benutzer, Verwendung von Methodenbanken und eine eigene Datenbasis. Diese Datenbank setzt sich aus internen und externen Datenbeständen einer Unternehmung zusammen. Bei solchen Anforderungen bietet sich der Einsatz von Personal Computern geradezu an. Endbenutzerprodukte wie Spreadsheetprogramme, Graphiksysteme, Kommunikationspakete und Datenbankprogramme erleichtern wesentlich den Aufbau von Prototypen. Aus derartigen Prototypen kann durch ständige Weiterentwicklung das endgültige Programmsystem entstehen. Diese Art der Systementwicklung ermöglicht eine rückgekoppelte Entwurfs- und Realisierungsphase, in die der Endbenutzer seine Fachkenntnisse einbringen kann. Anforderungen an DS-Systeme auf PC-Basis werden in dem vorliegenden Beitrag diskutiert. Dabei werden insbesondere die Problematiken natürlichsprachliche Kommunikation, Copy- und Extract-Management sowie Zugriff auf Hostsysteme erläutert.

Als Anwendung wird ein realisiertes Entscheidungsunterstützungssystem für die Controlling-Abteilung eines deutschen Konzerns mit mehreren Geschäftsstellen beschrieben. Zum einen wird ein Kontrollinstrument vorgestellt, das ein natürlichsprachliches Interface an eine Vertriebsdatenbank enthält. Zum anderen wird ein Planungsinstrumentarium entwickelt, das ausgehend von einer Umsatzplanung entsprechende Finanzpläne, Personalpläne und Kostenpläne generiert. In diese Modelle, die mit kommerziellen Planungssprachen (wie z.B. PERSONAL W, IFPS und MICRO-FCS) erstellt wurden, laufen sowohl eigene Datenbestände, die auf dem Host gehalten werden, als auch externe Daten über die voraussichtliche Marktentwicklung.

Literatur

K. Huckert: Entwurf und Realisierung von PC-gestützten Decision Support-Systemen. Veröffentlichungen der Projektgruppe Praxisorientierte Informatik, Fachhochschule des Saarlandes, Saarbrücken, Heft 4, 1987.

AI Contributions to Decision Support: Where can They Really Help?

Matthias Jarke, Passau

Much has been written about the role of artificial intelligence, especially expert systems, in decision support systems (and vice versa). Recently, the term "expert support system" has been coined to illustrate the relationship between the two. But what is the "expertise" of decision makers and decision analysts?

Often, the answer to this question appears rather naive. AI fanatics attempt to replace carefully developed mathematical approaches by generalized "knowledge representations" that are inevitably less adequate to the problem at hand, and lead to combinatorial search algorithms. It must be recognized that AI is typically good at "weak methods" for search, and at representations that are close to human perception of a problem but less suitable for machine efficiency.

In my opinion, the real expertise of decision analysts, and the main role for AI support, lies in the development and maintenance of decision models where these models themselves can involve highly complex mathematics. The presentation will review a number of expert systems approaches to DSS, with an emphasis on model management systems, and will compare their effectiveness with respect to criteria such as:

- capability for explanation,
- capability for synthesis,
- capability for complexity control, and
- capability for maintenance.

Based on experience with a number of experimental model management systems [1], a normatively based model management environment is sketched. This environment, under development at Passau, centers around a view of models as reusable hierarchies of specifications, where each specification may be associated with corresponding methods and databases. Each such hierarchy should not only be reusable in its original form but adaptable to new applications and requirements, e.g., by renaming other minor changes in components. Interactive problem recognition, (re-) definition, and solution can be visualized as

- recognizing the applicability of existing models, using a conceptual theory of the application domain as a guide [2]
- choosing suitable paths through the specification hierarchy (=model formulations, methods, databases) by a decision-theoretically justified multiple criteria approach,
- evaluating choices against qualitative and quantitative constraints, and backtracking selectively on poor subdecision without having to repeat the modelling process.

Literature:

[1] M. Jarke: Kopplung qualitativer und quantitativer Theorien in der Entscheidungsunterstützung, *Proc. 2. GI-Kongreß Wissensbasierte Systeme*, München 1987.

[2] M. Jarke, T. Rose: Global KBMS Design and Development Strategy, Report Esprit Project 892 (DAIDA), Universität Passau 1987.

Decision support in scheduling

F. J. Radermacher, Passau

The talk deals with ongoing research concerning a DSS for scheduling that aims at an extended degree of support. Details of the system architecture in the present development phase (1986 - 1989) and experiences with already existing components are included as are some general design ideas for advanced DSS in general:

For further information and references cf.:

Radermacher F.J.: Entwicklungsperspektiven rechnergestützter Entscheidungsfindung, erscheint in: Wolff J. (Hrsg.): Proceedings des IBM-Symposiums "Entscheidungsunterstützende Systeme", Oldenbourg Verlag, München 1987.

INTEGRAL LOGISTICS IN CENTRALISED WOODPROCESSING

M.P. Reinders, Wageningen

In the research we deal with integrated logistics in centralised woodprocessing. Goal of the research is to build a decision support system (DSS) with which we are able to design and control woodprocessing centres.
Techniques to improve logistic design, productionplanning and technology as well as inventory control systems are linked to a highly integrated system to increase flexibility and reliability of the overall production system. The DSS will be able to tackle problems occuring on several levels of aggregation.

First there is the operational level. On this level we make use of techniques such as dynamic programming to overcome production efficiency problems.
On this level our goal is to maximize the "relative value" of the logs, as a function of market prices and inventory levels.

On the tactical level in which we are confronted with the masterproductionplan we make use of the multiple goal programming technique. On this level we optimize a set of logs, anticipating on future demand and capacity shortages (production smoothing). Also harvesting and buying of logs is taken into account.

The third level is the level of strategic planning. Managers have to deal with investment decisions and financial analysis of the company to improve its effectiveness. At this level of high aggregration the DSS will provide simulation based facilities to design the manufacturing system optimally.

All three levels are integrated in one DSS because they heavily interact with another.

Entscheidungsunterstützung für die europäische Umweltpolitik Optimierung des Kapitaleinsatzes bei der Reduzierung von Luftschadstoffen

Günter Schmid, Stuttgart

Zur Vorbereitung und Unterstützung nationaler und europäischer Gesetzesinitiativen auf dem Gebiet der Emissionsreduzierung findet zur Zeit das von der EG-Komission initiierte und geförderte Projekt 'Energie und Umwelt' statt. Ziel dieser Studie ist es, Kosten und Effektivität verschiedener Schadstoffminderungstechnologien zu berechnen sowie die zukünftigen Auswirkungen verschiedener Strategien auf die jeweiligen Energieversorgungssysteme bei heutigem Wissensstand zu analysieren. Hierzu wird das modulare Energiemodell EFOM-12c eingesetzt, welches auf dem Ansatz der linearen Optimierung basiert und den optimalen Kapitaleinsatz garantiert. Im Rahmen dieses Modells werden alle energie- und umweltrelevanten Prozesse auf einem adäquaten Aggregationsniveau abgebildet. Gleichzeitig werden sektorale Umweltmodule hinzugefügt, die es erlauben, die Schadstoffe SO2, NOx, Staub, CnHm und CO zu bilanzieren und vorhandene Reduktionstechnologien einzufügen.

Im ersten Schritt werden bei zeitlich konstant angenommenem Nutzenergieverbrauch verschiedene Umweltszenarien untersucht: Zunächst wird die Entwicklung der Schadstoffe ohne Minderungstechnologien (doing-nothing case) und bei Durchsetzung nationaler Gesetze (legal case) erforscht. Danach werden die Schadstoffe sukzessiv soweit wie möglich vermindert und schließlich werden die Auswirkungen europäischer Gesetzesvorlagen analysiert. Gleichzeitig werden sektorale und volkswirtschaftliche Trade-off-Kurven entwickelt, die intersektorale und internationale Vergleiche ermöglichen. In einem weiteren Schritt wird der Einfluß verschiedener Energiebedarfsentwicklungen auf die optimalen Strategien ermittelt. Die in diesen Schritten erarbeiteten und hier vorgestellten Ergebnisse sollen sich an der politischen Umsetzbarkeit orientieren. Sie dienen der Entscheidungshilfe in den Gremien der europäischen Gemeinschaft bei der aktuellen Diskussion um eine umweltverträgliche Energiepolitik.

Ein wissensbasiertes System zur strategischen Planung auf Basis von PROLOG

Reinhart Schmidt, Kiel

Die Anwendung von quantitativ und qualitativ orientierten Verfahren der strategischen Planung kann gefördert werden, wenn ein wissensbasiertes System dem Anwender aufgrund seiner speziellen Probleme Antworten bezüglich des adäquaten Modell- und Methodeneinsatzes und bezüglich der Datenbeschaffung gibt. Deshalb wird zunächst der Rahmen eines allgemeinen Problemlösungssystems in TURBO-PROLOG beschrieben.

Die Wissensbasis in dem Teilberich "Probleme" enthält Problemmerkmale sowie Problemerfordernisse im Hinblick auf Methoden und Daten. In dem Teilbereich "Methoden" sind Angaben über Methodenmerkmale sowie über Methodenerfordernisse im Hinblick auf die Daten gespeichert. Schließlich sind in dem Teilbereich "Daten" Datenmerkmale vorhanden. Das System prüft, ob die für aus einem Problem abgeleiteten Erfordernisse entsprechdende Methoden und Daten vorhanden sind.

Im Rahmen einer strategischen Unternehmensplanung wird das Planungsproblem über Teilprobleme abgebildet, die den Phasen des Planungsprozesses entsprechen. Es wird gezeigt, wie der Systembenutzer aufgrund einer Problemartikulation zunächst zu einer Antwort von einzusetzender Methode und anzuwendenden Daten kommt. Das System ist auch in der Lage, anschließend entsprechende, in anderen Programmiersprachen geschriebene, kompilierte Programme aufzurufen und damit den Vollzug der Systemempfehlung vorzunehmen. Danach kehrt das System auf die Ebene der Problemartikulation zurück.

EXPERTISE IN ENVIRONMENTS FOR INFORMATION SYSTEMS DESIGN

Henk G. Sol, Delft

For problem-solving in general, and thus e.g. for simulation modelling, for decision support and for information systems design, the development of conceptual models is of great importance. We observe that the relevance of application domain models is greatly overlooked, mainly because in many decision-making situations emphasis has been on the exploration of alternative solutions, not on the activities of problem identification and problem specification.

Effective and efficient modelling requires a lot of expertise. The consultation of a base of conceptual models from various application areas may speed up the specification of an appropriate model of the specific application domain. This brings us to some requirements on how expert systems could be used and what their contents should be to support effective modelling during the design of decision support and information systems.

DESCRIPTIVE MODELING AND EXPERT SYSTEMS

C.A.Th. Takkenberg Utrecht

Designing a Descision Support System requires a detailed understanding of decision making in organizations.
In the diagnostic approach the designer first observes the process of decision making aiming at a clinical description rather than searching for normative models to improve a presumed actual decision making. Designers first need a descriptive model as the basis for indentifying a normative direction.
These ideas were advocated ten years ago by Peter Keen. At that time the world was quite unaware of the phenomenon we now call expert systems. However the similarity between a formalized description of a certain kind of problem solving and the contents of a knowledge base of an expert system is striking.
This paper concerns a case study of the descriptive phase in designing a DSS. We give explicit attention to the methodological aspects answering for scientific standards. This leads to a model cycle debouching into a verified or validated descriptive model.
Such a model is a good starting point for solving the real problems. Outcomes of this study corresponds with well known descriptive studies of Newell, Simon and Bowman.
The problem field is the capacity planning in a factory of a multinational company.

SOME PRINCIPLES OF DESIGNING CONFIGURATION CONTROL SYSTEMS

Eelco A. van Veen and Johan C. Wortmann
Eindhoven

In assemble-to-order industries Product Design often has created a large number of products which are basically similar, but differ on a number of engineering characteristics. For every customer order the functional specifications (e.g. voltage, torque etc.), which define a product that suits the customers needs, have to be determined. In order to define a Bill-of-Material for that specific product, the functional specifications have to be related to the engineering characteristics (e.g. a specific type of motor). The task of determining appropriate specifications and creating a corresponding Bill-of-Material can be very complex and time-consuming and is often subject to errors and inconsistencies.
These problems are due to the complex interactions that can exist between functional specifications and engineering characteristics and between functional specifications mutually.
For these problems the concept of the configuration control system could be helpful. A configuration control system is a software product, which supports the configuring of complex products in terms of functional or other specifications. Some of the systems currently known are based on a formal description -by means of decision trees- of the process of configuring a product. We will show that this approach often leads to very complex and redundant decision structures, which are therefore very vulnerable to (engineering) changes. A better approach would be not to describe the way a product must be configured, but merely to describe the constraints that must be met by any configuration from a given range of products. Two typical configuration constraints are the exclusion and the inclusion. The first represents the fact that certain specifications are not to be combined. The latter represents the fact that if a certain combination of specifications has been selected, some other specification must be selected in addition. A disadvantage of this approach seems to be that a product must be fully specified before its feasibility can be evaluated. We will explore an architecture for configuration control systems which copes with this disadvantage. Because of the nature of the configuration constraints the application of AI-programming techniques may prove to be helpful within this architecture.

ASPECTS OF A DSS FOR BIDDING PRICE CALCULATIONS

A.H. Vellekoop & G. van der Hoek
Rotterdam

The building of a DSS for the calculation of bidding prices is a complex multidisciplinary task. For the problems at hand in the dredging company Royal Boskalis Westminster a project group was established of people from cost accounting, planning and information systems.

In this presentation aspects of data modeling and the analysis of the mathematical structure of the iterative method developed will be focused.

A DECISION SUPPORT SYSTEM FOR MANPOWER PLANNING

Diederik J.D. Wijnmalen, The Hague

We have developed a decision support system for medium and long range manpower planning and policy analysis, which is currently being implemented in two (large) organisations. The system consists of a data-base with several interactively accessible files, a model-base with several models, and a dialogue component.
The following model types are included:

* steady-state model for calculating equilibrium manpower systems (markov model),
* multi-period simulation models for forecasting the time-dependent behaviour of manpower systems (stochastic and deterministic models, markov and mixed markov-renewal models).

All model types can be run in batch and interactive mode, depending on the problem size. An interactive run of the multi-period simulation models can be interrupted at the end of each time period in order to analyse intermediate results or to change parameter values.
The system can handle the manpower in an aggregate or individual manner and stores it dynamically, using the method of linked allocation of PASCAL records.
Although external masterfiles can be read from, all input data can be prepared and altered conversationally: the manpower career streams, five types of push and pull flows between grades, flow-restrictions, required strengths, financial data, wastage, etc. New grades may be inserted and existing ones deleted; in fact, an altogether new career structure may be defined interactively. An algorithm sorts out binary order relations between flows and computes the complete final order in which the flows are to be performed, taking into account any (feasible) order that the user may wish to previously arrange for.
Basically the same set of data and parameter values may be input to different model types. The system will check on the feasibility and completeness of the input data when a specific model type has been chosen.
The system supports decisions on the policy making level, where current manpower policies are assessed and new ones are formulated (e.g. as to career schemes, age of retirement, long term manning levels), and on the planning level, where recruitment, promotions, training, etc. are planned within the bounds set by current manpower policy of the organisation.

Ein Ansatz zur Operationalisierung des Technologie-Managements

Reinhard Alves
Jülich

Ein erfolgreiches Technologie-Management setzt sowohl die Umsetzbarkeit strategischer Zielvorgaben in operationale Definitionen entsprechender Projekte als auch die Begründbarkeit von Technologie-Strategien durch vorhandenes Know How voraus. Das gegenwärtig zur Verfügung stehende Instrumentarium ist für derartige Aufgabenstellungen nicht hinreichend geeignet.

Es wird deshalb zunächst ein methodologischer Rahmen zur Ermittlung des "Eigenwertes" von Technologie-Projekten beschrieben, wobei der Eigenwert über die Verfügbarkeit von wissenschaftlich/technischem Know How in Hinblick auf projektspezifische Problemstellungen definiert wird. Konstituenten des methodischen Ansatzes sind einerseits Verfahren zur Problemcharakterisierung (Kontext-Analyse, Separierung) und zur Problemkonzeptualisierung (Strukturierung, Partition, Vereinfachung), andererseits Verfahren zur Formalisierung (Modellbildung, Interpretation) und Organisation von Know How.

Zur Implementation der Methodologie werden ausgewählte automatentheoretische und verwandte Verfahren zur Diskussion gestellt, deren Realisierung prinzipiell über EDV erfolgen kann. Damit soll der erforderliche Methodenverbund mit Verfahren des CAE garantiert werden.

Abschließend werden Überlegungen skizziert, den dargestellten Ansatz in Hinblick auf die Erfordernisse einer Prozeßsteuerung von Technologieprojekten zu erweitern, die die Grundlage für Entscheidungen unter Wirtschaftlichkeitsaspekten bilden kann.

STRATEGISCHE ENTSCHEIDUNGSMODELLE: WO BLEIBT DIE EMPIRISCHE FORSCHUNG?

F. Hanssmann, München

Einleitung

Das zentrale Instrument des Operations Research ist das quantitative Entscheidungsmodell. Wir verstehen darunter eine nicht-definitorische funktionale Beziehung zwischen gestaltbaren Managementvariablen (Entscheidungsvariablen) und relevanten Konsequenzen. Die Rolle des Operations Research im strategischen Management hängt davon ab, ob solche Zusammenhänge auf strategischer Ebene existieren. Strategische Entscheidungsvariable sind abstrakter und unspezifischer als ihre operativen Gegenstücke, weil die strategischen Entscheidungen ihrer operativen Umsetzung in konkretere Maßnahmen zwangsläufig erheblich vorausgehen müssen. GÄLWEILER (1983) sah die Aufgabe der strategischen Planung als "Vorsteuerung des Erfolgs". Ob auf der "Vorsteuerungsebene" mit ihrer sehr viel größeren Abstraktheit, Unschärfe der Begriffe und Unsicherheit nützliche quantitative Modellzusammenhänge existieren, kann nur durch empirische Forschung entschieden werden, die solche Zusammenhänge nachweist und validiert. Meine These ist, daß die einschlägige empirische Forschung erheblichen Rückstand und empfindliche Lücken aufweist, so daß die Frage teilweise offen bleibt. Andererseits liegen auch ermutigende Ergebnisse vor. Ich möchte in beiden Richtungen einen kurzen Überblick geben und damit zukünftige Entwicklungen anregen.

Orientierungsrahmen

Zur Einordnung einschlägiger Modellierungsversuche benutze ich ein allgemeines strategisches Entscheidungsmodell der Unternehmung, wie ich es auch in meiner "Quantitativen Betriebswirtschaftslehre" (1982) vorgeschlagen habe. Das Modell ist modular aufgebaut. Meine These lautet: Wir stehen in der Modellierung der Umweltentwicklung und der interessierenden Marktpotentiale einigermaßen gut da, in der Modellierung der Interaktion konkurrierender Marketingstrategien und der daraus resultierenden Aufteilung des Marktpotentials in Marktanteile dagegen recht schwach. Hier im Marketingbereich - sehr viel weniger in der Produktion und Finanzierung - liegt das Haupthindernis für eine strategieabhängige Projektion des Cash Flow und eine moderne quantita-

Operations Research Proceedings 1987

tive Bewertung von Strategien, etwa im Sinne des Shareholder Value nach RAPPAPORT (1986). Im folgenden wird daher ein gewisser Schwerpunkt bei den strategischen Marktanteilmodellen liegen.

Stand der Kunst in der Modellierung des Marktpotentials

Dennoch soll der Stand der Kunst auch in den vorgelagerten Teilmodellen kurz gewürdigt werden. Charakteristisch wäre beispielsweise eine Modellierung des Automobilmarktes, gemessen an der Zahl der Neuzulassungen, als Funktion sozialer, ökonomischer und technischer Umweltbedingungen. In einem kürzlichen Ansatz von KELLNER (1987) wurden die fahrfähige Bevölkerung, das Einkommen (Nettolohn- und Gehaltsumme) und die Autokosten (Kraftfahrerpreisindex) als unabhängige Variable herangezogen und über eine exponentielle Sättigungsfunktion mit dem Zulassungsvolumen verknüpft. Der statistische Fit über 30 Jahre, die die Motorisierungswelle mit der entsprechenden Veränderung der Größenordnung des Marktes einschlossen, erreichte ein Bestimmtheitsmaß von 97 % (Bild 1). Begriffe wie "Autokosten" sind genügend unspezifisch und abstrakt (insbesondere wird von Einzelheiten der Produktgestaltung abgesehen), um auf der strategischen Vorsteuerungsebene brauchbar zu sein. Quantitative Relationen zwischen solchen unspezifischen, abstrakten strategischen Größen sind also grundsätzlich möglich.

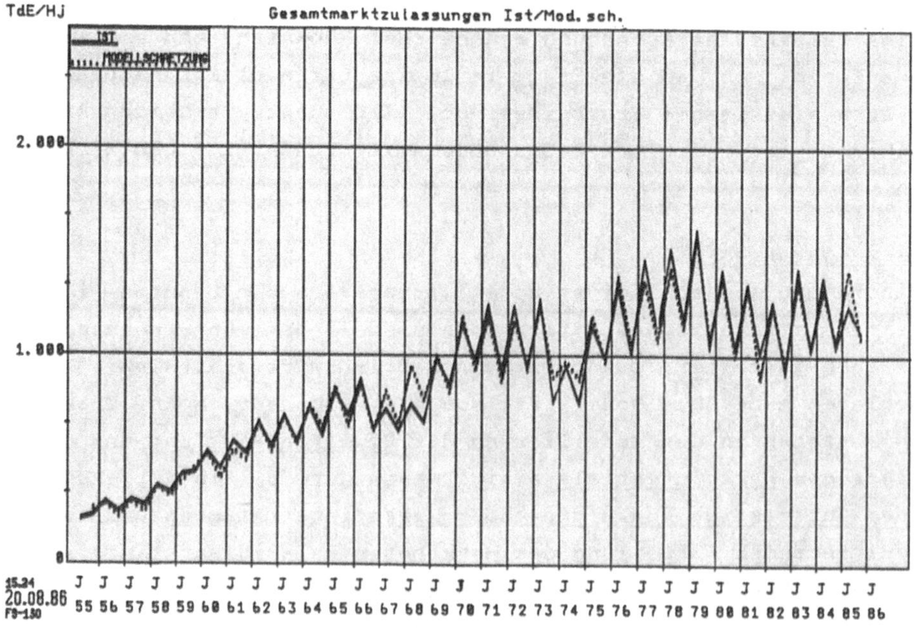

Bild 1: Modellierung der Gesamtzulassungen

Auch die modellgestützte Segmentierung des Marktpotentials weist einen guten Stand auf. Bei produktseitiger Segmentierung des Automarktes ergibt sich die Frage nach dem Anteil eines definierten Produktsegments am gesamten Marktpotential.

Ein Modellansatz von KELLNER (1987) für den Segmentanteil benutzte als entscheidende Einflußgröße die Segmentattraktivität. Diese Größe wurde operationalisiert durch Zuweisung einer subjektiven Punktzahl für jede Neuerscheinung in diesem Segment, die dann (mit Rücksicht auf den Produktzyklus) zeitlich abgewertet wurde. Das auf Segmentattraktivitäten basierende Segmentanteilmodell erreichte ein Bestimmtheitsmaß von 97 %.

Stand der Kunst in der Modellierung der Marktanteile

Eine (segmentweise oder globale) Modellierung des Marktanteils muß die Interaktion der strategischen Marketingvariablen der konkurrierenden Firmen in den Mittelpunkt stellen. Denn Strategie ist Gestaltung des Konkurrenzkampfes, möglichst mit dauerhaftem Konkurrenzvorteil! Es gibt nur wenige Modelle, die die Konkurrenten und ihre strategischen Marketingvariablen explizit einbeziehen.

Vielversprechend sind Modelle, die von einer Operationalisierung des Kundennutzens der konkurrierenden Angebote ausgehen. Im Falle eines elektronischen Geräts mit bestimmter technischer Funktion und gewissen "features" konnte der Kundennutzen durch die quantitative Größe "meßbare Ersparnisse + Zahlungsbereitschaft des Marktes für Features - Kaufpreis" operationalisiert werden (OREN, 1980). Dieses Konzept des Kundennutzens wurde in einen Logit-Ansatz für den Marktanteil eingebracht: jedem der konkurrierenden Anbieter wird eine "Attraktionskraft" zugeordnet, die exponentiell mit dem angebotenen Kundennutzen wächst. Der Marktanteil wird als das Verhältnis der eigenen Anziehungskraft zur Summe aller Anziehungskräfte modelliert. Das Modell konnte mit Hilfe von Erhebungsdaten erfolgreich validiert und anschließend zur Unterstützung einer gravierenden Produkteinführungsentscheidung strategischen Ausmaßes eingesetzt werden.

Das Modell ~~ist~~ vom "Attraktionstyp". Zahlreiche weitere Ansätze vom Attraktionstyp wurden formuliert und harren der empirischen Validierung. Erfolgreich war die Modellierung des Marktanteils eines Autoherstellers in einem bestimmten Produktsegment (KELLNER, 1987). Es handelt sich um einen Attraktionsansatz aufgrund grober Produkteigenschaften (etwa so grob wie in einer Automobilzeitschrift zu finden, daher vom strategischen Standpunkt durchaus geeignet). Aus einer subjektiven Punktbewertung jeder Eigenschaft wurde eine gewichtete Punktsumme für jedes konkurrierende Angebot hergeleitet, die wiederum mit dem Alter des Modells zeitlich abklang. Weitere Variable traten hinzu. Die wichtigste Variable war aber die (in diesem Fall sogar linear formulierte) relative Attraktionskraft des Herstellers. Der Ansatz ergab ein Bestimmtheitsmaß von 95 % (Bild 2).

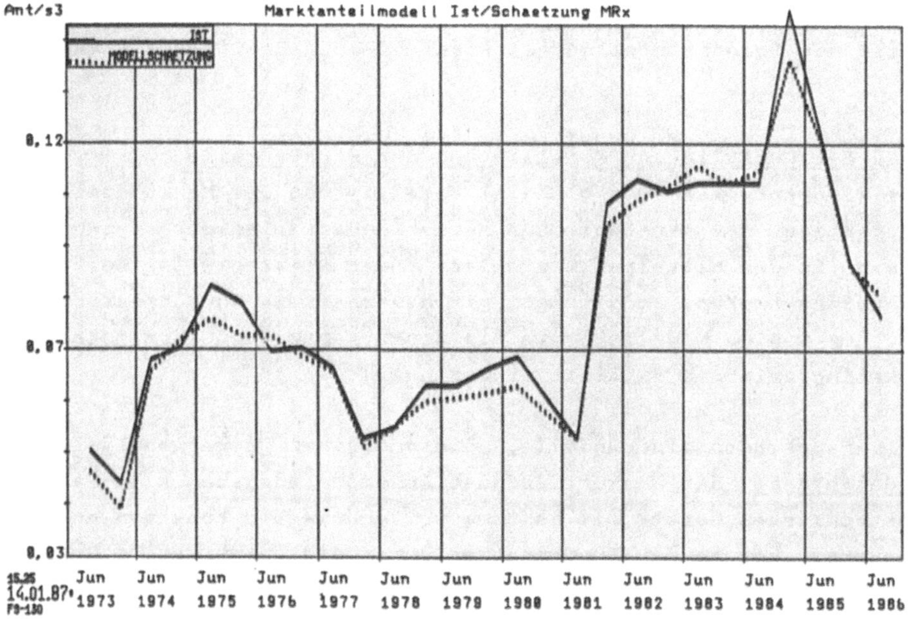

Bild 2: Modellierung des Marktanteils

Sehr aussichtsreich erscheinen auch <u>hierarchische Erfolgsfaktorenmodelle.</u> Hier wird der Marktanteil abhängig von der relativen Stärke der Firma in einzelnen Erfolgsfaktoren gesehen, die ihrerseits einer Gewichtung vom Standpunkt des Kunden unterliegen. Wir illustrieren durch ein Beispiel aus dem politischen Bereich. Gewisse politische Problemkomplexe, die vom Wähler gewichtet werden, dienen als Erfolgsfaktoren. Der Wähler gewichtet ferner die konkurrierenden Parteien bezüglich jedes einzelnen Erfolgsfaktors und schafft so ein "Konkurrenzprofil" jeder Partei (Bild 3). Wird dieses Profil mit dem Vektor der gewichtete te Erfolgsfaktoren multipliziert, so entsteht ein theoretisches Maß der "Wettbewerbsstärke" einer Partei. Die gute Übereinstimmung mit den ebenfalls erhobenen empirischen Wählerpräferenzen für Parteien (Größenordnung des Fehlers 1 %) zeigt, daß damit ein gut validiertes Marktanteilmodell auf Grundlage von Erfolgsfaktoren gewonnen ist. Die Übertragung auf Produktmärkte und kommerzielle Markterfolgsfaktoren liegt auf der Hand.

Bild 3
WETTBEWERBSSTÄRKE POLITISCHER PARTEIEN UND EMPIRISCHE WÄHLERPRÄFERENZEN

Politisches Problem ("Erfolgsfaktor")	Gewicht	Konkurrenzprofil der Partei				Summe
		CDU	FDP	SPD	Grün	
Arbeit und Soziales	.359	.455	.038	.461	.046	1.000
Umwelt	.085	.303	.051	.222	.424	1.000
Außenpolitik	.120	.383	.058	.534	.025	1.000
Wirtschaft	.186	.624	.048	.314	.014	1.000
Innenpolitik	.250	.390	.099	.367	.144	1.000
Wettbewerbsstärke	1.000	.449	.058	.399	.094	1.000
Empirische Präferenz	-	.464	.051	.404	.081	1.000

Die geschilderten Modelle reichen natürlich nicht aus, aber sie beweisen grundsätzliche Möglichkeiten. Nur angemerkt werden kann, daß speziell in Commodity-Märkten der Preis oft keine frei gestaltbare strategische Variable ist, sondern aufgrund der konkurrierenden Kapazitätsangebote zustande kommt. Die Industriekostenkurve präzisiert diesen Zusammenhang und leistet nützliche prognostische Hilfe bei Veränderungen der Kapazitätsstrategie.

Gesamtbewertung einer Strategie

Ich übergehe die Untermodelle für Produktion und Finanzierung und mache noch einige Anmerkungen zur Gesamtbewertung von Strategien. Es wäre ideal, wenn wir quantitative Erfolgsmaßstäbe - z. B. ROI (return on investment) oder noch besser "economic value" nach RAPPAPORT (1986) mit strategischen Entscheidungsvariablen modellmäßig verknüpfen können. In diesem Zusammenhang muß unbedingt das PIMS-Projekt erwähnt werden, das mit Hilfe einer firmenübergreifenden großen Datenbank empirisch-quantitative Zusammenhänge zur Unterstützung der strategischen Planung liefern will. Leider liefert PIMS keine Entscheidungsmodelle. Dies wird illustriert an der bekannten Tabellierung des ROI in Abhängigkeit von Marktanteil und Qualität. Die Tabelle gestattet nicht, die aus einer qualitätsverbessernden Strategie resultierende Verbesserung des ROI zu prognostizieren, weil die Qualitätsveränderung ihrerseits auch auf den Marktanteil einwirkt, jedoch in unbekannter Weise. Der Marktanteil ist eben keine Entscheidungsvariable, sondern selbst das Resultat strategischer Entscheidungsvariablen.

Was benötigt wird, ist ein kausales Entscheidungsmodell, das die PIMS-Resultate allein aufgrund gestaltbarer Managementvariablen erklärt! Es besteht Hoffnung, daß ein solches Modell entwickelt und an den PIMS-Daten geeicht werden kann. Ein in meinem Institut getesteter Ansatz ist wiederum das Logit-Modell, das den Marktanteil als Funktion von je drei Managementvariablen der Firma und der Konkurrenz erklärt: Preis, Qualität und Marketingaufwand (Bild 4). Hinzu kommen einige Untermo-

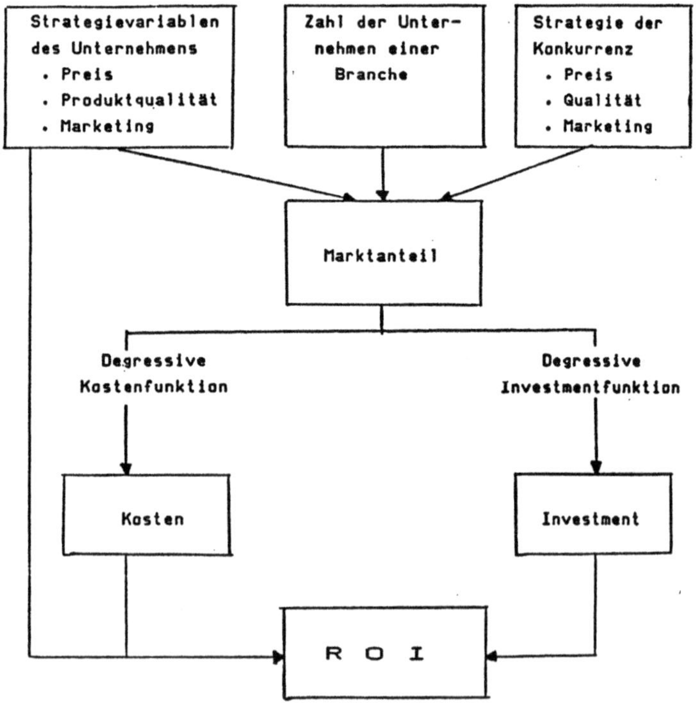

Bild 4: Kausales PIMS-Modell

delle auf der Kosten- und Investitionsseite, die Größendegressionen mit dem Marktanteil beinhalten. Zusammengenommen entsteht ein ROI-Modell, das den ROI allein durch strategische Entscheidungsvariable erklärt. Es gelang, das Modell so zu eichen, daß alle empirischen PIMS-Resultate mit hoher Rangkorrelation reproduziert wurden (Bild 5). Weitere Verbesserungen stehen in Aussicht, so daß wir in Kürze über echte PIMS-Entscheidungsmodelle mit empirischer Validierung verfügen dürften.

Auswirkung von Marktanteil und Produktqualität auf den ROI

		Produktqualität		
	Marktanteil	niedrig	mittel	hoch
PIMS	unter 12 %	4.5 %	10.4 %	17.4 %
	12 - 26 %	11.0	18.1	18.1
	über 26 %	19.5	21.9	28.3

		Produktqualität		
	Marktanteil	niedrig	mittel	hoch
KAUSAL-MODELL	unter 12 %	15.7 %	15.8 %	19.6 %
	12 - 26 %	18.1	19.7	21.0
	über 26 %	20.4	29.2	33.1

Bild 5: Gegenüberstellung von PIMS-Ergebnis und Ergebnis des kausalen Modells (I)

Zu den erfolgreichen Bewertungsmodellen gehört auch das Wertlinien-Modell, das abschließend wenigstens erwähnt werden soll. Es verknüpft die (strategieabhängige) Eigenkapitalrentabilität mit dem Börsenwert der Firma, gemessen als Vielfaches des Buchwertes (market-to-book ratio). Die entsprechende Wertlinie wurde mit Hilfe der Daten zahlreicher Firmem empirisch erforscht.

Zusammenfassung und Schlußfolgerungen

1. Empirische Forschung hat die Existenz quantitativer Zusammenhänge auf der Ebene der strategischen "Vorsteuerung" erwiesen und damit gewisse grundsätzliche Antworten gegeben.

2. Es existieren jedoch empfindliche Forschungslücken.

3. Insbesondere sollte die empirische PIMS-Forschung in Richtung kausaler strategischer Entscheidungsmodelle umgesteuert werden.

4. Das noch sehr kleine Arsenal strategischer Marktanteilmodelle, die die Interaktion konkurrierender Marktstrategien abbilden, sollte dringend vergrößert werden.

5. Von beiden empirischen Forschungsrichtungen kann man eine erhebliche Bereicherung und Unterstützung der strategischen Planung erwarten.

Literatur

Gälweiler, A.: Erfahrungen mit der strategischen Planung, in: ZfbF, Sonderheft, 1983

Hanssmann, F.: Quantitative Betriebswirtschaftlehre, München 1982

Kellner, M.: Marktpotential- und Marktanteilsprognose in der Automobilindustrie, Dissertation, München 1987

Oren, S.S., Rothkopf, M.H., Smallwood, R.D.: Evaluating a New Market: A Forcasting System for Nonimpact Computer Printers, in: Interfaces, Vol. 10, Nr. 6 (Dez. 1980)

Rappaport, A.: Creating Shareholder Value, New York 1986

GESTALTUNG STAATLICHER RAUMFAHRTPROGRAMME MIT ASPEKTEN DER UNTERNEHMENSPLANUNG

Joachim F. Majus, Köln

Das neue Weltraumprogramm und die Rolle des Staates

Auf der politisch-programmatischen Ebene sind die anstehenden Entscheidungsprobleme über ein neues europäisches Raumfahrtprogramm:
- der Gesamtumfang des Engagements der Bundesrepublik Deutschland und
- das Setzen von Schwerpunkten.

Hierbei sind neben den allgemeinen staatlichen Vorsorge- und Fürsorge-Aufgaben unternehmerische Aufgaben wahrzunehmen und entsprechend zu berücksichtigen. In der Raumfahrt haben Staaten weltweit unterschiedliche Aufgaben, die - obwohl abhängig voneinander - unterschiedlich zu begründen und zu beurteilen sind (s. Tabelle 1). In der Raumfahrt ist der Staat überwiegend sein eigener "Kunde".

Staatsaufgabe:	Nutzer	Betreiber
staatliche Dienstleistung	♦ Nachrichten-Übertr. ♦ Meterologie ♦ Umwelt-Überwachung	♦ Satelliten, Plattf. ♦ Transportsysteme ♦ Betriebseinrichtgn.
Förderung von Wissenschaft, Technik und Wirtschaft	♦ Erforschung des Weltraums ♦ Erforschg. d. Erde ♦ Forschg. unter Mikrogravitation	♦ Raumfahrt-Technik ♦ Transfer von Hochtechnologie ♦ Internationale Kooperation

Tabelle 1: Die Rollen des Staates in der Raumfahrt

Gemeinsamkeiten staatlicher und unternehmerischer Planung

Staat und Unternehmen müssen grundsätzlich Policy-Entscheidungen treffen hinsichtlich der Schwerpunktsetzung zwischen zwei einander zunächst entgegenstehenden Zielen:
- Erzielung eines maximalen kurzfristigen Gewinns auf der Basis des vorhandenen Kapitals
- Sicherung der zukünftigen Erwerbsfähigkeit durch Schaffung neuer / weiterer Ressourcen (Kapital/Rücklagen, Anlagen, KnowHow, Image, Kundenstamm).

Es gibt viele Ziele, die auf dieser Ebene kein Entweder-oder zulassen: Zunächst gehört dazu das o.g. Problem der Konkurrenz zwischen Verbrauch und Rücklagen, das genauso einen Finanzminister bewegen muß gegenüber seinen Steuerzahlern wie einen Konzernchef gegenüber seinen Aktionären. Akzeptanz und Konsensbildung sind zur Entscheidungsfindung bei dieser Art Probleme gefragt. Einzelmaßnahmen können zwar jeweils das eine oder das andere Ziel oder beide erfüllen - die getrennte Bewertung nach beiden wäre sinnlos. Solche Verteilungsziele sind die internationalen Kooperationsbeziehungen, die zu fördernden Forschungsdisziplinen, die Industrie-Beteiligung und die Schwerpunkte für die Technologie-Entwicklung.
Angesichts der Natur der Ziele und der geringen Quantifizierbarkeit ihrer kurz- bis mittelfristigen Erfüllung durch neue Maßnahmen sind eindimensional maximierende Bewertungsverfahren nicht anwendbar.

Auf technischer Ebene gibt es die gleichen Relationen von Angebot-Nachfrage bzw. Anforderung-Leistung bzw. Kosten-Preis(Gebühren) wie in der gewerblichen Wirtschaft. Hier sind die Methoden und Verfahren der Planung und Entscheidungsvorbereitung noch stärker verwandt.

Entscheidungshierarchie und Dienstleistungsstruktur

Für die Programme der Raumfahrt und ihre Projekte und Systeme wie für Programme in der Wirtschaft mit ihren Anlagen und ihrer Infrastruktur gibt es ein streng hierarchisches, aber nicht baumförmiges Zielsystem entsprechend den Anforderungs-Leistungs-Beziehungen. Teilt man das Entscheidungsproblem auf die einzelnen Ebenen auf, kann einem mehrdimensionalen Ziel auf einer einzelnen solchen Ebene ein Anforderungsprofil zugeordnet werden, für das auf der nächsten Ebene eine oder mehrere sich ergänzende Maßnahmen definiert und analysiert und hinsichtlich bekannter Vorgaben beurteilt werden können.

Auf höheren Ebenen findet das Verfahren seine quantitative Grenze, wo ein erwarteter langfristiger wirtschaftlicher Nutzen nicht bestimmbar ist. Bereits die analytischen Verfahren sind weitgehend auf die rückschauende Betrachtung beschränkt, mit denen für Folgen vergangener Aktivitäten der Nachweis erbracht werden sollen. Deren Übertragbarkeit in die Zukunft generell und erst recht für andere, neue Aufgaben ist für komplexere Wirkungsmechanismen nicht realisierbar. An der daher unzureichenden Kenntnis der zu erwartenden Zielerfüllung scheitert dann die Anwendung komplexer monetär quantifizierbarer Bewertungsverfahren.

Eine Entscheidung wie die anstehende Programmentscheidung ist als "Schichtenmodell" darstellbar, das sich selbst auf jeder Ebene noch detaillieren läßt. Von Entscheidungsprozessen auf hoher (politisch-programmatischer) Ebene sind viele darunterliegende Entscheidungsebenen mitbetroffenen. Angesichts deren Vielzahl ist der kommunikative Aspekt des Entscheidungsprozesses von besonderer Bedeutung. Ihm ist durch organisatorische Maßnahmen Rechnung zu tragen. Diese Hierarchie enthält drei Ebenen stark unterschiedlicher Entscheidungsproblematik, die der Staatshaushalte und Ressorts, die der Programme und deren nutzungsorientierter Projekte und die der Raumfahrtsysteme und -Technik.

Auf den beiden unteren Ebenen lassen sich Auftrags-Leistungs-Beziehungen identifizieren, wie sie bei anderen technischen Dienstleistungen wie im Verkehr oder bei der Fernmeldetechnik bekannt sind. Konsequente technische Anwendung eines solchen Anforderungs-Leistungs-Prinzips ist das Schichtenmodell in der Telekommunikation, bei dem jede funktionale Ebene die Leistungen aller darunter liegenden mitrepräsentiert.

Das Grundprinzip der funktionalen Hierarchie in der Raumfahrt wird durch die Elemente Anwendungsaufgabe/Nutzer - Nutzungsaufgabe/Nutzlast - Orbitalsystem - Transportsystem - Betriebsinfrastruktur gebildet. Für jede Beziehung gibt es eine Relation Anforderungen-Leistungen/Kosten, die sich als Planungs-Auftrag von oben nach unten und als Realisierungskonzept von unten nach oben darstellt. Jede Leistung einer Stufe ist dabei gleichzeitig Repräsentant der Anforderungen an die zu ihrer Realisierung nötigen Teilaufgaben auf der nächstniedrigeren Stufe - jede Anforderung repräsentiert für die darüberliegende Aufgabe eine Leistung.

Entscheidungsverfahren

Für die auf unmittelbare Anforderungs-Leistungs-Relationen ausgerichtete Auswahl-Entscheidung ist die Kostenwirksamkeitsanalyse das geeignete Verfahren. Bei ihr wird die jeweils kostengünstigste Variante der die einzelnen Anforderungen eines mehrdimensionalen Anforderungsprofils erfüllenden Lösungen ausgewählt. Übererfüllung einzelner Anforderungen kann nur durch einen neuen, strengeren Satz von Anforderungen erfaßt werden. Dessen Notwendigkeit und Kosten sind auf der höheren Ebene zu beurteilen. Dort besteht nämlich die Notwendigkeit, wieder die Anforderungen als Repräsentation einer erwarteten Leistung hinsichtlich der darüberliegenden Aufgabe zu beurteilen.

Bei diesem Verfahren müssen keine einzelnen Nutzenbeiträge aggregiert werden. Das erlaubt die Bewertung von Verteilungsvorgaben, entbindet jedoch nicht von der Notwendigkeit, auf einer höheren Entscheidungsebene dafür begründete Vorgaben zu machen. Außerdem wird das Problem vermieden, in einem Entscheidungsschritt gleichzeitig aus Sicht der Kosten und des "Nutzens" (Zielerfüllung) zu bewerten.

Die Anwendung von Nebenzielen zur eigentlichen Anwendungsaufgabe führt in der Regel zu Mehrkosten, die nach einem Ursachen-Wirkungs-Prinzip zu behandeln sind. Dieses Prinzip ist für eine saubere Planung zumindest rechnerisch anzuwenden und transparent zu machen, auch wenn eine Budgetierung nach anderen Gesichtspunkten erfolgt. Dafür wird das Anforderungsprofil um die Nebenbedingungen erweitert. Die Kosten können nicht mehr allein mit der Anwendungsaufgabe gerechtfertigt werden. Für die Erfüllung eines solchen Nebenziels ist sind Mehrkosten vorzugeben, indem den Nebenzielen eigene Budgets zugewiesen werden. Dies würde auch die Kostentransparenz von Programmen verbessern und wäre Basis für Gebühren-Modelle.

Beispiel meteorologischer Satellit

Die Anwendungsaufgabe eines meteorologischen Satellitensystems ist wieder staatlich: für die Wetterdienste und ergänzend für wissenschaftliche Aufgaben der Atmosphärenphysik. Sie kann als Datendienst beschrieben werden: Das Satellitensystem liefert Bilddaten und andere Meßdaten von der Erde in einem bestimmten Umfang mit einer bestimmten Güte. Die Spezifikationen dafür repräsentieren das Anforderungsprofil - zunächst als Planungsvorgabe in mehreren Leistungsstufen. Als Dienst könnten ihm Gebühren zugeordnet werden.

Eine zeitlichen Entwicklung läßt sich durch weitere Parametrisierung der Leistungsdaten auf einer Zeitachse einbringen. Die zu beurteilenden technischen Systeme können dann bzgl. ihrer Leistungsentwicklung mitbeurteilt werden. Schwieriger ist ein darüberhinausgehendes Entwicklungspotential zu bewerten: Hierfür ist eine Policy-Vorgabe nötig, die den Unwägbarkeiten der Anforderungsentwicklung Rechnung trägt und in das Nebenziel Vorsorge gehört.

Auf der Ebene des Anwendungsprogramms ist zu entscheiden, wie die Wetterdaten beschaffen sein müssen, mit denen eine bessere Wettervorhersage erzielt werden soll. Als Varianten können mehrere Szenarien unterschiedlicher Leistung und unterschiedlicher möglicher Parameterkombination festgelegt werden. Für jedes dieser Szenarien ist dann zu fragen, wie ein Satellitensystem aussehen muß, das diese Leistung möglichst kostengünstig erbringt. Diese Frage erzeugt wieder eine Reihe von Detail-Planungs- und Entscheidungsprozessen. Sie sind aber wieder von der Planung aller übriger Anwendungsaufgaben mit abhängig und kännen nicht für diesen Anwendungsfall isoliert betrachtet werden.

Als Ergebnis des 'Top down'-Planungsprozesses erhalten wir dann ein kostengünstigstes System je Anforderungsprofil (häufig nicht linear!) und damit in Stufen Leistungs-Kosten-Aussagen für bessere Wetterdaten. In der darüberliegenden Entscheidungsebene sind das Ausmaß, in dem Wettervorhersage verbessert werden soll, und die dafür aufzuwendenden Kosten mit analytischen Methoden allein nicht zu messen: Die Folgen besserer Wettervorhersage für Landwirtschaft, Seefahrt, Freizeit-Branche usw. sind von soviel weiteren Einflußgrößen abhängig, daß hier in der Regel nur politische Entscheidungen möglich sind.

Organisatorische Voraussetzung

An dem Beispiel wird deutlich, daß die Komplexität des Gesamtprozesses ein globales Entscheidungsverfahren verbietet. Vielmehr müssen die Einzelentscheidungen auf hoher Ebene in Wechselwirkung zueinander und zu den Planungsergebnissen auf den niedrigeren Ebenen stehen. Eine Koordinierungsfunktion für den Gesamtentscheidungsprozeß muß die Kommunikation für die in ihrem Bereich eigenverantwortlich planenden und ausführenden funktionalen Einheiten sicherstellen. Immer wo Beziehungen zu mehreren Auftraggebern und zu mehreren Kunden bestehen, sind sogenannte Profit-Center einzurichten - selbst wenn der "Profit" nur in der Minimierung der Staatsaufgaben für ein Bündel aus vorgegebenen Aufgaben besteht.

ELECTRONIC DATA PROCESSING IN STRATEGIC PLANNING

Jacob de Smit Rotterdam

The paper aims at explaining the developments of electronic data processing for decision making processes related to the formulation and implementation of policy and/or strategy for organizations.

In strategy making usually four phases of development are recognized:
Business policy-making, this is a rather intuitive process in which the major and often only actor is the corporate executive officer. Decision support or, more general, staff support plays hardly a role in this process.
Long range planning, the main emphasis is on forecasting, the role and importance of staff support is increased considerably. The use of mathematical models and electronic data processing emerges.
Strategic planning, a systematized process of decision making in which a number of participants are involved. Most important group of participants is formed by the line managers. Strong focus on formulating and choosing objectives, the role of forecasting diminishes. Corporate staff for strategy making becomes more process consultant than expert consultant.
Strategic management, involves a broadening of scope of strategic decision making, including the organizational structure and culture, as well as the political dimension of a company's societal behavior.

The last two phases show a redefining of the decision support from forecasting to the use of (interactive) databases and a broadening of scope of these databases. The increasing capacity for data processing, the potential for local data manipulation and the increased scope of the policy making process provide an interesting potential for the use of electronic data processing in order to improve policy making.

Aircraft-stand allocation at Schiphol Airport: a decision support system

Ko Anthonisse, Ben Lageweg, Amsterdam

The allocation of flights to gates at Schiphol Airport is preceded by the development of three plans, which are based upon different sets of information. Airlines have different schedules for different periods of the year. For each season a plan is developed and distributed. This plan consists of an allocation of flights to gates for each day of a typical week of the season. Each day in the evening a plan for the next day is developed. This plan contains all aircraft which will arrive or depart during the next day. In this plan the availability of gates is taken into account. The plan should not deviate too much from the seasonal plan. Copies of this plan are distributed to various services at the airport. During the operational day deviations from the scheduled arrival and departure times will occur. These, and the actual allocations, may necessitate modifications in the plan for the remainder of the day. However, it should stay close to the original plan for the day.

Plans are readily represented with the help of planning-boards. The use of these boards however, does not encourage the planners to develop alternative plans. The first goal of the decision support system is to store plans in the computer, to show them on a colour display and to provide the planner with interactive tools to develop and to modify a plan. Two representations of plans are useful: a planning-board and the lay-out of the airport. The planner can easily switch between those two representations. The second goal is to provide the planner with information on the quality of a plan, determined by many criteria which are difficult to quantify. To show the quality both numerical and graphical representations are available. The third goal is to provide procedures which will generate a plan. Such a plan should score not too badly on the main criteria, to serve as a starting point for further development by the planner.

Aircraft-stand allocation at Schiphol Airport: an optimization procedure

Ko Anthonisse, Ben Lageweg, Amsterdam

The allocation of flights to stands at an airport is a quadratic multi-criteria programming problem under uncertainty. The problem should not be restricted to flights and gates but should include the stationing of aircraft at other locations of the airport. These other locations can replace gates for the handling of the aircraft, with bus service to the terminal building, but can also be used to park airplanes between arrival and departure. The allocation problem is quadratic due to relations between flights which involve the distance between the locations of the aircraft. The passengers' walking distance between connecting flights is an example. With a different metric, the distance also is relevant to the apron-men and, with yet another metric, to crews working inside the building. There also are linear criteria. Gates and flights and aircraft have several properties which determine the suitability of a gate to accommodate a flight. The uncertainty in the problem is due to the stochastic nature of the arrival and departure times of flights.

We describe a procedure to generate allocations which at least can serve as a starting point for later improvements by the planner. In this procedure, the quadratic criteria and the stochastic elements are disregarded. Moreover, the linear criteria are jointly expressed as a preferential order of stands for each flight. The procedure first determines an allocation by minimizing the number of allocations at the least preferred level. Then we move up, step by step, to the next higher level of preference and minimize the number of allocations at that level, while maintaining the number of allocations at each previous level.

A METHOD FOR DATA COLLECTION FOR CAR NAVIGATION

Bart J. Beers, 's-Hertogenbosch

Over the last years car navigation systems have become a topic of major interest. A large category of these systems depend on what is called an electronic map, which can be seen as a digital copy of a road map.
TELE ATLAS is developing and producing electronic maps in cooperation with manufacturers of car navigation systems. It became clear that just copying existing maps and adding collateral information would not lead to a satisfactory database.
Indispensible information is not available in a standardised manner, and if it is, about 15% leads to contradictions.
This means that field surveys are inevitable.

For this purpose TELE ATLAS have developed a video car which will record information needed for car navigation databases. Amongst others this results in:

- an optical database, stored on common videotape
- digital recordings of distances, slopes, heights and magnetic anomalies.

In the video car this information is linked to a metric database, in a way that it is possible to easily retrieve the data and convert it to car navigation databases.
The field surveys are done on beforehand, which means that all possible roads are recorded, ensuring a well to plan production flow in the office.

Tests have shown that there would have been discrepencies from reality in the car navigation database, which would never have been revealed without such a thorough field survey with the video car.

A Decision Support System for a Location-Allocation-Routing Problem

A.J.M. Beulens, A.W.J. Kolen and E.G. Niekamp

In this paper we describe a decision support system that was developed for supporting strategical and tactical decisionmaking with respect to the determination of the distribution structure of companies. It deals with problems where the locations of warehouses have to be determined such that fixed and operational distribution- and routing costs are minimized subject to amongst others capacity-, vehicle fleet-, client and service degree constraints. The system, still under development, is used by consultants to perform distribution studies. The DSS contains a heuristic to find the "optimal" locations of Wp warehouses out of a possible set of warehouses W. At the beginning of each iteration we have Wp warehouses and the assignment of clients to one of these warehouses. We then solve the corresponding Wp single warehouse vehicle routing problems. The routes found are used to cluster the clients into groups. Next we solve a p-median problem defined on the clusters, which constitutes the starting point for the next iteration of the heuristic.

The software of our DSS consists of own made modules and of the Mover vehicle routing package. Mover and its associated road network have been used because of it being available, the capability to solve routing problems, the possibility to interface with own made software and the resulting development efficiency.
The system has been used to solve some practical problems. Problems involving three factories with associated main depots, up to 10 possible warehouse locations, and up to 3000 client orders. The use of the system for a set of practical problems showed that the system and its heuristic are practically feasible. The convergence of the heuristic takes place in two to three steps, independent of the starting solution. The processing times take up to 5 hours on an IBM RT, but are perfectly feasible. For one case, with up to 7 warehouses and about 650 clients, we obtained further results:

- the warehouse selection didn't change for client selections containing 50% ,60%, 70%, 80%, 90%, and 100% of the 650 orders,
- the costs per delivery increased considerably as the density of the distribution of the clients decreased,
- the studies give an indication of the expected inter client distance and driving time that can be obtained depending on the density of the distribution of the clients

Further research is necessary to improve the efficiency of the system and to get a better insight into the characteristics and behavior of it.

Zielsetzungen bei der Modellierung von Standortproblemen

Hans Ulrich Buhl, Karlsruhe

Zusammenfassung: Im Gegensatz zum Inhalt der reichhaltigen Standorttheorie - Literatur ist der Schwerpunkt der vorliegenden Arbeit die Diskussion ökonomisch relevanter Zielfunktionen für Vielziele - Optimierungsprobleme. Im besonderen werden einige Theoreme abgeleitet, in welchen einfache Zielsetzungen charakterisiert werden, die Probleme mit mehrfachen Zielsetzungen reduzieren.

Summary: Contrary to most of the rich literature existing on location theory the paper emphasizes the discussion of adequate objective functions for multi-objective-approaches. In particular, a number of theorems are presented characterizing single objective reductions of multi-objective problems.

1. EINLEITUNG

Die ökonomische Bedeutung von Standortentscheidungen nimmt sowohl im privatwirtschaftlichen Bereich als auch bei öffentlichen Einrichtungen aus mehreren Gründen zu: Marktwirtschaftliche Wettbewerbsprozesse geben den Wettbewerbern, die das ökonomische Prinzip konsequent - also auch bei der Standortwahl - anwenden, die besseren Wachstums- und Überlebenschancen. Die national und international zunehmende Arbeitsteilung und die mit ihr einhergehende Zunahme der Lieferbeziehungen verstärkt die Auswirkungen guter oder schlechter Standortentscheidungen bezüglich der Beschaffungs- und Absatzmärkte. Aufgrund prohibitiv hoher Kosten und einer Vielzahl nicht-monetärer Hindernisse, die der Verlagerung, Aufgabe oder Neuerrichtung eines Standorts entgegenstehen können, sind Standortentscheidungen zudem - zumindest kurzfristig - häufig irreversibel.

Bereits aus diesen Bemerkungen wird die herausragende Stellung von Standortproblemen bei einer Vielzahl ökonomischer Aktivitäten deutlich. (Empirische) Umfragen bei staatlichen und privatwirtschaftlichen Entscheidungsträgern stützen diese Aussagen. Es ist daher nicht überraschend, daß die Standorttheorie - insbesondere in den letzten 25 Jahren - (mit über 2000 wissenschaftlichen Publikationen) ein vielbeachteter Forschungsschwerpunkt geworden ist.

Es besteht jedoch eine auffällige Diskrepanz zwischen der Vielzahl entwickelter Theorien und Verfahren einerseits und deren tatsächlicher Anwendung bei praktischen Problemstellungen andererseits, wo nur in Ausnahmefällen quantitative Entscheidungsverfahren verwendet werden. Ein Grund für diese Diskrepanz dürfte darin liegen, daß in der Praxis häufig eine Vielzahl unterschiedlichster Zielsetzungen als für Standortentscheidungen bedeutsam angesehen wird, während die Anwendung der o.g. Theorien und Verfahren meist die Vorgabe eines Zielkriteriums erfordert. Die Analyse mehrfacher Zielsetzungen, die in weiten Teilen der Entscheidungstheorie eine große Rolle spielen, ist in der Standorttheorie nur wenig untersucht worden. Hier dominieren einfache Zielsetzungen.

Während in mikroökonomischen Problemstellungen - z.B. bei der Bestimmung transportkostenminimaler Standorte von Auslieferungslägern - meist Median-Ziele wie Minisum verwendet werden, unterstellt man in vielen volkswirtschaftlichen, insbesondere regionalwissenschaftlichen, Problemen - z.B. bei der Bestimmung zentraler Standorte wie Feuerwachen, Blutbanken, Rettungs-Depots, Rathäuser, Schulen - ausschließlich Zentren-Ziele wie das Rawls- Kriterium Minimax. Auf die genaue Definition dieser Zielsetzungen und deren ökonomische Bedeutung wird später noch eingegangen.

Operations Research Proceedings 1987

2. ZIELE UND WERTURTEILE BEI STANDORTENTSCHEIDUNGEN

Im einfachsten Fall möchte man bei Standortentscheidungen nach dem Median-Prinzip die Aggregate Summe der Entfernungen bzw. "Durchschnittsentfernung" minimieren, bei solchen nach dem Zentrums-Prinzip ist der individuell schlechteste Wert "Maximalentfernung" Ziel der Minimierung. Man kann deshalb etwas vergröbernd sagen, der Median sei kollektivbezogen, das Zentrum individualbezogen.

Standortentscheidungen volks- und insbesondere regionalwissenschaftlicher Art können schon aus politischen Gründen meist weder rein kollektivbezogen (z.B. wegen Minderheitenschutz) noch rein individualbezogen (z.B. wegen kollektiver Mehrheitsinteressen) gefällt werden. Es ist daher zweifelsohne sinnvoll, bei der Entscheidungsvorbereitung mittels quantitativer Methoden mehr als nur eine Zielsetzung zu betrachten. Dabei ist es übrigens keinesfalls ein Nachteil, wenn unterschiedliche Lösungen, die alle bezüglich mindestens eines der betrachteten Ziele optimal sind, dargestellt und erläutert werden müssen. Gerade das Verständnis, welche Zielsetzung welchen Einfluß auf die Optimallösung(en) hat, dürfte einen äußerst positiven Einfluß auf die Akzeptanz quantitativer Methoden bei Entscheidungsträgern haben. Insofern ist die Analyse mehrfacher Zielsetzungen vielleicht nicht nur eine Ergänzung zur einfachen Zieloptimierung, sondern schafft auch Anwendungsbereiche und Akzeptanz für diese. Im folgenden werden wir uns zunächst auf zwei Zielsetzungen nach den Median- und Zentrum-Prinzipien beschränken. Bevor wir jedoch auf die Vorstellung der bisherigen sog. Cent-Dian und Medi-Center Formulierungen sowie einen axiomatischen Ansatz kommen, ist etwas Notation notwendig.

Seien
x ein Standort aus der Menge O der potentiellen Orte,
u_i, $i = 1, \ldots, n$ ein vorgegebener Bezugspunkt,
b_i, $i = 1, \ldots, n$ die Gewichtung dieses Bezugspunkts,
$d_i(x)$ die Entfernung zwischen Bezugspunkt u_i und Standort x

Dann nennt man

$$(1)\quad S(x) = \sum_{i=1}^{n} d_i(x) b_i$$

die <u>Summe der gewichteten Entfernungen</u> zwischen Standort x und allen Bezugspunkten $u_1, \ldots, u_n$,

$$(2)\quad D(x) = S(x) \,/\, \sum_{i=1}^{n} b_i$$

die <u>gewichtete Durchschnittsentfernung</u> und

$$(3)\quad M(x) = \max_{i=1,\ldots,n} \{d_i(x)\}$$

die ungewichtete <u>Maximalentfernung</u>. Für den Median m und das Zentrum z gelten

$$(4)\quad S(m) \leqq S(x),\ D(m) \leqq D(x) \qquad \forall\, x \in O$$

bzw.

$$(5)\quad M(z) \leqq M(x) \qquad \forall\, x$$

3. CENT-DIAN und MEDI-CENTER Ansätze

Bisher in der Literatur beschrieben sind - soweit mir bekannt - die folgenden Ansätze für simultane Berücksichtigung von Median- und Zentren-Zielen.

3.1 Vektoroptimierungs-Ansatz

Lowe [1978] wählte diesen Ansatz für die Bestimmung der paretooptimalen Lösungen auf einem Baumgraphen. Von Ross/Soland [1980] wurde eine Verallgemeinerung mit einen iterativen Auswahlalgorithmus angegeben. Vorteil: Man erhält alle interessanten Standorte (zur Endauswahl). Nachteil: Insbesondere bei Mehr-Standort-Problemen kann die Menge effizienter Lösungen prohibitiv groß werden, so daß doch die Auswahl mithilfe von (einer Folge von) Ersatzproblem(en) erforderlich ist.

3.2 Restriktions-Ansatz

Dieser auf Toregas, Swain, ReVelle und Bergmann [1971] zurückgehende Ansatz wurde später von Church [1976], Church und Meadows [1975], Gleason [1974], Khumawala [1973], [1975], Levin [1977] und Odoni [1974] verwendet. In diesen Arbeiten sind heuristische und für kleinere Probleme - auch exakte Lösungsverfahren für den folgenden Ansatz beschrieben

(6) Min $S(x)$ u.d.N. $M(x) \leq \bar{M}$, $x \geq 0$.

Halpern [1977] betrachtete das Problem

(7) Min $M(x)$ u.d.N. $S(x) \leq \bar{S}$, $x \geq 0$.

und stellte interessante Dualitätseigenschaften für die Probleme (6) und (7) bei Zugrundelegung eines Baumgraphen dar.

Der Vorteil dieser Ansätze liegt klar in der Reduktion auf ein Problem mit einfacher Zielsetzung und Einschränkung des Lösungsraumes; den rechentechnischen Vorteilen steht jedoch ein gewichtiges Formulierungsproblem gegenüber: Insbesondere bei Unkenntnis der Werte $S(m)$ oder $M(z)$ dürfte es bei vielen Problemen wie unserem Feuerwachen-Beispiel schwierig sein Werte für $\bar{M}$ oder $\bar{S}$ festzulegen, die gleichzeitig

- klein genug sind, um das Optimierungsproblem zu beeinflussen,
- groß genug sind, damit die Lösungsmenge nicht leer ist,
- die Bedeutung der beiden Zielsetzungen widerspiegeln.

3.3 Straffunktions-Ansatz

Bei Verwendung der Zielfunktion

(8) $P(x) = \sum_{i=1}^{n} f(d_i(x))b_i$,

wobei $f: \mathbb{R}_+ \to \mathbb{R}_+$ eine monoton wachsende, konvexe Straffunktion ist, wird häufig mit $f(d) = d^2$ gearbeitet (z.B. Wolf [1977]). Begründet wird dieser Ansatz häufig mit der Beobachtung, daß die Maximalentfernung des bzgl. Ziel (8) optimalen Standorts meist deutlich besser als bei reiner Minisum-Zielsetzung ist. Mir erscheint dieser Straffunktions-Ansatz jedoch nur dann adäquat, wenn die gewählte Funktion $f: \mathbb{R}_+ \to \mathbb{R}_+$ tatsächlich z.B. entfernungsabhängige Kosten widerspiegelt. Dies mag im Feuerwachen-Beispiel der Fall sein, wenn die Brandschäden in der relevanten Anfahrzeit mit $f(d) = d^2$ wachsen. Im normalen Cent-Dian Problem ist i.a. jedoch die Festlegung der adäquaten Straffunktion schwierig.

3.4 Konvexkombinations-Ansätze

Diese gehen auf Halpern [1976], [1980] und Handler [1976], Handler und Mirchandani [1979] zurück. Verwendet wurden die (Minimierungs-) Zielfunktionen

(9) $Z_S(S(x),M(x);\lambda) = \lambda S(x) + (1-\lambda)M(x),\ \lambda \in [0,1]$

sowie

(10) $Z_D(D(x),M(x);\lambda) = \lambda D(x) + (1-\lambda)M(x),\ \lambda \in [0,1]$

Vor- und Nachteile dieser Zielfunktionen werden im folgenden (Kapitel 4) mithilfe von Charakterisierungen und Beispielen verdeutlicht.

4. CHARAKTERISIERUNG VON ZIELFUNKTIONEN

Seien a und b zwei Zielkriterien und μ_1, μ_2 die zugehörigen Gewichtungsfaktoren. Halperns und Handlers Zielfunktionen lassen sich mithilfe von (A1) und (A2) nach Aczel folgendermaßen charakterisieren:

(A1) Homogenität: Für alle $\rho > 0$ gilt:

$$Z(\rho a,\rho b;\mu_1,\mu_2) = \rho\ Z(a,b;\mu_1,\mu_2)$$

(A2) Translationsbedingung: Für $\mu_1 + \mu_2 > 0$ gilt:

$$Z(a+c,b+c,\mu_1,\mu_2) = Z(a,b;\mu_1,\mu_2) + c$$

Theorem 4.1 (in Anlehnung an Aczel)

Sei $Z_1: \mathbb{R}_+^4 \rightarrow \mathbb{R}_+$ monoton wachsend in den ersten beiden Argumenten. Dann gilt:

Z_1 erfüllt (A1) und (A2) <=>

$$Z_1(a,b;\mu_1,\mu_2) = \frac{\mu_1}{\mu_1 + \mu_2}\cdot a + \frac{\mu_2}{\mu_i + \mu_2}\cdot b\ .$$

Man beachte: Mit $\lambda = \mu_1/(\mu_1+\mu_2) \in [0,1]$ ergeben sich Handlers und Halperns Zielfunktionen Z_S und Z_D.

Sicherlich wird nicht jedermann die Eigenschaften (A1) und (A2) als unbedingt notwendig für Zielfunktionen einstufen. Das ihren Zielfunktionen innewohnende Werturteil kommt m.E. in der folgenden Charakterisierung deutlicher zum Ausdruck.

(AR) Normierung: Für $\mu_1 + \mu_2 > 0$ gilt:

$$Z(a,a;\mu_1,\mu_2) = a$$

(F1) Kompensationskriterium 1:

Für $\mu_1, \mu_2 \in [0,\infty)$ und $\mu_1 + \mu_2 > 0$ gilt:

$$\mu_1(a_2-a_1) + \mu_2(b_2-b_1) = 0 \iff$$

$$Z(a_1,b_1;\mu_1,\mu_2) = Z(a_2,b_2;\mu_1,\mu_2)$$

Die Normierungseigenschaft (AR) ist sicherlich natürlich und wenig einschränkend, während (F1) ein subjektives Werturteil enthält, welches die zulässigen Zielfunktionen erheblich einschränkt. Es gilt:

Theorem 4.2

Für $Z_1: \mathbb{R}_+^4 \to \mathbb{R}_+$ gilt: Z_1 erfüllt (AR) und (F1)

$$\iff \quad Z_1(a,b;\mu_1,\mu_2) = \frac{\mu_1}{\mu_1+\mu_2}\cdot a + \frac{\mu_2}{\mu_1+\mu_2}\cdot b \;.$$

Ändert man das Werturteil (F1) aufgrund der oben erwähnten Subjektivität etwas ab (zu (F2)), so wird eine Cobb-Douglas-Funktion charakterisiert.

(F2) Kompensationskriterium 2:

Für $\mu_1, \mu_2 \in [0,\infty)$ und $\mu_1 + \mu_2 > 0$ gilt:

$$\mu_1(\ln a_1 - \ln a_2) + \mu_2(\ln b_1 - \ln b_2) = 0$$

$$\iff (a_1/a_2)^{\mu_1} (b_1/b_2)^{\mu_2} = 1$$

$$\iff Z(a_1,b_1;\mu_1,\mu_2) = Z(a_2,b_2;\mu_1,\mu_2)$$

Theorem 4.3

Für $Z_2: \mathbb{R}_{++}^2 \times \mathbb{R}_+^2 \to \mathbb{R}_{++}$ gilt: Z_2 erfüllt (AR) und (F2)

$$\iff Z_2(a,b;\mu_1,\mu_2) = a^{\frac{\mu_1}{\mu_1+\mu_2}}\, b^{\frac{\mu_2}{\mu_1+\mu_2}}$$

Beweise zu den beiden Theoremen 4.2 und 4.3:

4.2 "=>" Für c: $\mu_1 c + \mu_2 c = (\mu_1 + \mu_2)c = \mu_1 a + \mu_2 b$ (1)

gilt wegen (F1) $Z(c,c,\mu_1,\mu_2) = Z(a,b,\mu_1,\mu_2)$ (2)

und wegen (AR) $Z(c,c,\mu_1,\mu_2) = c$. (3)

(1) - (3) implizieren zusammen:

$$Z(a,b,\mu_1,\mu_2) = c = \frac{\mu_1}{\mu_1+\mu_2}\cdot a + \frac{\mu_2}{\mu_1+\mu_2}\cdot b \;. \tag{4}$$

"<=" trivial.

4.3 "=>" Für c: $c^{\mu_1} c^{\mu_2} = c^{(\mu_1 + \mu_2)} = a^{\mu_1} b^{\mu_2}$ (1)
gilt wegen (F2) $Z(c,c,\mu_1,\mu_2) = Z(a,b,\mu_1,\mu_2)$ (2)
und wegen (AR) $Z(c,c,\mu_1,\mu_2) = c$. (3)

(1) - (3) implizieren zusammen

$$Z(a,b,\mu_1,\mu_2) = c = a^{\frac{\mu_1}{\mu_1 + \mu_2}} b^{\frac{\mu_2}{\mu_1 + \mu_2}} . \qquad (4)$$

"<=" trivial.

Die Kompensationskriterien (F1) und (F2) und mithin die charakterisierten Zielfunktionen Z_1, Z_2 erfordern explizite Werturteile in Gestalt der Gewichtungsparameter μ_1 und μ_2. Daß bei Verwendung der Konvexkombinations-Zielfunktionen von Halpern/Handler und Mirchandani die präferenzgerechte Angabe der Gewichtungsparameter Probleme aufwerfen kann, zeigen die folgenden Beispiele. In diesen bezeichnen die eingekreisten Zahlen die Einwohnerzahlen dreier Teilorte, die Zahlen an den Verbindungslinien deren Entfernungen. Es ist der "beste" Standort für ein öffentliches Gebäude zu bestimmen.

Beispiel 1:

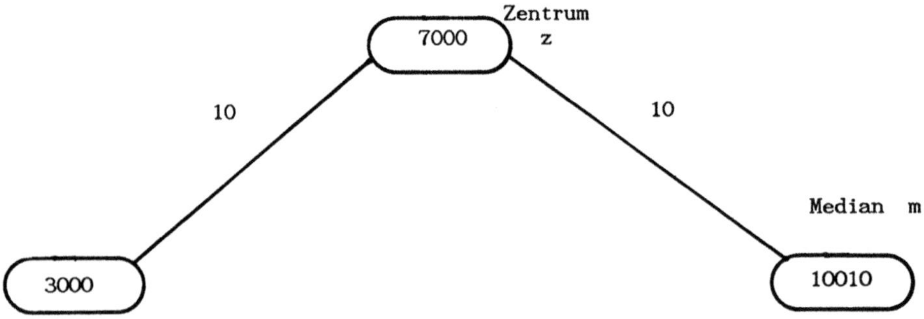

S(z) = 130100
M(z) = 10

S(m) = 130000
M(m) = 20

Beispiel 2:

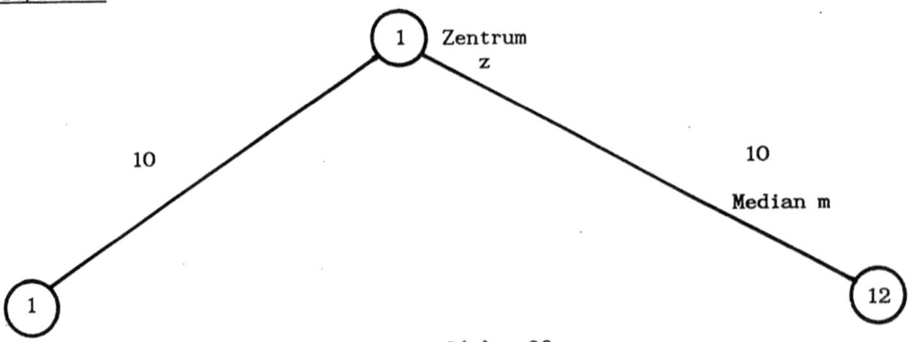

S(z) = 130
M(z) = 10

S(m) = 30
M(m) = 20

Als besten Standort für ein öffentliches Gebäude wählte eine vom Autor befragte (nicht ganz unverzerrte) Stichprobe von 10 Personen in Beispiel 1 das Zentrum z, in Beispiel 2 den Median m. Interessant ist daher der folgende

Satz: Für die Zielfunktion

$$Z_S(S(x),M(x);\lambda) = \lambda S(x) + (1-\lambda)M(x)$$

existiert kein Gewichtungsparameter $\lambda \in [0,1]$ so daß in Beispiel 1 das Zentrum Z dem Median m und im Beispiel 2 der Median m dem Zentrum Z vorgezogen wird.

Begründung: Die "intuitiv richtige Standortentscheidung" weist einen besseren Zielfunktionswert auf dann und nur dann wenn

- in Beispiel 1 $\lambda < 1/11$ ist,
- in Beispiel 2 $\lambda > 1/11$ ist.

Als Fazit aus den Beispielen können wir festhalten: Ohne Kenntnis der einfachen Zielfunktionswerte (S(m),M(Z),D(m)) ist eine "präferenzgerechte Gewichtung" bei den Konvexkombinations-Zielfunktionen schwierig - wenn nicht unmöglich!

Wenn man aber die einfachen Optima benötigt, bietet es sich an, diese auch in die Zielfunktion zu integrieren, z.B.

Ausweg: Da die Betonung absoluter Abweichungen in den Zielfunktionen Z_S und Z_D deren Anwendung begrenzen, kann man die Zielfunktionen Z_1, Z_2 bspw. von den Quotienten

$$a := \frac{S(x)}{S(m)} = \frac{D(x)}{D(m)}, \quad b := \frac{M(x)}{M(Z)}$$

als Maße für die relativen Abweichungen von den einfachen Optima abhängen lassen.

Verwendung der Eigenschaften (AN), (F1) oder (F2) und der Charakterisierungs-Theoreme ergibt dann die Zielfunktionen

$$Z_3(D(x),D(m),M(x),M(z);\lambda) = \lambda\frac{D(x)}{D(m)} + (1-\lambda)\frac{M(x)}{M(z)}, \qquad \lambda \in [0,1]$$

sowie

$$Z_4(D(x),D(m),M(x),M(z);\lambda) = \left[\frac{D(x)}{D(m)}\right]^{\lambda}\left[\frac{M(x)}{M(z)}\right]^{1-\lambda}, \qquad \lambda \in [0,1]$$

Bemerkung: "ZF - Werte - 1" geben durchschnittliche relative Abweichungen von einfachen Optima an.

Wenden wir nun die Zielfunktion Z_3 auf die Beispiele 1 und 2 an, so ergibt sich jeweils die "intuitiv richtige Standortentscheidung".

Bei Verwendung der Zielfunktion

$$Z_3(x) = \lambda\frac{S(x)}{S(m)} + (1-\lambda)\frac{M(x)}{M(Z)},$$

mit z.B. $\lambda = 0.5$, wird z $(Z_3(z) = 1{,}00038)$ gegenüber m $(Z_3(m) = 1.5)$ in Beispiel 1 klar bevorzugt!

Bei Verwendung der selben Zielfunktion Z_3 in Beispiel 2

$$Z_3(x) = \lambda\frac{S(m)}{S(x)} + (1-\lambda)\frac{M(Z)}{M(x)},$$

mit z.B. $\lambda = 0.5$ wird m $(Z_3(m) = 1.5)$ gegenüber z $(Z_3(z) = 2.67)$ klar bevorzugt!

Natürlich gibt es auch Beispiele, in welchen Zielfunktionen, die absolute Abweichungen betonen, als genausogut oder besser empfunden werden als solche, die relative Abweichungen betonen. Insofern wäre eine allgemeine Propagierung der "relativorientierten" Zielfunktionen sicher verfehlt. Gleichwohl bin ich der Ansicht, daß deren Anwendungsbereich wesentlich breiter als der der kritisierten ist. Im allgemeinen ist aber folgende Vorgehensweise zu empfehlen:

1. Bestimmung der einfachen optimalen Standorte und deren Zielfunktionswerte

2. Entscheidung über Zielfunktion unter Kenntnis von 1.

3. Entscheidung über Gewichtung unter Kenntnis von 1 + 2.

4. Lösung des Ersatzproblems mit 2 + 3.

Diese Vorgehensweise bringt mindestens 3 (nicht notwendig verschiedene) interessante und effiziente Standorte / bzw. Standort - Kombinationen.

5. Schlußbetrachtungen

Bisher wurden lediglich die beiden Zielsetzungen Minisum und Minimax und deren Kombinationsmöglichkeiten betrachtet. Bereits einleitend wurde aber darauf hingewiesen, daß bei vielen volkswirtschaftlichen Standortproblemen eine Vielzahl von Zielsetzungen für bedeutsam gehalten wird. Beispielsweise kann man in unserem Feuerwachen-Beispiel zusätzlich zu den bisher untersuchten Zielen auch die Zielsetzung "Maximierung der Anzahl der in einem (für wünschenswert gehaltenen) δ-Radius (Zeit/Entfernung) um x gelegenen Bezugspunkte (-gewichte)" für bedeutsam halten.

Naheliegend ist daher die Frage, ob und inwieweit sich die o.g. Ansätze (aus Kapitel 4) auf $n \in \mathbb{N}$ Ziele verallgemeinern lassen. Erfreulicherweise ist dies für die charakterisierten Zielfunktionen Z_1 und Z_2 sehr einfach möglich; wenn man die Forderungen für n Ziele verallgemeinert, erhält man auf gleichem Wege Zielfunktionen $Z(a_1, \ldots, a_n; \mu_1, \ldots, \mu_n)$ mit n Gewichtungsparametern μ_i, $i=1, \ldots, n$.
Gegen all die diskutierten Überlegungen könnte man zwei grundsätzliche Einwände erheben: erstens, daß aufgrund der mangelnden Quantifizierbarkeit vieler bedeutsamer Ein-flußfaktoren eine Optimierung der Standortwahl unter Berücksichtigung aller Ziele sowieso ausscheidet; zweitens, daß überhaupt die auf Optimierungsgedanken beruhende Standorttheorie keinerlei "Erklärungswert" für reale Entscheidungen hat, solange nicht gezeigt werden kann - und das ist nicht geschehen - daß tatsächliches Verhalten durch solche Optimalitätskriterien bestimmt wird.

Dem Standorttheoretiker, der aufgrund dessen und der eingangs erwähnten Diskrepanz zwischen Theorie und Wirklichkeit zweifelt oder gar verzweifelt, sei Trost gespendet mit einem Zitat von August Lösch [1943] zum Sinn der Frage nach dem Standort:

"Nein, das eigentliche Geschäft des Ökonomen ist nicht, die miserable Wirklichkeit zu erklären, sondern zu verbessern. Die Frage nach dem besten Standort ist ungleich würdiger als die Feststellung des tatsächlichen".

Literatur:

Aczel, J.: "Lectures on Functional Equations and Their Applications", Academic Press, New York, London, 1966.

Church, R.L.: "Absolute Medians of a Graph and the p-Median Problem with the Maximum Distance Constraints", Department of Civil Engineering, University of Tennessee, Knoxville, Tenn, 1976.

______/Meadows, M.E.: "An Optimal Approach to an Expanded Form of the Location Set Covering Problem", presented at the Las Vegas ORSA/TIMS meeting, Nevada, November, 1975.

Gleason, J.M.: "Integer Programming Models for Bus Stop Location", presented at the Puerto Rico ORSA/TIMS meeting, San Juan, Puerto Rico, 1974.

Halpern, J.: "The Location of a center-median combination on an undirected tree", Journal of Regional Science, pp. 237- 245, 16, 1976.

Halpern, J.: "Finding center-median convex combination (cent- dian) of a graph", Management Science, pp. 535-544, 24, 1978a.

Handler, G.Y.: "Medi-Centers of a Tree", Working Paper 278/76, Tel Aviv University, 1976.

______/Mirchandani, P.B.: "Location on Networks: Theory and Algorithms", MIT Press, Cambridge, 1979.

Khumawala, B.M.: " An Efficient Algorithm for the P-Median- Problem with Maximum Distance Contraints", Geographical Analysis, pp. 309-321, 5, 1973.

______/"Algorithm for the P-Median-Problem with Maximum Distance Constraints: Extension and Reply", Geographical Analysis, pp. 91-95, 7, 1975.

Levin, N.: "Location Problems on Weighted Graphs", unpublished M.S. Thesis, Tel-Aviv University, Israel, 1977.

Loesch, A.: "Die räumliche Ordnung der Wirtschaft", Fischer, Jena, 1943.

Lowe,T.J.: "Efficient Solutions in Multi Objective Tree Network Location Problems", Transportation Science, pp. 298-316, 12, 1978.

Odoni, A.R.: "Location of Facilities on a Network: A Survey of Results", Technical Report 03-74, OR Center MIT, Cambridge, 1974.

Ross, G.T./Soland, R.M.: "A Multicriteria Approach to the Location of Public Facilities, European Journal of Operational Research, pp. 307-321, 4, 1980.

Torregas, C./Swain, R./ReVelle, Ch.S./Bergman, L.: "The Location of Emergency Service Facilities", Operations Research, pp. 1363-73, 19, 1971.

Verfahren zur Lösung von vehicle scheduling-Problemen im Öffentlichen Personennahverkehr

Joachim R. Daduna, Hamburg

Das Problem des vehicle schedulings nimmt bei der Planung des Betriebsmitteleinsatzes in Betrieben des öffentlichen Personennahverkehrs eine zentrale Stellung ein. Dies bezieht sich sowohl auf die Bereitstellung von Fahrzeugen als auch auf den Bedarf an Fahrpersonal. Aus der maximalen Anzahl von Fahrzeugumläufen in den Verkehrsspitzen ergibt sich eine Untergrenze für den benötigten Fuhrpark. Die Gesamtumlaufdauer determiniert in einem wesentlichen Umfang den Bedarf an Fahrpersonal zur Erbringung der angebotenen Verkehrsleistung.

Ausgangspunkt für das vehicle scheduling ist die durch einen Fahrplan vorgegebene Fahrtenmasse. Da aufgrund der zeitlichen und räumlichen Komponenten eines Fahrplanes ein Fahrzeug in einem Planungszeitraum mehr als eine Fahrt leisten kann, läßt sich die Zuordnung von Fahrzeugen zu den Fahrten als ein spezielles Reihenfolgenproblem beschreiben. Zur Lösung dieses Problems sind in den vergangenen Jahren, meist eingebunden in DV-gestützte Planungssysteme, eine Reihe von Konzeptionen entwickelt worden. Diese lassen sich nach verschiedenen Kriterien unterteilen, wie:

° Minimierung des maximalen Fahrzeugbedarfs oder Minimierung der nichtproduktiven Zeiten als Zielsetzung.

° Anwendung einer "cluster first - schedule second" - oder "schedule first - cluster second"-Vorgehensweise.

° Beschränkung des Planungsproblems auf die Fahrten mit Personenbeförderung oder Einbeziehung der Ein- und Aussetzfahrten (Fahrten vor der ersten bzw. nach der letzten Fahrt mit Personenbeförderung).

° Einsatz von Heuristiken oder Optimierungsverfahren zur Lösung des kombinatorischen Problems.

° Automatisierungsgrad bei der Erstellung eines ersten zulässigen Umlaufplanes.

Ein exakter Vergleich der verschiedenen Konzeptionen auf der Grundlage parallel durchgeführter Tests existiert nicht. Ebenso sind Aussagen über die Effizienz mit Hilfe des Umlaufwirkungsgrades (Summe der Fahrten mit Personenbeförderung zu Gesamtumlaufdauer) nicht möglich, da die Rahmenbedingungen in den Verkehrsbetrieben sehr uneinheitlich sind. Verschiedentlich durchgeführte Vergleichstests zwischen manueller Arbeitsweise und dem Einsatz DV-gestützter Planungssysteme lassen allerdings eine deutliche Tendenz erkennen. Die Anwendung einer "schedule first - cluster second"-Vorgehensweise, verbunden mit dem Einsatz von exakten Optimierungsverfahren, bei denen ohne manuelle Eingriffe eine zulässige Lösung erstellt wird, bringt die deutlichsten Verbesserungen der Planungsergebnisse.

MEHR-DEPOT-TOURENPLANUNG

Ulrich Eberhard, Hans-Joachim Vaterrodt, Mannheim

Seit den fünfziger Jahren erscheinen in der Operations Research Literatur Veröffentlichungen zu Problemen der Tourenplanung. Der überwiegende Teil dieser Arbeiten beschränkt sich aber bei der Entwicklung von Lösungsverfahren auf die Betrachtung lediglich eines Depots. Werden in einem vielgliedrigen Logistiksystem jedoch umfassende Lösungen angestrebt, so ist eine simultane Betrachtung aller Depots, also eine Mehr-Depot-Tourenplanung erforderlich.
Neben einem kurzen Überblick über die bekannten Ansätze steht die Entwicklung eines neuen Verfahrens zur Lösung des Mehr-Depot-Tourenplanungsproblems im Zentrum der Betrachtung.

Bei der neuen Heuristik erfolgt die Bestimmung eines Tourenplans in drei Phasen:

- Zusammenfassung der Kunden zu Gruppen
- Verbesserung der Gruppeneinteilung
- Bestimmung der Kundenreihenfolge innerhalb einer Gruppe.

In der ersten Phase des Verfahrens werden Kunden aufgrund ihrer räumlichen Nähe zueinander zu Gruppen zusammengefaßt.
In der zweiten Phase wird die so erhaltene Gruppeneinteilung durch den Austausch von Kunden zwischen Gruppen verbessert. Da in dieser Phase noch kein Reihenfolgeproblem gelöst werden soll, wird die Vorteilhaftigkeit für das Entfernen eines Kunden aus einer Gruppe und dessen Einfügen in eine andere Gruppe nach einer Hilfszielfunktion bestimmt. Dabei hat sich die Verwendung der mittleren quadrierten Abstände zum Schwerpunkt einer Gruppe als günstig erwiesen.
Erst in der dritten Phase des Ansatzes wird mit den bekannten 2-opt/3-opt Verfahren für jede Gruppe das Travelling Salesman Problem gelöst.
Die Ergebnisse von Testrechnungen und deren Vergleich mit Resultaten aus der Literatur demonstrieren die Leistungsfähigkeit der neuen Heuristik.

Lit.: Eberhard, Ulrich, Mehr-Depot-Tourenplanung, München 1987
(erscheint voraussichtlich Ende November 1987)

EIN VERGLEICH DER FLEXIBILITÄT VON VERKEHRSMITTELWAHLMODELLEN

Jürgen Gerken, Bremen

Zur Beschreibung der Verkehrsmittelwahl (Modal Split) werden fast ausschließlich Probit- und Logitmodelle verwendet. Während im Falle der binären Wahl Probit- und Logitmodelle im wesentlichen äquivalent sind, ergeben sich für mehr als zwei Wahlmöglichkeiten wesentliche Unterschiede.

Das Probitmodell ist sehr flexibel, da es unterschiedliche Ähnlichkeiten zwischen den Alternativen zuläßt. Ein Nachteil ist jedoch, daß die auftretenden Integrale numerisch ausgewertet werden müssen. Dadurch wird das Probitmodell schwer handhabbar. Beim Logitmodell hingegen sind die Wahrscheinlichkeiten analytisch auswertbar. Dies ist sowohl für den Rechenaufwand, als auch für die Transparenz der funktionalen Abhängigkeiten von großer Bedeutung. Es ist jedoch nicht möglich, die Wahlwahrscheinlichkeiten mit hinreichender Genauigkeit zu bestimmen, wenn sehr unterschiedliche Ähnlichkeiten zwischen den Alternativen bestehen.

Eine Möglichkeit zur partiellen Überwindung dieser mangelnden Flexibilität stellt das hierarchische Logitmodell (nested logit model) dar. Das hierarchiesche Logitmodell ist bei richtiger Wahl der Hierarchie flexibler als das Logitmodell, jedoch weist es noch nicht die gleiche Flexibilität wie das Probitmodell auf.

Durch Berechnung der Funktionalmatrix (Matrix der partiellen Ableitungen) in einem ausgewählten Punkt lassen sich die verschiedenen Modelle hinsichtlich ihrer Flexibilität beschreiben. Ferner können funktionale Abhängigkeiten zwischen den erklärenden Variablen und zugehörigen Wahlwahrscheinlichkeiten durch Berechnung der Funktionalmatrix transparent gemacht werden.

Dadurch ist es möglich, eine Antwort auf die folgende Frage zu geben: In welcher Situation ist welches Modell geeignet?

The problem of "fuzzy" constraints in computerised planning

Pieter Klapwijk
AMSTERDAM

Computerised planning techniques are applied in a wide variety of environments. Their success has been proven for strategic and tactical planning problems.
For operational problems, however, successfull implementations are very few indeed.
It is argued that the lack of success does not lie in the inability to cope with the mathematics required to solve the problem, but in the inability to define the problem accurately. This is mainly caused by the fact that constraints are often "fuzzy" in nature, implying that they cannot be expressed in mathematical terms.
Computerised operational planning systems should therefore focuss primarily on problem manipulation-, rather then problem solving techniques.
It will be argued that once we are able to manipulate the problem definition effectively, it is relatively easy to find satisfactory solutions.

TOURENPLANUNG MIT EINEM PERSONAL COMPUTER

W.G. Kolenbrander, Eindhoven

Beim letzten Glied der Logistik-Kette - dem externen Transport - steckt die praktische Anwendung von automatisierten Systemen noch in den Kinderschuhen. Das Interesse wächst allerdings ständig. Das Tourenplanungssystem, das auf diesem Kongreß behandelt wird, basiert auf der besonders effektiven Operations Research-Technik: dem Savings-Algorithmus von Clarke und Wright. Dieser Algorithmus löst nach dem heuristischen Prinzip das klassische Transportproblem, nämlich die Verteilung einer Menge Waren aus einem Depot mit Hilfe eines gegebenen Fahrzeugparks und unter Berücksichtigung bestimmter Einschränkungen.

Die schnelle Entwicklung der Mikroelektronik ermöglichte es, diese OR-Technik von einem Großrechner auf einen Mikrocomputer zu verlagern, wobei angesichts der doch begrenzten Verarbeitung folgende Faktoren entscheidend sind: 1. Durchlaufzeit des Planungsprozesses, 2. MS-DOS-Standard, 3. Einpassung von PC und Software in bestehende Computerkonfigurationen, 4. Anwenderfreundlichkeit.

Der rechenintensive Teil des Modells erfolgt auch weiterhin auf einem Großrechner: 1. das Einrichten einer Abstandsmatrix (Dekompositionstechnik), 2. das Einrichten einer Savings-Datei für das Depot. Auf dem PC ist entwickelt:

- ein Informationseingabe- und änderungssystem
- Systeme für das Einrichten von einem Subset der Savings-Datei und anschließend das Einrichten einer begrenzten Abstandsmatrix
- Tourenplansystem, in das ggf. vom Planer eingegriffen wird
- Planungswahl-Menü für manuelle Anpassung und Displays
- Programme für wechselweises Laden von Eingabe und Planungsergebnis

Anwendung: Landesweite Distribution per firmeneigenem und gewerblichem Güterverkehr.

WAGENUMLAUFPLANUNG IM ÖPNV BEI BEGRENZTEN BETRIEBSHOFKAPAZITÄTEN
EIN GANZZAHLIGES MEHRGÜTERFLUSSMODELL MIT LÖSUNGSANSÄTZEN

Achim Lamatsch, Karlsruhe

Zusammenfassung: Die Bildung von Wagenumlaufplänen im ÖPNV läßt sich als ganzzahliges Mehrgüterflußproblem darstellen. Das Problem kann gelöst werden, indem ein Eingüterflußproblem als Lagrange-Relaxation des Mehrgüterflußproblems benutzt wird. Da die Bestimmung der Lagrange-Multiplikatoren im allgemeinen relativ lange Rechenzeiten erfordert, wird eine Heuristik vorgestellt, die bei einer beliebigen Lösung der Lagrange-Relaxation die Verletzungen der Mehrgüterflußbedingungen erkennt und sie durch Änderungen der Flußwerte schrittweise beseitigt.

Summary: The vehicle scheduling problem in public transport can be represented as a multicommodity network flow problem. This problem can be solved using a singlecommodity network flow problem as a lagrangean relaxation. For the determination of the lagrangean multipliers usually takes a long computation time, a heuristic is presented that reduces step by step the number of violations of the multicommodity flow constraints of any given solution of the Lagrangean relaxation.

1. Einleitung

Die Bildung von Wagenumlaufplänen im öffentlichen Personennahverkehr (ÖPNV) läßt sich überführen in das Problem, einen kostenminimalen Fluß in einem Netzwerk zu bestimmen (siehe z.B. Carraresi/Gallo 1984). Für die Wagenumlaufplanung mit einem Betriebshof mit einem Fahrzeugtyp reicht die Darstellung als Eingüterflußproblem aus, um auch bei begrenzter Betriebshofkapazität einen optimalen Wagenumlaufplan zu erhalten (siehe Rousseau et al. 1987). Wird jedoch ein Eingüterflußalgorithmus zur Wagenumlaufplanung mit mehreren Betriebshöfen eingesetzt, ergeben sich im Normalfall unzulässige Wagenumlaufpläne (siehe Lamatsch 1987).

In dieser Arbeit wird ein Mehrgüterflußmodell vorgestellt, das es erlaubt, unter bestimmten Nebenbedingungen (siehe unten) einen optimalen Wagenumlaufplan zu bestimmen. Das (ganzzahlige) Mehrgüterflußproblem kann durch eine Lagrange-Relaxation auf ein Eingüterflußproblem zurückgeführt und so gelöst werden. Ergänzt wird das Verfahren durch eine Heuristik, die eine dual zulässige Lösung der Lagrange-Relaxation in eine gute zulässige Lösung des Mehrgüterflußproblems umwandelt.

2. Problemstellung

Die Wagenumlaufplanung faßt die durch den Fahrplan vorgegebenen Fahrten zu Umläufen zusammen und ordnet die Umläufe Betriebshöfen zu. Die Wagenumläufe wiederum stellen die Vorgaben für die Dienstplanung dar. Rückwirkungen von der Wagenumlaufplanung auf den Fahrplan, die sich bei einer Sensitivitätsanalyse ergeben (vgl. Daduna 1986), oder Rückwirkungen von der Dienstplanung auf die Wagenumläufe werden nicht betrachtet. Um eine einfachere Problemformulierung zu erhalten, wird davon

Operations Research Proceedings 1987

ausgegangen, daß in jedem Betriebshof genau ein Fahrzeugtyp beheimatet ist. Sind in einem Betriebshof k (k>1) Fahrzeugtypen beheimatet, so wird ein Transformationsschritt durchgeführt, bei dem dieser (reale) Betriebshof durch k (Modell-) Betriebshöfe mit je einem Fahrzeugtyp ersetzt wird. Die sich durch den Transformationsschritt ergebende Anzahl an (Modell-) Betriebshöfen werde mit H bezeichnet.

Betrachtet werden n fahrplanmäßige Fahrten f_1 bis f_n, die zu Umläufen zusammengefaßt und den H (Modell-)Betriebshöfen h_1 bis h_H zugeordnet werden sollen. Jede Fahrt f_i wird beschrieben durch ihren Abfahrtszeitpunkt ABF_i, die Dauer der Fahrt (Fahrzeit) FZ_i, die Endstelle AP_i, bei der die Fahrt f_i beginnt, und die Endstelle EP_i, bei der sie endet. Hinzu kommt für jede Fahrt f_i die Angabe der Betriebshöfe, denen die Fahrt zugeordnet werden kann, durch die Zuordnungsmenge Z_i. Durch den Transformationsschritt sind damit auch die Fahrzeugtypen festgelegt, die Fahrt f_i durchführen können.

Die maximale Anzahl an Fahrzeugen, die vom Betriebhof h_l aus eingesetzt werden kann (die Betriebshofkapazität), werde mit KAP_l bezeichnet. Die Leerfahrzeit vom Endpunkt EP_i der Fahrt f_i zum Anfangspunkt AP_j der Fahrt f_j sei δ_{ij}. Mit $\delta_{h_l i}$ werde die Leerfahrzeit vom Betriebshof h_l zum Anfangspunkt AP_i der Fahrt f_i bezeichnet, mit δ_{ih_l} die Leerfahrzeit von EP_i nach h_l. Es wird angenommen, daß die Dauern der Leerfahrzeiten vom Fahrzeugtyp unabhängig sind.

Ein zulässiger Wagenumlaufplan muß die folgenden Bedingungen B1 bis B6 (vgl. auch Lamatsch 1987) erfüllen:

B1 : Alle fahrplanmäßigen Fahrten müssen durchgeführt werden.

B2 : Jeder Wagenumlauf endet in dem Betriebshof, von dem aus er begonnen wurde.

B2': Die Anzahl der Fahrzeuge muß in jedem Betriebshof am Ende eines Betriebstages genauso groß sein wie am Anfang.

B3 : Die Belastung der Betriebshöfe darf die vorgegebene Kapazitätsgrenze für keinen Fahrzeugtyp überschreiten.

B4 : Ein Fahrzeug kann erneut eingesetzt werden, nachdem es in den Betriebhof zurückgekehrt ist.

B5 : Es dürfen nur Fahrten zu einem Umlauf zusammengefaßt werden, wenn der ganze Umlauf von einem Fahrzeugtyp durchgeführt werden kann.

B6 : Jede Fahrt kann (vor der Optimierung) einem Betriebshof oder einer Betriebshofgruppe zugeordnet werden.

Bemerkung: Die Bedingung B2' stellt eine Abschwächung der Bedingung B2 dar.

Das vorgestellte Modell erlaubt sowohl die Minimierung der Anzahl der maximal eingesetzten Fahrzeuge als auch die Minimierung der Fahrzeugeinsatzkosten, jeweils unter den Bedingungen B1 bis B6. Die Fahrzeugeinsatzkosten ergeben sich als

Summe der Kosten für Leerfahrten und Pausenzeiten. Die Kosten lf_l für eine Minute Leerfahrzeit und die Kosten p_l pro Minute Pausenzeit für ein Fahrzeug vom Typ l ergeben sich aus Betriebs- und Personalkosten und lassen sich als kalkulatorische Kosten monetär angeben. Die Kostenkoeffizienten für die Verknüpfung der Fahrten f_i und f_j berechnen sich als mit lf_l bzw. p_l gewichtete Summe von Leerfahrzeit und Pausenzeit, wenn f_i und f_j mit einem Fahrzeug aus Betriebshof h_l durchgeführt werden. Zu den Kostenkoeffizienten können noch Strafkosten für Linienwechsel o.ä. hinzuaddiert werden.

3. Ein Mehrgüterflußmodell zur Wagenumlaufplanung

Bei dem im folgenden vorgestellten Mehrgüterflußmodell wird zur Darstellung der Wagenumläufe für jeden der H Betriebshöfe ein Gut l verwendet. Die Kapazitätsbeschränkungen der einzelnen Güter werden durch eine spezielle Wahl der Kostenkoeffizienten ersetzt.

Jede Fahrt f_i wird durch zwei Knoten i_1 und i_2 und den Pfeil $\langle i_1, i_2 \rangle$ repräsentiert. Die Minimalkapazität $\lambda_{i_1 i_2}$ und die Maximalkapazität $\kappa_{i_1 i_2}$ werden auf 1 gesetzt. Dies bewirkt, daß die Bedingung B1 eingehalten wird. Die Bewertung $c^l_{i_1 i_2}$ wird auf 0 gesetzt, wenn die Fahrt f_i von einem Fahrzeug aus Betriebshof h_l durchgeführt werden kann, d.h. wenn $l \in Z_i$ ist, sonst auf ∞. So werden die Bedingungen B5 und B6 eingehalten.

Jede mögliche Verknüpfung der Fahrten f_i und f_j wird durch einen Pfeil $\langle i_2, j_1 \rangle$ dargestellt. Die Bewertung $c^l_{i_2 j_1}$ kann für die einzelnen Güter l ($l = 1,\ldots,H$) unterschiedlich gewählt werden, sie wird in Abhängigkeit vom Fahrzeugtyp l auf die Kosten für die Verknüpfung von f_i und f_j gesetzt. Ist die Verknüpfung für Fahrzeugtyp l nicht erlaubt, wird $c^l_{i_2 j_1}$ auf ∞ gesetzt. Die Minimalkapazität $\lambda_{i_2 j_1}$ wird auf 0, die Maximalkapazität $\kappa_{i_2 j_1}$ auf 1 gesetzt.

Die Betriebshöfe h_1 bis h_H werden durch Knoten $b_{l,t}$ mit $l \in \{1,\ldots,H\}$ und $t \in \{0,\ldots,1440\}$ und Pfeile $\langle b_{l,t}, b_{l,t+1} \rangle$ ($t \in \{0,\ldots,1439\}$) dargestellt. Der Knoten $b_{l,t}$ entspricht dem Betriebshof h_l zur Tageszeit t (t wird in Minuten angegeben). Die Pfeile $\langle b_{l,t}, b_{l,t+1} \rangle$ erhalten die Bewertung $c^l_{b_{l,t} b_{l,t+1}}$ von 0 und die Minimalkapazität $\lambda_{b_{l,t} b_{l,t+1}}$ von 0. Für alle anderen Güter l' wird $c^{l'}_{b_{l,t} b_{l,t+1}}$ auf ∞ gesetzt. Dies bedeutet, daß Fahrzeuge nur in ihren Heimatbetriebshof einfahren dürfen. Die Maximalkapazität $\kappa_{b_{l,t} b_{l,t+1}}$ wird auf die Betriebshofkapazität KAP_l gesetzt. Diese Festlegung der Kapazitäten bewirkt, daß die Bedingung B3 eingehalten wird.

Ist vom Betriebshof h_l aus eine Einsetzfahrt zur Fahrt f_i möglich, enthält der Modelldigraph einen Pfeil $\langle b_{l,t}, i_1 \rangle$ mit $t = ABF_i - \delta_{h_l i}$. Entsprechend wird die mögliche Aussetzfahrt von der Fahrt f_i zum Betriebshof h_l durch den Pfeil $\langle i_2, b_{l,t}\rangle$ dargestellt mit $t = ABF_i + FZ_i + \delta_{ih_l}$. Auf diese Weise wird erreicht, daß die Bedingung B4 eingehalten wird. Die Minimalkapazität ist jeweils 0, die Maximalkapazität 1. Der Kostenkoeffizient $c^l_{b_{l,t}, i_1}$ bzw. $c^l_{i_2, b_{l,t}}$ ergibt sich zu $lf_l * \delta_{h_l i}$ bzw. zu $lf_l * \delta_{ih_l}$, für alle anderen Güter sind die Kostenkoeffizienten ∞.

Durch Hinzunahme weiterer Pfeile kann das Minimieren der Zahl der maximal gleichzeitig eingesetzten Fahrzeuge in das Modell eingefügt werden. Hierzu wird für jeden Betriebshof h_l ein Pfeil $\langle b_{l,0}, b_{l,1440}\rangle$ erzeugt mit einer Maximalkapazität $\kappa_{b_{l,0}, b_{l,1440}}$ von KAP_l und einer Minimalkapazität $\lambda_{b_{l,0}, b_{l,1440}}$ von 0. Wird die Bewertung $c^l_{b_{l,0}, b_{l,1440}}$ auf 0 gesetzt, wird der Wagenumlaufplan mit den geringsten Fahrzeugeinsatzkosten ermittelt. Wählt man $c^l_{b_{l,0}, b_{l,1440}}$ kleiner als 0, dann gibt $-c^l_{b_{l,0}, b_{l,1440}}$ an, um wieviel sich die Fahrzeugeinsatzkosten erhöhen dürfen für das Einsparen eines Fahrzeugs des Typs l. Für alle anderen Güter wird die Bewertung des Pfeils $\langle b_{l,0}, b_{l,1440}\rangle$ auf ∞ gesetzt.

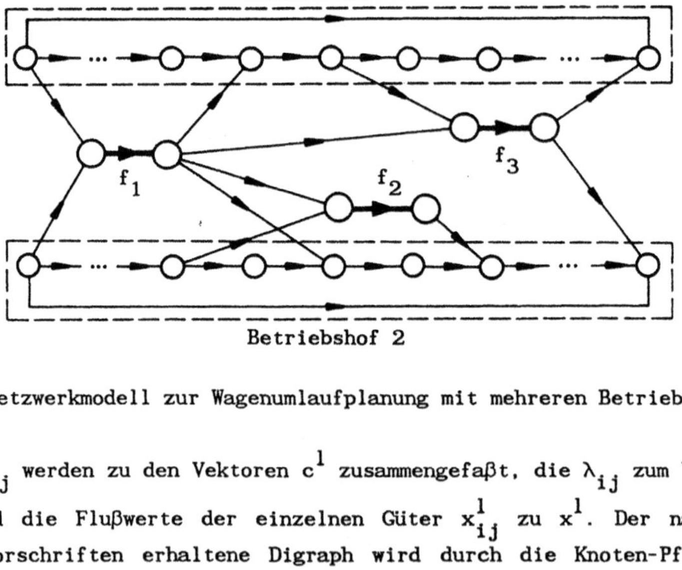

Abb. 1 Netzwerkmodell zur Wagenumlaufplanung mit mehreren Betriebshöfen

Die c^l_{ij} werden zu den Vektoren c^l zusammengefaßt, die λ_{ij} zum Vektor λ, die κ_{ij} zu κ, und die Flußwerte der einzelnen Güter x^l_{ij} zu x^l. Der nach den oben angegebenen Vorschriften erhaltene Digraph wird durch die Knoten-Pfeil-Inzidenzmatrix A dargestellt. Die Höhe a_i^l des Angebots an Fluß des Gutes l betrage KAP_l für den Knoten $b_{l,0}$ und $-KAP_l$ für den Knoten $b_{l,1440}$. Alle anderen Knoten sind Umladeknoten mit einem Angebot a_i^l von 0. Das Angebot der einzelnen Knoten wird zusammengefaßt zum Vektor a^l. Damit ergibt sich das Mehrgüterflußmodell (MGF):

$$
\text{(MGF)}\quad
\begin{array}{lllr}
\text{Min.} & (c^1)^T x^1 + \ldots + (c^H)^T x^H & & (1)\\
\text{u.d.N.} & -x^1 - \ldots - x^H & \leqq -\lambda & (2)\\
 & x^1 + \ldots + x^H & \leqq \kappa & (3)\\
 & A\,x^1 & = a^1 & (4.1)\\
 & \ddots & \vdots & \vdots\\
 & A\,x^H & = a^H & (4.H)\\
 & x^1_{ij} \in \mathbb{Z}_+ & & (5)
\end{array}
$$

Die Nebenbedingungen (4.1) bis (4.H) erzwingen das Einhalten der Bedingung B2. Daraus ergibt sich, daß auch die Bedingung B2' eingehalten wird.

Wird in dem so festgelegten Modell ein kostenminimaler ganzzahliger Mehrgüterfluß bestimmt, kann die Lösung folgendermaßen in einen Wagenumlaufplan umgesetzt werden :

Ist der Fluß $x^1_{i_2 j_1}$ gleich 1, so werden die Fahrten f_i und f_j miteinander verknüpft. Ist der Fluß $x^1_{b_{1,t} i_1}$ gleich 1, so enthält der Wagenumlaufplan eine Einsetzfahrt vom Betriebshof h_1 zur Fahrt f_i, ist der Fluß $x^1_{i_2 b_{1,t}}$ gleich 1, wird an die Fahrt f_i eine Aussetzfahrt zum Betriebshof h_1 angehängt. Der Fluß $x^1_{b_{1,t} b_{1,t+1}}$ gibt an, wieviele Fahrzeuge sich im Zeitintervall $[t,t+1)$ im Betriebshof h_1 befinden.

Das Problem (MGF) kann wegen der Ganzzahligkeitsbedingung (5) nicht mit den üblichen Lösungsverfahren für Mehrgüterflüsse, wie z.B. der Dantzig-Wolfe-Dekomposition (siehe z.B. Kennington/Helgason 1980) bearbeitet werden, da für mehr als zwei Betriebshöfe die optimale Lösung eines Mehrgüterflußproblems nicht notwendig ganzzahlig ist.

4. Ein Lösungsansatz für das Mehrgüterflußmodell mittels Lagrange-Relaxation

Die Nebenbedingungen (4.1) bis (4.H-1) von (MGF) werden zu (4.H) addiert und zusätzlich bei der Lagrange-Relaxation (LGR1) des Problems (MGF) in die Zielfunktion übernommen.

$$
\text{(LGR1)}\quad
\begin{array}{lllr}
\text{Min.} & (c^{1\,T} - u^1 A)x^1 + \ldots + (c^{H-1\,T} - u^{H-1} A)x^{H-1} + c^{H\,T} x^H & & (6)\\
\text{u.d.N.} & -x^1 - \ldots - x^{H-1} - x^H & \leqq -\lambda & (7)\\
 & x^1 + \ldots + x^{H-1} + x^H & \leqq \kappa & (8)\\
 & A\,x^1 + \ldots + A\,x^{H-1} + A\,x^H & = a^1 + \ldots + a^H & (9)\\
 & x^1_{ij} \in \mathbb{Z}_+ & & (10)
\end{array}
$$

(LGR1) kann zurückgeführt werden auf die Bestimmung eines kostenminimalen Eingüterflusses (wodurch auch die Ganzzahligkeitsbedingung (10) entfällt) durch die Festlegung

$$x_{ij} := x_{ij}^1 + \ldots + x_{ij}^H \qquad \text{für alle } \langle i,j\rangle \qquad (11)$$

$$\bar{c}_{ij} := c_{ij}^{l^*} - u_i^{l^*} + u_j^{l^*} := \min_{l=1,\ldots,H} c_{ij}^l - u_i^l + u_j^l \qquad \text{für alle } \langle i,j\rangle \qquad (12)$$

(Hierbei werden alle u_i^H und u_j^H als Null angenommen.) Damit ergibt sich (LGR2)

$$\text{(LGR2)} \qquad \begin{array}{lll} \text{Min.} & \bar{c}^T x & (13) \\ \text{u.d.N.} & -x \leq -\lambda & (14) \\ & x \leq \kappa & (15) \\ & Ax = a^1 + \ldots + a^H & (16) \end{array}$$

Eine optimale Lösung x des Problems (LGR2) wird in eine optimale Lösung x^1 bis x^H des Problems (LGR1) umgesetzt durch die Vorschrift

$$x_{ij}^{\;l} := \begin{cases} x_{ij} & \text{falls } l = l^* \quad \text{(aus (12))} \\ 0 & \text{sonst} \end{cases} \qquad (17)$$

5. Ein Algorithmus zum Korrigieren der Lösung der Lagrange-Relaxation in eine zulässige Lösung des Mehrgüterflußproblems

Die bei der Lagrange-Optimierung erhaltene Lösung stellt eine dual zulässige Lösung des Mehrgüterflußproblems (MGF) dar, d.h. die Lösung ist (sofern sie nicht optimal ist) nicht primal zulässig und kann deshalb nicht in einen (unter den Bedingungen B1 bis B6) zulässigen Wagenumlaufplan umgesetzt werden. Deshalb wurde ein Algorithmus entwickelt, der durch Änderungen der Flußwerte die Lösung der Lagrange-Relaxation (LGR1) in eine zulässige Lösung des Mehrgüterflußproblems (MGF) umwandelt.

Alle ganzzahligen Lösungen des Mehrgüterflußproblems (MGF) lassen sich erzeugen aus einer Lösung x der Lagrange-Relaxation (LGR2), die durch Flußänderungen entlang von Zyklen im Inkrementdigraphen zu x modifiziert und so in eine Lösung x^1 bis x^H von (MGF) zerlegt wird, daß die Knotenbedingungen (4.1) bis (4.H-1) erfüllt sind (siehe Meinhardt 1987). Jedoch ist bereits für relativ kleine Probleme die Anzahl der Zyklen derart groß, daß die Untersuchung aller Zyklen und deren Linearkombinationen nicht mehr in vernünftiger Zeit möglich ist. Für praktische Anwendungen muß deshalb die Menge der zu untersuchenden Zyklen reduziert werden.

Hierzu werden die Knoten i identifiziert, für die die Flußbedingungen (4.1) bis (4.H-1) nicht erfüllt sind. Nur von diesen Knoten i aus werden im Inkrementdigraphen zur aktuellen Lösung x Zyklen gesucht, so daß nach Flußänderung entlang eines solchen Zyklus für den Knoten i die Flußbedingungen eingehalten werden, ohne die Flußbedingungen für einen anderen Knoten zu verletzen. Existieren mehrere solcher Zyklen, wird derjenige mit geringster Kostenerhöhung ausgewählt. Da jeweils nur Zyklen im Inkrementdigraphen gesucht werden, wird die Betriebshofkapazität nie überschritten. Die sich nach der Flußänderung ergebende Lösung wird als neue aktuelle Lösung x verwendet und auf Verletzungen der Flußbedingungen hin untersucht.

6. Ergebnisse

Liegt ein Problem mit einem Betriebshof und einem Fahrzeugtyp vor, läßt sich die ermittelte optimale Lösung von (LGR2) direkt in die optimale Lösung des Wagenumlaufplanungsproblems umsetzen. Liegt ein Problem mit mehreren Betriebshöfen und/oder mehreren Fahrzeugtypen vor, so kann es sein, daß diese Lösung die Flußbedingungen des Mehrgüterflusses nicht erfüllt, so daß die Lösung nicht in einen zulässigen Wagenumlaufplan umgesetzt werden kann. Sichergestellt ist jedoch, daß die Bedingungen B1, B2', B3 und B4 eingehalten werden. Die Bedingungen B2, B5 und B6 können also verletzt sein (vgl. Lamatsch 1987). Das Einhalten der Bedingungen kann durch Anwenden des Korrekturalgorithmus oder durch Änderung der Lagrange- Multiplikatoren erreicht werden. Geeignete Strategien zur Wahl der Lagrange-Multiplikatoren werden zur Zeit untersucht, ein geeigneter Ansatz scheint die Subgradientenoptimierung zu sein (siehe Held et al. 1974).

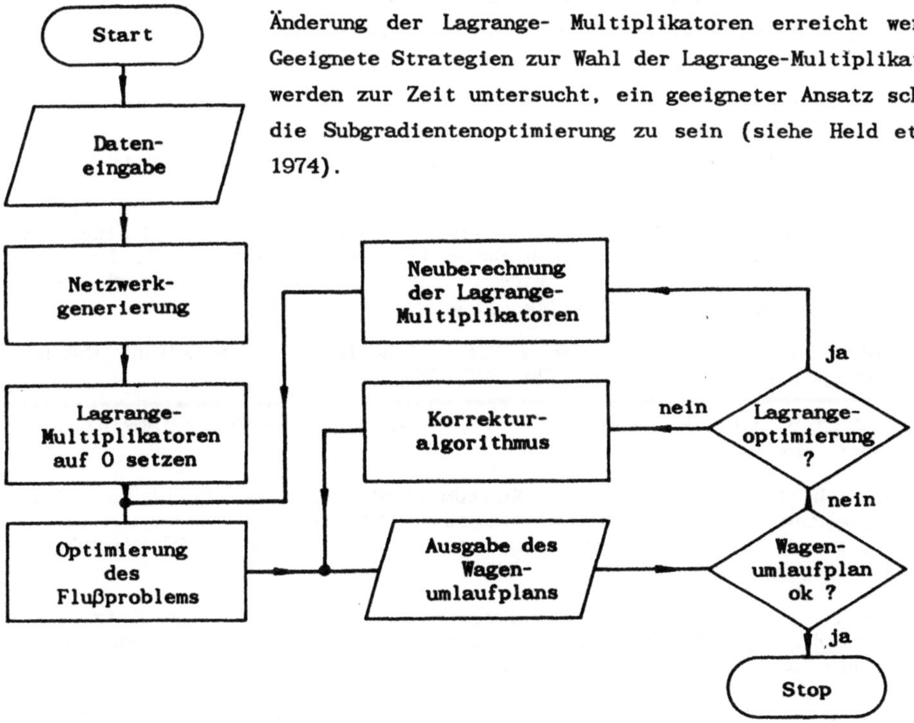

Abb. 2 Das Zusammenwirken der vorgestellten Algorithmen

Die Lösungen der Lagrange-Relaxation konvergieren gegen die optimale Lösung des Mehrgüterflußproblems (MGF), wobei die Anzahl der Verletzungen der Bedingungen B2, B5 und B6 abnimmt. Da jedoch die Rechenzeit für große Probleme bis zum Erhalten der optimalen Lösung des Mehrgüterflußproblems (MGF) sehr lang sein kann, gibt es die Möglichkeit, die Rechenzeit vorzugeben und dann zu entscheiden, ob neue Lagrange-Multiplikatoren berechnet werden sollen oder ob der Korrekturalgorithmus eingesetzt wird, der eine gute zulässige Lösung erzeugt.

Der beschriebene Korrekturalgorithmus findet in vielen Fällen schnell eine gute zulässige Lösung, die relative Abweichung des Zielfunktionswertes der Lösung des Korrekturalgorithmus vom Optimum verringert sich mit wachsender Problemgröße. Bei ungünstiger Wahl der Zyklen besteht jedoch die Möglichkeit, daß der Algorithmus keine weitere Korrekturmöglichkeit findet, obwohl eine zulässige Lösung existiert. Dieser Fall muß durch Eingriffe von Hand bearbeitet werden, indem eine Änderung der Flußwerte durchgeführt wird, die keine Verringerung der Anzahl der Verletzungen der Flußbedingungen bewirkt.

Soweit sich bisher absehen läßt, eignet sich das vorgestellte Verfahren insbesondere für die Wagenumlaufplanung im ländlichen Raum mit wenigen, dafür langen Fahrten und großen Leerfahrzeiten zwischen den Endstellen.

Literaturverzeichnis

[1] Carraresi, P.; Gallo, G.: Network models for vehicle and crew scheduling. EJOR 16, S.139-151, 1984

[2] Daduna, J.: Anwendung einer Dialog-gesteuerten Sensitivitätsanalyse bei Fahrzeugumlaufbildung in Verkehrsbetrieben. Aus : Streitferdt, L.; Hauptmann, H.; Marusev, A.W.; Ohse, D.; Pape, U. (eds.): Operations Research Proceedings, DGOR, Vorträge der 14. Jahrestagung, Springer, Berlin, 1986

[3] Held, M.; Wolfe, P.; Crowder, H.P.: Validation Of Subgradient Optimization. Mathematical Programming 6, 1974, S.62-88

[4] Kennington, J.L.; Helgason, R.V.: Algorithms for Network Programming. John Wiley & Sons, New York, 1980

[5] Lamatsch, A.: Modelle zur Wagenumlaufplanung im ÖPNV bei begrenzten Betriebshofkapazitäten. Aus : Isermann, H.; Merle, G.; Rieder, U.; Schmidt, R.; Streitferdt, L. (eds.): Operations Research Proceedings, DGOR, Vorträge der 15. Jahrestagung, Springer, Berlin, 1987

[6] Meinhardt, A.: Entwicklung eines Korrekturalgorithmus für verletzte Boolesche Nebenbedingungen bei Wagenumlaufplänen. Diplomarbeit, Institut für Wirtschaftstheorie und Operations Research der Universität Karlsruhe, Karlsruhe, 1987

[7] Rousseau, J.-M.; Lessard, R.; Désilets, M.: ALIAGES : A System For The Assignment Of Bus Routes To Garages. Publication #531. Centre de recherche sur les transports, Université de Montréal, Montréal, 1987

EIN MODELL FÜR DIE PLANUNG VON VERKEHRSBEGRENZUNGSMASSNAHMEN

Gert Marte, Bremen

Zusammenfassung: In der vorliegenden Arbeit wird ein Verkehrsmodell beschrieben, das ein stationäres Staumodell enthält und die Rückwirkung des Straßennetzes auf das Verkehrsaufkommen als elastische Nachfrage berücksichtigt. Es läßt sich mit diesem Modell berechnen, wie sich verschiedene Maßnahmen zur Begrenzung des Autoverkehrs auf die Verkehrsflüsse und Staus auswirken.

1. Einleitung

In der Broschüre STADTVERKEHR IM WANDEL, 1986, die vom Bundesminister für Raumordnung, Bauwesen und Städtebau herausgegeben wurde, wird auf die methodischen Probleme bei der Begrenzungsplanung eingegangen.
Es wird darauf hingewiesen, daß die bisher für die Entwicklung von Generalverkehrsplänen benutzte Methodik von der Überzeugung ausgeht, daß der Autoverkehr weiter wachsen muß. Außerdem wird erwähnt, daß die üblichen Modelle an der Grenze der Überlastung unbrauchbar sind.
Modelle für die Planung von Verkehrsbegrenzungsmaßnahmen müssen die Rückwirkung des Verkehrsnetzes auf das Verkehrsaufkommen erfassen. Außerdem wird ein Staumodell benötigt, da Staus verkehrsbegrenzend wirken und die Verhinderung von Staus durch Ausbaumaßnahmen zu einer Steigerung des Verkehrsaufkommens führt.
Das Verkehrsaufkommen kann sich durch eine Änderung der Erzeugung, der Verteilung und der Verkehrsmittelwahl verringern. Schon BECKMANN, 1956 hat ein exaktes Gleichgewichtsverfahren für die Wegewahl mit elastischer Nachfrage angegeben. Änderungen der Wegewahl und der Erzeugung in Abhängigkeit vom Verkehrsnetz lassen sich also durch ein exaktes Gleichgewichtsverfahren berechnen. Zur Berücksichtigung der Verkehrsmittelwahl ist eine heuristische Erweiterung des Gleichgewichtsverfahrens notwendig. LUTTER, 1987 hat gezeigt, daß simultane Verfahren einen größeren Konvergenzbereich besitzen als sequentielle Verfahren. Da die Konvergenz von der Steilheit der Reisezeitkennlinien abhängt, und die Reisezeitkennlinien an der Grenze der Überlastung stark ansteigen, ist für Stauanalysen ein simultanes Verfahren sinnvoll. Die Änderung der Verkehrsverteilung in Abhängigkeit von Netzänderungen wird bisher nicht berücksichtigt.

SCHWERDTFEGER, 1987 berücksichtigt die Rückwirkung des Straßennetzes auf die Verkehrsmittelwahl. Er berechnet die Umlegung und die Verkehrsmittelwahl sequentiell, was zur Analyse der flächenhaften Verkehrsberuhigung in der Regel ausreichen wird. MARTE, 1986 berücksichtigt die Verkehrsmittelwahl in einem simultanen Verfahren, so daß die Analyse stationärer Staus möglich wird. INOYE, 1986 hat ein exaktes Lösungsverfahren für die Berechnung stationärer Staus für den Fall angegeben, daß die Reisezeitfunktionen an der Kapazitätsgrenze senkrecht ansteigen. Er berücksichtigt aber nur die Wegewahl.
In der vorliegenden Arbeit wird ein stationäres Staumodell benutzt und in einem simultanen Gleichgewichtsverfahren die Rückwirkung des Straßennetzes auf die Wegewahl und die Erzeugung erfaßt. Als Gleichgewichtsverfahren wird wie in MARTE,1986 die sukzessive Gleichgewichtsumlegung benutzt. Durch Beispiele wird gezeigt, daß sich die in der Broschüre STADTVERKEHR IM WANDEL, 1986 und von KNOFLACHER, 1985 beschriebenen Konzepte für die Verkehrsbeschränkung analysieren, vergleichen und integrieren lassen.

2. Stationäres Staumodell

In Bild 1 ist das benutzte stationäre Staumodell dargestellt. Die Annahme von stationären Flüssen und stationären Staus vereinfacht die Berechnung ganz wesentlich. Es wird deshalb angenommen, daß sich eine konstante Zahl von Fahrzeugen im Stau befindet und sich die Staulänge durch die Rückwirkung der Stauzeit auf den Fluß stabilisiert. Die Rückwirkung wird bei dem Modell durch längere Alternativrouten oder durch eine elastische Nachfrage verursacht. Da sehr geringe Flußänderungen die Reisezeiten stark ändern, muß eine sehr genaue Einpegelung der Flüsse angenommen werden. Bestätigt wird das durch eine relativ stabile Staubildung während der Hauptverkehrszeiten. Tägliche Schwankungen der Stauzeiten treten natürlich ein. Die Verkehrsteilnehmer können sich dann nur an mittleren Staulängen orientieren.

3. Elastische Nachfrage

Durch die Annahme einer elastischen Nachfrage soll nach Bild 1 berücksichtigt werden, daß die Flüsse zwischen Quellen und Senken bei wachsender Reisezeit geringer werden. Um das Modell leicht auf mehrere Verkehrsmittel erweitern zu können, wird die elastische Nachfrage durch ein Logit-Modell beschrieben

$$F\,[I,J] = F1\,[I,J]\;\frac{2}{1 + \exp\,(B(T[I,J] - T1\,[I,J]))} \qquad (1)$$

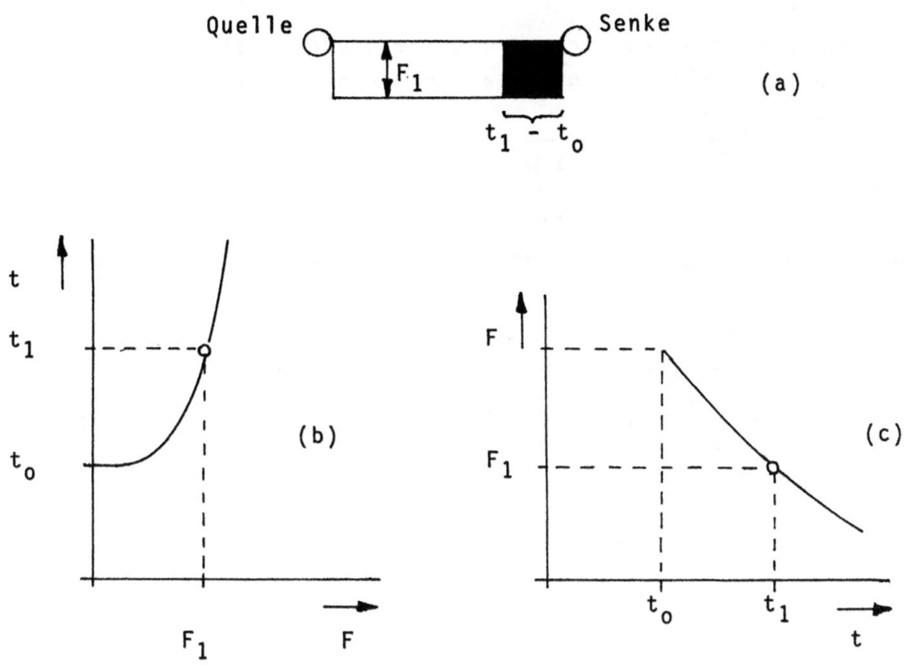

Bild 1 Stationäres Staumodell
a) Straßenverbindung von einer Quelle zu einer Senke
b) Reisezeitfunktion
c) Nachfragekurve

T1 ist die Reisezeit im gemessenen Nullfall und F1 der gemessene Quelle-Ziel-Fluß. Die Konstante B bestimmt die Stärke der Rückwirkung. Wenn man annimmt, daß Netzdaten und eine OD-Matrix gegeben sind, dann kann man zur Kalibrierung für fest gegebene OD-Flüsse eine Umlegungsrechnung machen und daraus die Zeiten T1 [I,J] berechnen. Bei der Analyse von Netzänderungen wird dann die elastische Nachfrage nach G1 (1) berücksichtigt, die für T [I,J] = T1 [I,J] auch F [I,J] = F1 [I,J] liefert.

Für die Konstante B kann man einen Anhaltspunkt gewinnen, wenn man für die Fahrer ein konstantes Zeitbudget annimmt. Nach GOODWIN, 1981 muß man bei einem Zeitbudget mit einem Variationskoeffizienten von 30% rechnen. Für eine Abschätzung des gesamten Verkehrsaufkommens wäre diese Unsicherheit sehr groß. Da aber das Zeitbudget nur benutzt wird, um Abweichungen vom gemessenen Jetztzustand zu ermitteln, reicht die Genauigkeit für eine Abschätzung aus.

Bei einem als konstant angenommenen Zeitbudget ist die Häufigkeit der Reisen umgekehrt proportional zur Reisedauer.

$$\frac{F\ [I,J]}{F1\ [I,J]} = \frac{1}{\dfrac{T\ [I,J]}{T1\ [I,J]}} \qquad (2)$$

In erster Näherung ergibt sich für kleine $\Delta T = T\ [I,J] - T1\ [I,J]$

$$\frac{F\ [I,J]}{F1\ [I,J]} \approx 1 - \frac{\Delta T\ [I,J]}{T1\ [I,J]} \qquad (3)$$

Aus Gl. (1) folgt für kleine $\Delta T\ [I,J]$

$$\frac{F\ [I,J]}{F1\ [I,J]} \approx 1 - \frac{B}{2}\ \Delta T\ [I,J] \qquad (4)$$

woraus

$$B = \frac{2}{T1\ [I,J]} \qquad (5)$$

folgt (Bild 2).

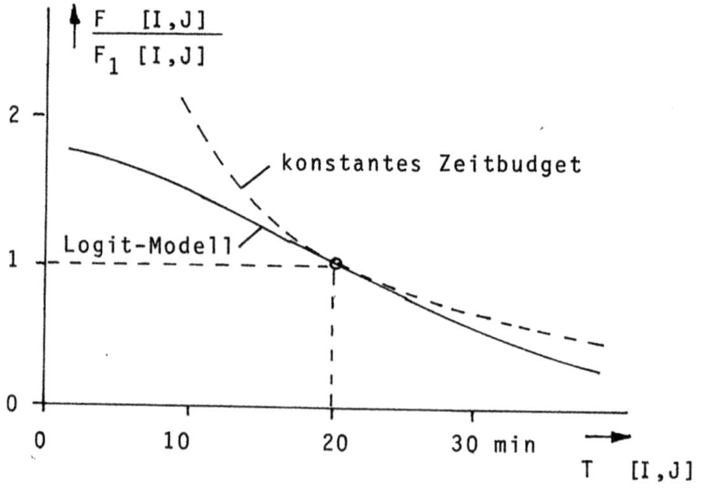

Bild 2 Elastische Nachfrage

4. Beispiel

In den Bildern 3 und 4 ist ein einfaches Beispiel dargestellt. Die Knoten 1, 2 und 3 sind Quellen und Senken, die durch fiktive Kanten mit dem Straßennetz verbunden sind. Durch Reisezeiten auf diesen fiktiven Kanten lassen sich Fußwegzeiten erfassen. Der Weg 4, 5, 6 soll eine Verbindung vom Außenbereich zum Zentrum darstellen, wobei nur der Verkehr zum Zentrum während der Morgenspitze dargestellt wird. Bei der Nullversion sind zwei Strecken 4, 5 und 5, 6 über-

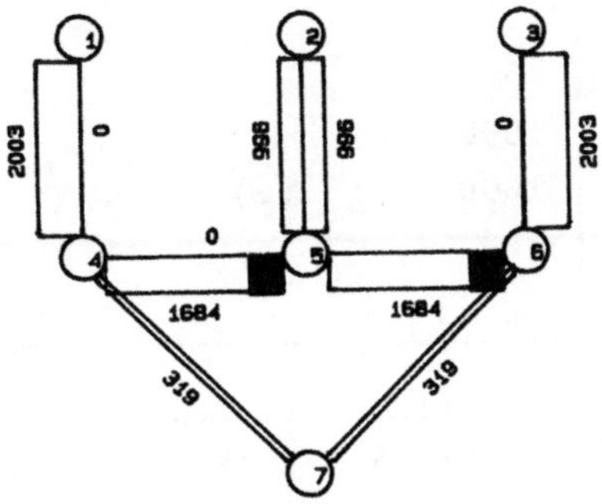

(0)

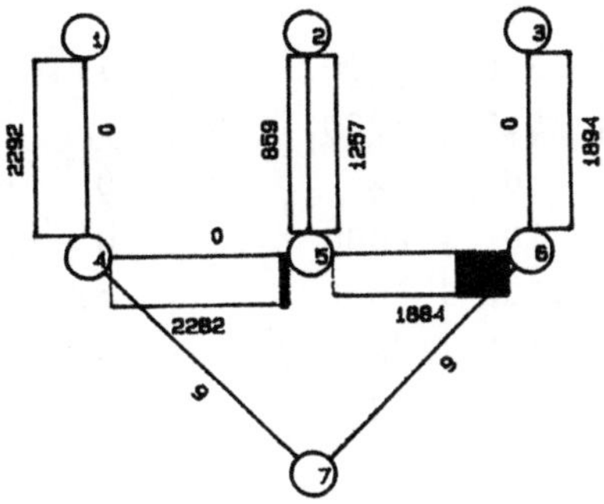

(1)

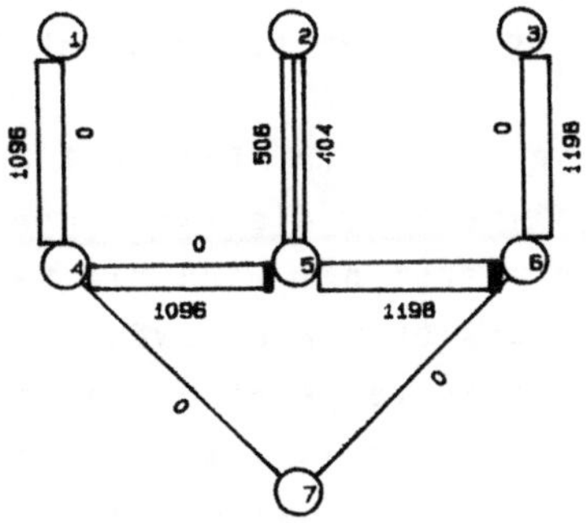

(2)

<u>Bild 3</u> Graphische Darstellung des Beispiels

0) Jetztzustand

1) Anpassungsplanung

2) Beschränkungsplanung

von \ nach	1	2	3
1	0	1000	1000
2	0	0	1000
3	0	0	0

(a)

Version		0			1			2		
AK	EK	K	T0	T1	K	T0	T1	K	T0	T1
1	4	∞	0	0	∞	0	0		5	5.0
2	5	∞	0	0	∞	0	0	∞	5	5.0
3	6	∞	5	5.0	∞	5	5.0	∞	10	10.0
4	1	∞	0	0	∞	0	0	∞	5	5.0
4	5	2000	10	15.0	4000	10	11.1	2000	14	14.1
4	7	2000	15	15.0	2000	15	15	2000	21	21.0
5	2	∞	0	0	∞	0	0	∞	5	5.0
5	4	2000	10	10	2000	10	10	2000	14	14.0
5	6	2000	10	15.0	2000	10	17.9	2000	14	15.9
6	3	∞	5	5.0	∞	5	5.0	∞	10	10.0
7	6	2000	15	15.0	2000	15	15	2000	21	21.0

(b)

Bild 4 Daten für das in Bild 3 dargestellte Beispiel
a) Quelle-Ziel-Matrix für die Nullversion
b) Netzdaten, wobei die gegenüber der Nullversion geänderten Eingabedaten hervorgehoben sind.

lastet mit Stauzeiten von je 5 min. Ein Teil des Flusses (319 Fzg/h) weicht auf eine Alternative aus, die als länger und nicht überlastet angenommen wird. Bei der Version 1 wird die Kapazität auf der Kante 4, 5 verdoppelt. Diese Version wurde gewählt, da sich die Kapazitäten im Außenbereich oft leichter vergrößern lassen als in Zentrumsnähe. Man erkennt, daß sich die Stauzeit auf der Kante 4, 5 von 5 min. auf 1,1 min. verringert. Der Stau wird aber in Richtung Zentrum verlagert. Auf der Kante 5,6 steigt die Stauzeit von 5 min auf 7,9 min an. Außerdem nimmt der Fluß auf Kante 4,5 zu, da sich der Quelle-Ziel-Fluß von Quelle 1 zur Senke 2 vergrößert und der Fluß von der Alternativroute 4,7,6 auf die Route 4,5,6 verlagert wird. Obwohl das Zeitbudget näherungsweise konstant gehalten wird, kann man auf der Kante 4,5, die baulich verändert worden ist, eine Zeitersparnis feststellen. Als Version 2 wird eine Verkehrsbeschränkungsplanung gewählt. Nach

KNOFLACHER, 1985 werden Fußgängerzonen in allen Stadtgebieten angenommen. Als mittlere Fußwegzeiten werden 5 min angenommen, wobei nach KNOFLACHER, 1985 Fußwegzeiten wenigstens um den Faktor 2 höher bewertet werden als Fahrzeiten, so daß reale zusätzliche Fußwegzeiten von 2,5 min ausreichen würden. Zusätzlich wurde entsprechend dem Vorschlag in der vom Bundesminister für Raumordnung, Bauwesen und Städtebau herausgegebenen Broschüre STADTVERKEHR IM WANDEL, 1986 eine Reduzierung der Höchstgeschwindigkeit auf einen Wert zwischen 20 km/h und 40 km/h angenommen. Es wurden die Grundreisezeiten um den Faktor 1,4 vergrößert. Wie Bild 3 zeigt, gehen Staus und Verkehrsflüsse entscheidend zurück. Die von KNOFLACHER, 1985 angegebene Verringerung der Staus läßt sich also mit Hilfe des Modells quantitativ nachvollziehen.

Literatur

BECKMANN, M., Studies in the economics of transportation. Cowles Commission Monograph, Yale University Press, 1956.

BUS DEMONSTRATION PROJECT, Summary Report No.8 Southampton, Department of the Environment, 1975.

GOODWIN, P.B., The usefulness of travel budgets, Transp. Res.-A Vol. 15A, pp. 97-106, 1981.

INOUYE, M., Traffic equilibria and its solution in congested road networks, 5 th IFAC/IFIP/IFORS, Control in Transportation Systems, Wien, 1986.

KNOFLACHER, M., Katalysatoren für Nichtmotorisierte, Wien, 1985.

MARTE, G., BOTHNER, P., HÖNERLOH, H., LUTTER, W., An equilibrium model for the environmental orientated design of traffic networks, Control in Transportation Systems, 5th IFAC/IFIP/IFORS, Wien, 1986.

LUTTER, W., Verfahren zur simultanen Berechnung der Verkehrsmittelwahl und der Wegewahl, Dissertation, Universität Bremen, 1985.

SCHWERDTFEGER, T., Planungsinstrumentarium für Verkehrsberuhigungsmaßnahmen, Heureka 87, Optimierung in Verkehr und Transport, Forschungsgesellschaft für Straßen- und Verkehrswesen, 1987.

STADTVERKEHR IM WANDEL, Bundesminister für Raumordnung, Bauwesen und Städtebau, 1986.

FAHRZEUGEINSATZPROBLEME IN STÄDTEREINIGUNGSUNTERNEHMEN

Heinrich Paessens, Flensburg

Die heutige Entwicklung in der Abfallwirtschaft ist unter anderem durch eine Zentralisierung der Abfallbehandlungsanlagen gekennzeichnet. Die Anzahl der Deponien ist in den letzten Jahren erheblich gesunken. Existierten 1972 in Baden-Württemberg noch 4000 Hausmülldeponien, so ist diese Anzahl bis heute auf 60 Zentraldeponien gesunken. In entsprechendem Maße haben sich die Transportentfernungen vergrößert. Man kann heute davon ausgehen, daß 70% der Gesamtkosten bei der Hausmüllbeseitigung auf den Bereich Sammlung und Transport entfallen. Der optimale Einsatz der Fahrzeuge ist daher von besonderer Bedeutung.

Folgende Problemstellungen bei der Planung des Fahrzeugeinsatzes (Ermittlung von Touren und Zusammenstellung der Touren zu einem täglichen oder wöchentlichen Fahrzeugeinsatzplan) in kommunalen und privaten Städtereinigungsunternehmen treten auf :

- regionale Hausmüll- und Wertstoffsammlung (knotenorientiert)
- innerstädtische Hausmüll- und Wertstoffsammlung (straßenabschnitts-, kantenorientiert)
- Krankenhausmüllentsorgung
- Entsorgung mit Container- (Mulden-) Fahrzeugen
- Straßenreinigung
- Schneeräum- und Streudienst
- Klärschlammentsorgung mit mobilen Pressen

Die praktischen Einsatzmöglichkeiten von Operations Research Verfahren als Planungshilfsmittel zur Lösung derartiger Probleme werden aufgezeigt .

GEOGRAPHICAL MARKET SEGMENTATION

C.J. van der Plas
G.J.R. Forch
J.J. Remmerswaal
G. van der Hoek
H.W. van den Meerendonk

Abstract: This paper discusses various mathematical modeling approaches to solve the geographical market segmentation problem of a specialized bank.

Keywords: geographical market segmentation, quadratic programming, mixed integer programming.

0. Introduction

A specialized bank - an affiliate of ABN Bank - wants to cover the total Dutch market with a limited number of branch-offices of equal size. Equal size is important for various reasons, e.g. comparability, automation and job rotation. Equal sizes can be obtained by giving the branches equal shares of the market. Of course, these shares of the market have to be geographically as compact as possible to minimize travelling distances. Customers are allocated to branches dependent on their addresses. In this paper we look at the problem of determining these compact equal geographical market segments.
In solving this problem, the exact location of the branch-offices is a second order problem. In general, customers do not visit offices; they are visited by branch agents.
We will present two methods which aim to solve the original problem of demarcating geographical market segments without locating branch-offices. These are a quadratic programming model and a heuristic method. A third method demarcates segments and locates offices simultaneously. This is a mixed integer programming model.

For practical purposes we have sought for a method that could be implemented on a IBM-PC.

This paper describes the three solving methods and compares quality of results and technical implementation.

Operations Research Proceedings 1987

1. Preliminaries

We can define our problem as follows:

"How can we divide the Netherlands into a number of geographical segments of equal size in such a way that the travelling distance within each segment is minimal?"

In solving this problem, we start from the following modeling assumptions:

- o The number of geographical segments is fixed to 37.
- o The size of the market is measured by number of inhabitants. The total population is scaled to 37, i.e. the number of segments.
- o The 80, so called, nodal regions are used as a base division of the Dutch market. This is a complete and exhaustive division made by the Central Bureau of Statistics. All economic activity in these regions centers round the nodal point, i.e. a town.
- o Distances are measured between nodal points; distances within nodal regions are zero.

For distances we have used two measures:

1) Euclidian distance.
   These distances have been calculated based on the coordinates of the nodal points. This measure has proven to be a good approximation of the actual - motorway - distance according to BERENS and KORLING (1985) in their study of the situation in the BRD. Their results are confirmed by COOPER (1983) and KOLEN & SNOEI (1986).
2) Actual - motorway - distance.
   BOM (1986) has developped the computer program Customat which calculates distances between ZIP-codes. For each nodal point a central ZIP-code has been determined.

A geographical segment is a combination of one or more - parts of - nodal regions.

In this paper the following definitions and notations will be used:

m : number of branches (37)

n : number of nodal regions (80)

p_i : size of region i, $\sum_{i=1}^{n} p_i = m$ because of the scaling of the total population to the number of inhabitants

d_{ij} : distance between the regions i and j, $d_{ii} = 0$ for $i=1,..,n$

x_{ij} : part of nodal region i covered by branch j

The x_{ij} are restricted by:

$$x_{ij} \geq 0 \qquad i=1,..,n;\ j=1,..,m \tag{1.1}$$

$$\sum_{j=1}^{m} x_{ij} = p_i \qquad i=1,..,n \tag{1.2}$$

$$\sum_{i=1}^{n} x_{ij} = 1 \qquad j=1,..,m \tag{1.3}$$

Restriction (1.2) states that every region has to be covered completely. Restriction (1.3) ensures branches of equal size.

2. Quadratic Programming Model

Our first approach to the problem has been the following formulation:

$$\text{minimize} \sum_{j=1}^{m} X_j^T\ D\ X_j \tag{2.1}$$

with $X_j^T = \{x_{1j},..,x_{nj}\}$

s.t. (1.1) - (1.3)

This formulation maximizes compactness of the segments, i.e. minimizes travelling distances within the segments. The location of the branch-office is left free. However, the matrix D is indefinite. So, many solving methods are not applicable or do not guarantee a minimum solution, neither global nor local. Other methods only guarantee a local extreme.

Another formulation of the objective function is:

$$\text{minimize } X^T\ D^*\ X \tag{2.2}$$

with $X^T = [X_1 \ldots X_m]$

$D^* = I_m \otimes D$

This formulation shows how sparse the D* matrix is. With n=80 and m=37 the total matrix has 8.7 mln elements of which .23 mln are non-zero. Sparsity also complicates optimization.

A third complication with this approach is the size of the model, especially in relation to the IBM PC.

These complications make is necessary to look for solving methods which use few intermediate results and can handle special structures (i.e. sparsity). Conjugated and reduced gradient methods seem suited for this job.

SANDGREN & RAGSDELL (1981) have compared 35 codes and conclude:

> "Of the nonlinear programming algorithms currently available for use, the generalized reduced gradient algorithms are computationally faster and are capable of solving a large variety of problems."

LOOTSMA (1981) confirms these conclusions. DEMBO (1981) reports good results with application of the reduced gradient method on large-scale, semi-definite problems. GILL ET AL. (1984) mention that many sparse problems are solved with standard methods; special software is underdeveloped.
Based on these experiences we have applied the reduced gradient method. We have used the code written by KRIBBE (1982) which is based on the algorithm of BENVENISTE (1979). This code is currently only implemented on a mainframe.

3. Mixed Integer Model

The quadratic programming approach causes severe complications. Therefore, we have sought for an alternative by using a linear objective function.
The formulation:

$$\text{minimize} \sum_{j=1}^{m} D\, X_j \qquad (3.1)$$

s.t. (1.1) - (1.3)

leads to a trivial solution.

We have to formulate our problem as a mixed integer model, that is:

$$\text{minimize} \sum_{j=1}^{M} D^{**}\, X_j \qquad (3.2)$$

$$\text{s.t.} \quad \sum_{j=1}^{M} x_{ij} = p_i \qquad i=1,..,n \qquad (3.3)$$

$$\sum_{i=1}^{n} x_{ij} = y_j \qquad j=1,..,M \qquad (3.4)$$

$$\sum_{j=1}^{M} y_j = m \qquad (3.5)$$

$$x_{ij} \geq 0 \qquad i=1,..,n;\ j=1,..,M \qquad (3.6)$$

$$y_j = [0,1] \qquad j=1,..,M \qquad (3.7)$$

with: d_{ij}^{**} distance between region i and location j

M number of locations

$$y_j \begin{cases} 0 & \text{if location } j \text{ is not used} \\ 1 & \text{if location } j \text{ is used} \end{cases}$$

Some regions are of such size that the location of more than one branch-office has to be considered. So, M is much bigger than n (180 versus 80).

In this approach, geographical segments are grouped around chosen locations for branch-offices.

The high number of variables and the special structure of our problem in combination with the limitation of the IBM PC make a straight forward application of standard - branch and bound - techniques inappropriate. We have implemented an exchange method as described by PAPADIMITRIOU & STEIGLITZ (1982). This method tries to improve a given solution by exchanging λ locations simultaneously. The resulting solution is called λ-optimal.
The higher the λ, the less important is the starting solution according to EILON, GALVAO (1978) and KAUFMAN, BROECKX, PAPADIMITRIOU (1982). However, a higher λ-value causes an increase of the number of substitutions. We have chosen $\lambda=1$.
Every exchange involves the solution of a transportation problem. This is done with a primal-dual algorithm, written by SYSLO & KICHARKYK (1975). Approximative solving methods for the transportation problem have been investigated, but they did not further speed up the model.
The construction of secondary clusters has been a successful improvement: instead of all possible exchanges, only exchanges to clusters of regions are considered. If a cluster does not look promising, this cluster is not investigated further.

4. Heuristic Algorithm

Another way of getting round the complications of the quadratic programming model is the development of a heuristic algorithm. We have investigated a method which demarcates geographical segments without locating the offices.
Roughly sketched, our approach in solving the problem heuristically is as follows:

a) choose a starting nodal region of the regions not already allocated
b) look for the best extension in neighbouring regions until the combined (parts of the) regions are of sufficient size for a branch-office.
c) repeat steps a and b until all regions are allocated to branch segments.

The selection of the starting point is a critical step. Firstly, we only look at regions on the border of the unallocated area. Otherwise, an algorithm is likely to split up the unallocated area in two or more subareas; this complicates the establishment of segments of equal size.
Furthermore we have to start at the most extreme region on the border. We have tried out several concepts of extremeness. The following solution has proven to be robust in the case of the Netherlands. Determine a central nodal region of such a size that it naturally will have an office, e.g. Amsterdam or Utrecht. Calculate the Euclidian distances of all regions to this centre. Choose the region on the border with the largest distance as a starting point.
Given a starting region segment, we have to expand it until the size of the segment

is equal to the size of a branch. We look for neighbouring regions which are relatively close to our starting segments (that is, relatively apart from the remaining unallocated regions).
At first, we have taken relative Euclidian distances as a measure. At a later stage we have replaced it by a measure based on common boundary. Of every neighbouring region we can determine the lenght of the boundary which is common with our starting segment. We relate this lenght to the total boundary of the region, that is boundary of the regions with other Dutch regions. We select the region with the relative largest common boundary.

An important advantage of this approach, especially the boundary measures, is that it is easy to take into account specific factors that influence practical distances, as for instance:
- cultural differences between regions
- quality and quantity of roads between neighbouring regions
- distances accross water.

5. Experiments, Results & Conclusions

Experiments:
In our experiments we have applied the algorithms to:
a) all nodal regions of the Netherlands at once
b) a partitioning of the Netherlands into 4 sets of nodal regions.
The quadratic programmic model has been applied only to b because of memory and CPU constraints; for this experiment we have used a VAX-11. The experiments with the other methods have both been carried out on a IBM/Olivetti-XT, for both a and b.

Results:
- Quadratic Programming Model
  Memory constraints have necessitated us to partition our problem, that is the Netherlands, into four subproblems. We have solved a problem with a X-vector of 280 elements and a D-matrix of size 280x280. A problem with respective sizes 490 and 490x490 could not be solved because of memory constraints.
  As for CPU, the VAX-11 solved the problem using a few minutes for each subproblem.
- Mixed Integer Problem
  This problem has been solved on a PC using the exchange methode. Even with our adjustments to speed up the algorithm, the program takes a considerable amount of

time. For the partitioned problem (b) some minutes, for the total problem (a) a few hours.

- Heuristic Algorithm
  This algorithm is very fast. However, it is recommendable to screen the solution on adjustments.

Conclusions:

If we review our results we conclude in moving from quadratic programming via exchange to heuristic algorithm we lose precision but we gain time. The Q.P. performs good on small problems but can not be applied on larger problems. The exchange method solves problems of considerable size within reasonable CPU-time. The heuristic algorithm is fast, so it is especially suited for generating a starting solution which is adjusted further by one of the other algorithms or by hand.

References

Benveniste, R., A quadratic programming algorithm using conjugate directions, Math. Programming 1979.

Berens, W., and F.J. Korling, Estimating road distances by mathematical functions, EJOR, 1985.

Bom, M.F.M., Het bepalen van afstanden met het basisnetwerk MSc Thesis, Erasmus University Rotterdam, 1986.

Cooper, J.C., The use of straight line distances in solutions of the vehicle scheduling problem, Journal of Operational Research Society, 1983.

Dembo, R.S., Large scale nonlinear optimization in: M.J.D. Powell, Nonlinear optimization 1981, Academic Press 1982.

Eilon, S., and R.D. Galvao, Single and double vertex substitution in heuristic procedures for the P-median problem, Management Science, 1978.

Gill, Ph.E., et al., Trends in nonlinear programming software, EJOR, 1984.

Kaufman, L., and F. Broeckx, The capacitated P-median model for location of facilities in the public sector.

Kolen, A., and M. Snoei, Interactieve routeplanning van bulktransport: een praktijktoepassing, Kwantitatieve Methoden nr 21, 1986.

Kribbe, W., Documentation of the Fortran subroutines for quadratic programming: Conqua and start, Econometric Institute, Erasmus University Rotterdam, 1982.

Kribbe, W., Nonlinear programming algorithms using conjugate directions in reduced dimensions, Econometric Institute, Erasmus University Rotterdam, 1982.

Lootsma, F.A., Performance evaluation of nonlinear optimization methods via multi-criteria decision analysis and via linear model analysis in: M.J.D. Powell, Nonlinear optimization 1981, Academic Press 1982.

Papadimitriou, C.H., and K. Steiglitz, Combinatorial optimization: Algorithm and complexity, Prentice Hall, 1982.

Plas, C.J. van der, Geografische gebiedsafbakening in het kader van een vestigingsplaatsonderzoek, MSc Thesis, Erasmus University Rotterdam, 1987.

Sandgren, E., and K.M. Ragsdell, On some experiments which delimit the utility of non-linear programming methods for engineering design in: M.D.J. Powell, Nonlinear optimization 1981, Academic Press, 1982.

Syslo, M.M. et al., Discrete optimization algorithms, Prentice Hall, 1978.

LOCATION OF ROTTERDAM FIRE STATIONS

Jan Schreuder, Enschede

ABSTRACT

In Rotterdam economizing should have its effects on the organization of the Fire Brigade. In order to deal with these problems the municipality decided to research the location of the fire stations. Based on earlier research by the OR-group in Twente, we were asked to investigate these problems in the following three fases. (I) is the determination of the number of fire stations needed. (II) is the number of fire stations needed with respect to the present locations. (III) is the generation and evaluation of alternative locations in view of safety.

A road netwerk approach (P-median/center) is used for determining the set of possible location sites. The set-covering formulation of the location problem is solved by a translation to a satisfiability problem.

KEYWORDS

Public Service, Location, Network.

INTRODUCTION

Since 1975 there exists a working relation between the Rotterdam Fire Department and the Operations Research Group of the Faculty of Applied Mathematics of the University of Twente. In the first decade the research was mainly devoted to reach such a location for the fire stations in order to fulfill the prescribed norms. The research carried out in 1987 was based on possible reductions of the Department while maintaining the present level of covering.

In this article we first describe some starting points and assumptions. Next we present the used network model: why choosen, how to construct it and how to determine the distances. At last we present some results and some covering measures.

STARTING POINTS AND ASSUMPTIONS

Starting point for the evaluation of different alternatives is the protection against the consequences of fire by the Fire Department. The importance of not exceeding the norms for the attendance times is reflected in the Figures 1-2. If the 'flash-over' point (spontaneous ignition) is passed, the damage to the house and the danger for the inhabitants are increasing very fast. In such a case a fast attendance is the more important

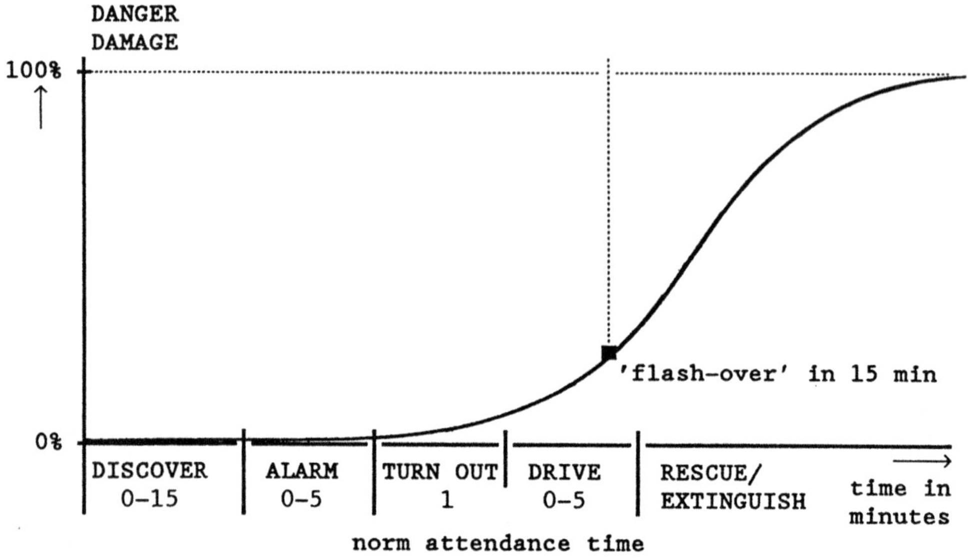

Figure 1 DEVELOPMENT OF DANGER AND DAMAGE IN TIME

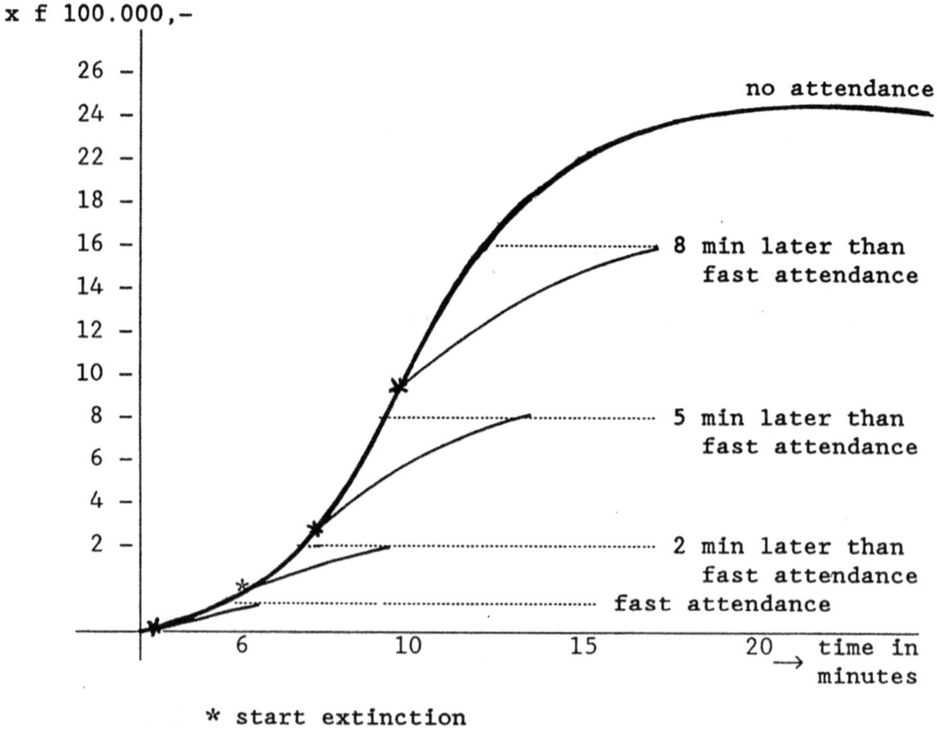

Figure 2 EXAMPLE OF DAMAGE IN RELATION TO ATTENDANCE TIMES

for saving human lives and protecting the surroundings for catching fire. The relieving of the norm for the first attendance could result in a disproportional increase of the danger and the damage! For example, an increase of the attendance time with 10% could result in an increase of the damage with 30%.

In order to maintain a reasonable level of safety in the different districts of the area to cover, the protection demands are translated into a risk classification, see Table 1. A is the highest and F the lowest risk The classification prescribes for each district the number of the first attendance pumpers and their time after the alarm in which the fire adress should be reached.
attendance

Data of turn outs in Rotterdam indicate an average speed of 36 km/hour by the pumpers.

Table 1 RISK CLASSIFICATION

RISK CATEGORIE	TYPE DISTRICT	MAXIMAL ATTENDANCE TIMES IN MINUTES PUMPER 1	PUMPER 2	PUMPER 3	PUMPER 4
A	heavy industry	6	6	8	8
B	industry	6	6	8	-
C	center	6	8	-	-
D	appartment	6	-	-	-
E	outskirts	8	-	-	-
F	rural area	15	-	-	-

MODEL

After a long history of testing different models mentioned in the literature and try-outs of ourselves, we finally choose a networkmodel to present the area to cover by the Fire Department.

The problem as posed here, is in reality a searching for a measurement for the effectivity of the repressive task of the Fire Brigade. We choosed a network approach because distances between two points in an area to cover, i.e., the fire station and the fire address, cannot accurate enough be calculated with distance functions or gravity models. Furthermore, the networkmodel is very useful in reducing the number of possible fire addresses and possible location areas.

A network consists of a number of nodes and edges. The nodes represent roadcrossings and the edges those roads which the Fire Brigade can use. The area to cover is divided in a number of districts, see Figure 3. These districts are represented by a central node-roadcrossing in the district (DC). Known of these nodes has to be the name (of the district), the risk, the correction factor (to compensate the difference in surface/inhabitants), the distance to the most difficult point (MDP), the travel speed in the district and the neighbour nodes.

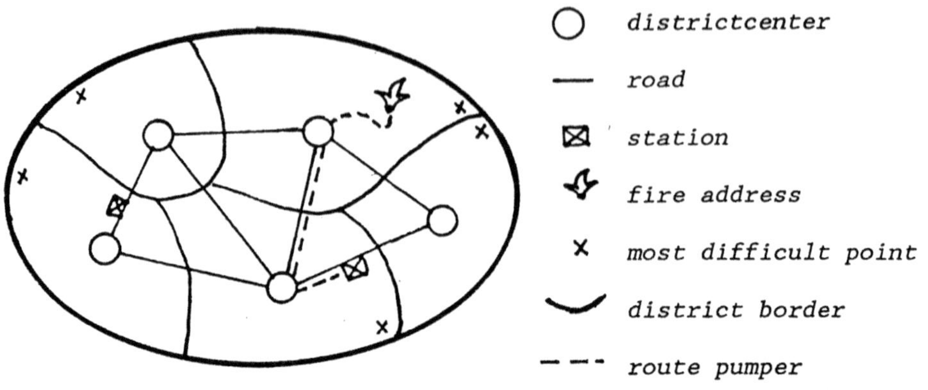

Figure 3 EXAMPLE NETWORK AREA TO COVER

After a fire call, the pumper of the nearest station turns out (STA), takes the shortest route over the network (RCR ... RCR) to the district center and from there to the fire adress. Determined for the attendance time of the whole district is the travel time to the MDP (no average, otherwise half of time too late!), see Figure 4.

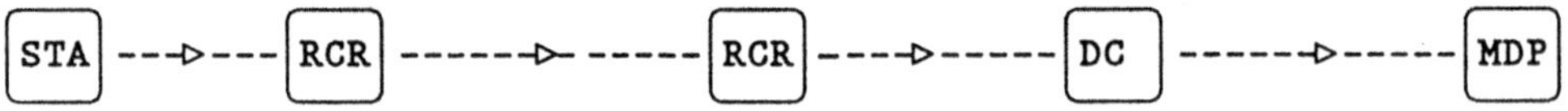

Figure 4 ROUTE PUMPER FROM STATION TO FIRE ADDRESS

A description of the mathematical model and its solution can be found in [4].

LOCATION STATIONS

In accordance with the research assignment of the Municipality of Rotterdam we detemined the number and location of the fire stations in the following three fases.

Fase I: The determination of the optimal location and the number of the fire stations needed. The present location of the fire stations is ignored in this fase. The calculations are based on the attendance time of 6, 8 and 10 minutes, see Table 2. The average travel speed of the first attendance cars is 36 km/hour. Double covering (pumpers from different fire stations to compensate traffic congestions) for the higher risk districts is prescribed. The results are given in Table 3 (one pumper in each station).

Table 2 VARIATION IN ATTENDANCE TIMES PUMPER 1 AND 2

risk category	type district	NORM 6 *	attendance times in minutes 6–8	8	8–10	10
A	heavy industry	6 6	6 8	8 8	8 10	10 10
B	industry	6 6	6 8	8 8	8 10	10 10
C	(shopping)center	6 8	6 8	8 8	8 10	10 10
D	appartment	6	6	8	8	10
E	outskirts	8	8	8	10	10
F	rural area	15	15	15	15	15

* prescribed national directive

Table 3 OPTIMAL NUMBER OF STATIONS FOR DIFFERENT ATTENDANCE TIMES
NO STATIONS FIXED

MAIN AREAS TO COVER	NORM 6	ATTENDANCE TIME IN MIN 6–8	8	8–10	10	REAL NUMBER OF STATIONS
Rotterdam-Noord (N)	10	8	7	5	5	7 *
Rotterdam-Zuid (Z)	8	6	5	4	4	5 **
Hoogvliet + industrie	9	6	4	3	3	2 ***
Maasvlakte + Rijnmond	25	16	12	10	10	1
Hoek van Holland	2	2	2	2	2	1 ****

* included 1 volunteers (Hilligersberg)
** included 1 professional partly occupied (Laagjes)
*** included 1 industry brigade (Heyplaat)
**** 1 volunteers (Hoek van Holland)

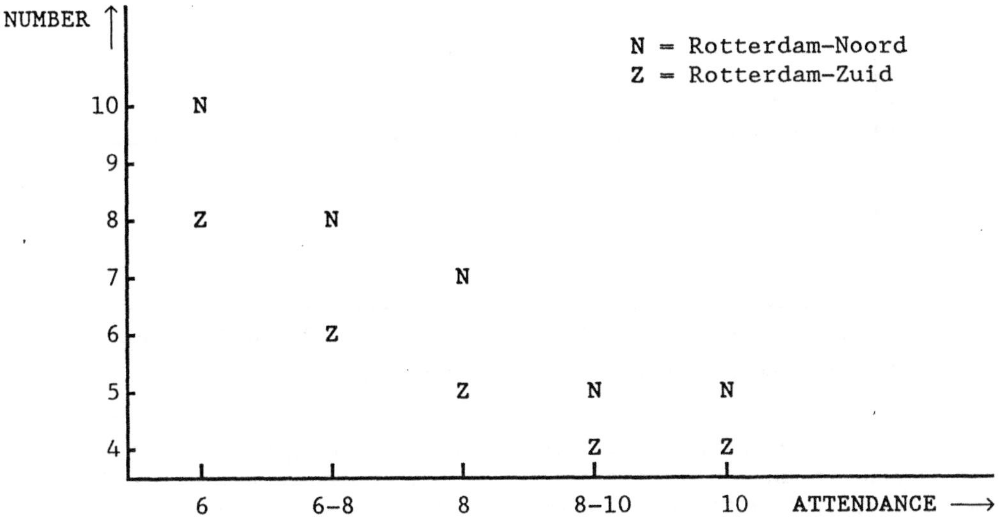

Figure 5 GRAPHICAL PRESENTATION OF TABLE 3

Fase II: The determination of the optimal location and the number of fire stations needed with respect to the present locations, see Table 4.

Table 4 OPTIMAL NUMBER OF STATIONS FOR DIFFERENT ATTENDANCE TIMES PRESENT STATIONS FIXED

	NORM	ATTENDANCE IN MINUTES				REAL NUMBER
	6	6–8	8	8–10	10	OF STATIONS
Rotterdam-Noord	16	14	11	10	10	7
Rotterdam-Zuid	12	8	7	6	6	5

The difference in the number of stations with Table 3 is explained by the bad location of the present stations.

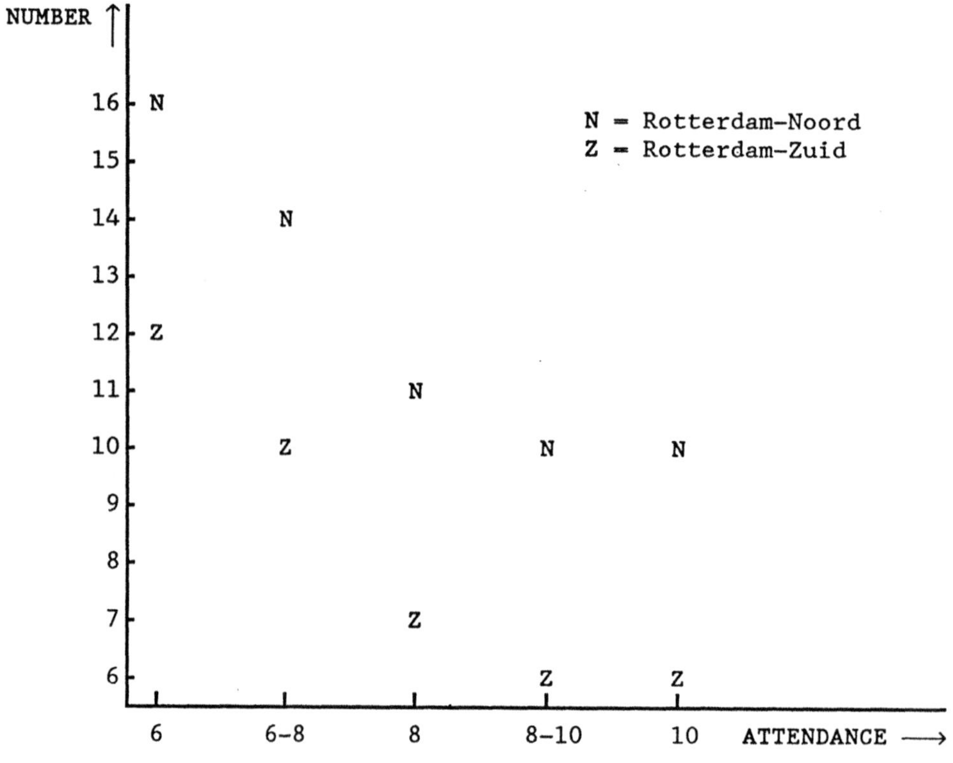

Figure 6 GRAPHICAL PRESENTATION TABLE 4

Based on the results of Fase I and II it appears that if the attendance is the only criterium, no station should be closed down. The prescribed attendance time is violated with 2–3 minutes or 35–50%.

A wider framework of criteria is used for the covering measures in Fase III and demonstrated in Table 5. Examples are the number of districts with a violated risk norm, the appreciation of the covering (mark) or the number of districts covered within a certain attendance time.

Final results of the project cannot be given in this place because the report (see [7]) is still in discussion with the city council.

TABLE 5 EXAMPLE CRITERIA ALTERNATIVES

ROTTERDAM-ZUID
NORMS ATTENDANCE WITH 8 MINUTES 8 8 8 8 8 8 8 8 15
32 33 34 35 37
1 1 1 1 1

DISTRICT NAME	NORM		STA COV	ATT MIN	LATE MIN	LATE PERC	MARK
AFK	B	8	34	5.3	.0	0.	7.3
BHF	B	8	34	4.2	.0	0.	7.9
CNS	B	8	34	4.3	.0	0.	7.8
VKO	C	8	32	6.3	.0	0.	6.8
VKO	C	8	37	9.0	-1.0	13.	5.5
ZDW	D	8	35	7.2	.0	0.	6.4

1	2	3	4	5	6	7	8	9	10	11	minutes
0.	0.	0.	9.	35.	49.	72.	88.	93.	93.	100.	% of districts

EXCEEDING ATTENDANCE TIMES IN PERCENTAGES: 25%
NUMBER OF DISTRICTS WHICH EXCEED : 5

REFERENCES

[1] Beasley, J.E. (1987), "An algorithm for set covering problem", EUROPEAN JOURNAL OF OPERATIONAL RESEARCH 31, 85-93.

[2] Loon, J.N.M. van & J.A.M. Schreuder (1976), "Number, location and capacity of fire stations in Rotterdam", RESEARCH REPORT UNIVERSITY OF TWENTE, in Dutch.

[3] Plesnik, J. (1987), "A heuristic for the p-center problem in graphs", DISCRETE APPLIED MATHEMATICS 17, 263-268.

[4] Schreuder, J.A.M. (1981), "Application of a location model to fire stations in Rotterdam", EUROPEAN JOURNAL OF OPERATIONAL RESEARCH 6, 212-219.

[5] Schreuder, J.A.M., J.H. Haken and W. Brinkhuis (1983), "Risk-covering fire department Hengelo", CONTRACT-RESEARCH UNIVERSITY OF TWENTE, (Enschede), in Dutch.

[6] Schreuder, J.A.M., R. Hommes, "Risk-covering and organisation fire department Lelystad", CONTRACT-RESEARCH UNIVERSITY OF TWENTE (Enschede), in Dutch.

[7] Schreuder, J.A.M. (1987), "Optimal location of fire stations Rotterdam", CONTRACT-RESEARCH UNIVERSITY OF TWENTE (Enschede), in Dutch.

Aircraft-stand allocation at Schiphol Airport: problem description

Idze Spilker

The basic airport process is a transfer process. A typical airport goal is providing and controlling that transfer process. The transfer has different faces. From a transportation point of view we can distinguish wheelmode at landside and wingmode at airside. The majority of the process steps take place in and around the terminal building between landside and airside. From a capacity point of view we must differentiate between flow capacity and stay capacity.

The allocation of aircraft to stands is a major decision problem. At airside the decision involves both stay capacity and flow capacity for aircraft. If we focus on passengers, the decision also involves the related flow and stay capacity in the terminal building. In general the aircraft, passenger and baggage handling processes are generated by time and place of actual aircraft arrival and scheduled departure. Deviations from the scheduled times may cause deviations in place, and thereby planning and actual allocation problems.

The growing number of aircraft, stands, objectives and limitations to be handled result in a rather complex decision situation. One of Schiphol's requirements for example is a stable seasonal allocation, which will cost some capacity under uncertain arrival and departure times. How should we plan and how can we make the proper decisions under time pressure in the daily operations when deviations occur, without disturbing the going process?

A strategic model for the solution of the location-allocation problem of a major oilcompany

Carlo F.M. Stokx, Rotterdam

The objective of the model was to give an answer to the following question: Which of the available gasoline depots in Belgium should be utilized and how should retail outlets be allocated to these depots and to trucks stationed at these depots, in order to supply the outlets with the required quantities of gasoline, minimize the total costs and satisfy the constraints ?

In order to find a solution to this problem, a simulation model was developed. The model calculates the minimal annual distribution costs for a given set of depots. The optimal combination of depots can then be found by applying the model to several feasible sets of depots and selecting the set which results in the lowest distribution costs.

The solution procedure for calculating the minimal total distribution costs is, briefly, as follows:

1. Develop mathematical functions representing the costs of each of the depots in terms of annual throughput.
2. Develop a mathematical function representing the costs of a trip using hired tanktrucks.
3. Develop a mathematical function representing the costs of a trip using a tanktruck owned by the oilcompany.
4. Allocate the outlets to a depot in such a way that distribution cost (the sum of depot costs and transportation costs) are minimized. Using the results of steps 1 to 3 this can be done in the following way:
   - Assess for each outlet from which of the available depots it has to be replenished in order to minimize annual replenishment costs, when using owned tanktrucks.
   - Assess for each outlet from which of the available depots it has to be replenished in order to minimize annual replenishment costs, when using hired tanktrucks.

With the information, resulting from these calculations, a function can be developed representing the total annual distribution costs as a function of the amount of owned transport capacity (in hours used annually) and the amount of hired transport capacity. Lineair programming is then used to establish the owned and hired transportcapacity, needed for each outlet, in order to minimize total annual distribution costs, while satisfying the constraints.

TRANSPORTATION PLANNING: RECENT DEVELOPMENTS IN THE NETHERLANDS

C. Bernhard Tilanus, Eindhoven

Abstract

In their Golden Age, the seventeenth century, the Dutch were the freighters and traders of Europe. They still are, to some extent. Then with ships, now with trucks.

This paper is an introduction to the section on Logistics and Traffic. Logistics has been restricted to physical distribution, which is logistics external to the firm. Traffic concentrates on transportation planning and vehicle routing. The Netherlands were canvassed more actively than West Germany, therefore the survey focuses on recent developments in transportation planning in the Netherlands.

Topics addressed are: the infrastructure for transportation planning, location-allocation case studies, algorithms for vehicle routing, and software packages for transportation planning.

1. Introduction

> "The Low Countries have as many ships and vessels as eleven kingdoms of Christendom, let England be one. They build every year near 1000 ships, although all their native commodities do not require 100 ships to carry them away at once."
>
> Walter Raleigh, "Observations touching trade and commerce with the Hollanders, and other nations, presented to King James. Wherein is proved that our sea and land commodities serve to enrich and strengthen other nations against our own", between 1604 and 1616. Quoted in C. Busken Huet, The Land of Rembrandt (in Dutch), Tjeenk Willink, Haarlem, p. 187.

In the seventeenth century, the Hollanders were the freighters and traders of Europe. But free trade was soon curbed by nationalist impediments, like Cromwell's Act of Navigation.

Today, there exists a foundation called "Holland Distribution-Land". Why? Because the Dutch still are traders and distributors, but legal obstructions in the European Community still abound and should be abolished (Van der Padt, 1986):

- Rotterdam is world harbour number one;
- Schiphol Airport is the fourth, and fastest growing, airport in Europe;
- the Netherlands have a 35 per cent share of total European inland shipping;
- the Netherlands have a 72 per cent share of non-bilateral, cross-EC, road transport;

Operations Research Proceedings 1987

- the Netherlands have a 75 per cent share of bilateral, Netherlands-Germany road transport - but this would have been 10 per cent higher if Germany had given more trip authorizations!

Thus there is a strong Dutch drive away from bureaucracy toward liberalization:

- early 1984: truckers blocked the Brenner Pass and the Mont Blanc Tunnel, as a retaliation against an Italian customs strike;
- May 1985: the European Court of Justice ruled that EC ministers should liberalize EC transport as yet, which they had been obliged to do as early as 1970;
- 1987: the average time loss for a truck border passage:
  * between the Netherlands and Germany is 45 minutes,
  * between Belgium and France is 2 hours,
  * between Austria and Italy is 6 hours;

  the costs of truck border passages in Europe (paper tariffs) are estimated at Dfl 100 million per diem;
- 1 January 1988: Benelux, Germany and France introduce the Single Administration Document (SAD); SAD replaces 30 different customs forms - but SAD still has 30 appendices!

This paper serves as an introduction to the section on "Logistics and Traffic." What do we understand by these terms? According to Oxford Advanced Learner's Dictionary of Current English, logistics is "supply, distribution and replacement of materials and personnel, e.g. for the armed forces", and traffic is "(1) (movement of) people and vehicles along roads and streets, of aircraft in the sky (2) transport business done by a railway, steamship line, airline, etc. (3) illicit trading". In the armed forces, personnel may be regarded as a peculiar kind of material (cannon-fodder: "men regarded as expendable material in war"). At any rate, I would like to drop "personnel" from the concept of "logistics".

For this section, it is appropriate to restrict the meaning of "logistics" still further, viz. to external logistics ("supply, distribution and replacement of materials outside the firm or organization"), since there are several other sections covering most of the internal logistics ("supply, distribution and replacement of materials within the firm or organization"), for instance the sections on:

- Production Planning and Inventory,
- Flexible Manufacturing Systems,
- Combinatorial Optimization.

When thinking of the subject of external logistics, many think of the sub-subject of transportation planning, and many even think of the sub-sub-subject of vehicle routing and scheduling. This is apparent from the contributions to this section.

Transportation planning is a lively subject nowadays, cf:

- in 1984, in the 26th International Meeting of The Institute of Management Sciences, the stream "Routing" got the highest average attendance in 40 streams;
- 1-2 June 1987, a special conference on "OR in Transportation" (NOAS '87) was held in Copenhagen;
- 16-19 June 1987, the "Third EURO Mini Conference on Transport Planning and Traffic Control" was held in Hercenovi, Yugoslavia;
- this section is the largest section of this conference, comprising 21 papers running in a stream from the first to the last session.

Why transportation planning is a lively subject, will be supported in Section 2 by an example. A survey of the papers on Logistics and Trafficfrom the Netherlands, to be presented in this conference, will be given in Section 3.

2. A problem and a conclusion

In the village where I lived as a boy, twice a day a horse-cart came along the single road collecting milk-cans for the local milk-factory. No vehicle-scheduling problem there!

Today, one dairy concern, DMV-Campina, processes all the milk produced in an area of about one quarter of the Netherlands, South of the river Meuse. The milk is collected (Bocxe and Tilanus, 1985):

- from 450,000 cows
- on 9000 farms
- for 14 plants
- by 250 tankers
- once every two or three days.

In that same area, in 1948, there were 41,000 farms and 119 independent factories. Between 1948 and 1983, milk production in that area increased five-fold, milk production per farm increased twenty-fold, and milk processed per dairy plant increased forty-fold. DMV-Campina has a large-scale, multi-depot vehicle-scheduling problem and the scale is ever increasing.

Vehicle scheduling is done in part by hand, in part, since 1975, by IBM's VSPX package. In 1982, Bocxe and Tilanus (1985) evaluated the nine vehicle-scheduling packages then available on the Dutch market, and found them all unsatisfactory to solve DMV-Campina's problems with VSPX, viz.;

- small changes in data input generating completely different routes, causing unwanted "unrest" with the farmers;
- bad scheduling of arrival times at the plants;
- no automatic assignment of farms to firms, hence no solving of the multi-depot problem;
- no scheduling of trailers as mobile depots; etc.

All programs in 1982 were batch programs. Since then, almost all newly developed software is interactive, often with visual display as well.
(I have been informed that professional route planners do not care for visual display of routes!) DMV-Campina is still busy developing satisfactory, interactive vehicle-scheduling software.

The persistent existence of this vehicle-scheduling problem and its increasing scale illuminate why transportation planning is such a lively subject nowadays.

Large scale is an important factor ("problem too small-scale" can be the cause of failure of projects, see Tilanus, 1985). A counter-example may illustrate this. At Van Gend & Loos, the largest transporter of goods in the Netherlands, transportation planning is done by hand. The computer is used for administrating the planning. The procedure is as follows. Van Gend & Loos has a 48-hours delivery service: the orders are received, say, today; the goods will be collected in one of fifteen regional depots tomorrow; inter-depot transportation will take place tomorrow-night; and the goods will be delivered the day after tomorrow. About 80 per cent of nightly inter-depot transportation is by fixed truck line service; 20 per cent is variable and planned by hand. Apparently, the scale of the problem at Van Gend & Loos is not yet large enough to necessitate computer scheduling!

From the growing scale of transportation problems we may conclude that interest in transportation planning will remain lively in the future.

3. A survey of contributions from the Netherlands

When organizing the section on Logistics and Traffic, I canvassed the Netherlands more actively than Germany. Therefore, only a survey of Dutch contributions is presented. They may be representative of ongoing developments.

First of all, there is the infrastructure. Data bases, etc., form the infrastructure of transportation planning, as roads form the infrastructure of transportation. We have the papers:

- "Scheduling the construction of Dutch roads", by G.T. Timmer;
- "The handling of road network data", by J.M. van Rooijen;
- "A method for data collection for car navigation", by B.J. Beers;
- "The problem of "fuzzy" constraints in computerised planning", by P. Klapwijk.

Case studies make up the meat of transportation literature. They show what the problems are and what approaches to solve them are followed in practice. We have the following case studies:

- "A strategic model for the solution of the location-allocation problem of a major oil-company", by C.F.M. Stokx;
- "Location of Rotterdam fire stations", by J. Schreuder;
- "Geographical market segmentation", by C.J. van der Plas, G.J.R. Förch, G. van der Hoek, H.W. van den Meerendonk, J.J. Remmerswaal;
- "Aircraft-stand allocation at Schiphol airport: Problem description", by J.I. Spilker;
- "Aircraft-stand allocation at Schiphol airport: A decision support system", by K. Anthonisse, B. Lageweg;
- "Aircraft-stand allocation at Schiphol airport: An optimization procedure", by K. Anthonisse, B. Lageweg.

Finally, at several universities and in software houses, work is done on developing new algorithms and software packages. Not all are represented (e.g., Fleuren, 1986). Some are presented in the section Combinatorial Optimization (see below).

The savings algorithm still plays an important part here (Paessens, to be published). One reason that DMV-Campina did not want to consider a package based on the sweep algorithm, was that they had invested 2Λ man-years constructing a road network database consisting of 9760 nodes and 13,953 links.

We have the following theory and software papers:

- "A location-routing problem", by J.A.M. Beulens, A.W.J. Kolen;
- "Tourenplanung mit einem personal computer", by W.G. Kolenbrander;
- "Transportation planning - as easy as 1-2-3", by H.J.J. Uyttenhove;
- "Local search for constrained routing problems", by M.W.P. Savelsbergh (in section: Combinatorial Optimization);
- "New exact and heuristic solution methods for the vehicle routing problem", by A.W.J. Kolen (in section: Combinatorial Optimization).

4. Round-up

A working group of the Netherlands Society for Logistics Management (NEVEM, 1985) made a checklist of items constituting vehicle scheduling problems, indicating whether they are easy, not-so-easy, or difficult to implement in algorithms and software. At first, it was thought that all items could be attributed to three basic entities:

- objects, to be transported;
- addresses, from, to, or through which the objects are to be transported;
- means, i.e., vehicles, by which the objects are to be transported.

After a while, other basic entities were added, like a (road) network, chauffeurs, trailers. Classification of vehicle scheduling problems proved to be a bigger problem than it first seemed. Ronen (to be published) distinguishes 65 characteristics for a classification of vehicle scheduling problems. He concludes that there is a vast variability in truck routing and scheduling problems.

From this vast variability of vehicle scheduling problems, which is quite unlike "pure" mathematical problems like the LP problem or the traveling-salesman problem, we conclude that there is a vast amount of work to be done, developing algorithms and programs as well as applying them, which will keep many people busy for a long time to come.

References

Bocxe, M.A.G., and Tilanus, C.B., (1985) "Testing vehicle scheduling programs for milk collection", European Journal of Operational Research 20, 25-33.

Fleuren, H.A., (1986) "Trip generation in a set partitioning approach for vehicle routing and scheduling", Twente University, Enschede.

NEVEM (Netherlands Society for Logistics Management)(1985) "Automated route planning: Results of NEVEM working group" (in Dutch), The Hague.

Padt, A. van der, (1986) "Entrepreneurship at the European level" (in Dutch), lecture held at Eindhoven University of Technology, Faculty of Industrial Engineering and Management Science, 30 January 1986.

Paessens, H., "The savings algorithm for the vehicle routing problem", to be published in European Journal of Operational Research.

Ronen, D., "Perspectives on practical aspects of truck routing and scheduling", to be published in European Journal of Operational Research.

Tilanus, C.B., (1985) "Failures and successes of quantitative methods in management", European Journal of Operational Research 19, 170-175.

SCHEDULING THE CONSTRUCTION OF DUTCH ROADS

Gerrit T. Timmer, Rotterdam

Several hundreds of millions of US dollars are spent each year on the construction of new national roads by the central Dutch government. At the moment, some 300 projects, concerning the construction of parts of a road, are under consideration. Characteristics of the projects such as expected duration (typically 5 years), expected costs, and relations between different projects, are known. A priority has been assigned to the projects using multi-criteria analysis. The projects are scheduled 10 years ahead. However, new political, technical and other developments may call for an update of the current planning. At least two updates are determined every year. Since it became practically impossible to update the planning by hand, ORTEC Consultants designed and implemented a computerized planning system.

The planning system must determine the starting year of every project, subject to bounds on these starting years, several types of precedence restrictions, and budget restrictions (lower- and upper bounds on the yearly expenditure). The objective has to take into account the priority of the projects, the wish to avoid huge changes in the schedule and the wish to stay as close as possible to certain target budgets. Ignoring the target budgets would lead to a linear integer programming model that is already unsolvable by existing standard software. Incorporating the target budgets worsens the situation even more, since the objective becomes nonlinear.

Therefore, a heuristic procedure has been designed. First a (continuous) linear programming relaxation is solved, after which an improvement scheme tries to determine the optimal schedule. This heuristic appears to produce schedules which are quite satisfactory and which are close to the optimal one.

The method has been programmed and incorporated in an easy to use system. The resulting planning instrument has been used sucessfully for more than a year now. It makes it possible to react more quickly on new developments and leads to better schedules.

TRANSPORTATION PLANNING- AS EASY AS 1-2-3

Hugo J.J. Uyttenhove
Eindhoven

As a special application of Linear Programming, a model is described that minimizes route mileage in a clear and understandable way. This application makes use of two inexpensive products, Lotus 1-2-3 and Whats´Best! which combined provide a user-friendly modelling technique for truck routing. An example will be discussed whereby a fleet of 135 trucks operate out of twenty terminals providing overnight delivery. Cutting out between 1000 and 2500 miles each night explains the large savings that can be reached. In addition to the description of the application, the author gives an explanation of how models using these two tools can be easily constructed. Some comments on limitation and problem size are discussed.

Werbeplanung in Theorie und Praxis

Ralph Berndt, Tübingen

Zusammenfassung:
Für drei bedeutsame Felder der Werbeplanung, der Zielplanung, der Werbebudgetierung und der Mediaselektion wird zum einen dargelegt, welche informations- und modellgestützte Konzepte aus theoretischer Sicht vorgeschlagen werden; zum anderen wird das tatsächliche Planungsverhalten in der Werbepraxis anhand empirischer Ergebnisse beschrieben. Auf der Basis einer Konfrontation theoretischer Konzepte und praktischer Verhaltensweisen werden Empfehlungen gemacht, die sich sowohl an den Theoretiker als auch an den Praktiker der Werbung richten.

Summary:
Theoretical concepts of advertising planning in the fields of goals, budgets and media are confronted with the planning behavior in practice. Recommendations are made for future advertising planning directed to theorists and practitioners of advertising.

A. Einführung

Die Bundesrepublik Deutschland wie die Niederlande zählen zu den führenden Werbenationen der Industrieländer. Ein internationaler Vergleich der Werbeausgaben (31) zeigt, daß - ausgehend vom jeweiligen Anteil an Gesamtwerbeaufkommen der Industrieländer - die Bundesrepublik Deutschland an 4. Stelle und die Niederlande an 9. Stelle steht; legt man hingegen den Anteil der Werbeausgaben am jeweiligen Bruttosozialprodukt zugrunde, so liegen die Niederlande (mit 0,91 %) noch vor der Bundesrepublik (0,75 %).

Die Werbung in der Bundesrepublik Deutschland ist charakterisiert durch

- hohe Gesamtwerbeaufwendungen, die in sehr unterschiedlicher Weise auf verschiedene Werbeträgergruppen entfallen,
- zum Teil sehr hohe betriebliche Werbebudgets,
- aber auch häufig hohe Einschaltkosten pro Belegung.

Die Gesamtwerbeaufwendungen im Jahre 1986 (13) beliefen sich auf 16,486 Mrd. DM. Knapp 40 % der Aufwendungen emtfielen auf Tageszeitungen, knapp 17 % auf Publikumszeitschriften und 12 % auf die Direktwer-

Operations Research Proceedings 1987

bung, gefolgt von der Fernseh- und Fachzeitschriftenwerbung (mit je 9 %) und der Adressbuchwerbung (mit 6,5 %); die restlichen Mediengruppen haben eine nur nachgeordnete Bedeutung.

Die größten Werbeanstrengungen im Jahre 1985 (20) unternahmen
- C & A Brenningmayer mit 212,5 Mio. DM,
- Volkswagen mit 134,6 Mio. DM und
- Karstadt mit 108,0 Mio. DM;

die durchschnittliche Budgethöhe der 25 größten Werbetreibenden belief sich auf 79,9 Mio. DM.

Die Belegungskosten (8) sind allgemein abhängig von der Werbemittelgröße (Insertionsmedien) bzw. -länge (Elektronische Medien); einmalige Schaltungen von ganzseitigen, schwarz-weißen Anzeigen kosten bei
- ADAC-Motorwelt 92.440 DM
- Hör Zu 72.160 DM
- Stern 51.072 DM
- Spiegel 37.554 DM;

bei vierfarbigen Anzeigen werden Preisaufschläge zwischen 55 und 60 % auf die Grundpreise verlangt. Die Kosten der Schaltung eines 15-Sekunden-Spots belaufen sich bei ARD (ZDF) auf 50.093 DM (34.633 DM); bei 30-Sekunden-Spots werden etwa um 70 % höhere Preise verlangt.

Aufgrund der damit hohen gesamtwirtschaftlichen und einzelwirtschaftlichen Bedeutung der Werbung wird im folgenden für drei typische und wichtige Planungsbereiche der Werbung, der Ziel-, Budget- und Mediaplanung zum einen dargestellt welches Planungsverhalten im Rahmen der Werbetheorie vorgeschlagen wird; zum anderen wird dargelegt, welche Vorgehensweisen tatsächlich in der Praxis herangezogen werden. Auf der Basis einer Konfrontation der Werbeplanung in Theorie und Praxis werden dann Empfehlungen gegeben hinsichtlich der zukünftigen Gestaltung der Werbeplanung; diese Empfehlungen sind sowohl an den Theoretiker als auch an den Praktiker gerichtet.

B. Werbepolitische Zielsetzungen

Theoretische Grundlage der Formulierung von Werbezielen sind individuelle Werbewirkungsmodelle, in denen der Prozeß vom Kontakt mit einem Werbemittel bis zur abschließenden Kaufhandlung beschrieben wird (ein Überblick über individuelle Werbewirkungsmodelle findet sich in (1, S. 139)). Nach dem Wirkungsmodell von Lavidge/Steiner (15) sind - ausgelöst durch Werbemittelkontakte - die fünf aufeinander aufbauenden psychischen Phasen Aufmerksamkeit, Bekanntheit, Positive Einschätzung, Bevorzugung, Kaufabsicht zu durchlaufen, bis es zu einer Kaufhandlung kommt. Dabei kann jede psychische Phase Bezugspunkt für ein psychologisches Werbeziel sein. Der Werbemittelkontakt selbst ist Basis für streutechnische Ziele; die abschließende Kaufhandlung Basis für ökonomische Werbeziele. Auskünfte über die praktische Relevanz der unterschiedlichen Arten von Werbezielen gibt eine empirische Erhebung aus dem Jahre 1985 (5). Befragt wurden insgesamt 38 Unternehmen der Bundesrepublik, unter denen 8 der 10 größten Werbetreibenden vertreten waren, nach den primär bzw. sekundär verfolgten Zielgrößen; Mehrfachnennungen waren möglich. Es zeigt sich, daß bei langlebigen Konsumgütern psychologische Werbeziele wesentlich häufiger verfolgt werden als ökonomische Zielgrößen; zwei Drittel der Befragten gaben an, insbesondere eine Steigerung der Aufmerksamkeit und der Bekanntheit anzustreben. Bei kurzlebigen Konsumgütern wird die psychologische Zielgröße "Aufmerksamkeit" am häufigsten genannt, gefolgt von der psychischen Zielgröße "Bekanntheit" und der ökonomischen Zielgröße "Absatzmenge".

	langlebige Konsumgüter			kurzlebige Konsumgüter		
	primär	sekundär	insgesamt	primär	sekundär	insges.
Aufmerksamkeit	44	22	66	74	7	81
Bekanntheit	67	0	67	56	11	67
Interesse	11	0	11	4	15	19
Kaufabsicht	22	22	44	15	15	30
Absatzmenge	22	0	22	56	11	67
Rentabilität	11	0	11	11	0	11

Abb. 1: Primär und sekundär verfolgte Werbeziele in der Praxis

Daß ein Heranziehen psychologischer Werbeziele in der Praxis nicht unproblematisch ist, wird deutlich, wenn die theoretische Fundierung psychologischer Werbeziele, also Strukturmodelle des Käuferverhaltens, näher analysiert wird (vgl. (2, S. 12 ff.), (18), (19, S. 30 ff.)). Bei Strukturmodellen des Käuferverhaltens handelt es sich um idealtypische Modelle, deren psychologische Fundierung umstritten ist. Sie sind elementenpsychologisch begründet; die These des "quasi-mechanischen" nacheinander Durchlaufens verschiedener psychologischer Phasen wird seit geraumer Zeit aus ganzheitspsychologischer Sicht bestritten. Hinzu kommt, daß eine Aggregation über verschiedene Individuen nicht möglich ist. Schließlich ist ein derartig extensiver Wirkungsprozeß nicht bei allen typischen Käuferverhaltensprozessen relevant, zwar bei extensiven Kaufentscheidungsprozessen, aber nicht bei Wiederholungskäufen bzw. Impulskäufen. Es ist daher zu empfehlen, von der Idee einer strengen Abfolge zwischen den psychischen Teilprozessen abzugehen und stattdessen auf der Grundlage des situativen Ansatzes im konkreten Fall jeweils einzelne relevante psychologische Zielgrößen herauszufinden und als unvollkommene Indikatoren für die individuelle Kaufentscheidung anzusehen.

Zu dieser Vorgehensweise existiert noch eine Alternative, die Erfassung des individuellen Werbewirkungsprozesses nicht in einem deterministischen Kontext, sondern als stochastischen Prozeß. Dabei ist es erforderlich, alternativen Ansprachenzahlen eines Individuums entsprechende Kaufwahrscheinlichkeiten zuzuordnen; dies kann auf der Basis des Markoffschen Markenwahlprozeßmodells oder des linearen Lernmodells erfolgen, wobei jeweils die Modellparameter in Abhängigkeit von der Zahl der Werbekontakte zu erfassen sind (2, S. 38 ff.). Im Gegensatz zu den Strukturmodellen des Käuferverhaltens ist hier eine Aggregation über die Zielpersonen möglich, wodurch einer geplanten Werbemaßnahme die zu erwartenden Mengen/Erlöse oder Gewinne zugeordnet werden können (s. (25), (4)). Offensichtlich ist, daß bei einem Heranziehen eines stochastischen Prozeßmodells alle Probleme entfallen, die bei dem zuvor erörterten deterministischen Ansatz anfallen; es muß aber beachtet werden, daß diese Vorgehensweise nur für kurzlebige Konsumgüter relevant ist, außerdem müssen die erforderlichen Meßmethoden zur Bestimmung der Kaufwahrscheinlichkeiten in Abhängigkeit von alternativen Ansprachenzahlen (vgl. (2, S. 45 ff.)) weiterentwickelt werden.

C. Werbebudgetierung

Eine Vielzahl an Modellen zur Werbebudgetierung (vgl. die Überblicke in (26), (16), (17), (10)) ist vorgeschlagen worden, die sich in vielen Modell-Charakteristika unterscheiden:

- Es sind Modelle sowohl für Einproduktunternehmen als auch für Mehrproduktunternehmen mit beliebigen absatzmäßigen Verflechtungen und beliebigen Werbeeffekten formuliert worden.
- Die Modelle beziehen sich sowohl auf monopolistische als auch auf dyopolistische Märkte.

- Es sind sowohl statische als auch dynamische Modelle entwickelt worden; bei den dynamischen Modellen sind beliebige time-lag Strukturen und carry-over Effekte beachtet worden.
- Den Modellen liegen verschiedene ökonomische Werbeziele zugrunde.
- Zur Ermittlung optimaler Werbebudgets sind diverse Optimierungstechniken herangezogen worden: von der einfachen und partiellen Differentiation über dynamische Simulationsmodelle bis hin zur dynamischen deterministischen Kontrolltheorie.

Fast allen Modellen liegt aber einhellig eine sehr restriktive Prämisse zugrunde: Es wird unterstellt, daß die entscheidungsrelevanten Daten vollständig und mit Sicherheit bekannt sind (Ausnahmen sind mit (12) und (27) gegeben).

Welche praktische Relevanz derartige Budgetierungsmodelle haben, lassen verschiedene empirische Erhebungen erkennen (vgl. Abb. 2).

- Eine willkürliche Festlegung von Werbebudgets ist stark zurückgegangen.
- Quantitative Modelle, also Optimierungsmodelle, haben eine geringe Verbreitung. Ein ergebnismäßiger Ausrutscher ist nur bei der USA 1981-Erhebung gegeben; die Erhebung zeichnet sich aber dadurch aus, daß bei fast allen Verfahren stark überdurchschnittliche Angaben gegeben sind, die Gesamtzahl der Nennungen ist weit überdurchschnittlich.
- Stattdessen weisen eine besondere Bedeutung sogenannte Praktikerverfahren auf, wie die Planung des Budgets als Prozentsatz des zukünftigen oder bisherigen Umsatzes oder Gewinnes, gemäß verfügbarer Mittel, in gleichem Maße wie die Konkurrenz bzw. gemäß der Zielsetzung und den einzusetzenden Mitteln (objective and task).

Eine theoretische Analyse der Praktikerverfahren zeigt nun aber, daß fast alle Praktikerverfahren unsinnig bzw. falsch sind; eine Ausnahme bildet nur das objective and task-Verfahren. Die Festlegung des Budgets als Prozentsatz des bisherigen bzw. erwarteten Umsatzes ist sachlogisch falsch, weist eine zyklische Wirkung auf und hat das Problem der Wahl des richtigen Prozentsatzes. Generell gültige bzw. branchentypische Prozentsätze existieren nicht; die empirische Erhebung (5) zeigt eine breite Streuung von 0,5 % bis mehr als 10 %. Zweifelsohne gibt es einen optimalen Prozentsatz, der mit einer Lösung z. B. eines deterministischen, marginalanalytischen Ansatzes mitgeliefert wird; doch diese ist selbstverständlich bei einem Verzicht auf eine Optimierung des Werbebudgets unbekannt. Eine entsprechende Beurteilung gilt für den Fall

Land Jahr (Fallzahl)	USA 1975 (50)	Europa 1977 (41)	GB 1977 (53)	D 1981 (269)	USA 1981 (54)	D 1985 (36)
Willkürlich	12 %	24 %	4 %	n. e.	4 %	n. e.
Quantitative Modelle	2 %	10 %	8 %	7 %	51 %	8 %
0/0 vom zukünftigen Umsatz / Gewinn	58 %	68 %	52 %	60 %	74 %	31 %
0/0 vom bisherigen Umsatz / Gewinn	20 %	24 %	28 %	16 %	20 %	n. e.
gemäß verfügbarer Mittel	30 %	37 %	4 %	41 %	20 %	25 %
wie die Konkurrenz	n. e.	n. e.	2 %	12 %	24 %	33 %
Objective and Task	6 %	15 %	n. e.	n. e.	63 %	69 %
Quellen	(24)	(22)	(9)	(14)	(21)	(5)

Mehrfachnennungen möglich
n. e.: nicht erhoben

Abb. 2: Werbebudgetierung in der Praxis (Konsumgüter-Industrie) einer Bestimmung des Budgets als Prozentsatz des Gewinnes. Eine Bestimmung des Werbebudgets gemäß verfügbarer Mittel ist sachlogisch fragwürdig, außerdem ist wiederum eine zyklische Wirkung gegeben. Eine Budgetfestlegung in gleicher Weise wie die Konkurrenz hat zumindest zwei Schwächen: das zukünftige Verhalten der Konkurrenten ist unbekannt, und Unterschiede bei den Ressourcen, der Kostenstruktur, den Zielsystemen werden nicht berücksichtigt. Zu akzeptieren ist nur die objective and task-Methode (vgl. z. B. (32)), welche folgende Schritte umfaßt:

- Festlegung der Werbeziele
- Festlegung der Mittel, um die Ziele zu erreichen
- Schätzung der damit verbundenen Kosten
- Summe der Kosten gleich dem erforderlichen Budget.
- Falls das Budget die verfügbaren finanziellen Mittel übersteigt, Änderung der Ziele und erneutes Budgetierungsverfahren von Anfang an.

Bei diesem Verfahren sind fehlerhafte Charakteristika nicht zu erkennen. Der zweite Schritt beinhaltet aber dieselben Prognoseprobleme wie bei den angeführten Optimierungsmodellen. Außerdem ist dieses Budgetierungsverfahren nur sehr allgemein beschrieben.

Für die Weiterentwicklung auf dem Feld der Werbebudgetierung sind grundsätzlich zwei Wege, ein induktiver und ein deduktiver Weg gegeben. Zum einen ist eine Neuorientierung an der <u>deskriptiven</u> Entscheidungstheorie wünschenswert, die im Marketingbereich schon gegeben ist mit dem induktiven Modell des Einkaufsverhaltens in Supermärkten von Bettman (6) und dem induktiven Modell der Bewertung und Auswahl industrieller Lieferanten von Vyas und Woodside (30). Erforderlich ist es, die einzelnen in der Praxis ablaufenden Werbebudgetierungsprozesse vollständig aufzuzeichnen und zu analysieren mit Hilfe von direkten Beobachtungen, Protokollanalysen, und Dokumentenanalysen, um typische Prozeßmodelle der Werbebudgetierung in Organisationen herauszufinden. Die bisher durchgeführten empirischen Erhebungen zur Werbebudgetierung sind zu einfach strukturiert; insbesondere Interaktionen zwischen verschiedenen Einflußfaktoren der Budgetermittlung werden nicht erhoben.

Zum anderen sollte auch im Bereich der Werbebudgetierung die sehr restriktive Prämisse vollständiger Kenntnis und sicherer Erwartungen bezüglich aller entscheidungsrelevanter Daten fallengelassen werden und - wie in anderen betriebswirtschaftlichen Planungsbereichen - die Theorie der Entscheidungsfindung in <u>Risikosituationen</u> herangezogen werden. Eine Werbekampagne ist als Marketing-Investition anzusehen, damit ihr eine spezifische Auszahlungs- und Einzahlungsreihe verbunden ist. Sämtliche typischen Analyse- und Entscheidungsverfahren für Risikosituationen wie z. B. risikoanalytische Verfahren oder e ne Entscheidungsfindung auf der Basis spezieller Entscheidungsregeln (vgl. (27), (11)) sind grundsätzlich geeignet. So kann mit Hilfe einer Risikoanalyse z. B. die Vorteilhaftigkeit einer dynamischen Werbestrategie beurteilt werden auf der Basis verschiedener zufallsabhängiger Inputdaten eines Kapitalwertmodelles wie Länge des Wirkungszeitraumes, Absatzmenge pro Periode und gegebenenfalls Verhalten von Konkurrenten. Auf der Grundlage eines flexiblen Planungsansatzes für Risikosituationen läßt sich ein optimales Werbebudget für eine Startperiode und optimale zufallsabhängige Budgets für Folgeperioden planen, was eine optimale Anpassung des Werbebudgets an die tatsächlich eintretende Umweltsituation erlaubt.

Bei einer Anwendung von Analyse- und Entscheidungsverfahren für Risikosituationen sind valide Wahrscheinlichkeitsschätzungen für die stochastischen Inputgrößen erforderlich; dies setzt voraus, daß Entscheidungsträger in der Praxis bereit und imstande sind, mehrwertige Erwartungen mit Hilfe subjektiver Wahrscheinlichkeiten zu spezifizieren (zur Ermittlung siehe z. B. den Überblick (11)). Ein Verzicht auf eine explizite Erfassung von Risikosituationen hat zur Konsequenz, daß eine Entscheidungsfindung entweder in Ungewißheitssituationen oder in Sicherheitssituationen vorgenommen werden muß. Für Ungewißheitssituationen fehlen allgemein akzeptable Entscheidungsregeln; die Sicherheitssituation ist eine Fiktion, die normalerweise in der Praxis nicht gegeben ist und die verlangt, mehrwertige Erwartungen auf einwertige Erwartungen zu reduzieren, wodurch die eigentlich gegebene Informationssituation unzureichend abgebildet wird.

D. Mediaselektion

Im Rahmen der Mediaselektion ist darüber zu entscheiden, welche Werbeträger in welchen Häufigkeiten belegt werden sollen. Auszugehen ist dabei vom verfolgten werbepolitischen Ziel; zu berücksichtigen sind Restriktionen in Form einer Budgetbedingung sowie maximaler Belegungszahlen für die einzelnen Werbeträger. In welcher Weise die Inputdaten einer Mediaplanung zu verknüpfen sind, um für alternative Belegungspläne Erfolgs-Kosten-Relationen bestimmen zu können, wird anhand des Ablaufdiagramms der Abb. 3 verdeutlicht. Bei den Inputdaten lassen sich harte

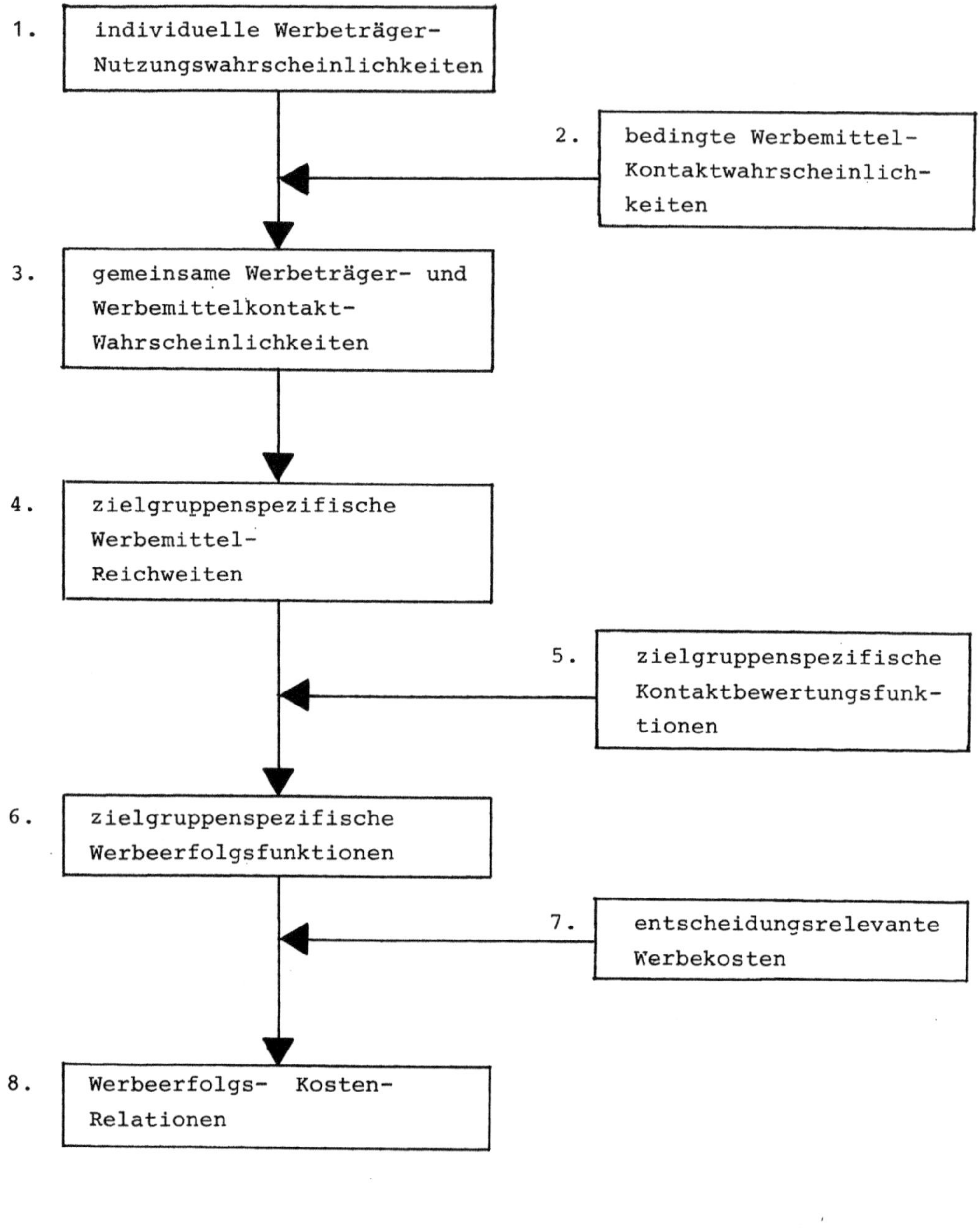

Abb. 3: Ablaufdiagramm zur Bewertung von Mediabelegungsplänen

und weiche Daten der Mediaplanung unterscheiden. Als harte Daten werden Inputdaten bezeichnet, die auf einer Befragung repräsentativer Stichproben mit anerkannten Meßmethoden beruhen oder mit Sicherheit bekannt sind. Zu den harten Daten zählen

- die individuellen Nutzungswahrscheinlichkeiten bezüglich der einzelnen Werbeträger, für deren Messung numerische Skalen entwickelt worden sind (s. (1, S. 30 ff.)),
- die entscheidungsrelevanten Werbekosten, die mit Sicherheit bekannt sind,
- und zielgruppenspezifische Kontaktbewertungsfunktionen beim Verfolgen streutechnischer Ziele, da diese aus der Zielformulierung deduzierbar sind, damit mit Sicherheit bekannt sind (vgl. (2, S. 83 ff.)).

Als weiche Daten sind hingegen anzusehen

- die bedingten Werbemittelkontaktwahrscheinlichkeiten (vgl. (1, S. 90 ff.)),
- die Kontaktbewertungsfunktionen bei Verfolgung psychologischer Werbeziele (vgl. (1, S. 117 ff.), (2, S. 66 ff.)),

da bis heute für beide Inputdaten allgemein akzeptierte, zuverlässige und valide Meßmethoden fehlen.

Grundbaustein eines jeden Mediaselektionsmodelles ist der Tausender-Kontaktpreis eines Werbeträgers. Hier handelt es sich um eine Wirtschaftlichkeitskennziffer, definiert als Quotient aus den Kosten einer Belegung eines Werbeträgers und der zielgruppenspezifischen Nutzerschaft des einzuschaltenden Werbemittels, multipliziert mit 1000. Der Tausender-Kontaktpreis alleine ist nicht geeignet, die Medien zu selektieren, da explizite Angaben von Ziel, Budget und Belegungsobergrenzen für die einzelnen Medien fehlen. Es ist aber möglich, das Kriterium zu einer Tausender-Kontaktpreis-Planungsrechnung zu erweitern, im Rahmen derer knappe finanzielle Mittel auf alternative Werbeträger unter Berücksichtigung von Belegungsobergrenzen und eines Rangfolgekriteriums zu verteilen sind (vgl. (3)). Eine solche Planungsrechnung ist sehr einfach, damit in der Praxis für beliebig umfangreiche Entscheidungssituationen anwendbar. Kritisch zu hinterfragen ist daher der Anwendungsbereich einer Tausender-Kontaktpreis-Planungsrechnung:

- Die Tausenderkontaktpreis-Planungsrechnung ist immer anwendbar, wenn das spezielle Werbeziel der Maximierung der Bruttoreichweite, d. h. der Zahl der Ansprachen, verfolgt wird (vgl. (3, S. 115 ff.)).
- Sie ist auch anwendbar bei einigen anderen streutechnischen Werbezielen (wie Maximierung der Nettoreichweite), falls keine externen Überschneidungen von Werbeträgern gegeben sind (vgl. (3, S. 118 ff.)).
- Falls ber externe Überschneidungen, d. h. Überschneidungen der

Nutzerschaften verschiedener Werbeträger, gegeben sind und nicht das Ziel der Maximierung der Bruttoreichweite verfolgt wird, versagt diese einfache Planungsrechnung (vgl. (3, S. 122 ff.)).

Will man eine derartige umfassende Entscheidungssituation exakt abbilden, so müssen Ansätze der binären Optimierung herangezogen werden, die jedoch für praktische Entscheidungssituationen nicht lösbar sind, da die Zahl der erforderlichen Variablen und Nebenbedingungen mit der Zahl der Werbeträger explodiert (vgl. (29)). Dies bedeutet, daß eine optimale Lösung des typischen Mediaselektionsproblems in der Praxis nicht möglich ist.

So ist es nicht überraschend, daß in der Praxis in den 70er Jahren heuristische Verfahren entwickelt worden sind, bei denen auf ein exaktes Erfassen externer Überschneidungen bei der Konstruktion von Mediabelegungsplänen verzichtet worden ist. Bei den heuristischen Verfahren der Mediaselektion lassen sich drei Modellarten unterscheiden (vgl. z. B. (32)):

- Rangreihungsprogramme, deren Funktion darin besteht, einzelne Medien zu bewerten und in eine Rangfolge zu bringen.
- Evaluierungsprogramme zur vergleichenden Bewertung alternativer vorgegebener Mediapläne.
- Konstruktionsprogramme, deren Funktion darin besteht, einen Streuplan selbständig aufzubauen.

Welche Akzeptanz diese Verfahren in der Praxis finden, ist von Simon und Thiel (28) im Rahmen einer empirischen Erhebung erfragt worden (vgl. Abb. 4 und 5). Es zeigt sich eine zunehmende Benutzung von Rangreihung- und Evaluierungs-Verfahren, ein starker Rückgang der Nutzung von Konstruktionsprogrammen um etwa 85 % zwischen 1971 und 1977. Das heißt, in deutlich stärkerem Maße werden Pläne nicht konstruiert, sondern mehr oder weniger willkürlich zusammengestelle Pläne werden vorgegeben und nur noch im Vergleich bewertet.

Programmtyp	1971	1972	1973	1974	1975	1976	1977
Rangreihung	826	1337	2515	2603	3012	3435	2513
Evaluierung	4230	5173	7075	7624	11915	12724	19817
Konstruktion	534	401	404	222	107	107	72

Abb. 4: Die Nutzung von Mediaselektionsprogrammen in Verlagen

Programmtyp	1971	1973	1975	1977
Rangreihung	110	231	303	365
Evaluierung	152	245	350	457
Konstruktion	146	77	52	21

Abb. 5: Die Nutzung von Mediaselektionsprogrammen in Werbeagenturen

Damit ist für den Bereich der Mediaplanung insgesamt eine äußert schwierige Situation zu konstatieren: Optimierungsmodelle können für Entscheidungssituationen in der Praxis nicht eingesetzt werden; vorhandene heuristische Verfahren, welche Lösungen konstruieren, finden eine recht geringe Akzeptanz. Wenn man sich nicht damit zufrieden geben will, daß für mehr oder weniger willkürlich erstelle Pläne nur eine modellgestützte Bewertung vorgenommen wird, so muß nach den Konsequenzen für die zukünftige Mediaplanung gefragt werden:
Erstens müssen heuristische Konstruktionsmodelle von vornherein so angelegt sein, daß Sensitivitätsanalysen für die weichen Inputdaten quasi enthalten sind. Dies ist erforderlich, da bei jedem denkbaren Mediaselektionsmodell - schon beim Tausenderkontaktpreis - weiche Inputdaten enthalten sind. Die Akzeptanz der Konstruktionsmodelle scheint umso größer zu sein, je mehr sie dem Werbeplaner Rechenarbeit abnehmen und Entscheidungshilfen anbieten, ohne scheinbar die Entscheidung vorwegzunehmen, je mehr dem Manager Möglichkeiten eingeräumt werden, in den Entscheidungsprozeß einzugreifen, z. B. durch Sensitivitätsanalysen in Zusammenhang mit weichen Daten.
Zweitens ist noch ein erheblicher Forschungsbedarf hinsichtlich der weichen Inputdaten zu konstatieren. Für die bedingten Werbemittelkontaktwahrscheinlichkeiten sind verschiedene unvollkommene Indikatoren (wie Seitenkontaktchance oder Leserblattbindung, vgl. (1)), vorgeschlagen worden; es fehlt eine vergleichende Analyse der Konzepte und eine Auswahl des vergleichsweise besten Konzeptes.
Zur Ermittlung der Kontaktbewertungsfunktionen für psychologische Werbeziele sollten nicht mehr Befragungen im Zusammenhang mit Labortests durchgeführt werden; vorzuziehen sind Feldexperimente, die gerade von den neuen Medien, den privaten Fernsehsendern, ermöglicht werden.
Drittens wäre es sehr wünschenswert, die heuristischen Konstruktionsmodelle so zu gestalten, daß sie für typische Planungssituationen in der Praxis bei Personal Computern eingesetzt werden können.

Literatur

(1) Bender, M., Die Messung des Werbeerfolges in der Werbeträgerforschung, Würzburg, Wien 1976
(2) Berndt, R., Optimale Werbeträger- und Werbemittelselektion, Wiesbaden 1978
(3) Berndt, R., Planungsrechnungen zur Mediaselektion, abgeleitet aus Optimierungsmodellen der Mediaplanung, in "Marketing - ZFP", 1981, S. 115 ff.
(4) Berndt, R., Stochastische Modelle des Käuferverhaltens als Grundlage für absatzwirtschaftliche Entscheidungsmodelle, in "Operations Research Proceedings 1982", Berlin, Heidelberg, New York 1983, S. 149 ff.
(5) Berndt, R., Werbeplanung in der Bundesrepublik Deutschland im internationalen Vergleich, Arbeitspapier Hamburg 1986
Ergebnisse einer empirischen Untersuchung, Arbeitspapier Hamburg 1986
(6) Bettman, J.R., An Information Processing Theory of Consumer Choice, Reading (Mass.) u.a. 1979
(7) Böcker, F., Gierl, H., Daten- und Verhaltensgestützte Mediaplanung, in "ZfbF", 38. Jg. (1986), S. 64 ff.
(8) Gesellschaft Werbeagenturen (Hrsg.), Zahlen und Daten für die Werbe-Planung, Frankfurt 1986
(9) Gilligan, C., How British Advertisers Set Budgets, in "Journal of Advertising Research", Vol. 17 (1977), No. 1, S. 47 ff.
(10) Hammann, P., Werbebudgetplanung, in "Werbung, Theorie und Praxis werblicher Beeinflussung", Hrsg. A. Kaiser, München 1980, S. 137 ff.
(11) Hampton, J.M., Moore, P.G., Thomas, H., Subjective Probability and its Measurement, in "Journal of the Royal Statistical Society", Serie A, 1973, Part 1, S. 21 ff.
(12) Holthausen, D.M., Assmus, G., Advertising Budget Allocation Under Uncertainty, in "Management Science", Vol. 28 (1982), No. 5, S. 487 ff.
(13) Kiefer, M.L., Der Werbemarkt 1986, in "Media Perspektiven", 1987, Nr. 6, S. 375 ff.
(14) Kleinert, H., Die Festsetzung des Werbe-Etats für eingeführte Produkte, Zürich 1981
(15) Lavidge, R.L., Steiner, G.A., A Model for Predictive Measurement of Advertising Effectiveness, in "Journal of Marketing", October 1961, S. 59 ff.
(16) Lilien, G.L., Kotler, P., Marketing Decision Making, New York 1983
(17) Little, J.D.C., Aggregate Advertising Models: The State of the Art, in "Operations Research", Vol. 27 (1979), No. 4, S. 629 ff.
(18) Mazanec, J., Strukturmodelle des Käuferverhaltens, Wien 1978
(19) Mühlbacher, H., Selektive Werbung, Linz 1982
(20) o.V., Die größten Werbeetats 1985, in "Werben & Verkaufen", 1986, Nr. 4
(21) Patti, C.H., Blasko, V., Budgeting Practices of Big Advertisers, in "Journal of Advertising Research", Vol. 21 (1981), No. 6, S. 23 ff.
(22) Permut, S.E., How European Managers Set Advertising Budgets, in "Journal of Advertising Research", Vol. 17 (1977), No. 5, S. 75 ff.
(23) Rogge, H.J., Planungs- und Informationsverhalten in Werbeagenturen, Ergebnisse einer empirischen Untersuchung, Osnabrück 1980
(24) San Augustine, A.J., Foley, W.F., How Large Advertisers Set Budgets, in "Journal of Advertising Research", Vol. 15 (1975), No. 5, S. 11 ff.
(25) Schaich, E., Eine Nachfragetheorie ohne Nachfragefunktionen, Diss. München 1967
(26) Schmalen, H., Kommunikationspolitik, Stuttgart u.a. 1985
(27) Schonfeld, E.P., Boyd, J.H., The Financial Payoff in Corporate Advertising, in "Journal of Advertising Research", Vol. 22 (1982), No. 1, S. 45 ff.

(28) Simon, H., Thiel, M., Die Anwendung von Mediaselektionsprogrammen, Erfahrungen und Schlußfolgerungen, in "ZfbF", 31. Jg. (1979), S. 238 ff.
(29) Theisen, P., ter Haseborg, F., Berndt, R., Optimierungsmodelle zur Mediaselektion unter Berücksichtigung von externen und internen Überschneidungen der Werbeträger, unveröffentliches Manuskript, Hamburg 1979
(30) Vyas, N., Woodside, A.G., An Inductive Model of Industrial Supplier Choice Processes, in "Journal of Marketing", Vol. 48 (Winter 1984), S. 30 ff.
(31) Waterson, M., Harrison, F., International Advertising Expenditure Data, European Tripartite Seminar, Rom 1986
(32) Zentes, J., Die Werbeentscheidungen und die Werbeoptimierungsmodelle, in "Die Werbung", Hrsg. B. Tietz, Band 3, Landsberg 1981, S. 2199 ff.

Schätzung von intervallskalierten Konkurrenzintensitäten aus subjektiven Rangordnungsurteilen

Sönke Albers, Lüneburg

Zusammenfassung: Es wird eine Methode zur Schätzung von intervallskalierten Konkurrenzintensitäten aus Rangurteilen vorgeschlagen, die es erlaubt, Konkurrenzintensitäten in Modellen zur Bestimmung von Umsatz-SOLL-Vorgaben zu berücksichtigen, was deren Akzeptanz deutlich verbessert. Das Verfahren erwies sich in einem Unternehmen der Versicherungs- und Finanzdienstleistungsbranche als gut anwendbar. Außerdem wird mit Hilfe eines Simulationsexperimentes gezeigt, daß die Methode eine hinreichend gute Reproduktionsgüte erreicht.

Summary: In this paper a method for the estimation of interval-scaled intensities of competition from rank-order data is developed. This method enables firms to take into account intensities of competition when establishing acceptable models for the determination of sales quotas. The procedure has been applied by an insurance company. In addition, a simulation experiment has been conducted to show sufficiently good results for the method in reconstructing true values.

1. Problemstellung

Bei geographisch gegliederten Verkaufsaußendiensten mit unterschiedlich "guten" Verkaufsgebieten können Umsatz-SOLL-Vorgaben zur Steuerung der Außendienstmitarbeiter nur dann durchgesetzt werden, wenn alle die Umsatztätigkeit beeinflussenden Faktoren berücksichtigt werden (Albers, 1987). Neben dem Umsatzpotential ist in vielen Fällen die Konkurrenzintensität in den einzelnen Verkaufsgebieten unterschiedlich und ein häufiges Argument für das Verlangen nach korrigierten Umsatzvorgaben. Eine direkte Operationalisierung von Konkurrenzintensität bestünde nun darin, die Wirkung der Marketing-Instrumente aller Wettbewerber zu messen. Da die Ausprägungen der Marketing-Instrumente der Konkurrenz jedoch kaum beobachtet werden können, insbesondere wenn es um den Verkaufsaußendienst geht, müßte man folgende indirekte Schätzung versuchen. Kennt man seinen eigenen Marktanteil und die Wirkung der einem selber bekannten Ausprägungen der eigenen Marketing-Instrumente, dann läßt sich die Gesamtwirkung der Marketing-Instrumente aller Anbieter am Markt aus folgender Definitionsgleichung ableiten:

Operations Research Proceedings 1987

$$\text{Marktanteil} = \frac{\text{Wirkung (Ausprägungen der eigenen Marketing-Instrumente)}}{\text{Wirkung (Ausprägungen der Marketing-Instrumente aller Anbieter)}}$$

In vielen Fällen existieren jedoch keine Branchen-Statistiken oder Paneldaten, aus denen Marktanteile errechnet werden können. In dieser Situation bleibt nichts anderes übrig, als die Konkurrenzintensitäten subjektiv zu schätzen, wenn man auf sie bei der Bestimmung von Umsatzvorgaben nicht verzichten will. Das Schätzproblem wäre trivial, wenn es in dem betrachteten Unternehmen eine Person gibt, die über einen hinreichend guten Marktüberblick verfügt, um auf einer fein abgestuften Intervall-Skala Schätzwerte pro Verkaufsgebiet abzugeben. Aber dies dürfte nur sehr selten der Fall sein. Eher hat man mit folgenden Schwierigkeiten zu kämpfen:

a) Typischerweise verfügt keine Person der Verkaufsleitung oder der Marktforschungsabteilung über das Vermögen, differenzierte Urteile auf einer fein abgestuften Intervall-Skala abzugeben. Erwarten darf man für das subjektive Urteil höchstens Rangreihen der Verkaufsgebiete nach Maßgabe ihrer Konkurrenzintensitäten.

b) In einem hierarchisch gegliederten Verkaufsaußendienst kann nur der unmittelbar vorgesetzte Regional-Verkaufsleiter verläßlich Konkurrenzintensitäten für die Verkaufsgebiete seiner Region schätzen, so daß man lediglich Rangreihen für Teilmengen der Verkaufsgebiete gewinnen kann. Allerdings können Verkaufsleiter höherer Hierarchiestufen zusätzlich Rangreihen über von zu Regionen aggregierten Verkaufsgebiete abgeben.

c) Schließlich kann der Fall auftreten, daß sich mehrere Personen (z.B. Verkaufsleiter, Marktforscher, Controller) berufen fühlen, verläßliche Urteile zu den Konkurrenzintensitäten abgeben zu können.

Unter diesen Bedingungen besteht das Problem darin, aus subjektiven Rangordnungsurteilen auf intervallskalierte Konkurrenzintensitäten zu schließen, deren Werte die einzelnen Rangordnungen am besten reproduzieren und bei divergierenden Urteilen eine Art Konsens herbeiführen. Dafür wird eine Methode entwickelt, deren Anwendbarkeit anhand eines Beispiels aus der Versicherungs- und Finanzdienstleitungsbranche demonstriert wird. Zusätzlich wird mit Hilfe einer deterministischen Simulation geprüft, ob hypothetisch angenommene wahre Konkurrenzintensitäten durch das vorgeschlagene Verfahren reproduziert werden können.

2. Schätzmethode

Die im folgenden vorgestellte Methode zur Ableitung von intervallskalierten Konkurrenzintensitäten aus subjektiven Rangordnungsurteilen basiert auf einer Idee von Srinivasan/Shocker (1973) zu einem ähnlichen Problem, nämlich das Auffinden von Schätzwerten unter der Zielsetzung der Minimierung der Verletzung von erfragten Rangordnungs-Urteilen als Goal-Programming-Ansatz der Linearen Programmierung zu formulieren. Bezeichne

S : Indexmenge der Personen, die Rangordnungsurteile zu Verkaufsgebieten abgeben,

I : Indexmenge der Verkaufsgebiete (n: Anzahl Verkaufsgebiete),

dann läßt sich jede erfragte Rangreihe von Verkaufsgebieten (Teilmengen von I) in eine Menge Q_s von geordneten Paaren $(i,j) \in Q_s$ $(s \in S)$ dekomponieren, wobei (i,j) bedeutet, daß die subjektiv empfundene Konkurrenzintensität des i-ten Verkaufsgebietes höher ist als die des j-ten Verkaufsgebietes. Für die zu schätzenden intervallskalierten Konkurrenzintensitäten x_i $(i \in I)$ soll dann gelten:

$$x_i - x_j > 0 \qquad ((i,j) \in Q_s;\ s \in S) \tag{1}$$

Sei a die Toleranzgrenze, bei deren Überschreiten x_i erst als wirklich größer als x_j angesehen wird, dann modifiziert sich (1) zu:

$$x_i - x_j \geq a \qquad ((i,j) \in Q_s;\ s \in S) \tag{2}$$

Sind die Urteile mehrerer Personen in sich inkonsistent, so können keine Werte für x_i $(i \in I)$ gefunden werden, die allen Bedingungen (2) genügen. Es gilt dann einen Konsens dadurch zu finden, daß man den Grad der Verletzung summiert über alle Bedingungen minimiert. Sei z_{ijs} der Grad der Verletzung des Urteils der s-ten Person zu dem i-ten und j-ten Verkaufsgebiet, dann kann folgender linearer Goal-Programming-Ansatz formuliert werden:

$$\sum_{s \in S} \sum_{(i,j) \in Q_s} z_{ijs} \Rightarrow \text{Min!} \tag{3}$$

$$x_i - x_j \geq a - z_{ijs} \qquad ((i,j) \in Q_s,\ s \in S) \tag{4}$$

$$x_i \geq 0 \qquad (i \in I) \tag{5}$$

$$z_{ijs} \geq 0 \qquad ((i,j) \in Q_s;\ s \in S) \tag{6}$$

Hierbei nimmt z_{ijs} aufgrund der Minimierungsvorschrift nur dann positive Werte an, wenn (4) nicht anders erfüllt werden kann, also eigentlich eine Verletzung von (2) vorliegt. Um eine triviale Lösung zu verhindern, in der alle $x_i=0$ $(i \in I)$ gesetzt werden, wird zusätzlich gefordert:

$$\sum_{i \in I} x_i = a * n * (n-1)/2 \qquad (7)$$

wobei $a*(n-1)/2$ den Mittelwert angibt, wenn man für x_i eine Folge von 0, a, 2a, 3a,... annimmt. Welcher Wert für $a > 0$ gewählt wird, hat aufgrund der Intervall-Skala-Eigenschaft von x_i $(i \in I)$ keinen Einfluß auf die Lösungsqualität.

Im Fall von zusätzlichen Rangordnungs-Urteilen über zu Regionen aggregierte Verkaufsgebiete sind folgende Nebenbedingungen hinzuzufügen:

$$(\sum_{i \in I_g} x_i)/m_g - (\sum_{i \in I_h} x_i)/m_h \geq a - y_{ght} \qquad ((g,h) \in R_t,\ t \in T) \qquad (8)$$

$$y_{ght} \geq 0 \qquad ((g,h) \in R_t,\ t \in T) \qquad (9)$$

T : Indexmenge der Personen, die Rangordnungsurteile zu Regionen abgeben,

I_g, I_h : Indexmenge der Verkaufsgebiete, die zu der g-ten bzw. h-ten Region gehören,

m_g, m_h : Anzahl der Verkaufsgebiete, die zu der g-ten bzw. h-ten Region gehören,

R_t : Menge der geordneten Paare von Regionen, abgeleitet aus dem Rangordnungsurteil der t-ten Person $(t \in T)$,

y_{ght} : Grad der Verletzung der Rangordnungsbeziehung der g-ten und h-ten Region im Urteil der t-ten Person $(t \in T)$.

Statt (3) ist nun folgende Zielfunktion zu minimieren:

$$\sum_{s \in S} \sum_{(i,j) \in Q_s} z_{ijs} + \sum_{t \in T} \sum_{(g,h) \in R_t} y_{ght} \Rightarrow \text{Min!} \qquad (10)$$

In dem bisher entwickelten Ansatz bietet sich auch grundsätzlich die Möglichkeit, die einzelnen Urteile nach der subjektiv empfundenen Schätzsicherheit und die die Urteile abgebenden Personen nach ihrer Bedeutung als Experte durch Multiplikation von z_{ijs} und y_{ght} mit entsprechenden Gewichten zu modifizieren.

3. Praxis-Erfahrungen mit einer Anwendung

Die in Abschnitt 2. beschriebene Methode wurde in einem Unternehmen der Versicherungs- und Finanzdienstleistungsbranche eingesetzt, in dem Außendienstmitarbeiter die Bestimmung von Umsatz-SOLL-Vorgaben kritisierten, weil der subjektiv als wichtig erachtete Einflußfaktor Konkurrenzintensität unberücksichtigt geblieben war. Ein höherer Grad an Verläßlichkeit der Schätzwerte konnte dadurch erreicht werden, daß neben den jeweiligen Rangordnungen der Verkaufsgebiete in den 5 Regionen durch die Regionalleiter und der Rangordnung der Regionen durch die Verkaufsleitung noch jeweils eine gesamte Rangordnung aller 22 Verkaufgebiete durch den Leiter der Marktforschung und den Vetriebs-Controller vorlagen. Die Nebenbedingungen (4) wurden über die urteilenden Personen so zusammengefaßt, daß jeweils nur eine Nebenbedingung pro Paar von Verkaufgebieten (i,j) formuliert wurde. Dazu wurde die Verletzungsvariable z_{ij} (der Index s entfällt bei Aggregation von (4) über Personen) mit der Anzahl der Personen gewichtet, die die Konkurrenzintensität des i-ten vor der des j-ten Verkaufsgebietes einschätzten abzüglich der Anzahl der Personen mit umgekehrtem Urteil. Auf diese Weise entstand ein linearer Goal-Programming-Ansatz mit 22 Variablen x_i, 231 Variablen z_{ij}, 10 Variablen y_{gh}, 231 Nebenbedingungen (4), 10 Nebenbedingungen (10) und der Beschränkung (7), in der a = 1 gesetzt war. Alle an dem Schätzprozeß Beteiligten empfanden die Schätzwerte für die Konkurrenzintensitäten der Verkaufsgebiete als plausibel. Bei einer Regressionsanalyse der Umsatztätigkeit mit Hilfe von erklärenden Einflußfaktoren (Albers, 1987) konnte dann entgegen den Erwartungen kein signifikanter Einfluß der Konkurrenzintensität festgestellt werden und damit die Vernachlässigung dieses möglichen Einflußfaktors bei der Bestimmung von Umsatz-SOLL-Vorgaben sachlich gut begründet werden, was die Akzeptanz von Umsatzvorgaben merklich steigerte.

4. Güte der Schätzmethode

Um die Güte des vorgeschlagenen Verfahrens besser beurteilen zu können, wurde getestet, inwieweit hypothetisch als "wahr" angenommene Konkurrenzintensitäten trotz der Unvollständigkeit der Rangurteile insgesamt durch das Verfahren wieder reproduziert werden können. Dazu wurden mit Hilfe eines Zufallszahlengenerators Konkurrenzintensitäten für eine Anzahl von Verkaufsgebieten bestimmt und dann als "wahr" unterstellt. Je nach Anzahl der Regionen wurden die Verkaufsgebiete dann gleichmässig auf die Regionen verteilt. Auf der Basis der so generierten hierarchischen Gliederung der Verkaufsorganisation und der Annahme, daß die Rangordnungs-Urteile der Verkaufsleiter mit den "wahren" Konkurrenzintensitäten korrespondieren, wurden dann die resultierenden Goal-Programming-Ansätze mit der Zielfunktion (10), den Beschränkungen (4), (8) und (7) sowie den Nichtnegativitäts-Bedingungen (5), (6) und (9) formuliert und optimiert. Da angenommen werden muß, daß die Reproduktionsgüte von der Größe und Struktur des Problems abhängt, wurde ein faktorielles Simulationsexperiment durchgeführt, bei dem die Ausprägungen folgender Faktoren systematisch miteinander kombiniert wurden:

Faktor	Ausprägungen
Größe des Problems (Anzahl Verkaufsgebiete)	24 ; 36
Anzahl Urteile (Anzahl Regionen)	1; 2; 3; 4; 5; 6
Struktur der Konkurrenzintensitäten	1. Methode: Zufällig verteilt über alle Verkaufsgebiete 2. Methode: Abgesehen von einem 50% Überlappungsbereich sind alle Konkurrenzintensitäten einer Region größer als die einer anderen Region

Tab. 1: Ausprägungen des faktoriellen Simulationsexperimentes

Zu jeder der insgesamt 2*6*2 = 24 Ausprägungskombinationen wurden 5 Replikationen gerechnet, so daß Ergebnisse für 120 Datensituationen vorliegen. Die Reproduktionsgüte wurde mit Hilfe des Rangreihenkorrelationskoeffizienten τ von Kendall (1955) gemessen, der aussagt, in wieviel Prozent aller Fälle (normiert auf den Bereich $-1 \leq \tau \leq +1$) Verletzungen der wahren Rangbeziehungen von Paaren von Verkaufsgebieten durch die geschätzten Werte auftreten. Über alle Fälle wurde ein mittlerer Wert von τ = 0,755 erzielt, der auf 12,25% Verletzungen hindeutet, was aufgrund der Unvollständigkeit der Urteile ein ausgezeichnetes Ergebnis darstellt. An dieser Stelle sei jedoch angemerkt, daß damit nur verfahrensbedingte Reproduktionsungenauigkeiten erfaßt werden, nicht jedoch das eventuell vorhandene Unvermögen, überhaupt subjektive Rangurteile über Konkurrenzintensitäten abgeben zu können. Welcher Einfluß von den Ausprägungen der experimentellen Bedingungen ausgeht, kann man an den Mittelwerten von τ für die einzelnen Zellen (siehe Tab. 2) ablesen:

Anzahl Regionen \ Struktur der Konkurrenzintensitäten / Anzahl Verkaufsgebiete	1. Methode		2. Methode	
	24	36	24	36
1	1,000	1,000	1,000	1,000
2	0,700	0,785	0,799	0,738
3	0,801	0,761	0,638	0,778
4	0,718	0,801	0,624	0,654
5	0,630	0,648	0,688	0,662
6	0,607	0,633	0,746	0,701

Tab. 2: Reproduktionsgüte gemessen als Kendalls für verschiedene Ausprägungen des faktoriellen Simulationsergebnisses

Eine Varianzanalyse dieses Datenmaterials zeigt nun, von welchen Faktoren der größte Einfluß auf die Reproduktionsgüte ausgeht und inwieweit die Einflüsse statistisch abgesichert sind (siehe Tab. 3):

Einflußfaktor	Erklärte Varianz in %	Signifikanzniveau nach Kruskal-Wallis-Test
Struktur der Konkurrenzintensitäten	0,0%	n.s.
Anzahl der Verkaufsgebiete	0,4%	n.s.
Anzahl der Regionen	56,7%	< 0,1%
Interaktionen	9,2%	n.s.= nicht signifikant auf dem 10%-Niveau
Nichterklärte Varianz	33,7%	

Tab. 3: Varianzanalyse des Einflusses der Ausprägungen der Simulationsbedingungen auf die Reproduktionsgüte

Wie man den Ergebnissen in Tab. 3 entnehmen kann, geht ein dominierender Einfluß von der Anzahl der Regionen aus (56,7% erklärte Varianz). Tab. 2 zeigt, daß die Reproduktionsgüte fast monoton mit der Anzahl von Regionen sinkt. Der Grund dafür liegt darin, daß mit immer weniger Regionen immer mehr Verkaufsgebiete paarweise verglichen werden, so daß die Vollständigkeit der Urteile mit abnehmender Anzahl von Regionen steigt. Von der Anzahl der Verkaufsgebiete geht praktisch kein Einfluß aus, lediglich daß bei 36 gegenüber 24 Verkaufsgebieten bei gleicher Anzahl von Regionen ein leicht vollständigeres Urteil gegeben ist. Schließlich kann man bei der Struktur der Konkurrenzintensitäten, jedenfalls nach Maßgabe der hier gewählten Operationalisierung, keinen Einfluß auf die Reproduktionsgüte feststellen.

5. Implikationen für die Unternehmenspraxis

Mit der hier vorgeschlagenen Methode zur Schätzung von intervallskalierten Konkurrenzintensitäten aus Rangurteilen wird es möglich, die Intensität der Anstrengungen der Wettbewerber in Modellen zur Bestimmung von Umsatz-SOLL-Vorgaben zu berücksichtigen oder "beweisen" zu können, daß die Konkurrenzintensität keinen nachweisbaren Einfluß auf die Umsatztätigkeit pro Verkaufsgebiet ausübt. Damit dürfte die Akzeptanz einer Außendienststeuerung mit Hilfe von Umsatzvorgaben verbessert werden. Eine Anwendung in einem Unternehmen der Versicherungs- und Finanzdienstleistungsbranche hat demonstriert, daß die geforderten Rangordnungsurteile zu den Verkaufsgebieten ohne größere Probleme erhoben werden können. Außerdem konnte mit Hilfe eines Simulationsexperimentes eine hinreichend gute Reproduktionsgüte des vorgeschlagenen Verfahrens gezeigt werden.

Literaturverzeichnis

Albers, S.: Steuerung von Verkaufsaußendienstmitarbeitern mit Hilfe von Umsatzvorgaben, erscheint in: W. Lücke et al. (Hrsg.): Betriebswirtschaftliche Steuerungs- und Kontrollprobleme, Göttingen 1987

Kendall, M.G.: Rank correlation methods (2nd ed.), London: Charles Griffin & Co., Ltd. 1955

Srinivasan, V.; Shocker, A.D.: Estimating the Weights for Multiple Attibutes in a Composite Criterion Using Pairwise Judgments, Psychometrika, Vol. 38 (1973), 473-493

MARKET RESEARCH SUPPORT BY DATA ANALYSIS TECHNIQUES: THEORETICAL DEVELOPMENTS VS. PRACTICAL APPLICATIONS

Wolfgang Gaul, Karlsruhe

There is an extensive and still growing variety of tools for the evaluation of marketing data.

The underlying paper reports on the results of a survey carried out among German market research institutes (see e.g. [5], [6]) in which - among others - questions concerning knowledge about and use of data analysis techniques have been asked. In contrast to these empirical findings, own research efforts with regard to clustering (see e.g. [2], [3], [7]), dual scaling (see e.g. [8]) and probabilistic multidimensional scaling (see e.g. [4]) are described.

Conclusions and an outlook for further research - e.g. the combination of decision support by data analysis techniques with possibilities which could arise by the use of knowledge-based systems (see e.g. [1]) - are also presented.

[1] Böckenholt, I., Both, M. and W. Gaul (1987), PROLOG-Based Decision Support for Data Analysis in Marketing, Diskussionspapier Nr. 118, Institut für Entscheidungstheorie und Unternehmensforschung, Universität Karlsruhe (Sept. 1987).

[2] Both, M. and W. Gaul (1986), Ein Vergleich zweimodaler Clusteranalyseverfahren, Methods of Operations Research 57, 593-605.

[3] Espejo, E. and W. Gaul (1986), Two-Mode Hierarchical Clustering as an Instrument for Marketing Research, in: Gaul, W. and M. Schader, Eds., Classification as a Tool of Research, North Holland.

[4] Gaul, W. (1987), Probabilistic Choice Behavior Models and Their Combination with Additional Tools needed for Applications to Marketing, Communication & Cognition, 20, 77-92.

[5] Gaul, W., Förster, F. and K. Schiller (1986), Typologisierung deutscher Marktforschungsinstitute: Ergebnisse einer empirischen Studie, Marketing ZFP 8, 163-172.

[6] Gaul, W. and Ch. Homburg (1987), The Use of Data Analysis Techniques by German Market Research Agencies: A Causal Analysis, to appear in Journal of Business.

[7] Gaul, W. and M. Schader (1987), Clusterwise Aggregation of Relations, Diskussionsbeiträge zur Statistik und Quantitativen Ökonomik Nr. 30, Universität der Bundeswehr Hamburg (Okt. 1987).

[8] Nishisato, S. and W. Gaul (1987), Marketing Data Analysis by Dual Scaling, Diskussionspapier Nr. 112, Institut für Entscheidungstheorie und Unternehmensforschung, Universität Karlsruhe (Sept. 1987).

DYNAMISCHE WERBEBUDGET-ABSATZREAKTIONSMODELLE

Fokko ter Haseborg, Graz

Zusammenfassung: Ziel des Beitrages ist es, einen Überblick über deterministische dynamische Werbebudget-Absatzreaktionsmodelle zu geben. Zunächst werden empirisch fundierte Absatzreaktionsphänomene der Werbung dazu herangezogen, um daraus Anforderungen an die Abbildungseigenschaften empirisch relevanter Absatzreaktionsmodelle herzuleiten. Verschiedene Möglichkeiten zur Modellierung von Absatzreaktionen auf Maßnahmen des Werbebudgeteinsatzes werden anschließend aufgezeigt und kritisch daraufhin verglichen, inwieweit sie die abgeleiteten Modellanforderungen hinsichtlich der empirischen Relevanz erfüllen. Abschließend werden mögliche Verallgemeinerungen und Weiterentwicklungen diskutiert.

Summary: The purpose of this study is to give a survey on deterministic dynamic advertising-sales-response models. First some empirical founded advertising-sales-effects are used to formulate modeling requirements which have to be fulfilled by empirical relevant sales-response models. Several alternative ways of modeling sales-advertising relationships are then discussed and critically compared whether they fulfil the derived modeling requirements for empirical relevance or not. Directions for potential generalizations are finally discussed.

A. Problemstellung und einige Abgrenzungen

In der Werbeliteratur sind zahlreiche dynamische Absatzreaktionsmodelle der Werbung entwickelt und analysiert worden (Siehe hierzu z.B. [1], [3] bis [20]). Häufig stehen dabei normative und schätztechnische Fragen im Vordergrund der Analysen. Ausgangspunkt dieses Beitrags ist dagegen die Frage, inwieweit die empirischen Strukturphänomene bei Absatzreaktionsprozessen der Werbung durch die Absatzreaktionsmodelle zur Werbebudgetierung modelliert werden können.

Der Untersuchungsgegenstand dieses Beitrages wird auf deterministische dynamische Reaktionsmodelle der Werbung für ein Konsumprodukt, welches auf einem monopolistischen, nicht-segmentierten (unvollkommenen) Markt angeboten wird, beschränkt. Als Instrumentalvariable wird das Werbe- bzw. - besser - Mediabudget des Produktes betrachtet, als Reaktionsvariable die Absatzmenge des Produktes. Von der Modellierung von Marketing-Mix-Effekten wird abgesehen. Die Beurteilung der Abbildungseigenschaften der Modelle findet im Rahmen einer Periodenanalyse statt, wobei die Periodenlänge auf ca. eine Quartalslänge nach oben begrenzt wird.

B. Empirisch fundierte Absatzreaktionsphänomene als Ausgangspunkt der Modellierung

Nimmt man eine Bestandsaufnahme der empirischen Befunde zur Absatzwirkung des Werbe- bzw. Mediabudgeteinsatzes vor, so lassen sich die Ergebnisse zu dem in Tabelle 1 wiedergegebenen Katalog von Abbildungseigenschaften, die von empirisch gehaltvollen Absatzreaktionsmodellen möglichst aufzuweisen sind, zusammenfassen (Vergleiche hierzu ausführlich [2]).

Operations Research Proceedings 1987

Statische Analyse

1. „Lang- und kurzfristiger Maximalabsatz"
1.1 Existenz von lang- und kurzfristigen Maximalabsatzwerten
1.2 Berücksichtigung von evtl. verschiedenen Maximalabsatzwerten in lang- und kurzfristiger Sicht
1.3 Berücksichtigung verschiedener lang- und kurzfristiger Maximalabsatzwerte für verschiedene Marken
2. „Lang- und kurzfristiger Minimalabsatz"
2.1 Existenz von lang- und kurzfristigen Minimalabsatzwerten
2.2 Berücksichtigung von evtl. verschiedenen Minimalabsatzwerten in lang- und kurzfristiger Sicht
2.3 Berücksichtigung verschiedener lang- und kurzfristiger Minimalabsatzwerte für verschiedene Marken
3. Verlaufsform der lang- und kurzfristigen statischen Absatzreaktionsfunktion
3.1 Abnehmende und gegen Null gehende Grenzabsatzwirkungen bei hohen Mediabudgets
3.2 Berücksichtigung vor allem von konkaven (weniger von s-förmigen) statischen Absatzreaktionsfunktionen

Dynamische Analyse

4. „Existenz und Anpassungsmöglichkeiten direkter Wirkungsabhängigkeiten"
4.1 Existenz verzögerter Mediabudget-Variablen (direkte Lag-Variablen)
4.2 Möglichkeit der Anpassung an verschieden lange direkte Lag-Zeiträume
4.3 Möglichkeit der Anpassung an verschiedene direkte Lag-Strukturen (-Funktionen, -Gewichte)
5. „Direkter Lag-Einfluß auf den Maximalabsatz" Abhängigkeit der periodenbezogenen Maximalabsatzwerte von Mediabudgets in Vorperioden
6. „Direkter Lag-Einfluß auf den Minimalabsatz" Abhängigkeit der periodenbezogenen Minimalabsatzwerte von Mediabudgets in Vorperioden
7. „Funktionale Ausgestaltung direkter Lag-Wirkungen"
7.1 Berücksichtigung maximaler Auswirkungen von früheren Mediabudgets auf den Absatz einer Periode
7.2 Berücksichtigung abnehmender Grenzabsatzwirkungen früherer Mediabudgets auf den Absatz einer Periode
7.3 Abhängigkeit der momentanen Grenzabsatzwirkungen einer Periode von den Mediabudgets in Vorperioden
8. „Existenz und Anpassungsmöglichkeiten indirekter Wirkungsabhängigkeiten"
8.1 Existenz verzögerter Absatzvariablen (indirekte Lag-Variablen)
8.2 Möglichkeit der Anpassung an verschieden lange indirekte Lag-Zeiträume
8.3 Möglichkeit der Anpassung an verschiedene indirekte Lag-Strukturen (-Funktionen, -Gewichte)
9. „Indirekter Lag-Einfluß auf den Maximalabsatz" Abhängigkeit der periodenbezogenen Maximalabsatzwerte von Absatzwerten in Vorperioden
10. „Indirekter Lag-Einfluß auf den Minimalabsatz" — Abhängigkeit der periodenbezogenen Minimalabsatzwerte von Absatzwerten in Vorperioden
11. „Indirekter Lag-Einfluß auf die momentane Grenzabsatzwirkung" Abhängigkeit der momentanen Grenzabsatzwirkungen einer Periode von den Absatzwerten in Vorperioden
12. „Vergessensprozesse" Möglichkeit der Abbildung zeitbeanspruchender Absatzabnahmeprozesse bei ausgesetzter Werbung

Tabelle 1: Empirisch fundierte Abbildungseigenschaften nach [2]

In die Bewertung des empirischen Gehalts der Absatzreaktionsmodelle gehen die verschiedenen Modelleigenschaften mit z.T. unterschiedlich hohem Beurteilungsgewicht ein. Einige Eigenschaften sind als Muß-Anforderungen aufzufassen (1.1, 2.1, 3.1, 7.1, 7.2); die anderen sind eher als Kann-Eigenschaften zu werten. Unter den Kann-Eigenschaften ist den Eigenschaften 1.3 und 2.3 sowie den Eigenschaften 4, 7.3, 8, 11 u. 12 ein höheres Bedeutungsgewicht zuzumessen.

C. Analyse der Abbildungseigenschaften der klassischen dynamischen Werbebudget-Absatzreaktionsmodelle

Im folgenden wird unterschieden zwischen Modellen mit zeitdiskreten Wirkungsrelationen und solchen mit zeitkontinuierlichen Wirkungsbeziehungen. Innerhalb dieser beiden Modellgruppen wird weiter danach klassifiziert, ob intermediäre Wirkungsbeziehungen abgebildet werden oder nicht. Um die Ansätze der verschiedenen Modellklassen vergleichend daraufhin untersuchen zu können, inwieweit sie die Abbildungseigenschaften 1 bis 12 aus Tabelle 1 aufweisen, wird die Beurteilung einheitlich auf der Basis von Periodenwerten vorgenommen. Zur Standardisierung der Zeitstruktur wird dazu für Modellle mit zeitkontinuierlichen Wirkungsbeziehungen davon ausgegangen, daß das Werbebudget einer Periode gleichmäßig über die Zeit verteilt wird. Ferner werden die resultierenden Absatzratenverläufe zu Periodenabsatzwerten integriert.

Für die 20 in Tabelle 2 konkretisierten Modelle werden die vollständigen Profile ihrer Abbildungseigenschaften in Tabelle 3 wiedergegeben. Sowohl in Tabelle 3 als auch im folgenden Kurz-Kommentar wird über die in runden Klammern angegebenen Referenznummern auf die verschiedenen Modelle Bezug genommen.

1. Modelle mit zeitdiskreten Wirkungsbeziehungen

a) Modelle ohne intermediäre Wirkungsbeziehungen

In den folgenden Modellen wird eine direkte Beziehung zwischen den Werbe- bzw. Mediaausgaben und den dadurch induzierten Absatzreaktionen hergestellt. Wichtige und in vielen empirischen Arbeiten verwendete Beispiele sind das variablen-lineare Modell (1), das semi-logarithmische und parameter-lineare Modell (2) und das multiplikative Modell (3). Gerade diese empirisch häufig überprüften Modelle erfüllen wesentliche Anforderungen hinsichtlich der Abbildungseigenschaften nicht. Die Anwendung dieser klassischen ökonometrischen Modelle stellt ein konventionelles Vorgehen dar, Absatzreaktionen zu prognostizieren für Werbeoperationen, die sich von historischen Werbemaßnahmen nicht sehr unterscheiden. Den Werbewirkungsprozeß vermögen sie - wenn überhaupt - höchstens auf einem sehr eng begrenzten Intensitätsbereich, nicht jedoch für die bei kurzfristigen Werbeplanungen relevanten Bereiche möglicher Werbeintensitäten angemessen zu repräsentieren.

Demgegenüber werden durch empirisch weniger überprüfte parameter-lineare bzw. parameter-linearisierbare Modelle - etwa in der Form von Exponentialansätzen nach (4) bzw. (5) - wesentliche Anforderungen an die Modelle bei Ausschöpfung schätztechnischer Vorteile erfüllt. Allerdings sind diese Modelle infolge der Restriktion, hinsichtlich der Parameter linear bzw. linearisierbar zu sein, bei der Anpassung an spezielle Absatzwirkungsverhältnisse nicht in dem Maße flexibel, wie es nach dem Profil der Abbildungseigenschaften als wünschenswert erscheint. Die nicht-parameter-linearisierbaren dynamischen Modelle vom Typ (6) und vom ADBUG-Typ (7) weisen dieses Ausmaß an Anpassungsflexibilität zusammen mit den anderen empirisch fundierten Abbildungseigenschaften auf.

b) Modelle mit intermediären Wirkungsbeziehungen

Einer ganzen Reihe von Arbeiten zur Werbebudgetierung liegt die Auffassung zugrunde, daß Werbemaßnahmen als Investition in eine Art Werbekapitalstock, der als Werbegoodwill bezeichnet wird und im Zeitablauf einer Abnahme- bzw. Vergessenswirkung unterliegt, anzusehen seien.

Referenz-Nr.	
	1. Modelle mit zeitdiskreten Wirkungsbeziehungen
	a) Modelle ohne intermediäre Wirkungsbeziehungen
(1)	$y_t=\alpha_o+\sum_{j=1}^{L_1}\alpha_j y_{t-j}+\sum_{i=0}^{L_2}\beta_i w_{t-i}$ $(L_1>0$ oder $L_2>0)$
(2)	$y_t=\alpha_o+\sum_{j=1}^{L_1}\alpha_j y_{t-j}+\sum_{i=0}^{L_2}\beta_i \ln(w_{t-i})$ $(L_1>0$ oder $L_2>0)$
(3)	$y_t = \alpha_o \prod_{j=1}^{L_1} y_{t-j}^{\alpha_j} \prod_{i=0}^{L_2} w_{t-i}^{\beta_i}$ $(L_1>0$ oder $L_2>0)$
(4)	$y_t=y_{t-1}-\sum_{j=1}^{L_1}\lambda_j(y_{t-j}-y^-)+[y^+-y_{t-1}+\sum_{j=1}^{L_1}\lambda_j(y_{t-j}-y^-)][1-\exp(-\sum_{i=0}^{L_2}w_{t-i})]$ $(L_1>0$ oder $L_2>0)$
(5)	$y_t = y^+ - (y^+-y^-)\exp(-\sum_{i=0}^{L}\beta_i w_{t-i})$ mit $L>0$
(6)	$y_t = \alpha_o + \sum_{j=1}^{L_1}\alpha_j\cdot y_{t-j} + \sum_{i=0}^{L_2}\beta_i\cdot\exp(\varphi_{i1}-\frac{\varphi_{i2}}{w_{t-i}}) + \sum_{i=0}^{L_1}\sum_{i'=i+1}^{L_2}\gamma_{ii'}\cdot\exp(\psi_{i1}-\frac{\psi_{i2}}{w_{t-i}})\cdot\exp(\eta_{i'1}-\frac{\eta_{i'2}}{w_{t-i'}})$ $+\sum_{i=0}^{L_2}\sum_{j=1}^{L_1}\delta_{ij}\cdot(y^+-y_{t-j})\cdot\exp(\mu_{i1}-\frac{\mu_{i2}}{w_{t-i}})$ wobei $L_1>0$ oder $L_2>0$.
(7)	$y_t = y_{t-1} - \sum_{j=1}^{L_1}\gamma_j(y_{t-j}-y^-) + (y^+-y_{t-1}+\sum_{j=1}^{L_1}\gamma_j(y_{t-j}-y^-))\cdot\frac{(\sum_{i=0}^{L_2}\beta_i w_{t-i})^\delta}{\gamma+(\sum_{i=0}^{L_2}\beta_i w_{t-i})^\delta}$ $(L_1>0$ oder $L_2>0)$
	b) Modelle mit intermediären Wirkungsbeziehungen
(8)	$g_t=\beta\cdot w_t+\gamma\cdot g_{t-1}$ mit $y_t=\alpha\cdot g_t^\varepsilon$
(9)	$g_t = \sum_{i=o}^{L_1}\beta_i w_{t-i} + \sum_{j=1}^{L_2}\gamma_j g_{t-j}$ mit $y_t = y^- + (y^+-y^-)\, g_t^\delta\cdot(\gamma+g_t^\delta)^{-1}$
(10)	$g_t=\sum_{i=o}^{L_1}\beta_i w_{t-i}+\sum_{j=1}^{L_2}\gamma_j g_{t-j}$ mit $y_t=\sum_{j=1}^{L_3}\alpha_j y_{t-j}+(1-\sum_{j=1}^{L_3}\alpha_j)(y^- + (y^+-y^-)\, g_t^\delta\cdot(\gamma+g_t^\delta)^{-1})$
(11)	$g_t = \gamma_o + \gamma_1 g_{t\cdot 1} + \beta w_t$ und $y_t = (1-\lambda)\cdot y_{t-1} + g_t\cdot(y^+-y_{t-1})$
(12)	$g_t=\gamma_o+\sum_{i=o}^{L_1}\gamma_j g_{t-j}+\sum_{i=o}^{L_2}\beta_i w_{t-i}$ und $y_t = \sum_{j=1}^{L_3}\alpha_j\cdot y_{t-j} + (1-\sum_{j=1}^{L_3}\alpha_j)\cdot\{(y_{t-j}-\sum_{i=0}^{L_4}\varepsilon_i(y_{t-i}-y^-)+$ $+[y^+-y_{t-1}+\sum_{i=0}^{L_4}\varepsilon_i(y_{t-i}-y^-)]\cdot f(g_t)\}$ mit $f(g_t) = g_t^\delta\cdot(\eta+g_t^\delta)^{-1}$
	2. Modelle mit zeitkontinuierlichen Wirkungsbeziehungen
	a) Modelle ohne intermediäre Wirkungsbeziehungen
(13)	$DY(z) = \alpha W(z)\,(Y^+-Y(z)) - \lambda\,(Y(z)-Y^-)$
(14)	$DY(z) = \alpha \log W(z) - \lambda Y(z)$
(15)	$DY(z) = W(z)^\delta\,(\gamma+W(z)^\delta)^{-1}\,(\alpha(Y^+-Y(z))+\lambda(Y(z)-Y^-))-\lambda(Y(z)-Y^-)$
(16)	$(D+\beta)Y(z) = \beta\{\sum_{j=1}^{L_2}\alpha_j\bar{Y}_{t-j}+(1-\sum_{j=1}^{L_2}\alpha_j)\cdot Y^g(W_t,\dots,W_{t-L_1})\}$ mit $Y^g(W_t,\dots,W_{t-L_1}) = Y^-+(Y^+-Y^-)(\sum_{i=o}^{L_1}\beta_i W_{t-i})^\delta\cdot$ $\cdot(\gamma+(\sum_{i=o}^{L_1}\beta_i W_{t-i})^\delta)^{-1}$
	b) Modelle mit intermediären Wirkungsbeziehungen
	$(D+\mu_1)(D+(\mu_1+\mu_2))G(z) = c\cdot\mu_2\,W(z)$ oder $(D+\mu)^r G(z) = \mu^r\cdot W(z)$
(17)	mit $Y(z) = Y^- + (Y^+-Y^-)\cdot G(z)^\delta\cdot(\gamma+G(z)^\delta)^{-1}$ oder
(18)	mit $(D+\alpha)Y(z) = \alpha\cdot[Y^-+(Y^+-Y^-)G(z)^\delta\,(\gamma+G(z)^\delta)^{-1}]$ oder
(19)	mit $DY(z) = \eta\, G(z)\,(Y^+-Y(z)) - \lambda(Y(z)-Y^-)$
(20)	$(D+\mu_1)(D+(\mu_1+\mu_2))G(z) = c\cdot\mu_2\sum_{i=0}^{L_1}\beta_i W_{t-i}$ oder $(D+\mu)^r G(z) = \mu^r\sum_{i=o}^{L_1}\beta_i W_{t-i}$ mit $DY(z) = \eta\cdot G(z)\,(Y^+-Y(z)-\lambda(\sum_{j=1}^{L_2}\alpha_j\bar{Y}_{t-j} + (1-\sum_{j=1}^{L_2}\alpha_j)Y(z)-Y^-)$

Tabelle 2: Ausgewählte Modellspezifikationen

Legende zu Tabelle 2

t bzw. z:	:	Perioden- bzw. Zeitpunkt-Index
y_t bzw. $Y(z)$	:	Absatz in Periode t bzw. Absatzrate zum Zeitpunkt z
y^+ bzw. Y^+	:	Maximalwert für y_t bzw. $Y(z)$
y^- bzw. Y^-	:	Minimalwert für y_t bzw. $Y(z)$
w_t bzw. $W(z)$	:	Mediabudget in Periode t bzw. Mediabudgetrate in z
D	:	Ableitungsoperator nach der Zeit
g_t bzw. $G(z)$	:	Goodwillbestand in Periode t bzw. zum Zeitpunkt z
W_t bzw. $\bar{Y}_t$	:	durchschnittliche Mediabudget- bzw. Absatzrate in Periode t
r; L, L_1,...	:	natürliche Zahlen
griech. Buchstaben	:	(vorzugebende oder zu schätzende) Funktionsparameter

Das hypothetische Konstrukt des Werbegoodwill kann charakterisiert werden als ein alternativer Mechanismus zur Modellierung von zeitlichen Carry-Over-Effekten.

Es stellt sich heraus, daß einfache Goodwillmodelle vom NERLOVE-ARROW-Typ (8) oder - erweitert - (9) mit monoton-steigenden, statischen Goodwill-Absatz-Transformationsfunktionen hinsichtlich der Abbildungseigenschaften im Vergleich zu direkten Absatzreaktionsmodellen ohne intermediäre Wirkungsbeziehungen keine Vorzüge zu bieten haben. Erst wenn auch dynamische Relationen zwischen den Absatzwerten mit jenen zwischen den Goodwillwerten kombiniert werden - wie dies bei Modellen vom BRANDAID-Typ nach (10) oder vom TURNER-NEUMAN-Typ nach (11) oder (12) der Fall ist -, ergeben sich Möglichkeiten zur Abbildung differenzierterer zeitlicher Carry-Over-Beziehungen, die in Modellen ohne intermediäre Goodwillvariablen nicht ohne weiteres nachgebildet werden können.

Während Modelle vom BRANDAID-Typ den meisten Anforderungen hinsichtlich der statischen und dynamischen Absatzreaktionsrelationen genügen, kann eine Abhängigkeit der momentanen Grenzabsatzwirkung vom erreichten Absatzniveau auf Grund der Separabilität der Goodwill-Absatzreaktionsfunktionen mit dieser Modellklasse nicht erfaßt werden. Dieser Einschränkung unterliegen die Modelle vom TURNER-NEUMAN-Typ nicht. Allerdings werden die in der Literatur bekannten Modelle dieses Typs gemäß (11) schon elementaren Anforderungen nicht gerecht. Mit der Erweiterung (12) können wieder die wichtigsten Abbildungseigenschaften sichergestellt werden.

2. Modelle mit zeitkontinuierlichen Wirkungsbeziehungen

Generell läßt sich bei einem Vergleich zwischen zeitdiskreten und zeitkontinuierlichen Modellen feststellen, daß einige empirische Strukturphänomene sich in zeitkontinuierlichen Modellen mit vergleichsweise einfachen Differentialgleichungen erfassen lassen. Ferner treffen die grundlegenden Unterschiede zwischen Modellen ohne bzw. mit intermediären Wirkungsbeziehungen, wie sie für zeitdiskrete Modelle beschrieben worden sind, in ähnlicher Weise auch für zeitkontinuierliche Modelle zu.

Abb.-eigenschaften nach Tab. 1			Modelle gemäß Tab. 2 mit																			
			zeitdiskreten Wirkungsbeziehungen												zeitkontinuierlichen Wirkungsbeziehungen							
			ohne Goodwill-Variablen							mit Goodwill-Variablen					ohne Goodwill-Variablen				mit Goodwill-Variablen			
			(1)	(2)	(3)	(4)	(5)	(6)	(7)	(8)	(9)	(10)	(11)	(12)	(13)	(14)	(15)	(16)	(17)	(18)	(19)	(20)
Statische Analyse	1.1		-	-	-	x	x	x	x	-	x	x	1	x	x	-	x	x	x	x	x	x
	1.2		-	-	-	-	-	x	-	-	-	x	-	-	-	-	x	x	-	x	-	-
	1.3		-	-	-	x	x	x	x	-	x	x	1	x	x	-	x	x	x	x	x	x
	2.1		x	-	x	x	x	x	x	x	x	x	x	x	x	-	x	x	x	x	x	x
	2.2		-	-	-	x	x	x	x	x	x	x	x	x	x	-	x	x	x	x	x	x
	2.3		-	-	-	x	x	x	x	-	x	x	x	x	x	-	x	x	x	x	x	x
	3.1		-	x	*	x	x	*	x	*	x	x	1	x	x	x	x	x	x	x	x	x
	3.2		-	x	*	x	x	*	x	*	x	x	1	x	x	x	x	x	x	x	x	x
Dynamische Analyse	4.1		x	x	x	x	x	x	x	-	x	x	-	x	-	-	-	x	-	-	-	x
	4.2		x	x	x	x	x	x	x	-	x	x	-	x	-	-	-	x	-	-	-	x
	4.3		x	x	x	-	x	x	x	-	x	x	-	x	-	-	-	x	-	-	-	x
	5		-	-	-	-	-	x	-	-	-	-	-	-	-	-	-	-	-	-	-	-
	6		x	-	-	*	x	x	x	-	x	x	-	x	-	-	-	x	-	-	-	x
	7.1		-	-	-	x	x	x	x	-	x	x	-	x	-	-	-	x	-	-	-	x
	7.2		-	x	*	x	x	x	x	-	x	x	-	x	-	-	-	x	-	-	-	x
	7.3		-	-	x	x	x	*	*	-	x	*	-	x	-	-	-	x	-	-	-	x
	8.1	y	x	x	x	x	-	x	x	-	-	x	x	x	x	x	x	x	-	x	x	x
		g								x	x	x	x	x					x	x	x	x
	8.2	y	x	x	x	x	-	x	x	-	-	x	-	x	-	-	-	x	-	-	-	x
		g								-	x	x	-	x					-	-	-	-
	8.3	y	x	x	x	x	-	x	x	-	-	x	+	x	+	+	+	x	-	+	+	x
		g								+	x	x	+	x					+	+	+	+
	9	y	-	-	-	-	-	x	-	-	-	x	-	x	-	-	x	x	-	x	-	-
		g								-	-	-	-	-					-	-	-	-
	10	y	x	-	-	x	-	x	x	-	-	x	x	x	x	-	x	x	-	x	x	x
		g								x	x	x	x	x					x	x	x	x
	11	y	-	-	x	x	-	x	x	-	-	-	x	x	x	-	x	-	-	-	x	x
		g								x	x	x	-	x					x	x	x	x
	12		x	x	-	x	+	x	x	x	x	x	x	x	x	-	x	x	x	x	x	x

Legende: x: ja *: möglich bei geeigneter Modifikation/Spezifikation +: ja, mit Einschränkungen
-: nein 1: nur in langfristiger Sicht
y bzw. g: kennzeichnet indirekte Lag-Struktur bezgl. des Absatzes (y) bzw. bezgl. des Goodwills (g)

Tabelle 3: Profil der Abbildungseigenschaften der Modelle aus Tab. 2

Sieht man einmal vom SETHI-Modell (14) ab, welches gegen eine ganze Reihe elementarer Anforderungen verstößt, so zeigt Tabelle 3, daß die gewöhnlichen Differentialgleichungs-Modelle in Anlehnung an VIDALE-WOLFE (13), SASIENI (15), NERLOVE-ARROW (17), TURNER-NEUMAN (19) sowie vom BRANDAID-Typ (18) zwar den Anforderungen in der statischen Analyse genügen, jedoch arm an dynamischen Effekten sind: direkte Lag-Strukturen werden überhaupt nicht abgebildet und indirekte Lag-Strukturen nur in sehr beschränktem Umfang.

Die in der Literatur vorgeschlagene Erweiterung der gewöhnlichen Differentialgleichungen zu Integro-Differentialgleichungen erweist sich aus planungspragmatischen Gründen als unbrauchbar. Die im Katalog der Abbildungseigenschaften erfaßten dynamischen Absatzreaktionseffekte können bei flexibler Anpassung an die jeweiligen Produktmarktverhältnisse modelliert werden, wenn unter Beibehaltung des Konzeptes zeitkontinuierlicher Wirkungsbeziehungen zeitdiskrete Lag-Strukturen in die Modelle - wie in (16) und (20) beispielhaft gezeigt - eingeführt werden.

D. Schlußbemerkungen

Für 20 ausgewählte klassische dynamische Werbebudget-Absatzreaktionsmodelle ist das Profil ihrer Abbildungseigenschaften analysiert worden (Eine differenziertere Analyse für insgesamt 200 Modelle findet sich [1]). Neuere Ergebnisse der empirischen Marktreaktionsmessung über zeitablaufbezogene, durch Mediabudget- und/oder Werbemitteländerungen induzierte Absatzreaktionseffekte und die Berücksichtigung von Konkurrenzwerbe- sowie Segmenteffekten bei nicht-monopolistischen und/oder segmentierten Marktstrukturen können dazu herangezogen werden, (a) den Katalog der Modellanforderungen in Tabelle 1 zu erweitern (Vergleiche [2]) und (b) - ausgehend von den sich als empirisch gehaltvoll erwiesenen klassischen Modellen - Modellspezifikationen zu entwickeln, die auch diesen zusätzlichen Anforderungen an die Abbildungseigenschaften genügen (Vergleiche [1]). Die Formulierung von in diesem Sinne empirisch gehaltvollen erweiterten Modellansätzen setzt umfassende und eingehende Analysen ihrer Abbildungseigenschaften voraus.

Literatur

[1] ter Haseborg, F.: Dynamische Marktreaktions- und Entscheidungsmodelle zur Zeitplanung des Werbemittel- und Mediabudgeteinsatzes, Habilitationsschrift, Hamburg 1983

[2] ter Haseborg, F.: Statische und dynamische Absatzreaktionsanalyse beim Werbemittel- und Mediabudgeteinsatz - Eine kritische Bestandsaufnahme der empirischen Befunde, in "der markt", 23. Jg., 1984, S. 104 - 117

[3] Lambin, J.-J.: Advertising, Competition and Market Conduct in Oligopoly over Time, Amsterdam, Oxford, New York 1976

[4] Little, J.D.C.: Models and Managers: The Concept of a Decision Calculus, in "Management Science", Vol. 16, 1970, S. B-466 - B-485; siehe auch die deutsche Übersetzung in Köhler, R.; Zimmermann, H.-J.: Entscheidungshilfen im Marketing, Stuttgart 1977, S. 122 - 147

[5] Little, J.D.C.: Brandaid: A Marketing-Mix Model, Part 1: Structure, in "Operations Research", Vol. 23, No. 4, 1975, S. 628 - 655

[6] Little, J.D.C.: Brandaid: A Marketing-Mix Model, Part 2: Implementation, Calibration, and Case Study, in "Operations Research", Vol. 23, No. 4, S. 656 - 673

[7] Little, J.D.C.: Aggregate Advertising Models: The State of the Art, in "Operations Research", Vol. 27, No. 4, 1979, S. 629 - 667

[8] Naert, Ph.A.; Leeflang, P.S.H.: Building Implementable Marketing Models, Leiden, Boston 1976

[9] Nerlove, M.; Arrow, K.: Optimal Advertising Policy under Dynamic Conditions, in "Economica", Vol. 29, 1962, S. 128 - 142

[10] Parsons, L.J.; Schultz, R.L.: Marketing Models and Ecomometric Research, North Holland, New York, Oxford 1976

[11] Sasieni, M.W.: Optimal Advertising Expenditure, in "Management Science", Vol. 18, No. 4, 1971, S. 64 - 72

[12] Schmalensee, R.: The Economics of Advertising, Amsterdam, London 1972

[13] Schmidt, B.; Topritzhofer, E.: Reaktionsfunktionen im Marketing: Zum Problem der Quantifizierung von Nachfrage- und Konkurrenzreaktionen, in Topritzhofer, E. (Hrsg.): Marketing, Wiesbaden 1978, S. 195 - 238

[14] Sethi, S.P.: Optimal Control of a Logarithmic Advertising Model, in "Operational Research Quarterly" , Vol. 26, No. 2, 1975, S. 317 - 319

[15] Sethi, S.P.: Dynamic Optimal Control Models in Advertising: A Survey, in "SIAM Review", Vol. 19, No. 4, Oct. 1977, S. 685 - 725

[16] Simon, H.: Goodwill und Marketingstrategie, Wiesbaden 1985

[17] Simon, J.L.; Arndt, J.: The Shape of the Advertising Response Function, in "Journal of Advertising Research", Vol. 20, No. 4, Aug. 1980, S. 11 - 28

[18] Turner, R.E.; Neuman, C.P.: Dynamic Advertising Strategy: A Managerial Approach, in "The Journal of Business Administration", Vol. 7, 1976, S. 1 - 21

[19] Vidale, M.L.; Wolfe, H.B.: An Operations-Research Study of Sales Response to Advertising, in "Operations Research", Vol. 5, 1957, S. 370 - 381; siehe auch die deutsche Übersetzung in Kroeber-Riel, W. (Hrsg.): Marketingtheorie, Köln 1972, S. 316 - 331

[20] Yon, B.; Mount, T.D.: The Response of Sales to Advertising: Estimation of a Polynomial Lag Structure, in Topritzhofer, E. (Hrsg.): Marketing, Wiesbaden 1978, S. 265 - 279

Multi-item production to order

Nico P. Dellaert, Eindhoven

In process industry we often meet a situation in which different types of products can be manufactured on one machine. There may be many possible varieties of each type of product, so that no safety stocks can be kept and we have to produce to order. Usually the machine has to be rebuilded, or some other work has to be done, before the production of another type can be started. This rebuilding-time is often quite large compared with the manufacturing-time of an order, whereas the rebuilding-time between the different varieties of the same type of product is rather small. Because of the large rebuilding-time one would like to manufacture a large lot of the same type without rebuilding. On the other hand there is the interest in short and accurate delivery-times for the orders as well as the capacity restrictions and storage costs that ask for rather small lotsizes.

For most firms it will be interesting to keep the delivery-times constant for certain groups of orders, for instance urgent orders are promised to be delivered 2 periods after ordering and non-urgent orders 5 periods after ordering. In order to take decisions about these delivery-times and about the capacity that should be available, we have to consider the percentages of the orders that will be delivered after their promised due-dates, the number of set-ups, the holding costs and the amount of overwork. A model is presented that considers all these aspects. In this model the costs of a certain delivery-time for a group of orders are rather independent of the delivery-times for other groups. This is realized by using a simple production-strategy, in which all known demand for a type of product is manufactured as soon as if there is enough priority-weighted demand for that type. Although, in practice, usually a more sophisticated production-strategy will be used, this model can be very useful in taking decisions about delivery-times and capacity.

OPERATIONAL RESEARCH IN PRACTICE - EXPERIENCES OF AN OR GROUP IN INDUSTRY

Leonard Fortuin, Eindhoven
Mynt Zijlstra, Eindhoven

Much has been written about the "failure of OR", its causes and possible remedies. The "practicality gap", with theoreticians in "ivory towers" on the one hand and practitioners on the other, is discussed at length. Literature on successful application of OR is less abundant.

We report on the experiences of an OR group in a large industrial organization, from 1981 to 1986. The adoption of a project approach, with clients asking for help of an OR character and paying for such help, turned out to be very fruitful.

Our analysis covers 140 projects. OR techniques applied are shown, as well as the areas of application and the level in the organization where the project originated. Most applications are in the departments for production and in the field of design and control of production systems. Discrete simulation and probability models appear as the most popular OR tools.

Operational Research is useless in practice - as long as it pursues the optimal solution of a problem that first was isolated from its original context. However, according to our experience, OR will encounter many satisfied clients, if it merely "seeks to produce an understanding of managerial problems and to develop models which will enable the consequences of decisions to be investigated" (EJOR).

ENTWICKLUNG VON TECHNO-ÖKONOMISCHEN STRATEGIEN ZUR MINDERUNG DER ANTHROPOGEN FREIGESETZTEN STOFFE SCHWEFELDIOXID UND STICKOXIDE IN BADEN-WÜRTTEMBERG MITTELS DES LP-ENERGIEMODELLS MESSAGE

Hans-Dietrich Haasis, Karlsruhe
Otto Rentz, Karlsruhe

Zusammenfassung: In diesem Bericht werden techno-ökonomische Strategien zur Minderung der anthropogen freigesetzten Stoffe Schwefeldioxid und Stickoxide in Baden-Württemberg unter Verwendung des, die realen Emissionsminderungsmaßnahmen mitberücksichtigenden LP-Energiemodells MESSAGE dargestellt und analysiert. Die dazu notwendigen Erweiterungen des Modells werden beschrieben und die Vorgehensweise zur Integration dieser Maßnahmen in das Modell dargestellt.
Die Ergebnisse basieren auf einer Untersuchung, welche die Autoren für das Projekt Europäisches Forschungszentrum für Maßnahmen zur Luftreinhaltung (PEF), Karlsruhe, durchgeführt haben.

Summary: In this report technical and economical strategies for reducing emissions of sulphur dioxide and nitrogen oxides in the Federal state Baden-Württemberg are described. The results were obtained by extending and applying the LP-energy model MESSAGE. The environmental orientated extension as well as the integration procedure for emission reduction measures are described.
The results are based on a study the authors have performed for the European Center for Air Pollution Control Measures, Karlsruhe.

1. Zum Zielkonflikt zwischen Energieumwandlung und Emissionsminderung

Zur Lösung von energiewirtschaftlichen Problemstellungen werden seit einigen Jahren mathematische, meist lineare Energieversorgungsmodelle entwickelt und auf unterschiedliche Szenarien angewandt /1-3/. Als Auslöser hierfür gelten die Ölkrisen in den 70er Jahren.
Wobei zunächst Versorgungsprobleme und Probleme des wirtschaftlichsten Einsatzes von Energieumwandlungsanlagen als Aufgabenstellungen zu Grunde lagen, treten seit einigen Jahren zunehmend umweltpolitische Fragestellungen bei der Entscheidungsfindung über künftige Energieversorgungsstrukturen in den Vordergrund, beispielsweise Fragen nach geeigneten Emissionsminderungsmaßnahmen und Grenzwerten /4-7/. Dies insbesondere vor dem Hintergrund des zunehmend ins Bewußtsein der Öffentlichkeit getretenen Phänomens neuartiger Waldschäden.

Aufgrund der Koppelproduktion von Endenergie und anthropogen freigesetzten Stoffen haben Maßnahmen zur Emissionsminderung Einfluß auf den optimalen Einsatz von Energieumwandlungsanlagen als auch auf die makroökonomische ausgerichtete optimale Energieversorgungs- und Industriestruktur. Dieser Einfluß macht sich deutlich in Verschiebungen der Energieträgereinsatz- und Technologiekapazitätsstruktur, beispielsweise im verstärkten Einsatz schwefelarmer Brennstoffe und/ oder dem verstärkten Zubau von Energieumwandlungsanlagen auf Basis nicht-fossiler Energieträger. Zur Lösung des Konfliktes zwischen Energieumwandlung und Emissionsminderung müssen die entwickelten Energiemodelle um einen technologieorientierten Umweltteil erweitert und angepaßt werden. Die erweiterten Modelle können als Entscheidungshilfe für die Analyse von Emissionsminderungsstrategien unter Berücksichtigung der realen Energieversorgungsstruktur herangezogen werden.

2. Methodik der Modellerweiterung und -anpassung

Die Energieversorgungsstruktur von Baden-Württemberg läßt sich mit der üblicherweise verwendeten Energieumwandlungskette beschreiben. Diese Kette, interpretierbar als gerichteter Graph, bildet die Basis für das Energie-Umwelt-Modell. Zur Analyse dieser Struktur lassen sich, entsprechend der Darstellung in Abbildung 1, zwei Vorgehensweisen unterscheiden: Der (eher deskriptive) Simulationsansatz und der (eher normative) Optimierungsansatz. Bei dem Konzept der Optimierung wird die künftige Energienachfrage deterministisch, beispielsweise direkt oder in Kopplung mit einem vorgelagerten Energienachfragemodell, vorgegeben, um hieraus retrograd die sich optimal einstellende Technologie- und Primärenergieträger-Verwendungsstruktur zu bestimmen; dies im Gegensatz zu der Verwendung von Simulationsmodellen, bei denen, ausgehend von vorgegebenen Technologiestrukturen, progressiv deren Nachfrageerfüllungspotential ermittelt werden kann.
Zur Ermittlung von optimalen Zuordnungen zwischen Umwandlungs- und Emissionsminderungstechnologien eignet sich der Optimierungsansatz. Aufgrund der Analyse zeitlicher Entwicklungen wird darüberhinaus ein dynamischer Ansatz verwandt, welcher es erlaubt den Planungszeitraum - gewählt werden 30 Jahre - in äquidistante Teilperioden (beispielsweise zu je 5 Jahren) zu unterteilen.
Zur Abbildung der realen Energieversorgungsstrukturen von Baden-Württemberg müssen im technischen Modellteil reale Umwandlungsanlagen bzw. Emissionsminderungsverfahren und Energieträger zu entsprechenden Typen aggregiert werden. Diese Aggregation hat sich an technischen, wirtschaftlichen und gesetzlichen Kriterien auszurichten, beispielsweise

- am Brennstoffband,
- am Emissionsverhalten von SO_2 und NO_x,
- an der Lastbereichszuordnung,

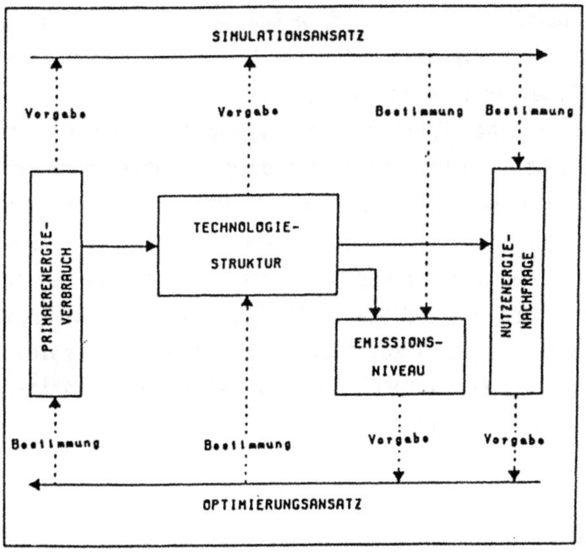

ABBILDUNG 1: Vereinfachte Darstellung der Vorgehensweise bei optimierendenund simulierenden Energie-Umwelt-Modellen

- an der Feuerungsbauart,
- an Größendegressionseffekten wirtschaftlicher und technischer Art,
- an gesetzgeberisch vorgegebenen Größenklasseneinteilungen.

Gerade die Abbildung des Emissionsverhalten beispielsweise von NO_x macht eine Disaggregation steinkohlebefeuerter Kraftwerke in Trocken- und Schmelzkammer-

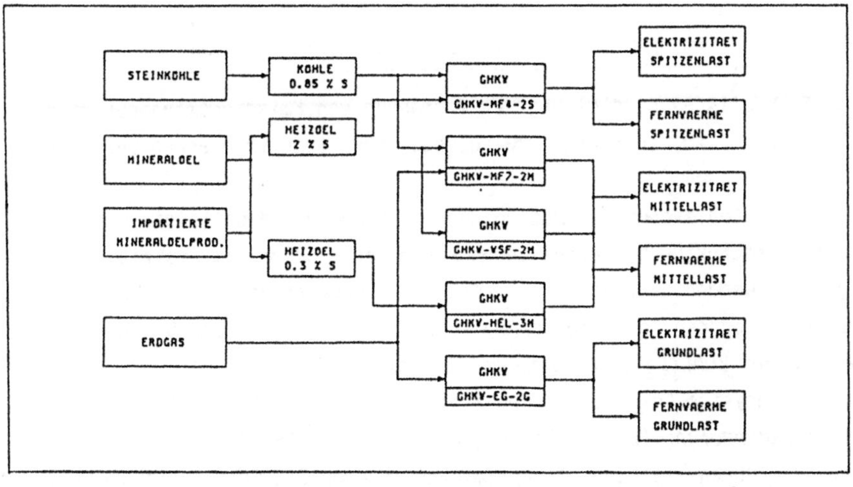

ABBILDUNG 2: Im Modell integrierte Technologiegruppen der Technologiegruppe "Gegendruck-Heizkraftwerke" (GHKW)

feuerungen erforderlich. Durch die so begründete Aggregation einzelner Anlagen erhält man die in Abbildung 2 exemplarisch für die Technologiegruppe "Gegendruck-Heizkraftwerke" dargestellten Technologietypen.
Zur Lösung des Konfliktes zwischen Energieumwandlung und Emissionsminderung müssen diesen Technologietypen, entsprechend Abbildung 3, unterschiedliche Emissionsminderungsmaßnahmen zugeordnet werden. In das Modell integrierte Maßnahmen zur SO_2-Minderung sind: Trockenadditivverfahren, Sprühabsorptionsverfahren, vereinfachte Kalkwaschverfahren und Kalkwaschverfahren. Verwendete Maßnahmen zur NO_x-Minderung sind: Primärmaßnahmen und SCR-Verfahren. Die Emissionsminderungsmaßnahmen werden charakterisiert durch zusätzliche Ausgaben und Energieverbräuche, durch veränderte Emissionsfaktoren für Schwefeldioxid und Stickoxide sowie durch veränderte Folgeproduktrelationen.

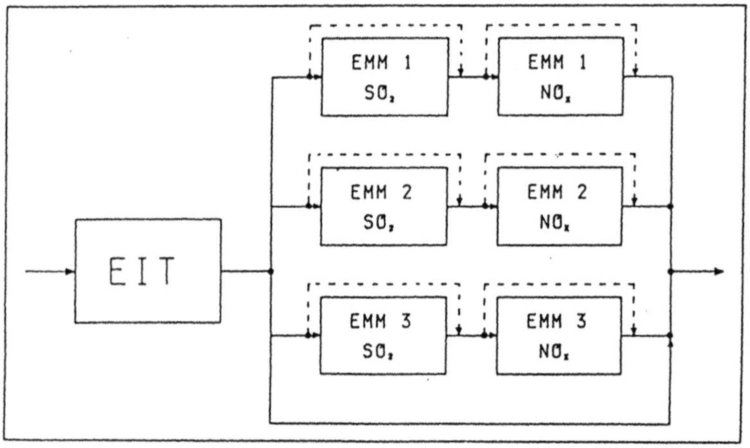

ABBILDUNG 3: Schematische Darstellung der Zuordnung von Emissionsminderungsmaßnahmen (EMM) zu Energieumwandlungs- bzw. Industrietechnologien (EIT)

Neben dieser Technologiestruktur, müssen zur Anwendung des Modells die die Energieversorgungsstruktur beeinflussenden Parameter a priori im Restriktionensystem vorgegeben werden, beispielsweise die Verfügbarkeit von heimischen und importierbaren Ressourcen, die Preise/Preisrelationen eingesetzter Brennstoffe, die gegenwärtig und künftig zu erwartenden Energienachfragen.
Das Zielfunktionensystem besteht aus den Zielfunktionen: Wirtschaftlichkeit (COST), Emissionsminderung (ENVT), Versorgungssicherheit (SCTY) und Technische Effizienz (EFCY). Durch individuelle Optimierung dieser Zielfunktionen stellen sich unterschiedliche optimale Energieversorgungsstrukturen ein. Es wird möglich, quantifizierte Zielkonflikt-Beziehungen, sogenannte Trade-Off-Funktionen zu bestimmen.

3. Ausgewählte Ergebnisse der Modellanwendung für Baden-Württemberg

Der o.a. methodische Ansatz wird in /6/ für das Bundesland Baden-Württemberg angewandt. Hierfür werden fünf Szenarien analysiert, bei denen das Prinzip zugrundegelegt wird, die Emissionsminderung durch anlagenspezifisch zugeordnete unterschiedliche Emissionsgrenzwerte sicher zu stellen. Szenario I dient als Referenzszenario ohne Berücksichtigung jeglicher Emissionsminderungsmaßnahmen für SO_2 und NO_x. In Szenario II sind die gesetzlichen Grenzwerte, wie sie sich aus der Großfeuerungsanlagenverordnung sowie der Technischen Anleitung zur Reinhaltung der Luft ergeben, einzuhalten. In Szenario III bzw. IV werden diese Emissionsgrenzwerte um in der Regel 25 % bzw. 50 % vermindert. Zusätzlich wird in Szenario V als Rahmenbedingung vorgegeben, daß künftig keine neuen Kraftwerke in Betrieb gehen und die derzeitig betriebenen Kernkraftwerke ihre Stromerzeugung bis zum Jahr 1992 auf Null zurückfahren. In diesem Bericht wird auf die Ergebnisse des Szenarios II eingegangen.

Die Ergebnisse des Szenario II spiegeln gemäß den Grenzwertanforderungen der aktuellen gesetzlichen Regelungen den derzeitigen Planungs- und Implementierungsstand von Emissionsminderungsmaßnahmen wieder. Impliziert durch eine Emissionsminderung von ca. 42 % im Jahre 2012 gegenüber 1982, für Schwefeldioxid dargestellt in der Abbildung 4, entstehen zusätzliche Betriebs- und Brennstoffausgaben in Höhe von ca. $330 \cdot 10^6$ DM/a. Die derzeit verfolgte Implementierungsstrategie in Baden-Württemberg wird damit im wesentlichen bestätigt, obgleich durch Umstrukturierungsmaßnahmen in der Primärenergieträgereinsatz- als auch Kapazitätsstruktur weitere Emissionsminderungen möglich wären, dargestellt in der Abbildung 5.

Als zusammenfassende Darstellung der Beziehungen zwischen den verwendeten Zielkriterien COST, ENVT, SCTY und EFCY können Trade-Off-Funktionen ermittelt werden. Bei der Analyse der Trade-Off-Funktionen für Szenario II, dargestellt in der Abbildung 6, sind drei Bereiche erkennbar:

* Bereich 1:
  Eine (gewichtete) Emissionsminderung während des Planungszeitraums von insgesamt $7{,}4 \cdot 10^6$ t auf $5{,}5 \cdot 10^6$ t geht zu Lasten einer alleinigen Bewertung des baden-württembergischen Energieversorgungssystems nach dem Kriterium COST, verbessert jedoch die Zielfunktionswerte der Kriterien EFCY und SCTY.
* Bereich 2:
  Eine (gewichtete) Emissionsminderung während des Planungszeitraums von insgesamt $5{,}5 \cdot 10^6$ t auf $5{,}1 \cdot 10^6$ t geht zu Lasten der Zielkriterien COST und SCTY, verbessert jedoch den Zielfunktionswert des Kriteriums EFCY.
* Bereich 3:
  Eine (gewichtete) Emissionsminderung während des Planungszeitraums von insgesamt $5{,}1 \cdot 10^6$ t auf $3{,}6 \cdot 10^6$ t geht zu Lasten aller übrigen Zielkriterien.

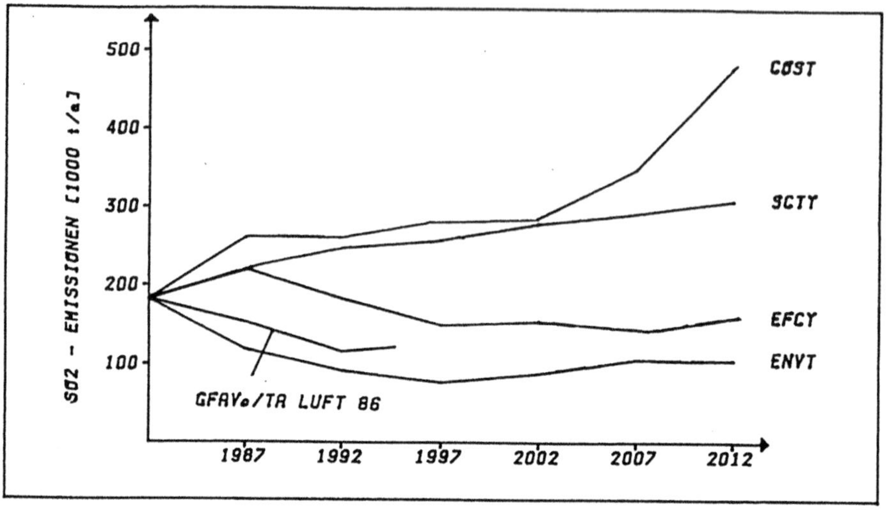

ABBILDUNG 4: Absolute Entwicklung der jährlichen SO_2-Emissionen des Energieumwandlungs- und Industriesektors für unterschiedliche Zielkriterien

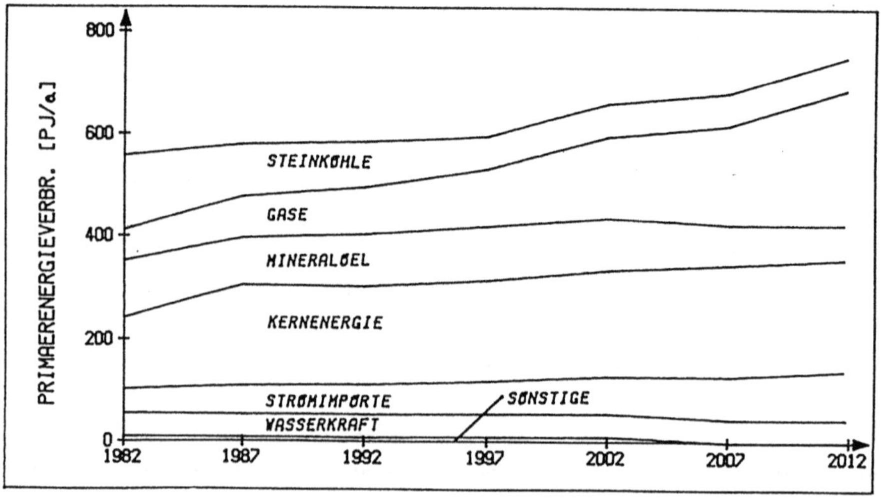

ABBILDUNG 5: Entwicklung des Primärenergieverbrauchs für das Zielkriterium ENVT

Den derzeitigen gesetzlichen Regelung entspricht eine (gewichtete) Emissionsminderung während des Planungszeitraums von insgesamt ca. $4{,}3 \cdot 10^6$ t (Bereich 3). Eine über diese Regelungen hinausgehende Emissionsminderung führt zu einer Verschlechterung der Bewertung des Energieversorgungssystems von Baden-Württemberg bezüglich COST, SCTY und EFCY.

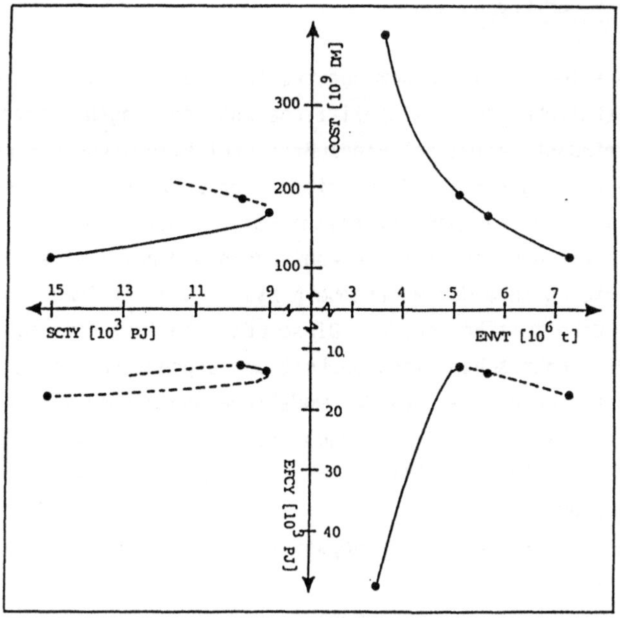

ABBILDUNG 6: Trade-Off-Funktionen für Szenario II

Eine Analyse der optimalen Zuordnung von Emissionsminderungsmaßnahmen und Umwandlungsanlagen ergibt für dieses Szenario folgendes Bild:

- Sofern zur Stromerzeugung steinkohlebefeuerte Technologien zugebaut werden, handelt es sich um Trockenfeuerungen, welche einen Emissionsgrenzwert für SO_2 bzw. NO_x von 400 mg/m^3 bzw. 200 mg/m^3 einzuhalten haben. Es werden vorwiegend Verfahren auf Kalkbasis bzw. SCR-Verfahren eingesetzt.
- Die Zunahme der kombinierten Strom-/Wärmeerzeugung resultiert bei Szenario II hauptsächlich aus dem Einsatz von i. a. mit Gas bzw. Heizöl S befeuerten Gegendruck-Heizkraftwerken. Hier werden zur Einhaltung der Grenzwerte für SO_2 von 1700 mg/m^3 und für NO_x von 200 (300) mg/m^3 vorwiegend Trockenadditiv- bzw. Sprühabsorptionsverfahren und Primärmaßnahmen eingesetzt. Daneben findet gerade die Wirbelschichtfeuerung hier ihr bevorzugtes Einsatzfeld.
- Im Industriesektor wird die bereits in Szenario I erkennbare Tendenz, Strom von öffentlichen Energieversorgungsunternehmen zu beziehen, hier fortgesetzt. Die Gründe für den Strom-Fremdbezug im Industriesektor liegen einerseits begründet in den für Emissionsminderungsmaßnahmen zusätzlich notwendigen Ausgaben als auch in der von den öffentlichen Energieversorgungsunternehmen verfolgten Strompreispolitik.
- Des weiteren greifen gerade im Industriesektor Brennstoffsubstitutionsmaßnahmen, d.h. Einsatz schwefelarmer Brennstoffe.

4. Umsetzbarkeit der Ergebnisse

Die Ergebnisse aller in /6/ zusammengestellten Szenarien können für künftige energiewirtschaftliche und umweltpolitische Entscheidungen ausgenützt werden. Sie fanden (zumindest teilweise) Einfluß in drei bisher von der Landesregierung Baden-Württemberg eingesetzten Kommissionen zur Entwicklung realer, der Energieversorgungsstruktur angepaßter, Emissionsminderungsstrategien. Obgleich dieser Annäherung zwischen Modellbauern/-anwendern und politischen/administrativen Entscheidungsträgern in Baden-Württemberg klafft derzeit immer noch eine nicht unbeachtliche Lücke zwischen beiden. Diese Lücke basiert u.a. auf Kritikpunkten hinsichtlich der praktischen Umsetzbarkeit der Ergebnisse. Bei der Bewertung der praktischen Umsetzbarkeit der aus der Modellanwendung resultierenden Ergebnisse muß berücksichtigt werden, daß die Ergebnisse - wie generell bei Systemstudien - eine Entscheidungshilfe darstellen. Derart interpretiert, lassen sich die Ergebnisse ausnützen, um

- Beziehungen des Energieversorgungssystems widerspruchsfrei darzustellen und zu analysieren,
- Trade-Off-Funktionen zu bestimmen,
- Anpassungsprobleme rechtzeitig aufzuzeigen,
- robuste nächste Schritte in Richtung einer "geeigneten" Strategie zu identifizieren,
- Unsicherheiten in den Einflußgrößen zu quantifizieren und deren Auswirkungen zu analysieren.

Durch eine enge Zusammenarbeit zwischen Modellanwendern und politischen/administrativen Entscheidungsträgern könnte es künftig möglich werden, die Lücke in der Akzeptanzbereitschaft weiter zu schließen.

Literaturverzeichnis

/1/ Häfele, W. (1981). Energy in a Finite World, Cambridge/Mass..

/2/ Van der Voort, E. et al. (1984). Energy Supply Modelling Package EFOM 12C, Mark I, Mathematical Description, Louvain-la-Neuve.

/3/ Schrattenholzer, L. (1978). The Energy Supply Model MESSAGE, IIASA RR-81-31, Laxenburg.

/4/ Haasis, H.-D.; Rentz, O. (1986). The Implementation of Emission Reduction Measures and their Effects on Optimal Future Energy Supply Structures, Hartmann, H.-F. (ed.): Proc. 7th World Clean Air Congress, Sydney, Vol. II, 166-173.

/5/ Hanicke, Th. (1985). Wirtschaftlich-technische Optimierung des Energieversorgungssystems der Bundesrepublik Deutschland anhand eines linearen multikriteriellen Optimierungsmodells, Dissertation, Karlsruhe.

/6/ Rentz, O. et al. (1987). Entwicklung von technisch-wirtschaftlichen Strategien für Emissionsminderungsmaßnahmen für Schwefeldioxid und Stickoxide aus stationären Produktionsanlagen in Baden-Württemberg, KfK-PEF Schriftenreihe Nr. 13, Karlsruhe.

/7/ Rentz, O. et al. (1988). Optimal Control Strategies for Reducing Emissions from Energy Production and Energy Use, Brussels, Karlsruhe.

QUANTITATIVE METHODS IN A MEDIUM/LARGE SIZE COMPANY

Jan F. van Haastrecht

Although this talk is given at an OR conference OR applications at Mars in Veghel are so closely connected with other quantitative methods that I rather like to use the wider setting.
So you will hear an overview of what has been done and what is being done in the company with the aid of quantitative methods.

Mars Veghel is producing more than 100.000 tons of chocolate products of which more than 80 percent is being exported.
Because both quality and costs are heavily influenced by raw materials, quite a few problems find their origin in the study of recipes and raw material prices.
This involves the divisions R&D (optimal mix, taste control, vendor assurance), Manufacturing (process control, technical restrictions) and Commercial (forecasting prices and buying raws).
Other applications related to physical processes are production planning and truck loading.
Marketing analysis and consumption forecasting are regular exercises which lead to a special regression approach. In the combined area of Marketing and Manufacturing several international "decomplexity" studies have been carried out, resulting in a reduction of the number of standard items produced. This is not meant to say that our efforts towards flexible production automation stopped.

Many of the methods developed for the problems above are used "off line" and often by specialized people.
Of course, many small problems are now solved by the responsible managers with the aid of personal computers (spreadsheets, statistical methods). The on-line solution however of larger problems proved to be not practical or not possible in most cases. Some reasons will be given.

Risikoanalyse für ein Investitionsvorhaben

Franz L. Klapp, Mülheim
Peter Müller, Düsseldorf
Wolfgang Wahls, Düsseldorf
Klaus Welters, Düsseldorf
Walter Griem, Ratingen

Ziel der durchgeführten Risikoanalyse war es, bessere Entscheidungsgrundlagen für ein bedeutsames Investitionsvorhaben der Mannesmannröhren-Werke AG zu gewinnen. Zusätzlich zu einer Wirtschaftlichkeitsrechnung, die den Rationalisierungserfolg der Investition aufzeigte, sollte ermittelt werden, wie die Ergebnislage des Produktes "Nahtlose Rohre" mittelfristig einzuschätzen ist. Außer von den sicher kalkulierbaren Kosten wird dieses Ergebnis hauptsächlich von der abgesetzten Menge und den Erlösen bestimmt, die - wie die Vergangenheit gezeigt hat - innerhalb weiter Grenzen schwanken können.

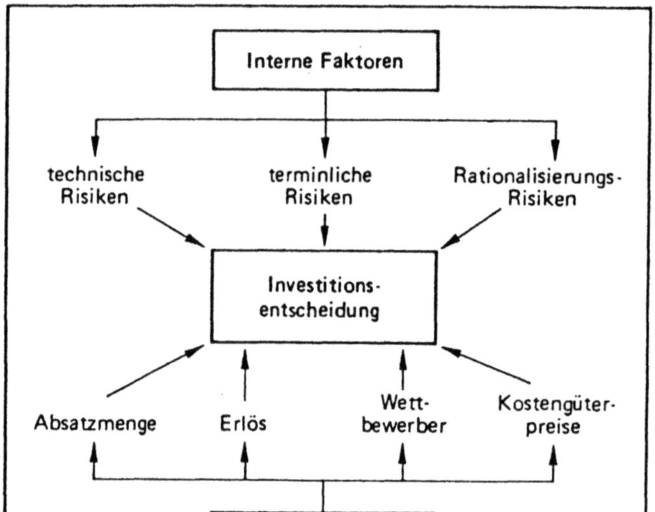

FAKTOREN FÜR DIE INVESTITIONSENTSCHEIDUNG

Da die Risiken im wesentlichen in den Marktdaten (Umsatzteil), also externen Faktoren liegen, war es erforderlich, ein Modell zu entwickeln, das die Bandbreite der Prognosedaten in ihrer Auswirkung auf das Ergebnis widerspiegelt. So ist z.B. die Verbrauchsmenge an Rohren in hohem Maße abhängig vom wirtschaftlichen Wachstum und vom Ölpreis. Das Erlösniveau wird darüber hinaus von der weltweiten Kapazitätsauslastung der Herstellerbranche bestimmt. Die Haupt-Einflußgrößen sind für die Zukunft so schwer zu schätzen, daß sie nur als stochastische Größen in die Modellbetrachtung einfließen können.

Die in den Kostenteil einfließenden Prämissen sind aufgrund vorliegender Erfahrungswerte kalkulierbar und deshalb im wesentlichen als deterministische Größen eingegeben worden. Der Umsatz- und der Kostenteil werden dann zur Hauptzielgröße Ergebnis zusammengeführt.

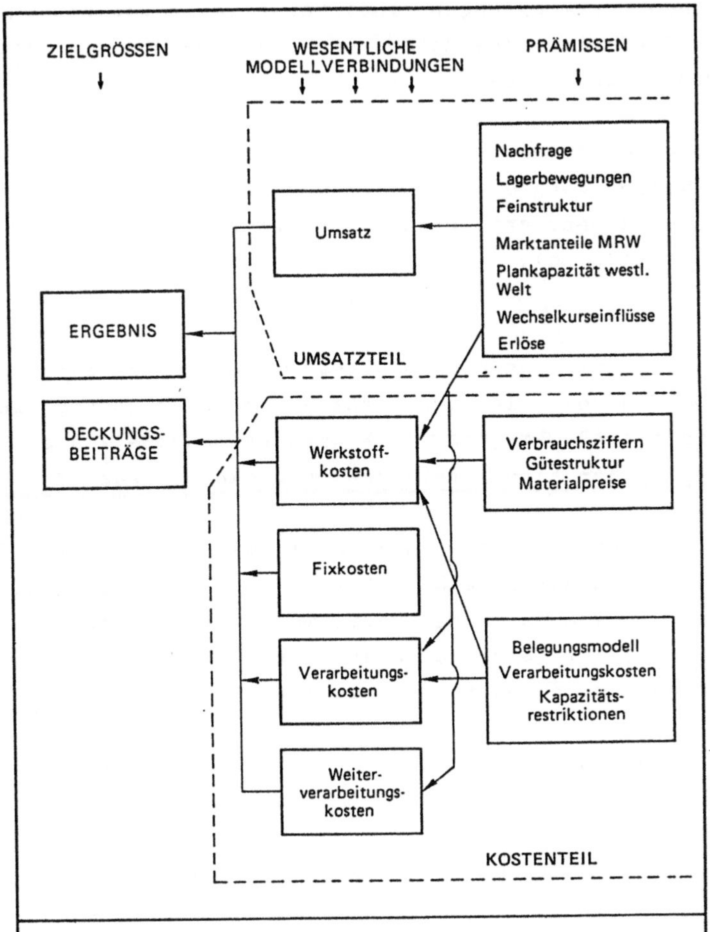

GROBSTRUKTUR DES ERGEBNISBAUMS

Das verwendete Risikoanalyse-Modell verarbeitet rund 300 stochastische Einzelgrößen unterschiedlicher Verteilungstypen (diskrete- sowie Dreiecksverteilungen, Polygonzüge; auch beliebige kontinuierliche Verteilungen, die durch Treppenfunktionen approximiert werden). Hinzu kommen etwa 1300 deterministische Eingabegrößen. Um eine übersichtliche Modellstruktur zu gewährleisten, wurden bis zu vierdimensionale Matrizen als Datentableaus verwendet. Die Modellergebnisse können grafisch oder tabellarisch ausgegeben und durch Sensitivitätsanalysen gestützt werden.

Das Modell ist in FORTRAN (etwa 3000 Statements) geschrieben und läuft auf einer IBM 3084Q. Die Programmstruktur umfaßt ein Hauptprogramm mit dem Ergebnisbaum und 20 Unterprogramme.

Die sachlichen Verknüpfungen zwischen den einzelnen Märkten und Produkten erforderten eine Methode, die Abhängigkeiten stochastischer Größen im Modell darzustellen.

Im folgenden wird ein Näherungsverfahren angegeben, wie es im Modell eingesetzt werden konnte (in der Statistik-Literatur sind solche Verfahren kaum zu finden, erst recht nicht dann, wenn die Verteilungen der X_i beliebig sind).

Sei (X_i, X_j) ein Paar von Einflußgrößen, dem die Korrelation ρ_{ij} mitgegeben werden soll. In der iterativen Modellrechnung werde (ohne Einschränkung der Allgemeinheit) X_i früher behandelt als X_j (d.h. aus X_i werde früher eine Zufallszahl gezogen als aus X_j).

1. Es wird eine über dem Intervall $[0,1]$ gleichverteilte Zufallszahl z_i vorgelegt und diese gemäß $x_i = F_i^{-1}(z_i)$ in eine entsprechende Zufallszahl x_i von X_i überführt, wobei F_i die Verteilungsfunktion von X_i darstellt (der Prozeß ist nicht immer eindeutig, da F_i zwar monoton aber nicht streng monoton ist).

2. Es wird eine weitere $[0,1]$ - gleichverteilte Zufallszahl z gezogen.
   Gilt

   $$|\rho_{ij}| \leq z ,$$

   so wird der Zufallszahlenziehungsprozeß für die Größe X_j nicht beeinflußt, d.h. es wird eine dritte $[0,1]$ - gleichverteilte Zufallszahl z_j gezogen und $x_j = F_j^{-1}(z_j)$ gesetzt (F_j sei die Verteilungsfunktion von X_j).

Gilt

$$|\rho_{ij}| > z ,$$

so setzt man

$$z_j = \begin{cases} z_i & \text{für } \rho_{ij} > 0 \\ 1 - z_i & \text{für } \rho_{ij} < 0 \end{cases}$$

und schließlich

$$x_j = F_j^{-1}(z_j) .$$

Das beschriebene Verfahren ist für Paare und damit auch für Ketten der Form $(X_i,X_j,X_k,\ldots)$ mit $i \neq j \neq k \neq i \ldots$ und $\rho_{ij} \neq 0$, $\rho_{ik} \neq 0, \ldots$ definiert.

Dieses Verfahren läßt sich nur in beschränktem Umfang auf Kreise von Zufallsgrößen wie z.B.

$$(X_i,X_j,X_k,X_i) \text{ mit } i \neq j \neq k \neq i \text{ und}$$

$$\text{mit Korrelationen } \rho_{ij} \neq 0, \quad \rho_{ijk} \neq 0, \quad \rho_{ki} \neq 0$$

erweitern.

Insgesamt stellt das benutzte Verfahren zur Berücksichtigung von Korrelationen zwischen stochastischen Einflußgrößen noch keine allen theoretischen Ansprüchen genügende Lösung dar; aber die Erfordernisse der Praxis wurden mit dieser Vorgehensweise hinreichend gut abgedeckt. Dabei halfen die nicht sehr starke Besetzung der Korrelationsmatrix (die positiv definit sein muß) wie viele Verifikationstests dieses Verfahrens.

Das Arbeitsteam "Risikoanalyse" setzte sich aus Mitarbeitern der Ressorts Datenverarbeitung, Technik, Absatz und Betriebswirtschaft zusammen.

SECURITY BUFFERS, THE CRYSTAL BALLS TO CONTROL AND IMPROVE YOUR BUSINESS

C. van Putten, Eindhoven

The bottom line results of a production business are not only determined by its constraints, such as resource capacity, material availability, market demand, but they are also crucially affected by disruptions, which always seem to destroy even the best prepared plans. Moreover, due to these disruptions, many times the results of improvement actions are less than expected.

However we are not as helplessly exposed to Murphy's law as we seem to be. Identifying the constraints correctly and safeguarding them from disruptions using properly located security buffers, pave the road to "what you get is what you plan".

If statistical fluctuations occur and e.g. may affect a capacity constraint, it should be protected by a time buffer placed in front of it, in addition to safety capacity in the preceeding process. Moreover, the goodsflow coming from this capacity constraint needs protection against inflowing disruptions, again by time buffers and additional capacity.

Thus security buffers, consisting of planned safety stock, time or capacity, help to achieve basic control of the production process, which is a prerequisite for its further improvement.

These buffers should not only be installed, but they should also be managed. Buffer management includes buffer sizing, i.e. the adjustment of the safety time to the safety capacity available and the disruptions present.

It also includes buffer monitoring by inspecting suitable perfomance indicators, like , e.g., the loss of the constraints' potential performance, the increment of the production lead time and the idle time of feeding operations.

Finally it includes productivity improvement analysis, based on the actual contents of the security buffers during production. It is the key to find which disruptive resource's improvement is the most productive one as a next step in a process of ongoing improvement.

Quantitative models, using inventory control theory, are under development to support the decisions to be made by buffer management.

STRUCTURING THE DEVELOPMENT OF A NEW PRODUCT AN APPLICATION OF OPERATIONS RESEARCH TECHNIQUES

Karl Spälti, Bern
Werner Popp, Bern

A manufacturer of electrical equipment intends to introduce a new product in a market within two years. The company devised a plan and devoted a substantial part of its R&D capacity to develop this new product. However, as the work continues, the plan must be reevaluated due to previously unforeseen technical difficulties. Furthermore, first contacts with future customers initiate additional changes in the technical design. The evolving nature of this development process necessitates a flexible planning procedure.

The process of developing the new product is split into various projects. Not all of them have to be completed but many of them are linked by logical conditions (e.g. "if...then" or "either...or" conditions are imposed on their selection). The selection of the projects must be based on an analysis taking these interdependencies into account.

To support the planning process, we formulate a mixed integer model maximizing the net present value subject to the following set of constraints:

- Logical conditions
- Capacities of R&D
- Production capacity
- Bounds on the sales

As opposed to the analysis traditionally used in the company, this approach allows the simultaneous evaluation of all projects which has, especially for strongly interrelated projects, surprising consequences.

Using the model as a tool in an interactive planning process, we derive solutions which are efficient with respect to a quantitative (i.e. financial) objective and qualitative goals.

INPUT/OUTPUT PLANNING IN MECHANICAL COMPONENT MANUFACTURING SHOPS

Ton M. van de Wakker/J. Will M. Bertrand
Eindhoven, Netherlands

Manufacturing firms can acquire a strong position in the market place by realizing short delivery times and a reliable delivery performance.
Many firms with high and medium volume assembly production have succeeded in shortening the delivery times by using flow-production techniques and applying the principles of Just-in-Time production.
However shortening flow times and improving delivery reliability is more difficult in a functionally organized mechanical, low volume component manufacturing environment.

The main characteristics of low volume mechanical component manufacturing shops are:
- A high variety of routings; a lot of different components are being produced, some completely to customer-specification,
- A high variety in work order processing times at the work centres,
- Irregular supply of work to the shop.
- Different work orders are related to each other because they are needed together at subsequent assembly stages. In essence, we are interested in the flow time and delivery performance of the networks of work orders.

Controlling the flow times and delivery performance of networks requires the maximal use of capacity flexibility and correct priority setting of work orders. Input/output planning has been demonstrated to be a useful tool to realise logistic performance goals for single component manufacturing shops. Therefore, also for network component manufacturing shops, emphasis must be put on flow time control by means of input/output planning. When applied consistently, these principles will provide us with a sound basis for the development of a decision support system for network flow time control.

Separate research has already been done on the issue of network flow time control in shops without input/output planning.
This research investigates the application of input/output planning techniques to mechanical network component manufacturing shops.

ÜBER DEN EINSATZ QUANTITATIVER METHODEN ZUR AKTIV-PASSIV-STEUERUNG EINER GROSSEN REGIONALBANK

Helmut Beeck, München

Kernaufgabe eines "asset and liability managements" einer Bank ist die Steuerung des Zinsänderungsrisikos (ZÄR). Wesentliche Komponenten dazu sind adäquate, ad hoc flexible auswertbare Informationen, Modelle zur quantitativen Bewertung von Szenarien und Entscheidungsalternativen ("what if"), und eine effiziente organisatorische Einbindung in den Management-Prozeß der Bank.
In dem Vortrag wird über Erfahrungen und Stand bei allen 3 Komponenten berichtet. Der Schwerpunkt liegt bei der Diskussion der verwendeten Modelle und einer Gegenüberstellung zu ähnlichen bzw. alternativen Modellen auf dem Markt bzw. in der Literatur (soweit dem Verfasser bekannt).

Zielsetzung u.a. ist auch, Faktoren für Akzeptanz/Nichtakzeptanz von OR-Methoden in der Bank-Praxis aufzuzeigen.

PERSONNEL PLANNING AND BUDGETING: the measurement of productivity in the operations division of a non-life insurance company

Hans van Gelder, Heiloo
Luc van Haastrecht, Groningen

Summary: The organisational context, the investigation of productivity and its determinants, a proposed formulation of the budget and directions for further research and implementation are discussed.

1) In the operations division of a non-life insurance company some 900 employees are engaged in underwriting, claims handling, document preparation and data entry.

2) An analysis has been performed on three and a half years of monthly production data to establish relationships between the dependent variable (productivity) and a number of independent variables (its determinants).

3) A Linear Program is proposed and tested as an alternative model of the planning and budgeting system that has been operative for the last eight years.

4) Further avenues for research and problems of implementation are discussed.

1. Introduction

The general or non-life insurance company sells its products through the intermediary of independent agents and brokers. The operations division (OD) liases with the agents and brokers, supplies them with the required information and quotations, and performs all the underwriting, claims handling, policy and document preparation and data entry activities.
In doing so the operations division performs two functions: sales and production. Notwithstanding extensive use of computers the process is highly labour intensive. Important issues are thus labour costs and quality of performance; one of the major quality criteria is processing speed.
The OD comprises of 4 similar regional and 2 specialized offices; each office is divided into a number of product groups each of which consists of an underwriting/policy preparation and a claims handling department. We distinguish two categories of personnel: direct, there is a more or less linear relationship between output and labour input, and indirect/supervisory, where there is no such relationship. In 1978 a planning and budgeting system was introduced in the OD based on concepts of queuing and inventory theory. For each line of business and each department workload is forecasted for each month of the year. Standard rates of production are forecasted on an annual basis. Work in progress (stock or backlog) at the beginning of the month, workload forcasted for that month, and labour available at the start of the month, all measured in mandays, are put into a model.

Required changes in the labour force were derived by bringing the end of month work in progress between pre-set control limits.
By a man-machine interactive process the labour force is scheduled to 'optimal'levels by scheduling hiring,'run off',training and special programs (holidays are scheduled as much as possisible in slack periods) taking into account seasonal variations in sick leave etc.

2. An investigation into the determinants of productivity

In this study we have attempted to explain fluctuations in direct labour productivity. Labour productivity per department within the product groups of each regional office has been estimated through ordinary least squares (OLS).
The independent variables employed in this are: for each month workload, backlog (inventory of work in progress)at the start of the month, the net labour capacity employed and a time(trend) factor(number of the month in the sample).
For each department an additive as well as a multiplicative model has been formulated. The latter has the advantage that the estimated coefficients may be interpreted as elasticities.

Model formulation.

$$PROD_{ijt} = u_0 + u_1 IN_{ijt} + u_2 WL_{ijt} + u_3 NC_{ijt} + u_4 T_t + r_t \qquad \text{(LINEAR)}$$

$$PROD_{ijt} = e^{(u_0 + r_t)} . IN_{ijt}^{u1} . WL_{ijt}^{u2} . NC_{ijt}^{u3} . T_t^{u4} \qquad \text{(MULT.)}$$

with:

$PROD_{ijt}$ =	productivity	department	i,	office	j	and period	t
IN_{ijt} =	work in progress inventory	"	"	"	"	" "	"
WL_{ijt} =	workload	"	"	"	"	" "	"
NC_{ijt} =	net direct labour capacity	"	"	"	"	" "	"
T_t =	time trend factor						
u_k =	coefficient k=(0,....,4)						
r_t =	residual error term						
e =	2.7181						

the number of observations is 42 (january 1983 - juli 1986).

The results of the estimation procedure were quite satisfactory. For all departments R^2 was reasonably high (0.80 on the average) and allmost all coefficients are significant at the 0.05 level. The outcomes are roughly identical for all departments. The elasticity of work in progress inventory (u_1) is positive for all cases and allmost always significantly different from zero at the 0.05 level; it ranges from 0.03 to 0.22. The effect of changes in net capacity (NC) and workload (WL) on productivity however is much stronger and allways significant. The mean elasticity of work load is +0.80. This implies that a high workload stimulates productivity while a low work load reduces productivity. This however is sometimes unavoidable when there is insufficient work to attain a high level of productivity.
The mean elasticity of net capacity (NC) is -0.74, which implies that a reduction in available net capacity (through sickleave for instance) will be compensated for by an allmost 75% higher productivity of the remaining (direct) clerical staff.

This also implies that an increase in available net capacity will be undone by lower productivity of a larger number of workers.
The increasing time trend is only significant in 7 out of 24 cases.

A second fase in our analysis consisted of pooling the data for similar departments in all four of the regional offices. Dummy variables were introduced to account for local differences. The assumption is that these dummy variables catch all the differences between various offices and that reaction to fluctuations in inventory of work in process, workload and nett capacity available will be identical.

Our model(s) now consist(s) of 6 equations

$$PROD_{it} = u_0 + u_{01}D_{1t} + u_{02}D_{2t} + u_{03}D_{3t} + u_1 IN_{it} + u_2 WL_{it} + u_3 NC_{it} + u_4 T_t + r_t$$

with $D_{1t} = 1$ for regional office number 1
$= 0$ everywhere else

etc.

The increased number of degrees of freedom make it now possible to test additional potential candidates to be included in the set of explanatory variables. The effect of including personnel mutations or the employement of temporary help turned out not to be highly significant however.

To test the predictive power of the model forecasts for 7 months (not in the sample) were made on the basis of the estimates of the pooled regression model. These forecast were compared to actually realized productivity. The mean relative absolute deviation is about 7% and the values for Theil's U - coefficient are quite low. In almost all cases the prediction of our model is better than the (predicted) standard rate of productivity employed in the budgeting system of the OD.

Our conclusion is that direct labour productivity is highly dependent on work in progress - backlog - at the start of each month, work load and net direct labour capacity for that month. With our model however productivity can be explained and predicted quite well.

3. A model for capacity planning

The second part of this study consists of an investigation into the feasibility of formulating the capacity planning model as a linear program. Our goal was to improve on the current budgeting model by employing LP- techniques while taking into account the fluctuations in productivity described in the previous chapter.

$$\text{MIN } a.SUM^n_{t=1} = nc_t + (1-a).SUM^n_{t=1} (su_t + sh_t)$$

subject to the following restrictions.

$$in_{t+1} = in_t + WL_t - nc_t . prod_t \quad (1)$$
$$prod_t = \underline{u}_0 + \underline{u}_1 . in_t + \underline{u}_2 . WL_t + \underline{u}_3 . nc_3 + \underline{u}_4 . T_t \quad (2)$$
$$in_t <= \overline{U}L_{in} \quad (3)$$

$in_t \quad >= LL_{in}$ (4)
$nc_t \quad = \quad NCB_t+sh_t-su_t$ (5)
$sh_t \quad <= UL_{sh}$ (6)
$su_t \quad <= UL_{su}$ (7)

nc_t, in_t, sh_t, su_t >=0 notation of exogenous variables by capitals

nc_t = required net direct labour capacity in period t
NCB_t = available net direct labour " "
sh_t = shortage in " " " " "
su_t = surplus in " " " " "
in_t = inventory of work in progress at the start of period t
WL_t = work load (inflow) in period t
$prod_t$ = productivity estimated with the model described in chapter 2
LL_{in} = lower limit inventory of work in progress
UL_{in} = upper limit " " "
UL_{sh} = upper limit shortage in direct labour
UL_{su} = upper limit surplus in direct labour

We will discuss the equations in the model briefly. Equation (1) describes the work in progess inventory balance over time taking cognizance of the variations in productivity. Direct labour productivity influences the required capacity; higher productivity reduces the number of men(days) required. Productivity is defined in eq.(2). The coefficients $\underline{u}_1,\ldots\ldots,\underline{u}_4$ have been estimated by linear regression on the pooled data(ch.2).
For quality of service considerations inventory of work in progress or backlog has been restricted by an upper limit(UL_{in}) that may vary between departments so that the insured do not have to wait too long for the receipt of their policy or before their claim has been handled. The inventory may not become too low either as this would result in excess capacity.
In the goal function the required capacity and consequently the associated costs will be minimised. Therefore the available capacity has to be adjusted to the required capacity. If the available capacity is too low it can be brought up to level by hiring of new personnel, by the employment of temporary help or by working overtime and by support from other departments if possible. Surplus capacity available can be reduced by 'firing'or 'laying off' people (in reality this is done by not filling vacancies created by employees leaving the company voluntarily or retiring) or by instituting training programs in slack periods and by offering support to other departments if required. This is described in eq.(5). By introducing su_t and sh_t in the goal function we try to minimize net capacity available whilst simultanously keeping a check on capacity surpluses and shortages. In the goal function the coefficient a (0<= a <=1) can be used to give priority to either a low nett capacity, regardless of hiring and firing, or small fluctuations in the labour force.
The model is non-linear and has been solved by an iterative method. The model has been tested for the first 10 months of 1986 and the results have been compared to actually realized values as well as to values derived from the current planning and budgeting system. The model proposes in many instances a lower required nett capacity than realized in actual practice.

This is possible because the model prevents the 'workload pressure'(the combined effect of inventory, work load and nett capacity available) from becoming too low which may cause productivity to become lower than the standard rate of productivity.
Because the model does take into account the increase in productivity in months with a high 'workload pressure', while the current system does not, the model does not always increase net direct labour capacity available in those months. The model only increases capacity to prevent excesses in inventory over the prescribed upper limit, something that occurs in actual practice.
In comparison with the budget over the first 10 months of 1986 employment of the model would have resulted in saving of 68 nett man days per month in the motor underwriting departments of the regional offices; this equals a reduction of the labour force by 4 clerical staff or f.250,000.- on an annual basis; this is a 5% reduction in the relevant labour force. Service levels are improved as well as the model gets rid of excessive backlogs and thus long delivery times.
If the solution of the model is not deemed satisfactory it is simple to change it in the desired direction by changing the upper limits for inventory or adjusting the value of the coefficient a in the goal function. Actually post optimality and sensitivity analysis are extremely useful techniques to help managers get a feel for the situation and strengthen their grip on the planning and budgeting process.

4. Problems of implementation and further avenues for research and development

Our study and research as reported here should be considered as a first step towards an improved planning and budgeting system for the operations division of a large non-life insurance company.

We have reasonably well succeeded in explaining and predicting the fluctuations in in direct labour productivity, however we still have to explain the differences in absolute levels of productivity between similar departments of the regional offices.
The currently employed planning and budgeting model as well as our improved model both depend crucially on the forecast of workload/inflow for each month. In our research project we investigated the quality of the currently employed (primitive) forecasting method which we found lacking in quality. A not yet explained fact is that in the early years of employing the present method the results were quite good while this has deteriorated over time.
A question remains however: what is a good forecast? There are the well known and obvious criteria that the forecast and the realization of the predicted event(s) should be close according to some statistical (distance measure) criterion; e.g. least squares. Furthermore we generally require forecasts to be unbiased.
Is it in actual practice really necessary to have an unbiased forecast if one has some idea about 'sign' and magnitude of the bias?

Our 'improved' model as well as the current system are deterministic while the processes that we model are highly stochastic. Most forecasting performed in actual practice leads to 'deterministic' point estimates. This is the only kind of forecast acceptable as input to a linear or non-linear programming model. A solution might be to resort to chance constrained programming or to programming under uncertainty.

The problem with these kind of models is that they labour under restricted size of computer programs whilst there are no standard programs readily available. Furthermore the interpretation of the solution is often quite obscure and at any rate rather difficult; in particular for practicing managers.

The processes of interest are:

- the inflow of workload; this processes may or may not display seasonal effects, a trend, effects of certain actions or campaigns instigated by the insurance company, and random elements; forecast should be formulated as probability distributions for each period;
- the availability of 'net capacity'; net capacity is the sum of gross capacity minus a number of deductions for sick leave, holidays, training, etc.; gross capacity is the product of the number of employees in the relevant category times the number of ours worked per head in the relavant period; the number of employees available depends on people leaving and entering the company and on their qualifications; the number of hours employees work depends on preferences for part time work; some of these variables are more under management control than others; mean percentage of absence due to sick leave for the operations division as a whole has fallen considerably over the last 10 years, but there are still very significant differences among offices and departments, while this fenomenon displays next to a marked seasonallity large random variation;
- productivity; at present productivity can be explained and predicted - in a way which suites our deterministic models - in a quite satisfactory manner by the model discussed in the second chapter of this paper; it seems however not unlikely that some additional explanatory variables may improve the explanatory and predictive power of the 'productivity model' we expect changes in personnel composition (inflow of new employees) to be a likely candidate as new clerical staff has to go through a period of training and familiarisation before they attain acceptable levels of skill and productivity; other candidates may be the changes in systems and procedures as staff has to unlearn old habits and learn new ones.

An other problem is the use to which the planning and budgetting system is put and consequently the time horizon of the forecasting and planning procedures. Until now the main purpose of the current system has been to produce estimates of required and available direct labour capacity in the framework of finanlcial planning and next year's budget. So at the end of the year forecasts were made of workload and percentual absence resulting from sick leave etc. for each month of next year while production standards were agreed upon among various layers of management. Then starting out from available manpower at the start of the year required additions and reductions in the labour force were calculated. This process operates in an iterative way by man-machine interaction. During the year actuals are compared to plan on a monthly basis and adjustments to the plan are made if required.
The latter is generally done in a more or less informal manner without going through the whole set of procedures of (re)forecasting and (re)planning.
An improved planning and budgeting procedure, employing more sophisticated models, however could be used to improve and extend the monthly evaluation procedure.

It could be employed to recalculate the plan for the next couple of months and in partcular it should provide management with distributions of the likelyhood of certain crucial service paramaters; e.g. the probability distribution of work in process inventory. On the basis of this information decisions could be made to adjust the value of process variables for the next period. In this respect one could wonder whether monthly periods would still be adequate or whether they should be replaced by weeks or even days.
If employed in the manner described above our models would be turned into a operations management decision support system.
The question then is whether we would need different models for different purposes. It is our feeling that the answer is affirmative.
The models, or improved versions thereof based on larger data sets and incorporating additional variables, that we discussed in this paper could be very well employed to assist in an annual or semi annual financial planning and budgeting process.
For use at the 'shop floor' as decision support for operational management however we feel that an additional set of models should be developed that do justice to the stochastic nature of the processes to be managed. We think in this respect either of 'risk analysis' types of models as they have been developed by David B. Hertz and others or of types of models as developed in research on reliability of production systems.

5.Concluding remarks

We have shown a strong relationship between productivity of direct labour in the operations division of a large non-life insurance company and 'workload pressure'. We have formulated a non-linear model for the annual planning and budgeting process, which was solved by an iterative application of linear progamming code.
Further research on the formulation of models and systems to support management at various levels in the organization has been advocated and discussed.
Until recently the application of quantitative analytical methods and techniques to modelling service processes have not been numerous. As services are playing a more and more important role in social and economic life it becomes crucial to manage service processes better. Services tend to be quite expensive as many are still labour intensive while quality assurance in service processes is as important as in manufactured goods but far more difficult. Even when large parts of many service processes may be automated and thus standardized the human touch often makes the difference. So the goal of management in service industries should be to provide high quality service at low cost. In order to achieve this careful and skilful management of the labour force is required to attain a high level of motivation and thus of quality and productivity.

Devisentermingeschäfte zur selektiven Absicherung offener Fremdwährungspositionen - ein portefeuilletheoretischer Ansatz

Holger Hinz, Kiel

1. Problemstellung

Wechselkursschwankungen beeinflussen den Erfolg und die Liquidität von Unternehmen. Empirische Untersuchungen haben gezeigt, daß dem Wunsch nach Absicherung von Transaktionsrisiken (" transaction risks")[1] traditionell mit Devisentermingeschäften (Outrights) nachzukommen versucht wird.[2]

Es soll daher ein Instrument entwickelt werden, das die Entscheidung über Höhe und Denominationswährung der Kurssicherungsaktivität mit Devisentermingeschäften unterstützt.

Insbesondere wird es dem Systembenutzer dem (μ,σ)-Prinzip entsprechend ermöglicht, eigene Vorstellungen über Wechselkursentwicklungen verlustmindernd auszunutzen ('selektive Absicherung'). Dazu wird im folgenden Abschnitt ein Modell in zwei Stufen entwickelt: Eine erste Formulierung zeigt, wie sich bei stochastischen Wechselkursen und deterministischen Fremdwährungszahlungen eine nichtlineare (quadratische) Nebenbedingung ergibt. Sodann wird die Prämisse deterministischer Exposure fallengelassen und man gelangt zu einer allgemeinen Formulierung, die das Problem der unternehmerischen Devisenkurssicherung bei Unsicherheit von Planzahlungen und zukünftigen Wechselkursen optimal mit Devisentermingeschäften zu lösen vermag.

Anschließend wird in Kapitel 3 zum besseren Verständnis dieser Modelle beispielhaft ein Absicherungsproblem dargestellt und für die unterschiedlichen Formulierungen gelöst. Ein Ausblick auf weitere Ausbaumöglichkeiten beschließt die Arbeit in Kapitel 4.

2. Modellierung

Sofern davon ausgegangen werden kann, daß die abzusichernden Fremdwährungszahlungen mit Sicherheit erfolgen werden und eine vollständige Absicherung angestrebt wird, ist das Entscheidungsproblem trivial:

Operations Research Proceedings 1987

Geplante Einzahlungen werden in gleicher Höhe auf Termin verkauft, Auszahlungen schon jetzt per Termin gesichert. Für den Fall jedoch, daß eine derart 'gesicherte' Zahlung ausfällt, stellt

Abb. 1: Bezeichnungen

n	Anzahl der zu betrachtenden Fremdwährungen
E	Erwartungswertoperator
Für jede Währung $i=1,..,n$ ist :	
WK_i	aktueller Wechselkurs in Heimatwährung für eine Fremdwährungseinheit
TK_i	Terminkurs in Heimatwährung für eine Fremdwährungseinheit am Planungshorizont
Z_i	geplanter Zahlungssaldo in Währung i Annahme : Z_i ist eine normalverteilte Zufallsvariable mit τ_i Erwartungswert der Zahlung in Währung i (Einzahlungen positiv, Auszahlungen negativ) Γ^2_i Varianz der Zahlung in Währung i $Z_1,..,Z_n$ stochastisch unabhängig
PK_i	Prognose für den Wechselkurs am Planungshorizont Annahme : PKi ist eine normalverteilte Zufallsvariable mit μ_i Erwartungswert der Prognose für Währung i σ_{ij} Kovarianz der Kursprognosen in den Währungen i und j
A_i	Entscheidungsvariable, die den gesuchten Absicherungsbetrag in Währung i angibt
VQUER	vorzugebende Maximalvarianz des Verlustes aus Devisengeschäften

Definition:

$$X := \sum_{i=1}^{n} (WK_i - PK_i) * Z_i + \sum_{i=1}^{n} (PK_i - TK_i) * A_i$$

Verlust aus Devisengeschäften	Verlust durch Kursschwankung	Verlust aus Absicherung

das Termingeschäft eine unerwünschte, selbstinduzierte Exposure dar. Die Abbildung dieser Problematik ist Gegenstand der zweiten Ausbaustufe des im folgenden zu entwickelnden Modells.

2.1. Absicherung bei unsicheren Kursprognosen

Zur Vereinfachung kann man zunächst versuchen, ein Modell unter Berücksichtigung normalverteilt schwankender Wechselkurse bei konstanten Zahlungsplänen zu entwickeln.

Abb. 2: MODELL I

Prämisse :

$Z_i = \tau_i$ und $\Gamma^2_i = 0$ für $i=1,..,n$

Zielfunktion : $\min ! \quad E(X)$ (I)

Nebenbedingungen :

Falls $Z_i \geq 0$: $A_i \geq 0$ (IIa)

Falls $Z_i \leq 0$: $A_i \leq 0$ (IIb)

für $i=1,..,n$

$$\sum_{\substack{i=1 \\ Z_i \geq 0}}^{n} WK_i * A_i \leq \sum_{\substack{i=1 \\ Z_i \geq 0}}^{n} WK_i * Z_i \qquad \text{(IIIa)}$$

$$\sum_{\substack{i=1 \\ Z_i \leq 0}}^{n} WK_i * A_i \geq \sum_{\substack{i=1 \\ Z_i \leq 0}}^{n} WK_i * Z_i \qquad \text{(IIIb)}$$

$$Var(X) \leq VQUER \qquad \text{(IV)}$$

Die Zielsetzung (I) des dargestellten MODELL I besteht - wie auch in der folgenden Ausbaustufe - in der Minimierung des erwarteten Verlustes. Wie aus der Definition von X hervorgeht, besteht durch die Wahl geeigneter Absicherungsbeträge A_i die Möglichkeit für das Devisenmanagement, diesen Verlust mit Hilfe einer geeigneten Absicherungsstrategie zu beeinflussen.

In den Bedingungen (II) spiegelt sich der Wunsch nach Vermeidung extrem spekulativer Engagements wider: Die Vorzeichen der Exposuredaten determinieren die Richtung der erlaubten Absicherung. Sofern beispielsweise eine Dollareinzahlung erwartet wird, stellt (IIa) sicher, daß nicht noch zusätzlich diese Währung auf Termin gekauft wird.

Da bis zu diesem Punkt nicht ausgeschlossen werden kann, daß die Position in Devisentermingeschäften die Planzahlungen übersteigt, ist in

(IIIa) die Forderung formuliert, daß die Summe der auf Termin verkauften Devisen nicht größer als der Gesamtwert der erwarteten Einzahlungen über alle Währungen sein soll. Entsprechendes wird in (IIIb) für Terminkäufe verlangt. So wird die Manövriermasse des Devisenmanagements durch eine Politikentscheidung festgelegt.
Bedingung (IV) sieht eine Risikobegrenzung vor, die bewirkt, daß die Varianz des Verlustes eine Obergrenze VQUER nicht übersteigt. Dadurch ergibt sich eine benutzerfreundliche Möglichkeit, die Risikoneigung des Anwenders zu erfassen.

Der Erwartungswert des Verlustes berechnet sich als gewichtete Summe der einzelnen Erwartungswerte :

$$E(X) = \sum_{i=1}^{n} (WK_i - \mu_i) * Z_i + \sum_{i=1}^{n} (\mu_i - TK_i) * A_i$$

Da die unsicheren Kursprognosen nicht stochastisch unabhängig sind, wirken sich bei der Berechnung der Varianz des Verlustes die vorgegebenen Kovarianzen aus:

$$Var(X) = \sum_{i=1}^{n} \sum_{j=1}^{n} (A_i - Z_i) * (A_j - Z_j) * \sigma_{ij}$$

Offensichtlich enthält Nebenbedingung (IV) Produkte von Entscheidungsvariablen. Zur Berechnung ist deshalb ein Algorithmus erforderlich, der ein Minimierungsproblem mit linearer Zielfunktion bei nichtlinearen Nebenbedingungen löst. Hier wurde das von SCHÜLER implementierte Schnittebenenverfahren von KELLEY verwendet.[3]

2.2. Absicherung bei Unsicherheit von Zahlungen und Wechselkursen

Für eine möglichst realitätsnahe Modellierung ist es notwendig, die Unsicherheit zukünftiger Zahlungen und Wechselkurse explizit zu berücksichtigen. Unter Berücksichtigung der Unabhängigkeit von PK_i und Z_i (d.h. $E(PK_i * Z_i) = E(PK_i) * E(Z_i)$) berechnet man

$$E(X) = \sum_{i=1}^{n} (WK_i - \mu_i) * \tau_i + \sum_{i=1}^{n} (\mu_i - TK_i) * A_i$$

Die Varianz des Verlustes ergibt sich als[4]

$$Var(X) = \sum_{i=1}^{n} \sum_{j=1}^{n} (A_i - \tau_i) * (A_j - \tau_j) * \sigma_{ij} + \sum_{i=1}^{n} \Gamma^2_i * ((\mu_i - WK_i)^2 + \sigma_{ii})$$

Im Vergleich zur Varianz im MODELL I repräsentiert der hinzugefügte Term die auf die Unsicherheit der Zahlungen zurückzuführende zusätzliche Schwankung. Auch hier ergibt sich in (IV) eine nichtlineare Nebenbedingung.

Abb. 3: MODELL II

Prämisse :

$(PK_1, .., PK_n), (Z_1, .., Z_n)$ stochastisch unabhängig

Zielfunktion	:	min ! E(X)	(I)
Nebenbedingungen	:		
Falls $\tau_i \geq 0$	:	$A_i \geq 0$	(IIa)
Falls $\tau_i \leq 0$	:	$A_i \leq 0$	(IIb)
		für i=1,..,n	
		$\sum_{i=1, \tau_i \geq 0}^{n} WK_i * A_i \leq \sum_{i=1, \tau_i \geq 0}^{n} WK_i * \tau_i$	(IIIa)
		$\sum_{i=1, \tau_i \leq 0}^{n} WK_i * A_i \geq \sum_{i=1, \tau_i \leq 0}^{n} WK_i * \tau_i$	(IIIb)
		$Var(X) \leq VQUER$	(IV)

3. Modellrechnungen

Alle relevanten Eingangsdaten für die nun folgenden Beispiele können Abbildung 4 entnommen werden. Liest man die Tabelle beispielhaft für den Dollar, so erwartet der Finanzmanager eine Auszahlung in Höhe von Eintausend bei einer Standardabweichung von Einhundert. Der Wechselkurs wird mit 2.07(DM/\$) prognostiziert und weist eine Standardabweichung von 0.02 auf. Außerdem zeigt die Kovarianzmatrix die Unabhängigkeit der Dollarkursprognose von der der anderen Währungen.

Abb. 4: Input der Modellrechnungen

WAEHRUNG	KASSAKURS	E(Prognosekurs)	TERMINKURS
DOLLAR	2.1100	2.0700	2.0800
YEN	1.1200	1.0800	1.1000
PFUND	3.4000	3.3800	3.3700
FRANC	0.3500	0.4500	0.4400

WAEHRUNG	E(Zahlung)	Var(Zahlung)
DOLLAR	-1000.00	10000.00
YEN	500.00	900.00
PFUND	400.00	10000.00
FRANC	-2000.00	90000.00

WAEHRUNG	KOVARIANZMATRIX			
	(D)	(Y)	(P)	(F)
DOLLAR (D)	0.0004	0.0000	0.0000	0.0000
YEN (Y)	0.0000	0.0009	-0.0005	-0.0003
PFUND (P)	0.0000	-0.0005	0.0004	0.0002
FRANC (F)	0.0000	-0.0003	0.0002	0.0004

VQUER : 1600.00

Die sich aus der Optimierung ergebenden Absicherungsbeträge zeigt die folgende Übersicht:

Abb. 5 Optimale Absicherungsstrategien

	Modell 1		Modell 2	
Währung	absolut	relativ	absolut	relativ
DOLLAR	0	0	-233	23
YEN	1267	253	963	193
PFUND	147	37	248	62
FRANC	-2570	129	-2344	117

Die Spalten 'absolut' beinhalten die Währungsbeträge, die per Termin zu verkaufen (positiv) oder zu kaufen (negativ) sind. Die relativen Werte stellen die prozentuale Beziehung von Absicherungsbeträgen und Exposure dar.

MODELL I erlaubt bereits eine zur Verringerung des erwarteten Verlustes führende Überabsicherung. Dies zeigt sich deutlich bei Betrachtung der Absicherungsvorschläge in Yen: Es wird ein Terminverkauf der Yen-Einzahlung in Höhe in Höhe von 253 Prozent vorgeschlagen. Einschränkend wirkt hier Bedingung (IV). Der in Pfund ausgewiesene Terminverkauf erscheint aus risikoneutraler Sicht ungünstig, erzeugt jedoch wegen der Korrelation der Kursprognosen eine Verminderung der Gesamtvarianz.

Die Ergebnisse von MODELL II zeigen deutlich die Wirkung der zusätzlichen Unsicherheit der Zahlungen: Die Absicherungsaktivität erstreckt sich nun auch auf Dollartermingeschäfte, die Schwankung erhöht sich und eine vollständige Absicherung ist nicht mehr erreichbar. Dieser Zusammenhang ist grafisch für parametrisch variierte Standardabweichungen verdeutlicht (Abb. 6).

Abb. 6: Effiziente Absicherungsstrategien

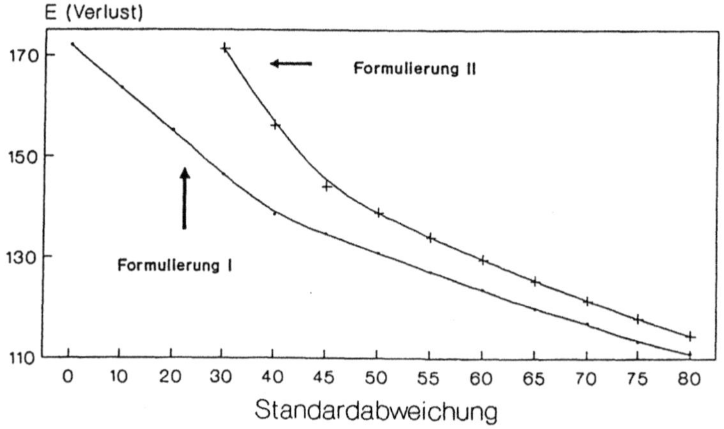

Stellt man bei den gegebenen Beispieldaten die in Abb. 5 vorgeschlagenen Absicherungsstrategien den Extrempositionen "nicht absichern" (d.h. $A_i = 0$ für $i=1,..,n$) oder "vollständig absichern" (d.h. $A_i = \tau_i$ für $i=1,..,n$) gegenüber, so ergibt sich das in Abb. 7 dargestellte Bild.

Abb. 7: Wirkung alternativer Absicherungsstrategien

	Modell 1		Modell 2	
Strategie	E(X)	√Var(X)	E(X)	√Var(X)
ohne Absicherung	188	154	188	157
vollständige Absicherung	172	0	172	31
optimale Absicherung	138	40	150	40

4. Ausblick

Die Übertragung der Modellstruktur auf andere Absicherungsinstrumente erscheint möglich: Während die Abbildung von (standardisierten) Devisenterminkontrakten zur Einführung von Ganzzahligkeitsbedingungen führt, erschwert sich bei Berücksichtigung von Währungsoptionen die Berechnung von Erwartungswert und Varianz des Verlustes.

Literatur

[1] vgl. zu gebräuchlichen Exposure-Definitionen:
Eiteman,D.K.; Stonehill,A.I.: Multinational Business Finance, 4. Aufl., Reading (Mass.) 1986, S.201-218.

[2] Müller,C.M., Foreign Exchange Management bei multinationalen Unternehmen Europas, Brugg 1979, S. 59.

[3] Albach, Horst, und Wolfgang Schüler, On A Method of Capital Budgeting Under Uncertainty, Journal of Mathematical and Physical Sciences 1970, S. 208 - 226.

[4] zur Berechnung der Varianz vgl:
Hinz,H.;Ralfs,D.: Optimale Sicherungsentscheidungen mit Devisentermingeschäften unter Unsicherheit, in: Manuskripte aus dem Institut für Betriebswirtschaftslehre der Universität Kiel, Nr. 194, Kiel 1987, S. 9-10.

PERFORMANCE-OPTIMIERUNG MIT HILFE DES PC- UND GROSSRECHNER-GESTÜTZTEN RENTENMARKT-ANALYSESYSTEMS "R E N S Y S"

Klaus P. L ü c k e, Eschborn

Noch in den siebziger Jahren wurde in der Bundesrepublik der Analyse festverzinslicher Wertpapiere im Rahmen der Anlagepolitik institutioneller und privater Anleger nur geringe Aufmerksamkeit geschenkt. Dies hatte seine Ursache einmal in dem vergleichsweise geringen Volumen des Sekundärmarktes, zum anderen waren die Zinsausschläge am Rentenmarkt zumindest in den 60er Jahren auch noch zu gering, um eine zyklische Anlagestrategie ausreichend erfolgreich erscheinen zu lassen. Mit der sukzessiven Ausweitung des Emissionsvolumens festverzinslicher Wertpapiere und einer seit 1973/74 deutlich grösseren Schwankungsbreite der Zinsen am deutschen Rentenmarkt nahm das Interesse an Rentenpapieren signifikant zu. Die Anlagebereitschaft ausländischer Investoren wurde durch den Abbau institutioneller Hemmnisse und die Währungsentwicklung unterstützt.

Das zunehmende Volumen und die gestiegene Attraktivität des bundesrepublikanischen Rentenmarktes liessen auf der anderen Seite die Forderung nach geeigneten Rentenmarkt-Analyse- und -Informationsinstrumenten, wie sie am Aktienmarkt verfügbar waren, immer lauter werden. Um diesem Bedürfnis entgegenzukommen, entwickelte die BHF-BANK in Zusammenarbeit mit dem Autor Anfang der achtziger Jahre das Rentenmarkt-Analyse-System 'RENSYS'. RENSYS beinhaltete gleichzeitig die Konzeption von Kurs- und Performance-Indizes, um Massgrössen für die Marktentwicklung und den Anlageerfolg bereitzustellen.

1. System-Konzeption

Konzeptionell weist RENSYS sowohl eine Fundierung auf der Grossrechnerseite als auch auf der Personal-Computer-Seite auf. Während der HOST durch die Bereitstellung umfangreicher Gattungs-, Portfolio- und Kursinformationen in erster Linie als voluminöser Datenspeicher dient, werden auf der PC-Seite von den Anwendern dezentral die eigentlichen Analyse- und Optimierungsprogramme "gefahren". Der Datenaustausch zwischen dem Grossrechner der BHF-BANK und den PCs erfolgt über File-Transfer. Im Hause der BHF-BANK verfügt jeder RENSYS-Anwender über einen Online-Anschluss zum HOST, der den Zugriff auf entsprechende IMS-Datenbank-Selektionsprogramme erlaubt. Externen Anwendern werden die sie interessierenden Datenbank-Segmente in täglichem und wöchentlichem Rhythmus als Updates über RJE zur Verfügung gestellt. Die übertragenen Informationen werden auf der PC-Seite des RENSYS-Systems in entsprechende lokale, relational organisierte Datenbanken eingestellt.

Folgende Informationen stehen auf der Grossrechnerseite zur Verfügung:

- Ausstattungsmerkmale aller in der Bundesrepublik gehandelten in- und ausländischen festverzinslichen Wertpapiere. RENSYS wird darüber hinaus gegenwärtig um Gattungsinformationen ausländischer Bondmärkte erweitert
- alle über die BHF-BANK gehandelten Namenspapiere

- aktuelle Kursinformationen auf täglicher Basis zu allen notierten Rentenpapieren
- historische Kurszeitreihen zu ausgewählten Wertpapieren, zum BHF-BANK Rentenmarkt-Kurs- und -Performance-Index sowie zu den entsprechenden Sub-Indizes
- historische Kurszeitreihen zu allen wichtigen Währungen
- zugriffsberechtigte Anwender sind darüber hinaus in der Lage, Portfolio-Informationen zu transferieren

Auf der PC-Seite läuft RENSYS unter dem Betriebssystem PC-DOS und belegt ca. 10 Megabyte an Plattenspeicherplatz. Das System kann im Rahmen eines lokalen PC-Netzwerkes eingesetzt werden. Dies hat den Vorteil, dass mehrere PC-Benutzer auf gemeinsame lokale Datenbestände zugreifen können.

2. Methoden und Instrumente

RENSYS umfasst alle für die Analyse festverzinslicher Wertpapiere relevanten Instrumente und Methoden. Zur Berechnung von Anleihen- oder Portfolio-Renditen sind alle wichtigen internationalen Rendite- und Stückzinsberechnungsverfahren implementiert. Die nachfolgende Abbildung vermittelt einen Überblick. Dabei ist jedem in der RENSYS-Datenbank gespeicherten Wertpapier eine spezifische Rendite-Methode zugeordnet. So werden Emissionen inländischer Emittenten vorwiegend mit Hilfe der Verfahren von Braess/Fangmeyer bzw. Moosmüller analysiert, während DM-Auslandsanleihen auf der Grundlage der AIBD-Formeln bewertet werden. Währungsanleihen werden die jeweiligen nationalen Renditemethoden zugeordnet.

KURS- UND RENDITERECHNUNG F F S

VERFÜGBARE RENDITEMETHODEN

METHODE LAND	STÜCKZINS-VERRECHNUNG bis 4 Jahre	alle Laufzeiten	nein	ZINSTAGE 360	365	PAPIER
Braeß	B1	B2	B3	X		DM Anl
Moosmülr	M1	M2	M3	X		DM Anl
AIBD	A1	A2	A3	X		DM Anl
AIBD		A4			X	WhrAnl
US-BONDS		U1			X	US-Dom
US-GLDM1			T1	X +	X	T-BILL
US-GLDM2			C1	X +	X	CDs
US-GLDM3			T2		X	T-NOTE
GB-BONDS		G1			X	BONDS
DISKONT			D1	X		BACCPT

C O D E : [B1] (Abbruch mit <ESC>) μP

Auch hinsichtlich der Stückzins-Verrechnung sind alle Alternativen implementiert. Im Rahmen von Markt- oder Portfolio-Analysen hat der Anwender die Möglichkeit, ein übergeordnetes Rendite- und Stückzinsverfahren zu wählen, um eine Vergleichbarkeit der Ergebnisse zu gewährleisten.

3. Performance-Masstäbe

Die Beurteilung des Anlageerfolgs bedarf geeigneter Masstäbe. Im Rahmen des RENSYS-Systems stehen dazu der BHF-BANK-Rentenmarkt-Kurs- und -Performance-Index zur Verfügung. Der BHF-BANK-Rentenmarkt-Kursindex basiert auf einem das Segment "Bund (einschliesslich Bundesobligationen), Bahn und Post" repräsentierenden Musterportfolio von 15 (fiktiven) Anleihen. Die Gewichtung dieser Anleihen entspricht der Verteilung des Emissionsvolumens im

Zeitraum 1969 bis 1983. Der Vorteil dieses Indexkonzeptes liegt im Unterschied zu einem auf effektiven Papiren basierenden Musterportfolio darin begründet, dass die aus der permanenten Laufzeit-Verkürzung der zugrunde liegenden Anleihen resultierenden verzerrenden Effekte vermieden werden. Die Anleihen des Musterportfolios werden auf täglicher Basis anhand effektiver Zinsstrukturen bewertet.

Der Performance-Index basiert auf der Zeitreihe des Kursindex und berücksichtigt zusätzlich laufende Kuponzuflüsse. Als Wiederanlage-Verzinsung wird standardmässig ein jährlicher Rhythmus zugrunde gelegt; die Zinseszins-Periodizität ist jedoch konfigurierbar und kann den Anforderungen des Vergleichs-Portfolios angepasst werden.

4. Portfoliotheoretische Ansätze

Im Rahmen der Modern Portfolio Theory (MPT) wird grundsätzlich zwischen aktiv und passiv orientierten Anlagestrategien unterschieden. Aktive Strategien verfolgen das Ziel, durch geeignete anlagepolitische Entscheidungen die zyklische Entwicklung der Kapitalmarkt-Zinsen systematisch zur Performance-Verbesserung zu nutzen. Passive Portfolio-Strategien gehen im Gegensatz dazu davon aus, dass eine erfolgreiche zyklische Anlagepolitik nicht möglich oder nicht erwünscht ist. Sie verfolgen vielmehr das Ziel, eine zu einem bestimmten Zeitpunkt gegebene Portfolio-Rendite von schwankenden Marktzinsen unabhängig zu machen, die Rendite des Portfolios also zu immunisieren.

RENSYS unterstützt sowohl aktiv als auch passiv orientierte Anlagestrategien am Rentenmarkt. Im Rahmen dieses Papiers stehen jedoch die aktiven Ansätze im Vordergrund.

5. Cash-Flow-Ansatz

Unabhängig von der verfolgten Anlagestrategie besteht die fundamentale Methode des RENSYS-Systems darin, jedes betrachtete Anlage-Instrument als Cash-Flow zu betrachten und zu bewerten. Damit verschwindet im Grunde eine Unterscheidung zwischen einem einzelnen Papier, einem Portfolio oder einem ganzen Marktbereich (z. B. dem Segment der Bundesanleihen). Ein Portfolio wird z. B. wie eine Anleihe mit unregelmässigen Zins- und Tilgungszahlungen betrachtet. Auf diese Weise wird eine direkte Vergleichbarkeit zwischen den Analyseergebnissen einzelner Anleihen und Portfolios hergestellt.

RENSYS beinhaltet darüber hinaus eine Reihe statistischer Instrumente zur Zinsstrukturanalyse (wie z. B. Regressions- und Interpolationsverfahren), zur Marktprognose (Glättungs- und Adaptionsverfahren) sowie zur Optimierung. Alle Ergebnisse können grafisch aufbereitet werden. Zeitreihen-Verwaltung und Graphik sind integriert. Da RENSYS nicht nur ein System zur Analyse und Optimierung von Portfolios und Anlagestrategien darstellt, sind darüber hinaus auch eher deskriptiv orientierte Informationsfunktionen implementiert.

6. Arbitrage-orientierte Performance-Optimierung

Arbitrage-orientierte Ansätze zur Portfolio-Optimierung basieren auf der Überlegung, dass Ungleichgewichte im Markt ausgenutzt werden können. Die Erfahrung zeigt, dass solche Ungleichgewichte tatsächlich vorhanden sind. Gerade in schwierigeren Marktsituationen können schon

Umsätze von nur wenigen Millionen DM Kursveränderungen bei einzelnen Papieren hervorrufen, die von RENSYS signalisiert und anlagepolitisch genutzt werden können.

Basis einer arbitrage-orientierten Performance-Optimierung mit RENSYS ist die Feststellung der Zinsstruktur. Zinsstrukturen werden nur für klar abgegrenzte, in sich homogene Marktbereiche ermittelt, um z. B. Bonitätseffekte oder die verzerrenden Auswirkungen von Kündigungsrechten zu vermeiden. Grundsätzlich wird bei RENSYS so vorgegangen, dass die Zinsstruktur für einen bestimmten Wertpapierbereich kalkuliert wird, da in der Regel Portfolio-Bestände allein nicht ausreichend sind, um Zinsstrukturen zu ermitteln. Hinzu kommt, dass es ja gerade darauf ankommt, Portfolio-Bestände mit dem "Markt" zu vergleichen. Auf diese Weise wird bei arbitrage-orientierten Portfolio-Optimierungen das Portfolio in mehrere Teilsegmente unterteilt. Für jedes dieser Segmente wird eine separate Zinsstrukturanalyse durchgeführt.

Eine Zinsstruktur stellt die Renditen der einzelnen Papiere eines Marktsegments oder Portfolios nach Laufzeiten (mittleren/rechnerischen Laufzeiten) gegliedert dar. Mit dieser Darstellungsweise kommt zum Ausdruck, dass die Mittelbindungsdauer ceteris paribus das entscheidende Kriterium für die Höhe der Renditen darstellt. Die wichtigsten verzerrenden Einflüsse sind durch die jeweilige Begrenzung auf ein bestimmtes Marktsegment eliminiert, so dass Abweichungen der Einzelrenditen von der Zinsstruktur als Basis der Optimierung herangezogen werden können. Diese Abweichungen werden als ausnutzbare Marktungleichgewichte interpretiert. Ihr Verschwinden, d. h. die Anpassung der Rendite an die Zinsstrukturkurve, würde Kursveränderungen in einem Umfang auslösen, die in der finanzmathematischen Mechanik des Rentenmarktes begründet sind.

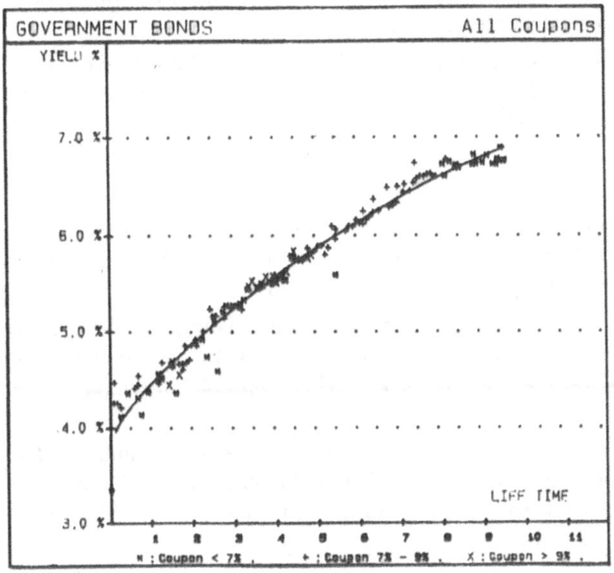

Zur Bestimmung der Zinsstrukturkurve selbst zieht RENSYS standardmässig einen Regressionsansatz mit linearem, quadratischem und logarithmischem Glied heran. Weitere Ansätze sind im Rahmen von RENSYS verfügbar, bleiben jedoch dem erfahrenen Anwender vorbehalten.

Die Kursveränderung, die bei einer Anpassung der Anleihen-Rendite an die Zinsstruktur resultiert, ist eine annähernd lineare Funktion der sog. "Duration". Die Duration ist ein Mass für die durchschnittliche Mittelbindung einer Anleihe, wobei im Unterschied zur Laufzeit auch Kuponzahlungen berücksichtigt werden. Salopp ausgedrückt stellt die Duration, die auf Macaulay zurückgeht, den mit den jeweiligen Barwerten der einzelnen Zins- und Tilgungszahlungen gewichteten Durchschnitt aller Zins- und Kuponlaufzeiten dar. Die Duration ist bei kupontragenden Anleihen daher immer geringer als die entsprechende mittlere oder Restlaufzeit. Je grösser die Duration, desto stärker ist tendenziell die Kursveränderung, die eine Renditeänderung auslöst. Der Grund hierfür liegt im Zinseszins-Effekt begründet.

Für die einzelnen Anleihen des betrachteten Portfolios werden die entsprechenden Renditeabweichungen von der Zinsstruktur und die Hebeleffekte auf die Wertpapierkurse, die beim

Verschwinden der "Ungleichgewichte" entstehen, vom RENSYS-System ausgewiesen. Sind die implizierten Hebeleffekte gross genug, können sie für eine arbitrage-orientierte Umschichtungspolitik oder für Neuanlagen herangezogen werden:

- Bei Neuanlagen sind Tausch-Gesichtspunkte nicht relevant. Der Anleger wählt in dem Laufzeitenbereich, in dem er seine Anlagen tätigen will, diejenigen Papiere aus, die gegenüber dem Marktdurchschnitt (der Zinsstruktur) bessere Renditen aufweisen. Die Entscheidung für den Laufzeitenbereich wird als gegeben vorausgesetzt und ist nicht Gegenstand einer arbitrage-orientierten Portfolio-Optimierung
- Bei Tausch-Operationen zwischen Markt und Portfolio (oder auch innerhalb von Portfolios) hängen die Arbitrage-Möglichkeiten vom gegebenen Portfolio-Bestand ab.

RENSYS bietet hier die Möglichkeit, dass der Anwender "sein" Portfolio bzw. Teile seines Portfolios für mögliche Tausch-Operationen deklariert. Alle vom System generierten Umtauschmöglichkeiten berücksichtigen dann diese Vorgabe. Um zu vermeiden, dass "Struktureffekte" hinsichtlich des Risikoverhaltens des Portfolios entstehen, kann der Anwender weiterhin eine zulässige Laufzeit-Abweichung und eine zulässige Kupon-Abweichung spezifizieren. Darüber hinaus ist es möglich, einen Mindest-Rendite-Vorteil bzw. eine Grösse für die Transaktionskosten zu spezifizieren. RENSYS prüft anschliessend für alle betrachteten Papiere, ob die gesetzten Bedingungen erfüllt sind und weist, soweit erfolgversprechend, Tausch-Vorschläge aus.

Diese Tausch-Vorschläge sind zunächst "nur" Ideen; die Realisierung steht auf einem anderen Blatt. Es zeigt sich jedoch, dass je nach Marktphase und Tendenz etwa zwischen 10 % und einem Drittel der Switch-Vorgaben effektiv umgesetzt werden können.

Die Erfahrung lehrt, dass neben Bonitäts- und Liquiditätsgesichtspunkten (Emissionsvolumen, Marktpflege) vielfach Kuponeinflüsse die Leistungsfähigkeit der zinsstruktur-orientierten Arbitrage-Strategie beeinflussen können. Aus diesem Grunde bietet RENSYS hier zwei weitere Möglichkeiten:

<u>Differenzierung nach Kuponklassen</u>

In diesem Falle, wichtig insbesondere seit etwa 1982, sind getrennte Umtausch-Analysen nach Kuponklassen durchzuführen. Tatsächlich können seit etwa 1982 systematische Kuponeinflüsse am Rentenmarkt festgestellt werden. Die Gründe hierfür sind vielschichtig:

Über die Zinserwartungen und eine gleichzeitige Betonung der Duration als Kursreagibilitätsmass verzerrt der Nominalzins das Ordnungskriterium "Restlaufzeit". So können z. B. sinkende Zinserwartungen zu einer Präferierung niedrig verzinslicher Papiere führen (und umgekehrt).

Nicht zu unterschätzen ist auch der Einfluss steuerlicher Gesichtspunkte bei Privatanlegern. Da bei Anleihen (Sonderregelungen z. B. bei Zero-Bonds einmal ausgeklammert) nur der Kupon, nicht jedoch Kursgewinne bzw. -verluste steuerlich relevant sind, geht von dieser Seite eine strukturelle Bevorzugung niedriger Nominalzinsen aus.

Ob eine Differenzierung nach Kuponklassen signifikant ist, kann im Rahmen von RENSYS über Signifikanztests der Regressionsparameter bei Zinsstrukturanalysen festgestellt werden. Werden für unterschiedliche Kuponklassen eines Marktsegments solche signifikanten Abweichungen ermittelt, wird empfohlen, arbitrage-orientierte Switch-Analysen nach Kuponklassen getrennt durchzuführen.

<u>Verwendung der Duration als Ordnungskriterium</u>

Im Rahmen von RENSYS können Zinsstrukturanalysen auch auf der Basis des Ordnungskriteriums "Duration" durchgeführt werden.

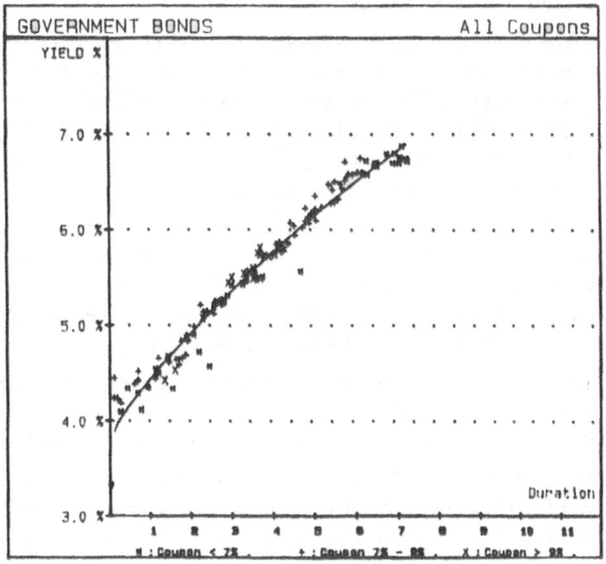

Da die Duration ein Mass für die Kursreagibilität darstellt, ist anzunehmen, dass hiermit Einflüsse von Zinserwartungen besser erfasst werden. Steuerliche Gesichtspunkte bleiben jedoch unberücksichtigt. Sie wären erst dann erfassbar, wenn in jeder Kuponklasse der Anteil inländischer "steuerehrlicher" Investoren bekannt wäre.

Aus den genannten Gründen wird im Rahmen praktischer Anwendungen daher vielfach einer nach Kuponklassen differenzierten Arbitrage-Optimierung der Vorzug gegeben.

Eine weitere, bei RENSYS implementierte Möglichkeit zur Arbitrage-Optimierung besteht darin, dass jede Anleihe in ein Portfolio fiktiver "Zero-Bonds" aufgelöst wird. Jeder Kupon und jede Tilgungsrate wird als ein separater, eigenständiger Zero-Bond betrachtet. Diese einzelnen Teil-Segmente einer Anleihe werden sodann mit einer relevanten Zinsstrukturkurve bewertet. Die Summe der Einzelpreise wird mit dem effektiven Kurs der Anleihe verglichen. Eine praktische Umsetzung dieser Arbitrage-Strategie scheitert allerdings am deutschen Rentenmarkt am fehlenden Material.

7. Szenario-Analyse

Während eine arbitrage-orientierte Anlagepolitik den risiko-orientierten Status quo eines Portfolios nahezu unverändert lässt, sind Veränderungen des Risikoprofils bei strategischen Anlageentscheidungen erwünscht. Eine strategisch orientierte Anlagepolitik besteht - lässt man Währungsgesichtspunkte einmal ausser Acht - im Grunde aus der Ausnutzung zyklischer Zinsschwankungen. Da sich im Zinszyklus in der Regel auch die Gestalt der Zinsstrukturkurve zum Teil dramatisch verändert, bedarf die anlagepolitische Analyse einer verbesserten Transparenz. Die Einflüsse, die auf eine Anleihe im Zinszyklus einwirken, lassen sich dabei theoretisch in 3 Komponenten unterteilen:

- Niveau-Effekt

  Veränderungen im Zinsniveau beeinflussen den Kurs eines Wertpapiers zweifellos am stärksten. Dieser Einfluss ist für ein einzelnes Wertpapier tendenziell um so ausgeprägter, je grösser seine Duration ist.

- Zinsstruktur-Effekt

  Da Veränderungen des Zinsniveaus in der Regel auch zu Verschiebungen der Zinsstruktur führen (der jüngste Zinsanstieg fand z. B. überwiegend - und untypisch - am langen Ende der Laufzeiten-Kurve statt), können sich je nach Laufzeit-Einordnung des einzelnen Papieres mehr oder weniger den Niveau-Effekt kompensierende oder verstärkende Einflüsse ergeben.

- Laufzeit-Verkürzungs-Effekt

  Die Verkürzung der Restlaufzeit während des Anlagehorizonts führt dazu, dass sich die Positionierung des betrachteten Papiers auf der Strukturkurve verändert. Bei einer steil ansteigenden Zinsstruktur, wie sie in Niedrigzinsphasen typisch ist, verstärkt dieser Effekt z. B. evtl. Einflüsse sinkender Zinsen bzw. schwächt Einflüsse steigenden Zinsniveaus ab. Bei einer inversen Zinsstruktur (Renditen im unteren Laufzeitenbereich sind höher als am oberen Ende der Zinsstruktur) sieht dies genau umgekehrt aus.

Im Rahmen von Szenario-Analysen berücksichtigt RENSYS diese Einflüsse simultan. Der Anwender hat die Möglichkeit, auf die Sicht eines beliebigen Anlagehorizonts seine Zinsstrukturerwartungen zu formulieren. RENSYS weist für die betrachteten Anleihen die erwartete Performance - bestehend aus induzierter Kursveränderung und laufender Verzinsung - in Form einer mit der Rendite vergleichbaren Grösse aus.

Auf diese Weise kann der Investor den erwarteten Anlageerfolg seines Portfolios oder einzelner Anleihen vor dem Hintergrund alternativer Szenarien überprüfen. Für einzelne Szenarien kann er darüber hinaus den performancemässig optimalen Laufzeitenbereich feststellen.

Für die einzelnen Papiere eines Portfolios oder Marktsegments kann im Rahmen von RENSYS wahlweise unterstellt werden, dass aktuelle Abweichungen der Einzel-Renditen von der Zinsstruktur bestehen bleiben oder verschwinden werden. Je länger der Anlage-Horizont, desto geringer sind die Einflüsse, die von der Anpassung an die erwartete Zinsstruktur ausgehen. Werden mehrere Kuponklassen mit einer Zinsstruktur gemeinsam betrachtet und kann unterstellt werden, dass auch zukünftig Kuponeffekte vorherrschen werden, wird mit der Option gearbeitet, die <u>keine</u> Anpassung unterstellt.

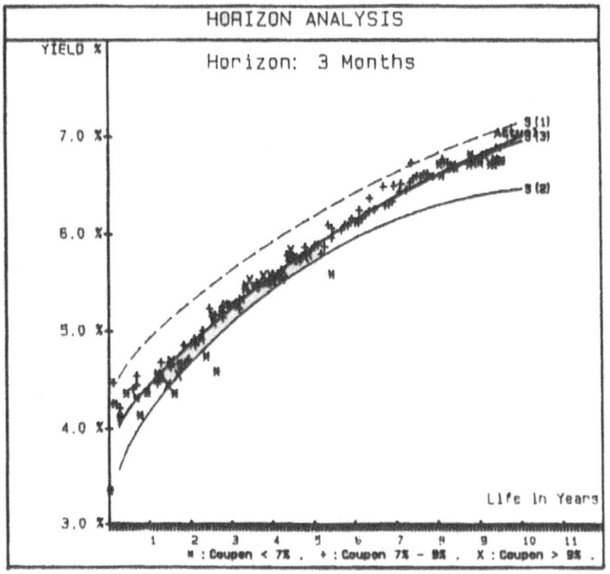

Mit Hilfe des RENSYS-Systems ist es auch möglich, mehrere Szenarien simultan zu betrachten. Jedem Szenario können dabei unterschiedliche Wahrscheinlichkeiten zugeordnet werden. Der "Erwartungswert" der Performance wird als - mit den Wahrscheinlichkeiten - gewichteter Durchschnitt ermittelt. Als Mass für den Unsicherheitsbereich wird die Standardabweichung herangezogen. Es zeigt sich in der Regel, dass die volatileren Anleihen (mit vergleichsweise grosser Duration) auch entsprechend höhere Risiko-Masse aufweisen.

Über die beiden Parameter "erwartete Rendite" und "Unsicherheitsbereich" (Standard-Abweichung) wird dem RENSYS-Anwender eine sogenannte Risk-Return-Grafik zur Verfügung gestellt. Bei überwiegend sinkenden Zinserwartungen ergibt sich ein positiv ansteigender Risk-Return-Tradeoff. Bei steigenden Zinserwartungen ist dieser Tradeoff negativ.

Als Entscheidungskriterium für die Wertpapieranlage gilt grundsätzlich, dass nur die Papiere als anlagewürdig gelten, deren Performance oberhalb des kongruenten Geldmarktsatzes liegt. Innerhalb der einzelnen Wertpapiere wird wie folgt selektiert:

- von zwei Papieren mit gleicher Performance-Erwartung wird dasjenige als anlagewürdig ausgewählt, das eine geringere Risikostreuung aufweist
- von zwei Anleihen wird diejenige als optimal ausgewählt, deren Performance-Erwartung abzüglich Standard-Abweichung grösser ist als die Performance-Erwartung der Vergleichs-Anleihe plus deren Standard-Abweichung. Dieser Test wird paarweise für alle analysierten Anleihen durchgeführt.

Es ist zweifellos richtig, dass über die Szenario-Analyse eine vergleichsweise effiziente Performance-Optimierung möglich ist. Es ist jedoch zu berücksichtigen, dass vom Anwender eine prognostische Leistung verlangt wird.

8. Timing-Instrumente

RENSYS verfügt über eine Reihe von das Timing unterstützenden Instrumenten. Diese Instrumente dienen der Verbesserung der Prognoseleistung des Anwenders:

- Technisches Indikatoren-System

  Das zum RENSYS-System gehörende technische Indikatoren-System erfasst Grössen, die dem Anwender Interpretationshilfen bei der Einschätzung des aktuellen Zinstrends geben. Bei der Auswahl und Entwicklung dieser Indikatoren, zu denen z. B. Volumensindikatoren - volumensgewichtete Momentumgrössen zum BHF-BANK Rentenmarkt-Index - gehören, wurde vor allem Wert auf signifikante Signaleigenschaften bei Trendwenden gelegt.

- Relative Price-Ratios

  Relative Price-Ratios geben eine vergleichende Auskunft über die Kursentwicklung unterschiedlicher Marktsegmente, z. B. Bund-Anleihen und Supranationale. Die Ratios geben Hinweise sowohl auf veränderte Marktsegment-Einschätzungen als auch auf evtl. vorzeitige Trendänderungen bei einzelnen Gattungsgruppen

- Spread-Analysen

  Hier erfolgt eine systematische zeitreihenanalytische Auswertung von Rendite-Abweichungen einzelner Papiere von den jeweiligen Zinsstrukturen

- Trend-Modul

  Ansätze zur automatisierten Prognose mit Hilfe von Verfahren des Exponential Smoothing (Winters Modell mit Saison).

- System zum Test technischer Anlagestrategien

  Darüber hinaus ist ein System zum Test technischer Anlagestrategien implementiert. Dieses System erlaubt es dem Anwender, selbst formulierte Regeln für Kauf- und Verkaufsentscheidungen anhand effektiver Kurszeitreihen bzw. der Zeitreihen des Rentenmarkt- Index zu überprüfen.

CREDIT SCORING SYSTEM

O. Naamani, Amsterdam

For centuries, before there was such a thing as creditscoring, credit grantors used their personal judgement to make credit decisions. In 1941 the first article about creditscoring was published. In this article Durand asserted the possibility to predict which applicants were most likely to go bad by systematically studying the characteristics of borrowers who had done so in the past. It was not until 1958, 17 years later, when the first creditscoring system for a finance company was launched in the USA.

Since the 70's, creditscoring systems are widespread used by the major credit and finance companies all over the world.

The yearly credit consumption in the Netherlands reaches an amount of some 7.5 billion Dutch guilders.

Algemene Bank Nederland developed during the last decade its own creditscoring system and gained a strong reputation in consumer lending through a direct marketing approach.

In recent years we are aware of two major changes which might have implications for the credit industry in the Netherlands. On one hand there is a growing use of sophisticated technology and of advanced statistical techniques which give new impulses for further research. On the other hand the Dutch parliament is expected to come in January 1, 1988, with a revised consumer credit legislation (WCK: Wet op het Consumptief Krediet). This new legislation will probably prescribe all credit institutions participation in a general credit registration bureau (BKR-toetsing). Furthermore the credit-supplier can be asked to show he was carefull in granting the credit. To applicants it probably will grant the right to know why an application eventually is rejected. A mental shift in consumer attitudes is expected as a consequence: applicants might from then on perceive credit as a right rather than a privilege. An increased exposure of data might also lead to an adverse publicity when portfolio registration is imperfect or absent.

All these factors combined make creditscoring systems not any more a luxury but a realistic necessity.

ABN works on some creditscoring variants for 3 of its finance companies (indirect marketing approach). We shall examine those on the background of the next models:

1. the traditional creditscoring model,
2. loglineair model,
3. advanced three-dimensional model,
4. decision-tree model,

and evaluate results of recent research on geo-demographic information as a component of the creditscoring system.

INTEREST MARGIN COMPUTER MODELS

Pier G. Sinia, Rotterdam

In banking it will be of utmost importance to face the uncertainties and the volatilities of the interest margin all the time. So management in banking is anxious to know the present and future developments of interest margin. There are several motives to face the interest margin conscientious, for instance:
- Profit in banking is chiefly determined by interest margin;
- Interest margin can be influenced by decisions about liquidity reserve level, capital adequacy level, strategy on matching the assets and liabilities, and so on;
- Interest margin is sensitive to and dependent on external phenomena such as structures of interest markets, volatility of foreign exchange rates, economic and political developments.

The question now is: Can an interest margin computer model (with or without use of operation research) serve as an effective tool controlling the interest margin?

Trying to answer this question, four financial planning computer models at Bank Mees & Hope in The Netherlands relating interest margin will be evaluated:
1. Capital adequacy model.
2. Linear programming model supporting the control of interest results.
3. Interest margin model.
4. Long-term financial planning model.

The idea of the capital adequacy model was to make a five year planning based on profit and solvency. The input consists of solvency facts at the beginning of the planning period, the percentage distribution of capital adequacy among the solvency-duty loan-groups, the average cost-price of liabilities and the target profit. The model calculates the break-even interest percentages of the assets belonging to the applied target profit and it gives other information. The objective of this model is not fulfilled because the planning horizon is too long and the outcomes are not useful as a management tool to take decisions.

The LP-model has a planning horizon of five quarters and the entire balance sheet is taken into account.The objective function is to maximize total interest profit, subject to many constraints about liquidity, solvency, limitations of growths, and so on. However the results are determined too much by the constraints, e.g. certain conflicting decisions are advised within the planning period. Maybe linear programming can be used more effectively in the field of interest margin if company plans are fully incorporated in the model in the place of some rather theoretical constraints.

The interest margin model is an interactive and deterministic model. The planning horizon is four quarters of a year; there is also a version with a planning horizon of five years. The balance sheet is split up in groups and subgroups by the criterion of interest behaviour. Some input figures are derived from the company plan and the budget; furthermore constraints about liquidity and solvency are build in and relationships of the interest market are taken into account. The interest margin model provides information of future interest margin based on the balance sheet classification and interest behaviours. It improves the insight into the future characteristics of the interest results such that the control of the interest margins gets improved significantly.The input figures are clear and the outcome is evident. The interest margin model is simple and valuable in practical sense. This model takes a place within the yearly planning cycle and the results are frequently used by several departments.

The long term financial planning model is mentioned here for the sake of completeness. The objective of this model is to get a global view about the consequences of some macro developments on the future profit & loss accounts and balance sheets. It's planning horizon is five years or beyond. This model is marginal with respect to the investigation of interest margin. It 's used incidentally; however the impact of this model is low.

From the evaluations of the four models above, following conclusion can be drawn. Interest margin computer models now can furnish already a valuable tool, but improvement is likely, perhaps by using operation research.

FINANCIAL MODELLING OF A LAND RECLAMATION PROJECT

J. Telgen and P. Verstraaten , Utrecht

For centuries now the Dutch are famous for their technical skills in land reclamation. The largest projects they undertook in this field are the so-called "Zuiderzeewerken", which started in 1918. These projects consisted of the construction of a dike to cut off a large inland sea and the reclamation of five large (± 500 km^2 each) pieces of this sea. Four of these five land reclamation projects have been finished and the fifth (Markerwaard) is about to start.

However, while the government funded the previous projects, there is a growing pressure to construct the Markerwaard without public funds, Thus a "Markerwaard Inc" has to be founded. Then after the project is finished the government buys all the land from Markerwaard Inc and sells it again to whomever the government deems appropriate. Subsequently Markerwaard Inc will be dismantled.

We were asked to assist with the preparation of a sound business plan for Markerwaard Inc by the government body that developes the reclaimed land, the Rijksdienst IJsselmeer Polders (RIJP). Part of this business plan is a (micro) economic analysis and evaluation of the technical plans (macro-economic effects may concern the government but not Markerwaard Inc).

This evaluation is complicated because:

- a 30 - year period is considered;
- there are many technical alternatives and possibilities, for which no clear choice has been made yet;
- there is a fair degree of uncertainty (inflation, interest rate, prices etc.) and risk (construction, geo-hydrology) involved;
- neither the mode of exploitation nor the way of funding are known in advance.

We developed a multi-level model for financial evaluation, in which the lower (more detailed) levels contribute information to the higher levels. The top level model consists of 5 items for costs and revenues each per year, resulting in a cash flow (before funding costs). From the cash flow the internal rate of return can be calculated, which gives the maximal return for funding. Then exploitation, profit and loss statements and balance sheets can be calculated and finally a complete cashflow forecast including interest results.

The emphasis in this paper is not so much on the technical model development, but on the use of modelling as a consultancy (intervention) instrument.

Regionale Standortplanung für medizinisch-technische Großgeräte

L. Bach, Bayreuth

In den letzten Jahren hat die Anwendung von medizinisch-technischen Großgeräten in Therapie und Diagnose eine rasche Verbreitung gefunden. Dabei sind neben Problemen in der Gesundheitsökonomie und -politik auch Probleme bei der regionalen Verteilung der Großgeräte entstanden. Beispiele für derartige Großgeräte sind Computertomographie-Geräte (CT-Geräte) oder Nierenlithotripsie-Geräte (ESWL-Geräte). Fragen, die im regionalen Zusammenhang zu lösen sind, beziehen sich darauf, wieviele Großgeräte für die angemessene Versorgung einer Region erforderlich sind und an welchen Standorten die Großgeräte lokalisiert sein sollen. Um für derartige regionale und gleichzeitig gesundheitsökonomische Fragen die notwendigen Informationen zu erhalten, die von Entscheidungsträgern im Gesundheitswesen benötigt werden, ist ein Modell für die regionale Standort- und Kapazitätsplanung von Großgeräten entwickelt worden. Mit dem Modell können alternative regionale Standortsysteme auf der Grundlage von Nutzungs- und Transportkosten analysiert und miteinander verglichen werden. Kriterien für den Vergleich sind die regionalen Gesamtkosten, die durchschnittlichen Kosten je Behandlung, die Zugänglichkeit zu den Großgeräten für die Bewohner, die Indikationsrate als Merkmal des medizinischen Versorgungsgrades und die Auslastung der einzelnen Großgeräte. Dabei können verschiedenartige Variablen der Nachfrage- wie der Angebotsseite Berücksichtigung finden. Das Modell entspricht dem Location-Allocation-Problem mit Kapazitätsrestriktionen. Mit der Anwendung des Modells bei Planungen für Großgeräte in Baden-Württemberg hat aufgezeigt werden können, daß im Fall von CT-Geräten mehr Geräte zu einer Senkung der Gesamtkosten in einer Region führen können (1), oder daß im Fall von ESWL-Geräten mit der preisgünstigen 2. Gerätegeneration ein erhebliches Dezentralisierungspotential für eine Region gegeben ist (2).

(1) Bach, L.; Hoberg, R.: A planning model für regional systems of CT Scanners, Socio-Economic Planning Sciences 19(1985)189-199.

(2) Bach, L.; Ehm, H.; Friedl, W.: Modellrechnungen zur Optimierung der Kapazität und Standortverteilung von ESWL- Geräten in Baden-Württemberg, Bayreuth 1987 (im Druck).

Ein flexibles Funktionalplanspiel zur ökonomischen Aus- und Weiterbildung von Krankenhauspersonal

Peter Hofweber, Nürnberg

Zusammenfassung: In Anbetracht der Kostensituation des Gesundheitswesens der Bundesrepublik wird immer häufiger mehr Management im Krankenhaus gefordert. Geeignete Fort- und Weiterbildungsmaßnahmen zur Verbesserung der Managementqualität fehlen jedoch weitgehend. Insbesondere Planspiele als aktive Lehr- und Lernmethoden können hier einen effektiven Beitrag zur Verbreiterung der Wissens- und Erkenntnisbasis leisten. Das dazu entwickelte Planspiel ARKTIS wird kurz vorgestellt, ein Problempunkt der Planspielkonstruktion angesprochen sowie ein Lösungsweg durch die Integration des Ansatzes von Expertensystemen aufgezeigt.

Summary: This paper introduces a new management game for training and educational purposes concerning the hospital pharmacy. Problems the design process entails are shown and some clues to surmount them are given.

1. Problemstellung

Das häufig diagnostizierte Managementdefizit in den Krankenhäusern der Bundesrepublik wird vielfach als wesentliche Ursache für deren Ineffizienz angesehen. Gezielte Schulung der Entscheidungsträger könnte hier sicherlich Abhilfe schaffen. Allerdings vernachlässigen herkömmliche, auf konventionelle Lehrmethoden wie z.B. Vorträge zurückgreifende Bemühungen die mit der Erwachsenenbildung verbundenen spezifischen Probleme. Häufig werden auch wichtige didaktische Aspekte, wie z.B. die Orientierung an den vorhandenen Fähigkeiten und Fertigkeiten der Schulungsteilnehmer oder deren Motivationsstruktur sowie die Bedeutung der individuellen Verwertbarkeit des Gelernten im Berufsleben, nicht berücksichtigt.

Aus motivationstheoretischer Sicht stellt dagegen das Planspiel eine Lehrmethode dar, die sich insbesondere auch für die Aus- und Weiterbildung von

Krankenhausmitarbeitern anbietet. Allerdings sind in dem weiten Feld von Managementspielen speziell für den Krankenhaussektor entwickelte und geeignete Spiele nur sehr spärlich vertreten. So ist auf nationaler Ebene lediglich das Planspiel KLIMA zu finden, welches auf das Trainieren von Managemententscheidungen auf höchster Ebene abzielt. Planspiele zur Verbesserung der Entscheidungsfindung auf unterer Ebene fehlen (z.B. für den Bereich der Materialwirtschaft oder der Ablaufsteuerung). Mit der Konzeption und Entwicklung des Planspiels ARKTIS (Arzneimittelversorgung im Krankenhaus - Training durch interaktive Simulation) wird gegenwärtig eine dieser Lücken geschlossen.

Mit traditionell konzipierten Planspielen können die angesprochenen didaktischen Mängel allerdings auch nicht völlig beseitigt werden. Insbesondere das Problem der Ausrichtung von Lerninhalten an der jeweiligen individuellen Wissensbasis der Spielteilnehmer bleibt dort unbefriedigend. Da mit der Entwicklung eines Planspiels ein hoher Zeitaufwand verbunden ist und diese damit sehr kostenintensiv sind, orientiert sich die Planspielkonstruktion am Einsatzpotential, d.h. Planspiele werden als Standardlösung entwickelt und tragen daher zwangsläufig das Stigma didaktischer Unvollkommenheit. Die Ursache dieses inadäquater Planspieldesigns sowie Ansatzpunkte zur Verbesserung dieser Situation werden nachfolgend aufgezeigt.

2. Realität und Abstraktion im Planspiel

Management-Planspiele werden einerseits in generelle und funktionale Spiele unterschieden, je nachdem ob die Struktur eines ganzen Unternehmens, mehrerer Unternehmensbereiche oder lediglich eines einzigen Unternehmensbereichs abgebildet wird. Daneben existiert eine Unterscheidung nach allgemeinen und betriebsspezifischen Spielen, die sich an der Abbildung irgendeines beliebigen, nicht näher definierten Unternehmens, der typischen Verhältnisse eines Wirtschaftszweigs, einer Branche, oder der konkreten Situation eines Betriebes anlehnt. Die Kombination dieser beiden Aspekte, Modellbreite und Modellbezug, ergibt die spezifische Art eines Spielmodells, wobei das Spektrum vom allgemeinen Unternehmensspiel bis zum betriebsindividuellen Einbereichsspiel reicht (Abbildung 1).

Bedeutung gewinnt die Art des Spielmodells infolge des engen Zusammenhangs zwischen der pädagogischen Effektivität von Planspielen und deren Realitäts- bzw. Abstraktionsgrad. Obwohl eine generelle Wertung hinsichtlich des Abstrak-

Modellbreite / Modellbezug	Unternehmen	Mehrere Bereiche	Ein Bereich
Allgemein	Allgemeines Unternehmens-spiel	Allgemeines Mehrbereichs-spiel	Allgemeines Einbereichs-spiel
Wirtschafts-zweig	Wirtschafts-zweigorien-tiertes Unter-nehmensspiel	Wirtschafts-zweigorien-tiertes Mehr-bereichsspiel	Wirtschafts-zweigorien-tiertes Ein-bereichsspiel
Branche	Branchen-orientiertes Unternehmens-spiel	Branchenorien-tiertes Mehr-bereichsspiel	Branchenorien-tiertes Ein-bereichsspiel
Betrieb	Betriebs-individuelles Unternehmens-spiel	Betriebs-individuelles Mehrbereichs-spiel	Betriebs-individuelles Einbereichs-spiel

Abb. 1: Arten von Spielmodellen

tionsniveaus kaum möglich ist, kann gesagt werden, daß die praktische Verwertbarkeit des im Spiel Gelernten umso größer ist, je konkreter sich der Modellbezug darstellt. Dahingehend ist ein allgemeines Unternehmensspiel in der Regel weniger brauchbar als ein betriebsindividuelles Unternehmensspiel. Daneben sind die Einbereichsspiele (Funktionalspiele) auch zur Wissensvermittlung besser geeignet als generelle Spiele [Wolfe 1985]. Daraus läßt sich die Forderung nach einem möglichst konkreten und realitätsnahen Modell ableiten.

Unabhängig von der Art des Spielmodells, ist ein realistisches Verhalten des Planspiels eine wesentliche Voraussetzung für einen erfolgversprechenden Planspieleinsatz. Diese "face validity" erfordert allerdings beim Funktionalspiel einen wesentlich höheren Realitätsgrad des Modells als beim generellen Spiel; ein an sich wünschenswerter Aspekt der jedoch erhebliche Probleme aufwirft:

"realism adds complexity, and as the game becomes more complex, the game becomes more difficult to learn, to play and to administer. Too complex a model does not permit the identification of the impact of important policies and decisions, thereby making ambiguous the structure underlying relationships and causal factors" [Norris 1986].

Als wesentliche Negativerscheinungen, die sich aus einem zu hohen Komplexitätsgrad ergeben, sind zu nennen:

- der Spielteilnehmer verliert den Bezug zu den zugrundegelegten Zusammenhängen, woraus eine verminderte Effektivität der Lehrmethode Planspiel resultiert.

- Der Spielteilnehmer wird Situationen ausgesetzt, die ihn überfordern, was Resignation und das Ausscheiden aus dem Spiel mit sich bringt.

Dem Streben nach Realität -und damit nach hohem Komplexitätsgrad- ist eine natürliche Grenze gesetzt, was Elgood [1984, S. 53] deutlich formulierte:

".... it still has to be accepted, that a simulation is useful precisely because it is in some respect not like real life. It is not physically or financially dangerous and it does not occupy so much time. These and other aspects of reality are deliberately excluded and it follows that total realism is an unattainable objective. The search for reality reaches its logical end when one considers war games: the final step towards making a war game real is to go to war - and when you have done that it is not a game anymore".

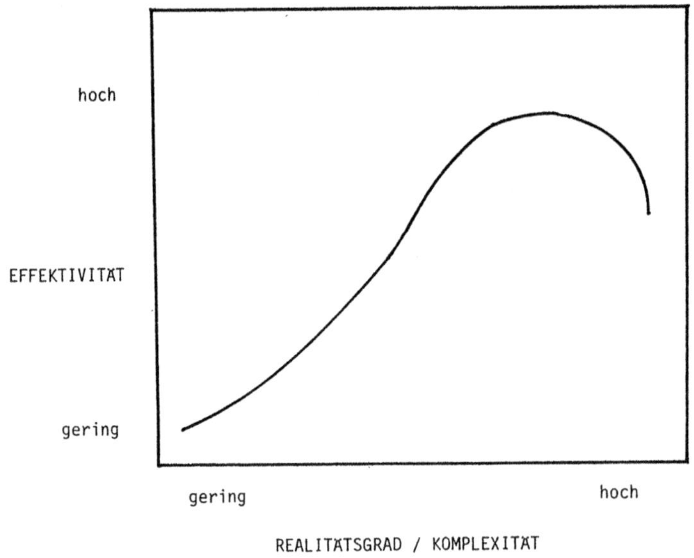

Abb. 2: Funktionaler Zusammenhang von Realitätsgrad und pädagogischer Effektivität

Andererseits bleibt angesichts des verbleibenden Spektrums an Abstraktionsmöglichkeiten die Frage nach dem anzustrebenden Realitätsgrad offen. Zwar hat Norris [1986] eine funktionale Beziehung zwischen dem Realitätsgrad und damit der Komplexität sowie der pädagogischen Effizienz eines Planspiels aufgestellt (Abbildung 2), jedoch skizziert diese den Zusammenhang lediglich exemplarisch und sie müßte, um aussagefähig zu sein, für jeden Spielteilnehmer individuell aufgestellt werden. Zudem ist ein Spiel immer die Darstellung der Realität aus der Sicht des Designers und selbst wenn sich dieser um Objektivität bemüht, werden sich unterschiedliche Wahrnehmungen der Realität aufgrund unterschiedlicher subjektiver Ansichten und Auffassungen seitens der Spielteilnehmer ergeben [Greenblat 1981]. Jeder Spieler spielt demnach ein spezielles Spiel, wobei es aus pädagogischer Sicht wünschenswert wäre, den Realitätsgrad und somit die Komplexität individuell anzupassen.

3. Konzeption eines flexiblen Planspiels

Im Planspiel ARKTIS, das den materialwirtschaftlichen Bereich einer Krankenhausapotheke zum Gegenstand hat, steht vor allem die medizinökonomische Ausbildung im Vordergrund. Hier werden die Spielteilnehmer mit materialwirtschaftlichen (oder nach derzeitigen Sprachgebrauch: logistischen) Problemen konfrontiert, an denen - unter Berücksichtigung medizinischer Aspekte - rationale Entscheidungsfindung zu trainieren ist. In diesem Zusammenhang bedingt die angestrebte rationale Entscheidungsfindung die Anwendung von Methoden des Operations Research, insbesondere Prognoserechnung und Modelle der Lagerhaltungstheorie.

Das Planspiel ARKTIS kann als Regelkreis beschrieben werden. Es existieren Ziele oder Sollvorgaben (Führungsgrößen) und es werden Entscheidungen getroffen (Stellgrößen), die sich unter dem Einfluß von Umwelteinwirkungen (Störgrößen) innerhalb der betrachteten Problemstellung (Regelstrecke) auswirken und zu bestimmten Istzuständen (Regelgrößen) führen. Eine sich anschließende Kontrolle beinhaltet den Vergleich von Führungs- und Regelgrößen und ermöglicht eine erneute Festlegung bzw. Korrektur der Stellgrößen. Im einzelnen können folgende Zuordnungen getroffen werden:

- Führungsgrößen sind die von den Spielern wählbaren Ziele, wie z.B. gewünschte Liefersicherheit, Art und Umfang des Artikelsortiments,

Unabhängigkeitsüberlegungen, Angebot von Dienstleistungen, Kostenaspekte etc.

- Stellgrößen sind die möglichen Spielerentscheidungen; zu diesen gehören z.B. Bestimmung von Bestellmengen und -zeitpunkten, Festlegung der Sicherheitsbestände, Aushandlung von Preisen und Zahlungskonditionen,Auswahl der Lieferanten, Einsatz von Generica, Tätigen von Investitionen, Wahl des Inventurmodus, Personalpolitik, Entscheidungen über Eigen- oder Fremdbezug etc.
- Störgrößen werden erfahren in den Ausprägungen der komplexen Umwelt, sowohl hinsichtlich der Ausgangsbedingungen als auch ihrer Veränderungen; z.B. Nachfrageverhalten der Stationen mit Strukturbrüchen, Entwicklung neuer Medikamente, Lieferzeitschwankungen, Preisänderungen, Ausfall von technischen Geräten etc.
- Regelstrecke ist der materialwirtschaftliche Teil des Apothekenbetriebs,
- Regelgrößenverkörperndie Erreichungsgrade der Führungsgrößen undinder
- Kontrolle werden die Abweichungen zwischen beiden durch den Soll-Ist-Vergleich gemessen, um dann schließlich durch eine Korrektur der Stellgrößen eine Annäherung an die festgelegten Ziele zu erreichen.

Die Software des Planspiels ARKTIS setzt sich aus vier Hauptmoduln zusammen, die die Problemkreise Lagerhaltung, Eigenfertigung, Investition und Personaldisposition umfassen. Diese Moduln sind nicht starr miteinander verbunden, sondern die Verbindung wird über Parameter gesteuert. Damit können durch entsprechende Modulkombinationen unterschiedliche Spielkonstellationen geschaffen werden. Insgesamt, nachdem das Lagerhaltungsmodul als Kern des Planspiels in der Kombination immer enthalten sein muß, lassen sich acht Alternativen konstruieren, jede davon mit einem anderen Komplexitätsgrad.

Leider läßt sich der Komplexitätsgrad eines Planspiels - anders als der von Softwareprodukten [vgl. z.B. Roth/Lesshafft 1987] - nicht messen und vermutlich korreliert die Komplexität je nach Anzahl der kombinierten Moduln stark, was zu einer beschränkten Anzahl an Komplexitätsalternativen führt, die zudem in Bezug auf das Schwierigkeitsniveau der einzelnen Möglichkeiten kaum eingeschätzt werden können. Wünschenswert wäre jedoch die Existenz einer Komplexitätsfunktion die sicherstellt, daß jedem Komplexitätsgrad des Kontinuums eine entsprechende Modulkombination zugewiesen werden kann.

Erreicht werden kann dies durch unterschiedliche Abstraktionsniveaus auf Modulebene. Damit kann die Modulkomplexität flexibel festgelegt werden, was im Zusammenspiel mit den verschiedenen Kombinationsmöglichkeiten der Moduln ein weites Variationsspektrum ermöglicht. In diesem Spektrum ist für jede Spielergruppe/jeden Spielteilnehmer die optimale Spielkonstellation in Bezug auf den Komplexitätsgrad enthalten. Um einen größtmöglichen Lernerfolg sicherzustellen wäre es nun konsequent, die Spielsituation den Fähigkeiten und Fertigkeiten der Spielteilnehmer anzupassen. Allerdings dürfte zu Beginn eines Planspiels kaum ein gesichtertes Bild über das Wissen der Spieler vorliegen und zudem ist auch der Komplexitätsgrad einer Spielsituation lediglich ein Erfahrungswert. Insofern ist die funktionale Beziehung

optimaler Komplexitätsgrad = f(Kenntnisse)

nicht handhabbar. Einen Ausweg aus diesem Dilemma stellt die intensive Betreuung der Spielteilnehmer während des Spielablaufs durch entsprechend geschultes Personal dar. Diese Fachleute könnten den Spielern mit Rat zur Seite stehen sowie bei offensichtlicher Über- oder Unterforderung eine entsprechende Anpassung der Spielkomplexität vornehmen. Andererseits kann diese Aufgabe auch von einem Expertensystem wahrgenommen werden. In dessen Wissensbasis sind die möglichen Handlungsalternativen und deren Angemessenheit in spezifischen Spielsituationen unterzubringen. Dadurch kann das System erkennen, inwieweit die Anforderungen der momentanen Spielsituation auf die Belange der Spielteilnehmer zugeschnitten sind. Mit Hilfe der Wissenserwerbskomponente können die Spielerentscheidungen in die Wissensbasis aufgenommen und durch den Vergleich von aktuellen mit früheren Entscheidungen kann der Lernfortschritt ermittelt werden. Falls ein Lernfortschritt ausbleibt oder sich die gestellten Anforderungen als unangemessen erweisen (zu hoch oder zu niedrig), könnte die Regelkomponente schließlich die flexible Anpassung der Spielsituation und damit des Komplexitätsgrades sicherstellen. In Anbetracht der unrealistischen Annahme, jedem Spieler bzw. Spielerteam einen Experten als Betreuer zur Seite stellen zu können, erscheint die Integration von Planspiel und Expertensystem der einzig gangbare Weg zu sein, will man optimale Lernerfolge im Rahmen funktionaler Planspiele erreichen. Durch die Konzeption von ARKTIS als flexibles Planspiel ist der erste Schritt in diese Richtung getan.

Literatur:

Elgood Ch., Handbook of Management Games, Aldershot/Hunt 1984

Greenblat C.S., Sociological Theory and the "Multiple Reality" Game, in: Greenblat C.S., Duke R.D., Principles and Practices of Gaming-Simulation, Beverly Hills, London 1981, S. 79-91

Norris D.R., The External Validity of Business Games, in: simulation and games 17 (1986), S. 447-459

Roth Ch., Lesshaft K., Ein Verfahren zur Quantifizierung von Software-Komplexität, in: Angewandte Informatik, Nr. 6, 1987, S. 227-239

Wolfe J., The Teaching Effectiveness of Games in Collegiate Business Courses, in: simulation & games 16 (1985), S. 251-288

OR IM GESUNDHEITSWESEN —

LOVE'S LABOUR'S LOST ?

EINE KRITISCHE BESTANDSAUFNAHME

Manfred Meyer, Nürnberg

In einer 1982 (nach den umfangreichen bibliographischen Arbeiten von Brant Fries) abgeschlossenen Auswertung englisch- und deutschsprachiger Literatur hat Bernd Page insgesamt 154 OR-Modelle für das Gesundheitswesen dokumentiert. Und vermutlich könnte diese Zahl mehr als verdoppelt werden, wenn man heute und unter Einbeziehung weiterer Länder eine entsprechende Untersuchung durchführen würde. Dennoch ist von solchen Modellen nichts in der Planungspraxis von Leistungsträgern, Finanzierungseinrichtungen, Behörden und Interessenverbänden des Gesundheitswesens der Bundesrepublik angekommen. Damit sieht das Bild dort weitaus trostloser aus als etwa in verschiedenen Branchen des privatwirtschaftlichen Bereichs, dessen Abstinenz gegenüber OR hin und wieder ja auch beklagt wird.

Andererseits bedarf aber gerade das Gesundheitswesen der Bundesrepublik zur Verbesserung seiner Effizienz dringend praxisgerechter Planungsinstrumente. Deshalb ist zu fragen, was die Ursache für die fehlende Innovationsbereitschaft z.B. in den Krankenhäusern, bei den Krankenkassen oder in den Gesundheitsministerien der Bundesländer ist. Hier bieten sich mehrere Aspekte für eine genauere Diagnose an, wovon die für den Gesundheitsbereich relevanten Gesetzes- und Verordnungstexte sowie die Ausbildungs richtungen der Entscheidungsträger im Gesundheitswesen die wichtigsten sind.

Desweiteren ist aber auch zu fragen, in welcher Weise OR selbst zu einem Wandel beitragen kann. Dabei ist von den diagnostizierten Systemdefiziten auszugehen, was zu der Erkenntnis führt, daß traditionelle Formen der Vermittlung von OR- Inhalten, etwa Vorträge und Zeitschriftenaufsätze, hier kaum zur Therapie taugen. Eine erfolgversprechende Alternative hält OR dagegen mit einem seiner klassischen Instrumente, der Simulation, bereit: Es kann als erwiesen gelten, daß ein Planspiel- und das ist nichts anderes als ein Simulationsmodell- gut dazu geeignet ist, z.B. medizinischem und Pflegepersonal in Krankenhäusern die betriebswirtschaftlichen Konsequenzen seines Handelns einsichtig zu machen. Da liegt wiederum der Gedanke nahe, in solche Planspiele selbst dann auch andere OR-Methoden zu verpacken, etwa EOQ-Modelle und die lineare Programmierung. In der Planspielfamilie KLIMA (Klinikmanagement), ARKTIS (Arzneimittelversorgung im Krankenhaus, Training durch interaktive Simulation) sowie ASTERIKS (Ablaufsteuerung im Krankenhaus) ist/wird dies realisiert.

NEUERE MODELLE UND SOFTWARE ZUR LINEAREN REGRESSION

Helmuth Späth, Oldenburg

Die Beschreibung von Daten (gegeben A und b für unabhängige und die abhängige Variable) durch lineare Regressionsfunktionen (Parameter x) für modellverifizierende oder prognostische Zwecke ist auch bei OR-Anwendungen ein sehr häufig eingesetztes Hilfsmittel. Die klassische Methode der kleinsten (Residuen-)Quadrate $\min_x \|Ax-b\|_2^2$ liefert jedoch nur beim Erfülltsein gewisser Voraussetzungen, die a priori kaum jemals nachprüfbar sind, gute Anpassung. Flexibler ist das Modell $\min_x \|Ax-b\|_p^p$ $(1 \leq p \leq \infty)$. An real world Beispielen wird gezeigt, daß datenabhängig unterschiedliche Werte von p, häufig solche mit $p \to 1$, günstig sind. Dies trifft ebenfalls bei der Elimination von Ausreißern, bei der Variablenselektion und der klassenweisen Regression zu. Robuste, Ridge und Average Regression werden behandelt. Für $p = 1,2,\infty$ werden lineare Nebenbedingungen $Cx = d$ und $Ex \geq h$ einbezogen; dies beinhaltet den ökonomisch sehr wichtigen Spezialfall $x \geq 0$ mit nichtnegativen Parametern. Hat nicht nur die abhängige, sondern haben auch die unabhängigen Variablen Fehler, sind also b und die Spalten von A gleich zu behandeln, so ist die orthogonale Regression angemessen, bei der die Summe der p-ten Potenzen der Beträge der orthogonalen Residuen minimiert wird. Schließlich genügen zur Anpassung manchmal statt Hyperebenen niedriger dimensionale lineare Mannigfaltigkeiten. Zu allen Anpassungskriterien werden die numerischen Verfahren skizziert, zugehörige, auf dem IBM PC AT 02 entwickelte FORTRAN 77 Subroutinen und damit durchgerechnete real world Beispiele vorgestellt.

Literatur:

H. Späth (unter Mitarbeit von J. Meier): Mathematische Software zur linearen Regression. R. Oldenbourg: München 1987

DIFFERENT APPROACHES TO COVARIANCE STRUCTURE ANALYSIS: A COMPARISON*)

Wolfgang Gaul und Christian Homburg
Karlsruhe

Zusammenfassung: Eine Kovarianzstrukturanalyse von beobachtbaren (Indikator-)Variablen, die zur Beschreibung nicht beobachtbarer oder latenter Variablen erhoben wurden, kann Hinweise auf Kausalzusammenhänge zwischen den latenten Variablen liefern. Für diesen Problembereich existieren mittlerweile unterschiedliche Modellansätze und Softwarepakete.
Der vorliegende Artikel beschreibt und vergleicht wichtige Entwicklungen und zeigt eigene Forschungsrichtungen auf.
Ein Beispiel zur Verdeutlichung einiger der angesprochenen Gesichtspunkte und ein mathematisch orientierter Anhang sind beigefügt.

Summary:
Covariance structure analysis with respect to observable (indicator) variables, collected for the description of inobservable or latent variables, can yield hints concerning causal relationships with respect to the latent variables. Meanwhile, for this problem area, different model approaches and software packages exist.
This paper describes and compares important developments and points out own research directions.
An example for clarification of some of the aspects addressed and a mathematically oriented appendix are included.

1. Introduction

Covariance structure analysis, often referred to as causal analysis, has become a prominent tool of data analysis in various fields of research, e.g. in social and behavioral sciences, psychology and marketing. Important reasons for the increased use of this new methodology are the recognition that many variables of interest cannot be observed directly (instead, multiple indicators for such inobservable or latent variables should be used to reduce biasing effects of any single indicator, which the new methodology allows for) and the fact that, meanwhile, different software tools are available for handling problems of the kind mentioned.
Historically, covariance structure modeling can be understood as a synthesis of confirmatory factor analysis and simultaneous equation modeling. These techniques were developed independently in the fields of psychometrics and econometrics and remained relatively unaware of each other until the 1960s (see e.g. Gaul and Homburg (1987) for a description of the historical background of this methodology).

*) Research has been supported by the Deutsche Forschungsgemeinschaft.

Operations Research Proceedings 1987

In most applications of covariance structure analysis, models are formulated in the frame of Jöreskog's model (see e.g. Jöreskog (1973, 1977)) and analyzed by either a ML (Maximum Likelihood) approach (based on the assumption of normally distributed variables) or a ULS (Unweighted Least Squares) approach both of which are available in the well-known LISREL (see e.g. Jöreskog and Sörbom (1984)) computer program. Besides ML- and ULS-approaches there is a class of GLS (Generalized Least Squares) approaches some of which are flexible in handling more general distribution models than those based on multivariate normality assumptions.

This paper gives a non-mathematical description of models and methods of covariance structure analysis and points out own research directions, focusing on comparative aspects some of which are demonstrated by means of a practical example.

Some more technical features of the methodology are given in a mathematically oriented appendix.

2. Description of Models and Methods: Comparative Aspects

Among the benefits of covariance structure analysis models, at least, the following two are worth mentioning: First, as mentioned in the introduction, the methodology allows to achieve a high quality in the measurement of relevant variables by explicitly taking into account measurement error and by using multiple indicators. Second, covariance structure modeling allows for the analysis of complex causal structures involving distinctions between direct and indirect effects or reciprocal causal relations between variables.

The flexibility within covariance structure modeling is documented by the fact that simultaneous equation models and, thus, standard regression models are special cases (see Table 1(a)).

In the following, we will describe a covariance structure model M in the form

$$M: \Sigma = \Sigma(\alpha) \qquad (1)$$

where Σ is the covariance matrix with respect to model M, based on p observed variables and expressed as a function of $t \leq p(p+1)/2$ parameters $\alpha = (\alpha_1,\ldots,\alpha_t)$. Usually, a model formulation as described by (1) is derived from a structural model given by a number of (linear) equations involving inobservable or latent as well as observed variables of the model.

The three most popular structural models are

- the confirmatory factor analysis model (see e.g. Jöreskog (1967));
- the Jöreskog model, sometimes referred to as "LISREL MODEL" (see Jöreskog (1973)), and
- the Bentler/Weeks model (see Bentler and Weeks (1979)).

The relations between these structural models are exhibited in Table 1(b), the mathematical formulations as well as corresponding covariance structures are given in the appendix (for a general moment structure model developed by Bentler (1983) see also appendix A.1.).

Table 1: Covariance Structure Analysis: Comparative Aspects*)

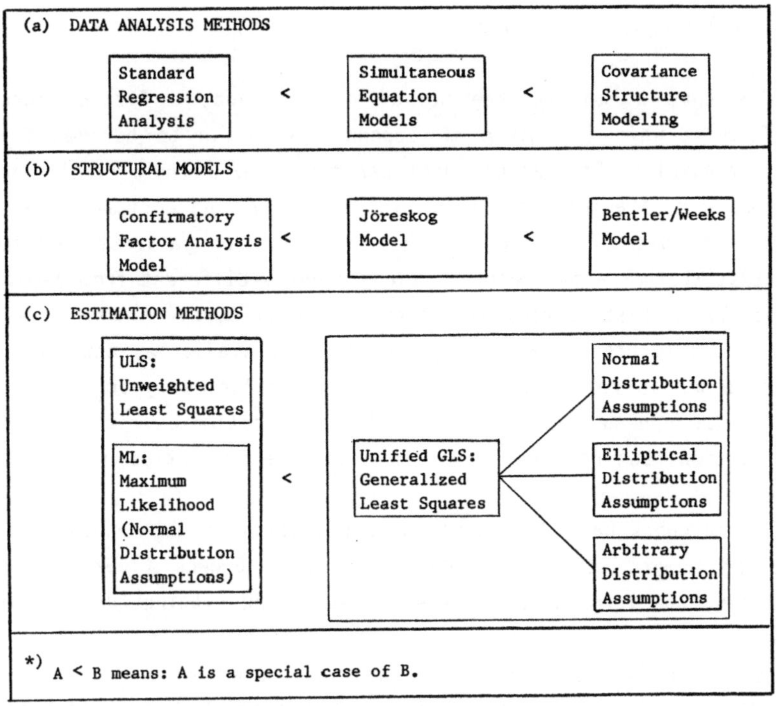

Table 2: Classification of GLS-Methods

Weight Matrix	Distribution Assumptions: Normal (N)	Elliptical (E)	Arbitrary (A)
Identity Matrix	NLS: Normal Least Squares	ELS: Elliptical Least Squares	ALS: Arbitrary Least Squares
Optimal Fixed Matrix	NGLS: Normal Generalized Least Squares	EGLS: Elliptical Generalized Least Squares	AGLS: Arbitrary Generalized Least Squares
Optimal Reweighted Matrix	NRLS: Normal Reweighted Least Squares $\sim$ ML	ERLS: Elliptical Reweighted Least Squares	ARLS: Arbitrary Reweighted Least Squares

Suppose, S denotes the given sample covariance matrix.
Then, the estimation problem consists in providing an estimate $\hat{\alpha}$ in such a way that $\hat{\Sigma} = \Sigma(\hat{\alpha})$ is - in some sense - close to the sample covariance matrix S (see e.g. Jöreskog (1977) for a discussion of problems concerning model identification). Usually, $\hat{\alpha}$ is obtained by iteratively minimizing some function $F(S, \Sigma)$ associated with the discrepancy between S and Σ, i.e.

$$F_S(\alpha) = F(S, \Sigma(\alpha)) \rightarrow \min. \tag{2}$$

If variables have a multivariate normal distribution, a fitting function, the specific form of which is given in appendix A.2, can be derived from maximum likelihood principles (Estimating model parameters by maximum likelihood has - as is well-known - desirable properties, e.g. the resulting estimates are asymptotically efficient, standard errors can be estimated for parameter estimates which allows to test whether a parameter is significantly different from zero, a measure (asymptotically χ^2-distributed) for testing the adequacy of the model used for representing the structure underlying the data is available, and so on. For details, the reader is referred to a whole set of articles by Jöreskog.). The ULS-method (see also appendix A.2.) yields consistent estimates without any distributional assumptions but does not have the desirable properties mentioned above for the ML-case.
Recently, researchers have discussed the application of GLS-methods to covariance structure and moment structure models (see e.g. Browne (1982, 1984), Bentler (1983)). Again, the formulation of the corresponding fitting function is given in appendix A.2.
GLS-methods can be classified into approaches based on normal, elliptical or arbitrary distribution theory. ULS and ML (under normal distribution assumptions) are special cases of GLS (see Table 1(c)).
A more detailed classification of GLS-methods is given in Table 2.
All methods in Table 2 yield consistent estimates under some very general regularity conditions (see Browne (1984), proposition 1). The methods in the first row of Table 2 (ordinary least squares methods) lead to estimates which are in general not efficient.
When methods as shown in the second row of Table 2 are used, optimal (in the sense of producing estimates with minimum asymptotic variances) fixed weight matrices are computed from the input data. In the case of normal or elliptical distribution assumptions these matrices can be computed from the second-order sample moments, under arbitrary distribution assumptions, additionally, fourth - order sample moments are needed.
The methods shown in the last row of Table 2 use optimal reweighted matrices. Here, explicitly, some kind of functional relationship between these matrices and the

model parameters is considered, and an updating of the weight matrices takes place within each iteration step of the fitting procedure, i.e. the matrices are re-weighted whenever the model parameter estimates are improved (see e.g. Bentler (1985) for details).

3. Example

As an example, an investigation concerning the causal structure between factors affecting the acceptance of computers is used. This problem was addressed because of empirical evidence (see e.g. Lange (1984), Müller-Böling and Müller (1986)) that "in the FRG attitudes towards computers are more negative than in the USA, the UK or France" and that "negative attitudes towards computers are part of a growing technology-future-pessimism, to which expectations of negative impacts of computer use contribute" (Lange (1984), p. 4).
In view of the high degree of complexity of the phenomenon to be explained, basic bivariate techniques as well as standard regression methods do not seem to be appropriate tools for analyzing such a problem so that the investigation was done by covariance structure modeling.
Data were collected from more than 500 students at Karlsruhe University. Because of page restrictions only some important features of the example can be described. The latent variables of the model are listed in Table 3.

Table 3: Latent Variables of Computer Acceptance Model

Variable	Meaning	Endogenous Variable	Exogenous Variable	Number of Indicator Variable
AT	Acceptance of Technological Innovations		X	2
E	Amount of Experience in Working with Computers		X	2
LM	Belief that Computers have negative Impacts on the Labour Market	X		2
WP	Belief that Computers have negative Impacts on the Working Place	X		4
AC	Acceptance of Computers	X		2

A variable is considered as endogenous if it is assumed to be causally affected by any other latent variable in the model.

The structural model shown in Figure 1 can be derived from substantive theory (see e.g. Lange (1984)). Here, $x \xrightarrow{+} y$ ($x \xrightarrow{-} y$) means that a positive (negative) impact of x on y is hypothesized.
With regard to confirmatory viewpoints, one can check whether the model of Figure 1 provides a satisfactory fit to the data.
With regard to exploratory viewpoints, one can search for a "best" model for fitting the given data (of course, restrictions based on known results from substantive theory can be taken into consideration). The model exhibited in Figure 1 was found to be the best model out of a set of more than 50 models by means of a cross-validation study (see Gaul and Homburg (1987)).

Figure 1: Specification of Relations Between Latent Variables of Computer Acceptance Model

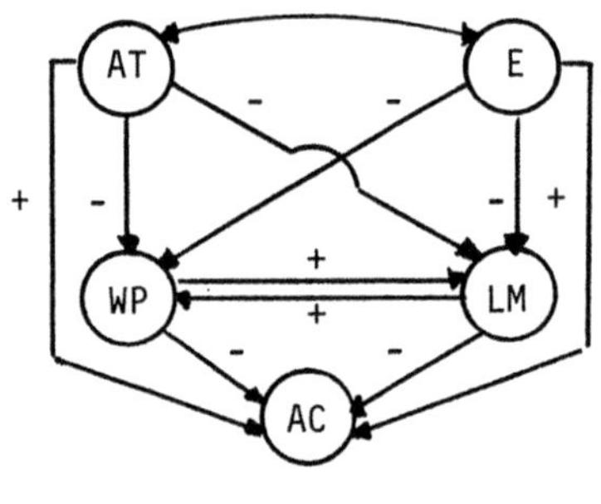

Table 4: Estimates for Causal Parameters (Standardized Results in Parentheses)

Parameters	ULS-estimates	AGLS-estimates
AT → LM	-.251 (-.257)	-.253 (-.265)
AT → WP	-.391 (-.432)	-.416 (-.442)
AT → AC	.260 (.271)	.272 (.254)
E → LM	-.087 (-.107)	-.174 (-.232)
E → WP	-.135 (-.178)	-.094 (-.127)
E → AC	.402 (.503)	.289 (.344)
LM → WP	.182 (.196)	.164 (.166)
LM → AC	-.113 (-.115)	-.115 (-.102)
WP → LM	.358 (.333)	.294 (.290)
WP → AC	-.419 (-.395)	-.471 (-.415)

In this example, it was found that the distribution of the observed variables deviated considerably from both multivariate normal and multivariate elliptical distributions. Therefore, estimation was performed by ULS (see Table 1) as well as by AGLS (see Table 2). For the AGLS-approach, the new computer programm EQS (see Bentler (1985)) was used.
Table 4 shows the results of the two different estimation methods. Both the ULS- and the AGLS-method yield estimates the signs of which are consistent with the directional hypotheses shown in Figure 1. It is also worth noting that the different estimates do not deviate considerably in magnitude. Moreover, both estimation methods indicated an excellent fit of the model to the data.
From a substantive point of view a distinction between direct, indirect and total effects (see Jöreskog and Sörbom (1984) for details) is of interest. Table 5 presents the total effects (direct and indirect effects) between latent variables.

Table 5: Total Effects (computed from unstandardized ULS-estimates)

on \ from	AT	E	LM	WP
LM	-.418	-.144	.070	.383
WP	-.467	-.161	.194	.070
AC	.503	.486	-.203	-.491

In Table 4, it can be seen that experience (E) has a stronger direct effect on the acceptance of computers (AC) than the acceptance of technoglogical innovations (AT). Yet, as the latter variable has stronger indirect impacts, the total effects are almost equally strong (see Table 5). Another result of interest is that beliefs concerning negative impacts of computers on the working place (WP) play a more important role than corresponding beliefs concerning the labour market (LM) which is true for both, the direct and the indirect effects. It is worth mentioning that the model we are presenting explains 99,8% of the variance of the variable AC (acceptance of computers).

4. Own Contributions and Conclusions

In this paper we gave a non-technical description of some important features of covariance structure analysis focusing on comparative aspects concerning data analysis methods, structural models, and estimation methods. Furthermore, part of the results of an empirical causal investigation of factors affecting the acceptance of computers were presented.

Own contributions have addressed the exploratory evaluation of covariance structure models which has experienced an increasing amount of interest in recent literature (see e.g. the references in Gaul and Homburg (1987)). Such exploratory evaluations may be based on modification indices, critical ratios, Lagrange multipliers and Wald tests or on general criteria which are not associated with a specific estimation method. We found some very encouraging results concerning the different methods' ability of detecting substantively meaningful models that fit the data in a satisfactory way (see Gaul and Homburg (1987) for more details). It goes without saying that a model that stands at the end of such a data-driven search process has not been "confirmed" in any real sense and that the exploratory approach to covariance structure modeling is not to be understood as an alternative to confirmatory analysis. In view of the fact, however, that in the social sciences, phenomena to be explained seldom seem to be so well understood that only one model may be derived from substantive theory, it seems that the exploratory evaluation of causal models is a very important supplement to the confirmatory approach.

A. Appendix

A.1. Structural Models, Covariance and Moment Structures

Confirmatory Factor Analysis Model

Model: $x = \Lambda\xi + \delta$

Notations:

- x: observed variables
- ξ: latent variables — ϕ: covariance matrix of ξ
- δ: residual variables — Θ_δ: covariance matrix of δ
- Λ: matrix of factor loadings

Covariance Structure: $\Sigma_x = \Lambda\phi\Lambda' + \Theta_\delta$
(of the form $\Sigma = \Sigma(\alpha)$)

Jöreskog Model

Model:

$$\eta = B\eta + \Gamma\xi + \zeta$$
$$y = \Lambda_y\eta + \varepsilon$$
$$x = \Lambda_x\xi + \delta$$

Notations:

- η: endogenous latent variables
- ξ: exogenous latent variables — ϕ: covariance matrix of ξ
- y:, x: observed variables
- ε:, δ:, ζ: residual variables — Θ_ε: covariance matrix of ε; Θ_δ: covariance matrix of δ; Ψ: covariance matrix of ζ
- B:, Γ: matrices of causal parameters — $\dot{B} = (I - B)^{-1}$

Covariance Structure:
(of the form $\Sigma=\Sigma(\alpha)$)

$$\Sigma_{\binom{y}{x}} = \left[\begin{array}{c|c} \Sigma_{yy} & \Sigma_{yx} \\ \hline \Sigma_{xy} & \Sigma_{xx} \end{array}\right]$$

$$= \left[\begin{array}{c|c} \Lambda_y\dot{B}(\Gamma\phi\Gamma' + \Psi)\dot{B}'\Lambda_y' + \Theta_\varepsilon & \Lambda_y\dot{B}\Gamma\phi\Lambda_x' \\ \hline \Lambda_x\phi\Gamma'\dot{B}'\Lambda_y' & \Lambda_x\phi\Lambda_x' + \Theta_\delta \end{array}\right]$$

Bentler/Weeks Model

Model: $\eta = \beta\eta + \gamma\xi$

Notations:

- η: dependent variables
- ξ: independent variables — ϕ : covariance matrix of ξ
- z: observed variables (among dependent and independent variables)
- β, γ: matrices of causal parameters — $B = \begin{pmatrix} \beta & 0 \\ 0 & 0 \end{pmatrix}$, $\Gamma' = (\gamma', I)$
- G: known selection matrix

Covariance Structure: $\Sigma_z = G(I - B)^{-1}\, \Gamma\phi\Gamma'(I - B)'^{-1}\, G'$

(of the form $\Sigma=\Sigma(\alpha)$)

General Moment Structure Model

Model: $z = A\xi$

Notations:

- ξ: independent variables
- z: observed variables
- A: parameter matrix
- $\boxtimes$: right Kronecker product
- ϕ_r, Σ_r: r-th order multivariate moment of ξ, z (respectively)

Moment Structure: $vec(\Sigma_r) = (\boxtimes_r A)\ vec(\phi_r)$

A.2. Estimation Methods

Method	Notations
Maximum Likelihood (normal theory)	
$F_{ML} = tr(\Sigma^{-1}S) - \ln\lvert\Sigma^{-1}S\rvert - p$	Σ : covariance matrix generated by a model
Unweighted Least Squares	S : sample covariance matrix
$F_{ULS} = 1/2\ tr[(S - \Sigma)^2]$	tr: trace operator
Unified Generalized Least Squares	
$F_{GLS} = (s - \sigma)'W(s - \sigma)$	s : vector of data to be modeled σ : corresponding values generated by a model W : weight matrix

References:

Bentler, P. M. (1983), Some Contributions to Efficient Statistics for Structural Models: Specification and Estimation of Moment Structures, Psychometrika, 48, 493 - 517.

Bentler, P. M. (1985), Theory and Implementation of EQS, A Structural Equations Program, Los Angeles.

Bentler, P. M and D. G. Weeks (1979), Interrelations Among Models for the Analysis of Moment Structures, Multivariate Behavioral Research, 14, 169 - 185.

Browne, M. W. (1982), Covariance Structures, in: Hawkins, D. M. (ed.): Topics in Applied Multivariate Analysis, Cambridge.

Browne, M. W. (1984), Asymptotically Distribution - Free Methods for the Analysis of Covariance Structures, British Journal of Mathematical and Statistical Psychology, 37, 62 - 83.

Gaul, W. and Ch. Homburg (1987), Marketing Research by Causal Modeling: Its Use as an Exploratory Method, Working paper, University of Karlsruhe.

Jöreskog, K. G. (1967), Some Contributions to Maximum Likelihood Factor Analysis, Psychometrika, 32, 183 - 202.

Jöreskog, K. G. (1973), A General Method for Estimating a Linear Structural Equation System, in: Goldberger, A. S. and O. D. Duncan (eds.): Structural Equation Models in the Social Sciences, New York.

Jöreskog, K. G. (1977), Structural Equation Models in the Social Sciences: Specification, Estimation and Testing, in Krishnaiah, P. R. (ed.): Applications of Statistics, Amsterdam.

Jöreskog, K. G. and D. Sörbom (1984), LISREL VI: Analysis of Linear Structural Relationships by Maximum Likelihood, Instrumental Variables and Least Squares Methods, Chicago.

Lange, K. (1984), Das Image des Computers in der Bevölkerung, St. Augustin.

Müller-Böling, D. and M. Müller (1986), Akzeptanzfaktoren der Bürokommunikation, München.

Absatzplanung durch Integration von Prognoseverfahren und empirischer Planung durch ein entscheidungsunterstützendes System

Wolfgang Martin
Darmstadt

Hochwertige Konsumgüter unterliegen in starkem Maße der Mode (z. B. "Designer-Qualität"). Die Absatzplanung solcher Konsumgüter birgt das Problem, daß die Produkt-Komponenten, die das modische Design wiederspiegeln, mathematisch sehr schwierig quantifizierbar sind. Das hat in vielen Fällen verhindert, daß brauchbare Prognosen mit Hilfe rein mathematisch-statistischer Prognoseverfahren ermittelt werden konnten.

Im Falle eines Unternehmens aus dem süddeutschen Raum wird gezeigt, wie man ein System zur Absatzplanung aufbauen kann, das mittels eines entscheidungs- unterstützenden Systems (hier System W von Comshare) Techniken von Standard-, Prognose- und -Hochrechnungsverfahren integriert mit dem Wissen von Experten aus der rein empirischen Planung heraus.

Die mathematisch-statistischen Verfahren sind unterschiedlich für Neu- und Altprodukte. Bei Neuprodukten - die Einführung neuer Produkte erfolgt z. Z. viermal pro Jahr - wird eine Hochrechnungstechnik als Prognose angewendet, die sich auf die einlaufenden Erst-Auftragseingänge stützt. Dieser Parameter spiegelt die produkttypische Markteinführungskurve wieder. Bei laufenden Alt-Produkten wird ein exponentielles Glätten unter saisonalen und Langzeittrend-Einflußfaktoren eingesetzt.

Zur Ermittlung der Parameter für die jeweiligen Prognoseverfahren wird in einer Datenanalyse eine Clusterung in homogene Gruppen pro Trend, Saison und Markteinführungskurve durchgeführt. Es hat sich gezeigt, daß Design-Merkmale der Produkte in Abhängigkeit vom jeweiligen Markt in dieser Klassifizierung sich wiederspiegeln. Die Clusterung wird in beliebigen Zeitabständen wiederholt, um die Klassifizierung der Produkte in homogene Gruppen aktuell dem Marktgeschehen anzupassen. Für die Experten bleibt also die Aufgabe, neue Produkte in das gefundene Raster einzuordnen aufgrund des Designs des jeweiligen Produktes. Die Datenanalyse dient so als Entscheidungsunterstützung für den Quantifizierungsprozeß. Auf Basis dieser Zuordnung kann ein maschineller Planvorschlag errechnet werden. Der Planvorschlag wird dann nochmals von den Marktexperten geprüft, die so markttypische externe Einflüsse noch korrigierend in den Absatzplan einbringen können. Hier ist das Prognoseergebnis die Entscheidungsunterstützung für den Marktexperten.

Der Einsatz des entscheidungsunterstützenden SYSTEM W ermöglicht die Integration der mathematisch-statischen Verfahren mit dem empirischen Expertenwissen in einer DV-Anwendung. Die speziellen Konstrukte von SYSTEM W erlauben, die Anwendung als "white box" für die Endanwendung zur Verfügung zu stellen, so daß ein interaktives Arbeiten mit dem Absatzplanungssystem, seinen Daten und seinen Strukturen ein endbenutzergerechtes Arbeiten wird.

Dieses Verfahren wird für ausgewählte Testmärkte angewendet. Die Ergebnisse werden dann nach einem Schlüssel auf das Konzernergebnis hochgerechnet und nach einem weiteren Schlüssel den Produktvarianten zugeordnet. So erhält die Logistik brauchbare Unterlagen für die Produktionsplanung.

Box-Jenkins Analysis of Air Pollution Data

Herbert Stahl, Mikael Weigelt, and Götz Wiegand; Berlin

Zusammenfassung: In einem Netz von Meßstellen, das sich über die Bundesrepublik Deutschland erstreckt, werden kontinuierlich Indikatoren der Luftqualität, wie z.B SO_2, NO_2, O_3 und Staub Konzentrationen gemessen.

1) Für Reihen aus diesem Meßnetz werden univariate und multivariate statistische Modelle nach der Box-Jenkins Methode gebildet.

2) Es wird gezeigt, daß sog. Standard-Modelle existieren, die ohne Strukturveränderungen auf ganze Klassen von Zeitreihen anwendbar sind, wodurch bei weiteren Analysen der Arbeitsaufwand ganz erheblich reduziert werden kann.

3) Bei den multivariaten Modellen werden verschiedene Ansätze (Transferfunktions-, ARMAX- und multiple Regressions-Modelle) miteinander verglichen, und es wird untersucht, ob und in welchem Grade mit steigender Komplexität auch die Erklärungsfähigkeit der Modelle zunimmt.

Summary: In a network of observation stations throughout the Federal Republic of Germany, SO_2, NO_2, O_3, and dust concentrations are continuously measured.

1) For a sample of time series from these measurements we fit univariate and multivariate Box-Jenkins models.

2) It is shown that so-called standard models exist, which have a good fit for all members of a whole class of time series. The knowledge of such models can shorten the model finding process considerably in further analyses.

3) In the multivariate case we compare different types of models (transfer function, ARMAX-, and multiple regressions models).

Keywords: Box-Jenkins Method, Transfer Function Models, Air Pollution.

A. Aims and Purpose of the Investigation

At different places throughout the Federal Republic of Germany indicators of air pollution, as for instance SO_2, NO_2, O_3 concentrations, or dust deposition are continuously measured. Based on a sample of time series from this measuring network, we have investigated and evaluated different types of statistical models. The aim was to find methods that are best suited for

- modelling monthly, daily, and hourly averages of measurements,
- fitting missing values in these time series,

Operations Research Proceedings 1987

- explaining connections between meteorological variables and air pollution data,
- forecasting or at least assessing the forecastibility of the considered series.

The investigation is part of a research project titled "Methoden zur Bewertung der Luftqualität" (Methods for Evaluating Air Quality Data), which has been commissioned by the Umweltbundesamt of the Federal Republic of Germany.

In the present paper we shall report about the modelling of monthly and daily data by the univariate and multivariate Box-Jenkins method. In the multivariate case we consider transfer function and regression models, i.e. models with only one output series and several input series, which in our case will be meteorological variables, as solarflux, temperature, wind intensity and direction, and dummy variables indicating specific weather situations.

The main aim of the paper is to communicate practical experiences with the Box-Jenkins technique in the area of ecological data. For a sample of monthly and daily averages of pollution measurements, we present concrete models. The percentage of variance explained by these models allows us to assess the forecastibility of the considered series.

A result of remarkable practical relevance are the so-called standard models. Since the fitting of Box-Jenkins models is a rather laborious task, the work can be speeded up considerably if there exists a standard model, which can be fitted to all series with a similar background and structure.

Finally, we report upon an empirical comparison of multivariate models of different degrees of complexity. We compare Box-Jenkins transfer function, ARMAX-, and multiple regression models. Limitations of space force us to be very restrictive in the selection of material. However, we hope that we are able to give an idea of the flavor and the type of results obtained in the whole investigation. A more detailed report can be found in [WWS1-4].

The outline of the paper is as follows: In the next section we review the models used in the Box-Jenkins method. This section is very condensed and can by no means be understood as an introduction to this technique. (For such an introduction we refere to [BoJe2] or [NGC]). In Section C we present results from the univariate analysis of monthly data, including the standard model for this type of data, in Section D we give analogous results for daily data, and finally in Section E we come to multivariate models.

B. ARMA- and SARIMA-Models

A basic building stone of the Box-Jenkins method is the so-called ARMA-(autoregressive-moving-average)-model, which is defined by the model equation

(B-1) $$X_t = \phi_1 X_{t-1} + \dots + \phi_p X_{t-p} + c + Z_t - \theta_1 Z_{t-1} - \dots - \theta_q Z_{t-q},$$

where the error term Z_t is assumed to be white noise. i.e. to consists of stochastically independent and identically distributed random variables. In order to simplify and shorten the notation, it is usual to introduce a so-called back-shift operator B, which moves a time series one step backwards, and is formally defined by

(B-2) $$BX_t := X_{t-1}, \quad t = 1,2,\dots$$

With its help we can rewrite the model equation (B-1) as

$$\text{(B-3)} \qquad (1 - \phi_1 B - \dots - \phi_p B^p) X_t = c + (1 - \theta_1 B - \dots - \theta_q B^q) Z_t.$$

which is the more common form in time series analysis. In short form this model is denoted by ARMA(p,q). (For a more detailed explanation of the ARMA class and related stochastic processes we refere to [NGC; Ch.VI]).

ARMA-models can be fitted to time series only if the series are stationary (or at least weakly stationary, cf.[[NGC; Ch.II]). However, most time series with an economical or ecological background have either a trend or a seasonal component or some other form of non-stationary behavior. In order to make the theory work for such series, Box and Jenkins introduced two extensions of the basic model (B-3), namely (i) differentiation and (ii) additional seasonal autoregressive and moving-average model factors. The extended models are called SARIMA- (seasonal-autoregessive-integrated-moving-average)-models, and they are defined by the model equation

$$\text{(B-4)} \qquad (1 - \phi_1 B - \dots - \phi_p B^p)(1 - \Phi_1 B^L - \dots - \Phi_P B^{LP})(1-B)^d (1 - B^L)^D X_t =$$
$$= (1 - \theta_1 B - \dots - \theta_q B^q)(1 - \Theta_1 B^L - \dots - \Theta_Q B^{LQ}) Z_t.$$

where again the error term Z_t is assumed to be white noise. The operations

$$\text{(B-5)} \qquad (1 - B) X_t = X_t - X_{t-1} \quad \text{and} \quad (1 - B^L) X_t = X_t - X_{t-L},$$

are called differentiation (ordinary and seasonal), and their reversion. the integration, is used in the naming of the model class. In short form the models (B-4) are denoted by SARIMA(p,d,q;P,D,Q;L), or ARIMA(p,d,q) if no seasonality is present. For reasons of simplicity we have dropped the constant term in (B-4). Of course, in every individual case it has to be checked wether or not a constant is necessary.

To complete the model class, we have to introduce a last building stone: the Box-Cox transformation. It depends on a parameter λ and can be defined as

$$\text{(B-6)} \qquad Y_t := \begin{cases} X_t^{\lambda} & \text{for } \lambda \neq 0 \\ \ln(X_t) & \text{for } \lambda = 0. \end{cases}$$

Most typical values for λ are $\lambda = 1$ (no transformation), $\lambda = 1/2$ (square root), $\lambda = 0$ (logarithm), and $\lambda = -1/2$ (reciprocal of a square root).

With the Box-Cox transformation all ingredients of the univariate Box-Jenkins method have been introduced. A second characteristic feature of the method is the organization of the model finding process. We will not elaborate on this, and refer to [NGC; Ch.X,XI]. In any case the model finding process causes the main work of the analysis, and sometimes it can really be very laborious. Maybe, this is one of the reasons, why applied researchers often hesitate to use the Box-Jenkins approach to time series analysis. But on the other hand, the procedure guaranties an optimal model within the SARIMA class.

C. Models for Monthly Data

In Table1 we give SARIMA-models that have been fitted to 10 time series of monthly averages of pollution measurements. The percentage of variance explained by every model is given in column 4 of Table 2. The length of these series varies from 132 to 400 months; the longest ones are from dust deposition measurements. In these latter two series there are missing values, which have been caused by interruptions in the measurement process. The gaps in the series have been filled by an iterative procedure based on forecasts by SARIMA-models. (For details see [WWS3]). The proportion of missing values will grow in the series of daily averages, which will be investigated in the next section.

Table 1

Results of univariate Box-Jenkins analysis for 10 time series of monthly averages

Series	Observations Missing Values	SARIMA-Model
O_3 Hohenpeissenbg Jan.71 - Feb.85	170 0	$(1-B^{12})(1-0.39\,B)\sqrt{X_t} = 0.003 + (1-0.95\,B^{12})Z_t$
SO_2 Waldhof Jan.73 - Dez.83	132 0	$(1-B)(1-B^{12})\ln(X_t) = (1-0.76\,B)(1-0.92\,B^{12})Z_t$
NO_2 Waldhof Jan.73 - Dez.83	132 0	$(1-B)(1-B^{12})\ln(X_t) = (1-0.83\,B)(1-0.94\,B^{12})Z_t$
SO_2 Schauinsland Jan.73 - Dez.83	132 0	$(1-B^{12})\sqrt{X_t} = -0.048 + (1-0.94\,B^{12})Z_t$
NO_2 Schauinsland Jan.73 - Dez.83	132 0	$(1-B^{12})(1-0.25\,B^4)\sqrt{\frac{1}{X_t}} = (1-0.93\,B^{12})Z_t$
SO_2 Deuselbach Jan.73 - Dez.83	132 0	$(1-B^{12})(1-0.24\,B^2)\ln(X_t) = (1-0.93\,B^{12})Z_t$
NO_2 Deuselbach Jan.73 - Dez.83	132 0	$(1-B^{12})(1-0.42\,B)(1+0.22\,B^{12})\sqrt{\frac{1}{X_t}} = (1-0.94\,B^{12})Z_t$
SO_2 Gelsenkirchen Jan.61 - Dez.80	240 0	$(1-B)(1-B^{12})(1-0.37\,B)\sqrt{\frac{1}{X_t}} = (1-0.94\,B)(1-0.93\,B^{12})Z_t$
Dust Gelsenkirchen Okt.56 - Sep.81	300 6	$(1-B)(1-B^{12})\ln(X_t) = (1-0.88\,B)(1-0.97\,B^{12})Z_t$
Dust Berlin Aug.52 - Nov.85	400 4	$(1-B)(1-B^{12})\ln(X_t) = (1-0.95\,B)(1-0.96\,B^{12})Z_t$

Although the 10 models in Table1 vary considerably in structure, it nevertheless turns out that the model

$$\text{(C-1)}\qquad (1 - B)(1 - B^{12}) \ln (X_t) = (1 - \theta_1 B)(1 - \Theta_1 B^{12}) Z_t$$

can be fitted rather well to all 10 series. Because of its general applicability it is called 'standard model'. Column 3 of Table 2 contains the percentage of variance explained by this model. The numbers show that the loss of performance is really minor if one takes the standard model instead of the optimally fitted one. An examination of other indicators, as for instance the autocorrelation function of the model residuals, show the same positive result.

Table 2

Coefficients and percentage of explained variance in the 'standard model' together with the percentage of variance explained by the optimally fitted model

Series		MA-Coefficients regular	MA-Coefficients seasonal	R^2 [%]	Optimal R^2 [%]
O_3	Hohenpeissenberg	0.83	0.95	82.5	84.7
SO_2	Waldhof	0.76	0.92	60.7	60.7
NO_2	Waldhof	0.83	0.94	84.5	84.5
SO_2	Schauinsland	0.89	0.94	37.1	46.5
NO_2	Schauinsland	0.89	0.94	24.2	24.6
SO_2	Deuselbach	0.83	0.94	49.1	50.7
NO_2	Deuselbach	0.72	0.93	65.6	67.9
SO_2	Gelsenkirchen	0.68	0.84	74.4	75.5
Dust	Gelsenkirchen	0.88	0.97	50.8	50.8
Dust	Berlin	0.95	0.96	17.6	17.6

Only the structure of the standard model is kept fixed, while the two parameters θ_1 and Θ_1 are re-estimated for every new time series. The estimated values are given in Table 2. They show that there is quite a variation. Of course, the use of a standard model implies a large reduction of work since parameter estimation is a purely numerical job that can be obtained by cheap computing.

Historically the Box-Jenkins method has superseded a number of special time series models, which were mostly connected with special purposes, as for instance smoothing or forecasting. The advantage of the new method was a systematic way for model selection (cf. the discussion in [BoJe1]). But this new philosophy does not exclude to look for models that can satisfactorily be fitted to a whole class of time series from a certain area of application. Moreover, the method allows to evaluate such a model in the class of all SARIMA-models, as has been done in Table 2. Of course, the set of series, on which the standard model is tested, must be sufficiently representative for the whole class.

D. Models for Daily Data

In Table3 we have assembled results from seven univariate Box-Jenkins models for daily averages. To keep the table small, the models are given only by their general structure. A full account can be found in Table3.1 of [WWS3]. In comparison to monthly averages, daily averages show a higher degree of random oscillation (compare the R^2 values of Table3 and Table2). In the next section we shall see that this phenomenon can be compensated by using multivariate models with weather data as external variables.

Two of the seven series in Table3 show a weak, but significant seasonal behavior. Seasonality means here a weekly pattern (L = 7).

Table 3

Results of the univariate Box-Jenkins analysis of daily averages

Series	Time Interval	Obs.	Missing Values	Model	Transf. λ	R^2 [%]
SO_2 Waldorf	Jan.81-Dez.83	1064	31	ARIMA(1,1,1)	0	38.3
SO_2 Gelsenkirchen	Jan.78-Dez.80	968	128	ARIMA(1,1,1)	0	51.4
O_3 Hohenpeissenbg	Jan.81-Dez.83	1084	11	ARIMA(3,1,1)	1	69.1
O_3 Deuselbach	Jan.81-Dez.83	1074	22	ARIMA(1,1,1)	1/2	69.8
O_3 Frankfurt	Jan.81-Dez.83	806	289	SARIMA(1,1,1;1,0,0;7)	1/2	68.1
SO_2 Frankfurt	Jan.81-Dez.83	849	246	ARIMA(1,1,1)	0	55.8
NO_2 Frankfurt	Nov.82-Okt.85	881	215	SARIMA(1,1,2;0,0,1;7)	0	59.3

In case of daily averages we can investigate time series that are considerably longer than the monthly ones. As a consequence we are confronted with the question of time-invariance of the models. Besides of the models given in Table3, we have also fitted models to earlier time intervals. All results are documented in Table3.1 of [WWS3].

Table 4

Coefficients and percentage of explained variance in the 'standard model' together with the percentage of variance explained by the optimally fitted model

Series	RA- coefficients	MA- coefficients	R^2 [%]	Optimal R^2 [%]
SO_2 Waldorf	0.52	0.95	38.3	38.3
SO_2 Gelsenkirchen	0.60	0.97	51.4	51.4
O_3 Hohenpeissenberg	0.60	0.94	68.0	69.1
O_3 Deuselbach	0.64	0.94	69.6	69.8
O_3 Frankfurt	0.45	0.89	66.4	68.1
SO_2 Frankfurt	0.46	0.91	55.8	55.8
NO_2 Frankfurt	0.66	0.95	54.3	59.3

It turns out that there exist small variations of the coefficients. but the model structure remain invariant, and the shift of the coefficients over time is negligible in comparison to the differences that exist between the seven time series of Table3.

For the univariate analysis of the daily averages, it was also possible to find a standard model. It is

$$\text{(D-1)} \qquad (1 - \phi_1 B)(1 - B) X_t = (1 - \theta_1 B) Z_t .$$

Table4 contains a comparison of the R^2 values of the standard model versus the optimally fitted models, and also the estimated values of the two coefficients in (D-1) for every one of the seven series.

E. Comparison of Multivariate Models

As we have seen in the last section, univariate models do not too well in case of daily averages of air pollution data. This, and also independently the interest to know more about the connection between meteorological variables and air pollution measurements suggests the use of transfer function models.

Let Y_t be one of the series of air pollution measurements, in the terminology of transfer function models it will be called output series, and let X_{kt}, $k = 1, \ldots, m$, be m input series, which are in our case meteorological variables as for instance: solarflux, temperature, wind intensity and direction, and dummy variables for different weather situations. Then the overall structure of the Box-Jenkins transfer function model is given by

$$\text{(E-1)} \qquad Y_t = F_1(B) X_{1t} + \ldots + F_m(B) X_{mt} + N_t ,$$

where the noise term N_t is modelled by a SARIMA-model in MA-form, and the lag-distribution-operators $F_k(B)$, $k = 1, \ldots, m$, are constructed in analogy to an ARMA- model (for details see [NGC; Ch.XI]). The structure of the model (E-1) is similar to that of a multiple regession model

$$\text{(E-2)} \qquad Y_t = \alpha_1 X_{1t} + \ldots + \alpha_m X_{mt} + Z_t ,$$

but there are two essential differences: (i) The input series X_{kt}, $k = 1, \ldots, m$, enter into the transfer function model (E-1) with a specific time-lag distribution that is governed by the factors $F_k(B)$, which are fitted to the individual set of time series Y_t and X_{kt}, $k = 1, \ldots, m$, while in the regression model (E-2), the coefficients α_k, $k = 1, \ldots, m$, are simply real numbers. (ii) The noise term N_t in the model (E-1) is autocorrelated in a specificly modelled way, whereas in the regression model the error term Z_t is assumed to be white noise, which is, however, rather unrealistic for most applications in time series analysis.

In order to simplify the model finding, ARMAX-models have been suggested. In the one equation case, which we have applied, it is a SARIMA-model extended by input variable terms as in the regression model (E-2).

For five time series of daily averages of air pollution measurements, we have simultaneously applied all three types of modelling. In Table5 the percentage of variance explained by each model is given for the five time series together with the values of R^2 for the univariate Box-Jenkins models. As was to be expected, the value of R^2 increases with the complexity of the model, but unfortunately the amount of work necessary for the modelling process increases in the same direction.

Table 5

Percentage of explained variance for different types of models. All models have been fitted for the time interval Jan.80 - Dez.83.

Series	R^2 [%]			
	Transfer function	ARMAX-Model	Multiple Regression	Univariate Box-Jenkins
SO_2 Langenbrügge	65.2	59.8	33.2	38.3
SO_2 Frankfurt	68.4	63.7	43.3	55.8
O_3 Frankfurt	77.7	75.5	66.9	68.1
O_3 Hohenpeissenberg	70.7	71.0	50.1	69.1
O_3 Deuselbach	74.7	73.5	48.7	69.8

The differences in R^2 values between transfer function and ARMAX-models are small, showing a uniform preference for the transfer function models. In a certain sense multiple regression and univariate Box-Jenkins models represent two opposite positions: In the regression model only external variables are used for explanation and the time series character of all variables is neglected, while in the univariate Box-Jenkins model the situation is reversed; now external variables are neglected and the time series characteristics of one single series are expained by the model. In this respect transfer function and ARMAX-models are a combination of the two special approaches, which also explains their rather good R^2 values.

F. References

[BoJe1] Box, G. E. P., and Jenkins, G. M. (1968): Some Recent Advances in Forecasting and Control. Appl.Stat., 17, pp 91 - 109.

[BoJe2] Box, G. E. P., and Jenkins, G. M. (1978): Time series analysis, forecasting and control. San Francisco; Holden-Day.

[Ha] Hannan, E. J. (1976): The Identification and Parametrization of ARMAX and State Space Forms. Econometrica, 44, pp 713 - 722.

[NGC] Nerlove, M., Grether, D. M., and Carvalho, J. L. (1979): Analysis of Economic Time Series. New York; Academic Press.

[WWS1] Wiegand, G., Weigelt, M., and Stahl, H. (1986): Box-Jenkins-Analyse der Monatsmittelwerte von Immissionsmessungen verschiedener Schadstoffe an unerschiedlichen Stationen. Monatsberichte aus dem Meßnetz. 7 (1986), Berlin; Umweltbundesamt II 1.6.

[WWS2] - - - : Multivariate Regressionsanalyse der Tagesmittelwerte von Immissionsmessungen ohne Berücksichtigung der Zeitreiheneigenschaft. Will appear in: Monatsberichte aus dem Meßnetz., Berlin; Umweltbundesamt II 1.6.

[WWS3] - - - : Box-Jenkins-Analyse der Tagesmittelwerte von Immissionsmessungen verschiedener Schadstoffe an unerschiedlichen Stationen. Will appear in: Monatsberichte aus dem Meßnetz., Berlin; Umweltbundesamt II 1.6.

[WWS4] - - - : Transfer-Funktionsmodelle der Tagesmittelwerte von fünf Immissions Zeitreihen. Will appear in: Monatsberichte aus dem Meßnetz., Berlin; Umweltbundesamt II 1.6.

Parameterschätzung bei differenzierbaren ergodischen Prozessen

Michael Weba, Hamburg

Zusammenfassung: Bei der Parameterschätzung ergodischer Prozesse werden zeitliche Mittelwerte anstelle von stochastischen Mittelwerten verwendet. Ist ein Prozeß in kontinuierlicher Zeit gegeben, so liegen diese Schätzer häufig in Form von (nicht unmittelbar beobachtbaren) stochastischen Riemann-Integralen vor, die man deshalb approximieren muß. Es wird eine Diskretisierungsmethode eingeführt, deren Approximationsfehler schnell gegen Null konvergiert.

Summary: Estimation theory of ergodic processes is usually dealing with time averages rather than ensemble averages. If a process in continuous time is considered estimates are fregently expressed as (nonobservable) stochastic Riemann integrals; therefore, these estimates must be approximated. A discretization method is introduced having the property that the approximation error rapidly tends to zero.

1. Einleitung: Es sei $X(t)$, $t \in (-\infty,\infty)$, ein stochastischer Prozeß mit reellem Zustandsraum und kontinuierlicher Parametermenge. Die Lösung zahlreicher Probleme erfordert brauchbare Schätzungen für gewisse Modellparameter des Prozesses $X(t)$. Drei Beispiele aus verschiedenen Anwendungsbereichen sollen dies verdeutlichen:
- die Spektralanalyse eines Elektroenzephalogramms,
- die Kohärenzschätzung zweier Signale,
- die Zukunftsprognose für eine Zeitreihe.

Zur Modellierung der hierbei auftretenden zufälligen Funktionen werden in der Regel - eventuell nach vorheriger Filterung - stationäre stochastische Prozesse $X(t)$ verwendet, und die Behandlung jedes der drei Beispiele ist nur möglich, wenn Näherungen für den Erwartungswert

$$\theta = E\,(X(t))$$

und die Kovarianzfunktion

Operations Research Proceedings 1987

$$C(s) = E((X(t_1) - \theta)\cdot(X(t_2) - \theta)),\ t_2 - t_1 = s$$

vorliegen.

In der Praxis ist man häufig auf eine einzige Realisierung von $X(t)$ angewiesen. Wird ein Pfad von $X(t)$ im Intervall $[0,T]$, $T > 0$, beobachtet, so finden zeitliche Mittelwerte als Schätzer Verwendung; beispielsweise wird zur Schätzung des Erwartungswerts θ die Größe

$$\hat{\theta}_T = \frac{1}{T} \int_0^T X(t)\, dt$$

herangezogen. Diese Vorgehensweise ist sinnvoll, falls der stochastische Erwartungswert durch den zeitlichen Mittelwert approximiert werden kann, d.h. falls der Prozeß $X(t)$ ergodisch ist und einer Beziehung der Form

$$\theta = \lim_{T\to\infty} \hat{\theta}_T$$

genügt. (Zur Ergodizität von Prozessen und der Formulierung hinreichender Bedingungen siehe Hannan [2], Koopmans [3], Priestley [4], Rozanov [6].)

Die Größe $\hat{\theta}_T$ kann - im Gegensatz zu Prozessen in diskreter Zeit, bei denen das Analogon zu $\hat{\theta}_T$ eine Summe ist - nicht unmittelbar beobachtet werden. $\hat{\theta}_T$ muß deshalb durch einen Näherungswert ersetzt werden, der auf endlich vielen Beobachtungen $X(t_0)$, $X(t_1), \ldots, X(t_N)$ mit Stützstellen $t_k \in [0,T]$ basiert. Es gelte z.B.

$$t_k = k \cdot h, \quad 0 \leq k \leq N$$

mit der Schrittweite $h = T/N$. Wählt man Approximationen $\hat{\theta}_{T,h}$ von $\hat{\theta}_T$ der Form

$$\hat{\theta}_{T,h} = \sum_{k=0}^{N} a_k \cdot X(t_k)$$

mit Gewichten $a_k \in \mathbb{R}$, so gilt:

1.) $$E\left(\hat{\theta}_{T,h}\right) = \left(\sum_{k=0}^{N} a_k\right) \cdot \theta$$

2.) $$(E|\theta - \hat{\theta}_{T,h}|^2)^{1/2} \leq (E|\theta - \hat{\theta}_T|^2)^{1/2} + (E|\hat{\theta}_T - \hat{\theta}_{T,h}|^2)^{1/2}$$

Im Gegensatz zum Abbruchfehler $E|\theta - \hat{\theta}_T|^2$ hängt der Diskretisierungsfehler $E|\hat{\theta}_T - \hat{\theta}_{T,h}|^2$ von den Gewichten a_k ab, und die Forderung

$$\sum_{k=0}^{N} a_k = 1$$

sichert die Erwartungstreue von $\hat{\theta}_{T,h}$. Es ist in der Literatur üblich, die Gewichte

$$a_k = 1/(N+1), \quad 0 \leq k \leq N$$

zu wählen. Die resultierende Näherung kann als Integral einer Treppenfunktion, d.h. als eine sehr primitive Quadraturformel zur Approximation des Integrals

$$\hat{\theta}_T = \frac{1}{T} \int_0^T X(t)\, dt$$

angesehen werden; aus diesem Grund konvergiert der Diskretisierungsfehler für fallende Schrittweiten h nur sehr langsam gegen Null.

Viele praktisch wichtigen Prozesse - etwa in den Bereichen Nachrichtentechnik, Signaltheorie, Meteorologie oder Enzephalographie - sind "glatt", d.h. stochastisch differenzierbar, beispielsweise Prozesse mit bandbegrenztem Spektrum oder Prozesse, welche implizit als Lösungen stochastischer Differentialgleichungen gegeben sind. Es liegt deshalb nahe, die Gewichte a_k so zu wählen, daß die resultierende Approximation formal einer Quadratformel entspricht, die für differenzierbare Integranden "schnell" gegen das Integral konvergiert. Als Beispiel soll im folgenden Abschnitt die Romberg-Quadraturformel untersucht werden.

2. Romberg - Schätzung des Erwartungswerts θ

Der Prozeß X(t) habe den konstanten Erwartungswert θ, und Differenzierbarkeit sowie Integrierbarkeit von X(t) ist stets im L^2-Sinne zu verstehen (siehe z.B. Ash und Gardner [1]). Es seien die natürlichen Zahlen m und N vorgegeben. Dann ist der Romberg - Schätzer $\hat{\theta}_{T,h,m}$ für θ zur Schrittweite $h = T/N$ wie folgt definiert:

1.) Man berechne die Trapezsummen

$$Q_i = h_i \cdot \left(\frac{1}{2} \cdot X(0) + X(h_i) + X(2h_i) + \ldots\right.$$

$$+ X(T-h_i) + \frac{1}{2} \cdot X(T)\Big)$$

für jedes $h_i = h \cdot 2^{-i}$, $0 \leq i \leq m$.

2.) Setze

$$\hat{\theta}_{T,h,m} = \Psi(0) / T,$$

wobei $\Psi(t)$ das Polynom vom Maximalgrad m (in t^2) mit

$$\Psi(h_i) = Q_i, \quad 0 \leq i \leq m,$$

darstellt.
Das Romberg-Verfahren ist ein Extrapolationsverfahren, und $\Psi(0)$ kann leicht rekursiv berechnet werden (Stoer und Bulirsch [5]). Der folgende Satz beschreibt das Verhalten des Diskretisierungsfehlers.

Satz

1.) Der Romberg-Schätzer $\hat{\theta}_{T,h,m}$ ist erwartungstreu für θ.

2.) Ist der Prozeß X(t) auf [0,T] (2m+2)-mal differenzierbar, so gilt für den Diskretisierungsfehler die Abschätzung

$$(E|\hat{\theta}_T - \hat{\theta}_{T,h,m}|^2)^{1/2} = O(h^{2m+2}).$$

Beweisskizze.

1.) $\hat{\theta}_{T,h,m}$ läßt sich als Linearkombination der Beobachtungen schreiben, wobei die Summe der Koeffizienten gleich 1 ist (Stoer und Bulirsch [5]).

2.) Sei Ω der zugrundeliegende Wahrscheinlichkeitsraum und $C_\Omega^{2m+2}[0,T]$ die Menge aller auf [0,T] (2m+2)-mal stetig differenzierbaren stochastischen Prozesse. Mit der punktweisen Addition und Skalarmultiplikation wird $C_\Omega^{2m+2}[0,T]$ ein linearer Raum, und

$$||X|| = \max_{0\leq\mu\leq 2m+2} \sup_{t\in[0,T]} (E|X^{(\mu)}(t)|^2)^{1/2}$$

definiert eine Norm auf $C_\Omega^{2m+2}[0,T]$. Die Menge $C^{2m+2}[0,T]$ aller "gewöhnlichen" deterministischen Funktionen ist ein Teilraum von

C_{Ω}^{2m+2} [O,T]. Werden Funktionen $f \in C^{2m+2}[O,T]$ mit Hilfe des Romberg-Verfahrens numerisch integriert, so läßt sich der Quadraturfehler wegen der üblichen Euler-MacLaurinschen Darstellung als beschränktes lineares Funktional auf C^{2m+2} [O,T] auffassen. Dieses Fehlerfunktional besitzt eine eindeutige Fortsetzung

$$\Phi : C_{\Omega}^{2m+2} [O,T] \rightarrow L^2(\Omega)$$

und stellt den Fehler bei der Romberg-Integration des stochastischen Integrals

$$\int_0^T X(t)\, dt, \qquad X(t) \in C_{\Omega}^{2m+2} [a,b]$$

dar. Hierbei bezeichnet $L^2(\Omega)$ den Raum aller quadratisch integrierbaren Zufallsvariablen über Ω. (Ein ausführlicher Beweis ist in [7] angegeben.) □

3. Verallgemeinerungen

1.) Bei der komplexen Demodulation von Zeitreihen sind Integrale der Form

$$\frac{1}{T} \int_0^T X(t) \cdot \exp(-i\lambda t)\, dt, \quad \lambda \in \mathbb{R}$$

auszuwerten (Koopmans [3]). Diese Berechnung kann ebenfalls mit dem Romberg-Verfahren durchgeführt werden.

2.) Die Resultate des Satzes gelten sinngemäß, wenn Kovarianzen geschätzt werden sollen.

3.) Der Beweis des Satzes stützt sich auf ein Fortsetzungsargument. Dieses Argument kann auch auf Fehlerfunktionale anderer Quadraturmethoden angewandt werden. Insbesondere kann man zur Schätzung stochastischer Modellparameter anstelle des Romberg-Schätzers auch Newton-Cotes-Schätzer heranziehen.

Literatur

[1] Ash, R.B. und Gardner, M.F. (1975). Topics in stochastic processes. Academic Press, London.

[2] Hannan, E.J. (1970). Multiple time series. Wiley, New York.

[3] Koopmans, L.H. (1974). The spectral analysis of time series. Academic Press, London.

[4] Priestley, M.B. (1981). Spectral analysis and time series. Academic Press, London.

[5] Stoer, J. und Bulirsch, R. (1981). Introduction to numerical analysis. Springer, New York.

[6] Rozanov, Y.A. (1967). Stationary random processes. Holden-Day, San Francisco.

[7] Weba, M. (1986). Quadrature of smooth stochastic processes. Preprint 86-7, Institut für Math. Stochastik, Universität Hamburg.

Conjoint Measurement: Eine Analyse der mit Hilfe des Schätzverfahrens LINMAP erzielten Ergebnisse für zufällig gezogene und empirisch erhobene Rangfolgen

Ursula Weisenfeld, Kiel

Zusammenfassung

Mit Hilfe der Conjoint-Analyse werden auf der Basis globaler Präferenzurteile bezüglich einer Anzahl Stimuli Nutzenfunktionen für die die Stimuli charakterisierenden Attribute geschätzt.
Ein mögliches Verfahren zur Schätzung der Parameter der Nutzenfunktionen ist LINMAP. Die durch LINMAP berechneten Anpassungsmaße zur Beurteilung der Güte der Schätzung und die aufgrund der geschätzten Parameter berechneten relativen Gewichte sind in starkem Maß von der Stimulus-Attribut-Matrix abhängig.
Am Beispiel von zwei verschiedenen Produktarten sollen die Auswirkungen der Stimulus-Attribut-Matrix auf die Ergebnisse demonstriert werden.

1. Problemstellung

Die Conjoint-Analyse ist ein dekompositionales Verfahren zur Schätzung von Nutzenfunktionen: Auf der Basis globaler Beurteilungen von n Objekten ("Stimuli") werden Nutzenfunktionen für die m die Objekte charakterisierenden Attribute geschätzt (vgl. GREEN/SRINIVASAN (1978)). Das Schätzverfahren LINMAP (vgl. SRINIVASAN/SHOCKER (1973)) ermittelt die Parameter mittels Linearer Programmierung und zeichnet sich gegenüber anderen Verfahren zur Schätzung der Parameter dadurch aus,

Operations Research Proceedings 1987

daß es globale Optima garantiert und Beschränkungen von Nutzenfunktionen erlaubt.
Die ordinal skalierten Präferenzurteile (Rangfolgen oder Paarvergleiche) werden in metrisch skalierte Stimuli-Nutzenwerte transformiert.

Es werden individuelle Nutzenfunktionen geschätzt, das heißt für jede Auskunftsperson $k \in K$ werden Parameter der Nutzenfunktionen ermittelt und zur Beurteilung der Güte der Schätzung zwei Anpassungsmaße berechnet:

* Der Rangkorrelationskoeffizient Kendall's τ mißt die Übereinstimmung von der empirisch erhobenen Rangfolge und der auf der Basis der geschätzten Parameter berechneten Stimulirangfolge.
* Der Index of fit $C^* = B^*/(1+B^*)$ mit B^* = Zielfunktionswert der Lösung des Linearen Programms liegt zwischen Null und Eins und nimmt den Wert Null an, wenn eine perfekte Übereinstimmung zwischen berechneten Stimulinutzen und beobachteten Präferenzen vorliegt.

Das Anpassungsmaß τ ist zur Beurteilung der Güte der Schätzung interessant, da es als Rangkorrelationskoeffizient die ordinal skalierten Inputdaten mit ordinal skalierten Outputdaten vergleicht.

Zur Schätzung der Parameter sind vorzugeben

* eine n*m - Stimulus-Attribut-Matrix,
* je Attribut eine Klasse von Nutzenfunktionen (Vektor-, Idealpunkt-, Teilwertnutzenmodell),
* gegebenenfalls Beschränkungen von Nutzenfunktionen.

Diese Vorgaben bestimmen die Zahl zu schätzender Parameter und die Zahl der Freiheitsgrade und legen zwei Probleme offen:

1.) Ist die Zahl der Freiheitsgrade gering, werden hohe Anpassungsmaße erzielt, und die Reliabilität der Parameter ist gering.
So berechneten MULLET und KARSON (1986) im Rahmen einer Simulationsstudie für verschiedene Experimentalanordnungen Index of fit - Werte und verdeutlichten die Problematik vergleichender Beurteilungen der Index of fit - Werte verschiedener Conjoint-Analysen.
Die Güte der Anpassung ist nicht nur absolut zu beurteilen: Die Einflüsse des jeweils eingesetzten Schätzverfahrens sind zu kontrollieren.

2.) Die geschätzten individuellen Nutzenfunktionen werden herangezogen, um die relativen Wichtigkeiten der Attribute zu berechnen; dabei hat jedoch die vorgegebene Anzahl an Ausprägungen je

Attribut Einfluß auf dessen berechnete relative Wichtigkeit.

2. Vorschläge zur Problemlösung

2.1 Das Anpassungsmaß τ

Um das Anpassungsmaß τ zu relativieren, können die Ergebnisse für empirisch erhobene Rangfolgen mit denen für zufällig gezogene Rangfolgen unter Verwendung identischer Stimulus-Attribut-Matrizen und Klassen von Nutzenfunktionen verglichen werden; dabei sind nicht nur Mittelwerte und Standardabweichungen der τ-Werte zu vergleichen sondern auch deren Verteilungen.

Ist L das Schätzverfahren LINMAP,
A eine Stimuls-Attribut-Matrix,
N Klassen von (gegebenenfalls beschränkten) Nutzenfunktionen,
R die Menge aller möglichen Rangfolgen (= n!),
K die Menge empirisch erhobener Rangfolgen,
Z die Menge aus R zufällig gezogener Rangfolgen und
r eine Rangfolge,

so liefert

$$L(A,N,r) \rightarrow \tau$$

das Anpassungsmaß Kendall's τ,

$$L(A,N,R)$$

die Häufigkeit des Auftretens einzelner τ-Werte bei Betrachtung der n! Rangfolgen,

$$L(A,N,K)$$

die Häufigkeit des Auftretens einzelner τ-Werte bei Betrachtung der empirisch erhobenen Rangfolgen und

$$L(A,N,Z)$$

die Häufigkeit des Auftretens einzelner τ-Werte bei Betrachtung von aus R zufällig gezogenen Rangfolgen.

Der Unterschied in den τ-Wert-Verteilungen für empirisch erhobene Rangfolgen einerseits und zufällig gezogene Rangfolgen andererseits kann mit Hilfe des Kolmogoroff-Smirnoff-Two-Sample-Tests festgestellt werden (vgl. KOLMOGOROFF (1941)): Dieser Test ermittelt die maximale Differenz D zwischen den beiden kumulierten Verteilungen; eine Gewichtung dieser Differenz mit den beiden Samplegrößen n_1 (= Anzahl empirisch erhobener Rangfolgen) und n_2 (= Anzahl zufällig gezogener Rangfolgen) liefert ein Abstandsmaß (vgl. SIEGEL (1956), S. 134)

$$Chi^2 = 4D^2 \frac{n_1 * n_2}{n_1 + n_2}$$

das den Vergleich von Conjoint-Analyse-Ergebnissen mit verschiedenen Stimulus-Attribut-Matrizen und Klassen von Nutzenfunktionen ermöglicht.

2.2 Die relativen Gewichte

Da die Anzahl der Ausprägungen je Attribut Einfluß auf dessen relative Wichtigkeit hat, ist eine Relativierung der Gewichte notwendig:
Die relativen Wichtigkeiten ergeben aufsummiert über alle Attribute den Wert Eins. Bei drei Attributen mit der gleichen Anzahl an Ausprägungen und einer Gleichgewichtigkeit bekommt jedes Attribut das Gewicht 0.33 zugewiesen; Werte, die deutlich darüber (darunter) liegen, weisen auf eine höhere (niedrigere) Wichtigkeit hin.
Ist die Anzahl der Ausprägungen für die Attribute verschieden, wird eine Gleichgewichtigkeit nicht durch den Wert 0.33 signalisiert, vielmehr sind als Vergleichsmaßstäbe die durchschnittlich erzielten Gewichte der Attribute bei Verwendung zufällig gezogener Rangfolgen heranzuziehen.

3. Empirische Analyse

Am Beispiel von zwei Produktarten ("Fleischsalat" und "Zigaretten") und unter Verwendung jeweils vier verschiedener Stimulus-Attribut-Matrizen (8*3, 12*3, 8*5, 12*5) soll

1.) die Relativierung der τ-Werte durch Berechnung von Chi^2-Werten und
2.) die Relativierung der Gewichte durch Vergleich mit für zufällig gezogene Rangfolgen berechneten Gewichten

gezeigt werden.

Dazu wurden globale Präferenzurteile über "Zigaretten" bzw. "Fleischsalat" erhoben: Die Probanden wurden gebeten, 8 bzw. 12 Stimuli, die auf Karten anhand von 3 bzw. 5 Eigenschaften beschrieben waren, in eine Rangfolge zu bringen.
Zur Relativierung der Ergebnisse wurden 389 Rangfolgen zufällig gezogen.
LINMAP ermittelte sowohl für die empirisch erhobenen als auch für die zufällig gezogenen Rangfolgen unter Verwendung des additiven Teilwertnutzenmodells Anpassungsmaße, Gewichte und Parameter der Nutzenfunktionen.

Tabelle 1 zeigt die erzielten τ-Werte, die D-Werte des Kolmogoroff-Smirnoff-Tests und die berechneten Chi^2-Werte.

Tabelle 1: Chi^2-Werte zur Relativierung von τ

Prod[a]	Stim[b]	Attr[c]	N1	N2	τ[d]	D[e]	Chi^2
FL	8	3	27	389	.923	.813	66.722
FL	12	3	27	389	.838	.890	78.145
FL	8	5	22	389	.974	.146	1.785
FL	12	5	20	389	.920	.835	53.111
ZI	8	3	24	389	.945	.753	51.298
ZI	12	3	24	389	.866	.896	72.608
ZI	8	5	20	389	.962	.179	2.432
ZI	12	5	20	389	.920	.805	49.292

[a] FL = Fleischsalat, ZI = Zigarette
[b] Anzahl zu beurteilender Stimuli
[c] Anzahl der Attribute
[d] Durchschnitt
[e] D-Wert des Kolmogoroff-Smirnoff-Tests

Obwohl für die 8*5 - Stimulus-Attribut-Matrix die höchsten durchschnittlichen τ-Werte mit .974 und .962 erzielt werden, sind die Chi^2-Werte und damit die Unterschiede zwischen den für empirisch erhobene und zufällig gezogene Rangfolgen ermittelten τ-Wert-Verteilungen gering (1.785 bzw. 2.432): Die Anzahl der Freiheitsgrade ist in diesem Fall so gering, daß auch für zufällig gezogene Rangfolgen gleichmäßig hohe τ-Werte erzielt werden und die für empirisch erhobene Rangfolgen ermittelten hohen Anpassungsmaße nicht den Anspruch erheben können, eine modelladäquate Beurteilung der Probanden zu bestätigen.

Die Auswirkungen der Anzahl der Ausprägungen auf die relativen Wichtigkeiten soll für die Stimulus-Attribut-Matrizen 8*3 und 12*3 demonstriert werden.
Beide Produktarten werden in diesem Fall durch 2 Attribute mit 2 Ausprägungen (Gewicht 1 und 2) und 1 Attribut mit 3 Ausprägungen (Gewicht 3) beschrieben.
Tabelle 2 zeigt die durch LINMAP berechneten durchschnittlichen Gewichte (Standardabweichungen) für empirisch erhobene und zufällig gezogene Rangfolgen.

Tabelle 2: Durchschnittliche berechnete Gewichte

Prod[a]	Stim[b]	Gewicht 1	Gewicht 2	Gewicht 3
FL	8	.171 (.176)	.142 (.169)	.688 (.204)
FL	12	.168 (.212)	.208 (.258)	.624 (.309)
ZI	8	.121 (.208)	.131 (.212)	.748 (.278)
ZI	12	.110 (.219)	.109 (.169)	.781 (.262)
ZG	8	.219 (.249)	.266 (.252)	.515 (.287)
ZG	12	.239 (.253)	.241 (.250)	.520 (.303)

[a]Produkt: FL = Fleischsalat, ZI = Zigaretten
ZG = zufällig gezogene Rangfolgen
[b]Anzahl der Stimuli

Beide Produkte weisen offensichtlich eine dominierende Eigenschaft auf: "Gewicht 3" beträgt für "Fleischsalat" .688 (8 Stimuli) bzw. .624 (12 Stimuli) und für "Zigaretten" .748 (8 Stimuli) bzw. .781 (12 Stimuli).
Als Vergleichsmaßstab sind die für zufällig gezogene Rangfolgen ermittelten durchschnittlichen Gewichte .515 (8 Stimuli) und .520 (12 Stimuli) heranzuziehen.
Die höheren durchschnittlichen Gewichte für empirische Daten weisen darauf hin, daß es sich bei dem 3. Attribut tatsächlich um einen dominierenden Beurteilungsfaktor handelt oder daß die im Vergleich zu den beiden anderen Attributen größere Anzahl an Ausprägungen eine größere Bedeutung dieses Faktors suggeriert.

Eine niedrige Anzahl an Freiheitsgraden bewirkt unabhängig vom Datenmaterial hohe Anpassungsmaße und führt dazu, daß die Aussagekraft der Ergebnisse von Conjoint-Analysen stark eingeschränkt wird.
Deshalb sind die für eine Conjoint-Analyse berechneten Anpassungsmaße und Gewichte anhand von zufällig gezogenen Daten zu relativieren: Die Einflüsse des Schätzverfahrens sind zu kontrollieren.

Literatur

Green, P.E.; Srinivasan, V. (1978): Conjoint Analysis in Consumer Research: Issues and Outlook, Journal of Consumer Research, Vol. 5, 1978, S. 103-123.

Kolmogoroff, A. (1941): Confidence Limits for an Unknown Distribution Function, The Annals of Mathematical Statistics, Vol. 12, 1941, S. 461-463.

Mullet, G.M.; Karson, M.J. (1986): Percentiles of LINMAP Conjoint Indices of Fit for Various Orthogonal Arrays: A Simulation Study, Journal of Marketing Research, Vol. 13, 1986, S. 286-290.

Siegel, S. (1956): Nonparametric Statistics for the Behavioral Sciences, New York, Toronto, London 1956.

Srinivasan, V.; Shocker, A.D. (1973): Linear Programming Techniques for Multidimensional Analysis of Preferences, Psychmetrica, Vol. 38, No. 3, 1973, S. 337-369.

The Development of GAMS and its Use on Personal Computers

J.J. Bisschop and C.A.C. Kuip
Enschede

This presentation will focus on the General Algebraic Modeling System GAMS. The GAMS system offers a symbolic modeling language for the representation of large-scale mathematical programming models. In addition, the GAMS system provides an automated interface with several solution algorithms.

The system has been designed and developed by the Analytic Support Unit of the World Bank. Initially, the system was only available for use on mainframes. The PC version has been introduced last year.

The purpose of the presentation is to familiarize the audience with the above development. Details about the GAMS system can be found in the following references.

[Biss82] Bisschop, J.J., Meeraus, A., "On the Development of a General Algebraic Modeling System in a Strategic Planning Environment", Mathematical Programming Study, Vol. 20 (1982) pp. 1-29.

[Four83] Fourer, R., "Modeling Languages Versus Matrix Generators for Linear Programming", ACM Transactions on Mathematical Software, Vol 9, No 2 (1983) pp. 143-183.

[Rose86] Rosenthal, R.E., "Review of the GAMS/MINOS Modeling Language and Optimization Program", OR/MS Today Volume 13, No.3, pp 24-32(even pages).

[Kend87] Kendrick, D.A., Meeraus, A., "Gams, an Introduction", The World Bank, 1818 H Street, N.W., Washington, D.C., 20433, U.S.A.

PC-PROG

A POWERFUL AND USER FRIENDLY MATHEMATICAL PROGRAMMING PACKAGE FOR PC's

Michel W.F.M. Draper
Erwin M.F. Kalvelagen
Amsterdam

Many universities are switching from minicomputers and mainframes to Personal Computers, particularly for educational purposes. Current computer packages for Mathematical Programming, however, are often mainframe oriented packages, even if they are adapted to PC's. This is for example the case for the well-known program LINDO.

PcProg is a package specially developed for the IBM-PC (or compatible). The shortcomings of related programs are avoided: it is an interactive system, where emphasis is put on user-friendliness. A full-screen editor (like Turbo-Pascal) is integrated into the package for comfortable model-entry. The correction of syntax-errors is as easy as possible: the cursor pin-points the exact location of the error and a plain English error-message tells you what is wrong.

An interactive environment is provided through the extensive use of windows which can display the progress of the algorithms, the final results, possible error-messages or On-line Help. Matrix-generation capabilities are added to generate large structered models in just a few key-strokes.

The following modules are included: Linear Programming (Revised Simplex method, using sparse matrix storage), Integer Programming, Mixed Integer Programming and Quadratic Programming. The LP-module can handle up to 500 variables and up to 250 constraints. The Mixed Integer Programming module has solved problems up to 230 integer variables. The implemented algorithms claim to be reliable, and are fast enough for interactive use even for large models.

The program is specially designed for educational purposes. It is easy to learn (almost no time investment to get started) and it is easy to use. This feature is particularly important for introductory O.R.-courses. Several Dutch universities have already adopted PcProg for computer-aided teaching.

Q-lib, a software package for the analysis of multiserver queues

M. H. van Hoorn

In many practical situations the phenomenon 'queueing' or 'waiting' plays an important role. Often, queueing introduces costs, inconvenients or waste. Therefore it is worth to analyse cause and effect of waiting and to seek for solutions to reduce or prevent it.

A well known example of a waiting situation is a post office where customers queue up when they cannot be served at once at one of the counters. Another example of queueing, involving high costs, takes place in the harbour when arriving ships do not find a free place on the quay for loading or unloading.

Other important applications of queueing models are found in telecommunications, computer systems, production/inventory systems, assembly lines, etc.

In the last ten years much effort has been spent on numerical work for queueing models. In the course of this work many mathematical approaches have been explored besides attempts to implement the existing theoretical results. However, the intrinsic complexity of queueing theory has been, and still is, an obstacle for its application in practice. Special mathematical knowledge and very skillful computer programming are required to obtain useful results.

For this reason a widely spread application of queueing models is feasible only if there is ready made software available.

With the completion of the software package Q-lib the most recent know-how on exact solutions for multiserver queueing systems becomes available. Q-lib is a valuable tool for design, dimensioning and performance evaluation of service and delay systems. Also, it can be used in commerce and industry e.g. to optimize production processes, to tune activities or to increase efficiency of expensive plants.

Q-lib has been achieved under supervision of Dr. M.H. van Hoorn and Drs. L.P. Seelen, both former members of the research group of professor H.C. Tijms at the Vrije Universiteit in Amsterdam.

WHAT TO COMPUTE WITH Q-LIB

Q-lib contains computer programs to analyse queueing systems with one or more servers and an infinite or finite waiting capacity. Q-lib offers a wide variety of characteristics for the service units and the arrival patterns of customers. Some performance measures that are computed:

* the probability that an arriving customer has to wait
* average and variance of the waiting time of a customer
* average and variance of the queue length
* the utilisation of the service facility
* the distributions of waiting time and queue length

In applications, for example, the following questions can be answered using these performance measures.

* what is the probability that a ship has to wait for a place at the quay for longer than half a day
* how many servers (machines, terminals,...) are needed to serve 90% of the customers within 10 minutes, given the offered traffic
* with how many terminals a computer configuration can be extended in order to reduce the waiting time of the users taking into account an increase of the response time

THEORETICAL BACKGROUND OF Q-LIB

Q-lib is a modularly structured software package for the exact analysis of G/G/c multiserver queueing systems. The underlying mathematical methods are based on the iterative solution of large sets of linear equations for suitably defined micro-state probabilities.
The following characteristics are computed

* p(n) steady state probabilities (time average)
* PW probability of waiting
* PB probability that all servers are busy
* ELQ mean queue length
* EWQ mean actual waiting time
* CLQ2 squared coefficient of variation of queue length
* Wq(t) waiting time distribution ($Wq(t) = Pr\{Wq \le t\}$)

The various modules of Q-lib analyse queueing models belonging to one of the following classes

* Ph/Ph/c queues
* Ph/D/c queues
* G/M/c queues

where Ph stands for a hyperexponential distribution or a mixture of Erlang distributions with the same scale parameter.

THE RANGE OF COEFFICIENTS OF VARIATION

Let CA2 and CS2 denote the squared coefficient of variation of the interarrival time and service time respectively. In Figure 1 the range of CA2 and CS2 is depicted that is covered by Q-lib. Clearly CA2 and CS2 are not the only parameters defining the corresponding distributions.

The shaded region reflects the models of type Ph/Ph/c. In view of the computing time and memory requirement, CS2 is restricted to CS2 $\ge$ 1/3. For CA2 the restriction CA2 $\ge$ 1/k holds, where k can be chosen, in principle, arbitrarily large. The lines at CS2 = 0 and CS2 = 1 reflect the classes Ph/D/c and G/M/c queues.

For the region of models with 0 < CS2 < 1/3 a 3-point interpolation is provided. For example, the results of a H2/E5/c queue are found by interpolation from the results of the H2/M/c, H2/E2/c and H2/D/c queues having the same arrival process. This interpolation procedure introduces errors in the order of one percent or less.

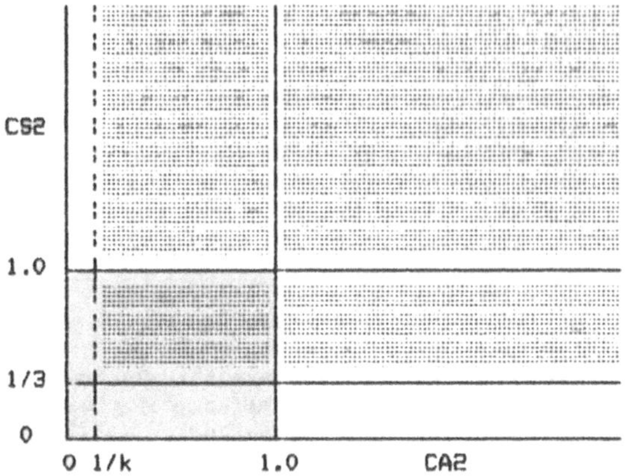

Figure 1. The range of squared coefficients of variation

REMARK

For the models of type Ph/Ph/c, the capacity of the waiting room may be chosesn finite or infinite. as is not allowed in the other classes. Also, in this class the machine repair model is included.

ORGANISATION AND USE OF Q-LIB

Q-lib has been designed as a subroutine library and is written in FORTRAN 77. Q-lib can be used as a program library in a user defined application program.

Another way to use Q-lib is in connection with the interactive, menu-driven input generator that has the following features:

* interarrival time or service time distribution of a model can be fully specified by all of its parameters
* a distribution can be specified by using some defaults, e.g. by giving only two moments
* accuracy and maximum number of iterations can be prescribed
* via a datagenerator mechanism a group of models can be specified at the same time (useful for parameter study or sensitivity analysis)
* help-functions are available to get more information on the screen and also a correction mechanism is provided
* a user may define his own defaults in order to accelarate the input

After having completed the input phase, a user can specify what should be done with the computed results. Besides printing in various lay-outs, Q-lib has a facility to display the results in a table.

The table lay-out definition program is very useful when the results of a group of models should be depicted in a conveniently organized form. Again by a menu-driven procedure a user can choose which parameter is to be varied horizontically/ vertically, which quantities are to be printed, etc. In Figure 2 an example of a table as produced by Q-lib is given. For various values of CA2 and CS2, the waiting probability PW and the mean queue length ELQ are depicted for a collection of G/G/c queues with c servers and an infinite waiting capacity.

ACCURACY AND COMPUTING TIME

The computing time for a queueing model depends on the number of equations for the micro-state probabilities to be solved and the required accuracy. Since the equations are solved iteratively, in principle any accuracy can be obtained.
Usually after about 5 iterations the first two digits of the solution are found. Hence by prescribing an accuracy of only two digits Q-lib provides very good approximations in relatively little computing time.
The computing time depends strongly on the computer system used. To give an indication, the 48 cases in Figure 2 required 9 CPU seconds on a CDC Cyber 175/750 and about 6 minutes elapsed time on a PC with a 8088 processor.

DISTRIBUTION OF Q-LIB

Q-lib is available for CDC systems under NOS/BE, for IBM mainframes under VM and IBM PC's or compatibles under MS=DOS. For the PC version MS-Fortran is used. For further information on Q-lib or an order form, apply to Dr. M.H. van Hoorn, Turfschip 195, 1186 XH Amstelveen, The Netherlands, telex 40038 VNWRK NL.

RO = 0.800 (utilisation)

		C= 1		C= 5		C= 10	
CA2	CS2	PW	ELQ	PW	ELQ	PW	ELQ
.2500	.0000	.5783	.3089	.2346	.1435	.1086	.0726
	.5000	.6609	1.044	.3830	.6354	.2412	.4130
	1.000	.6939	1.814	.4266	1.115	.2832	.7402
	2.000	.7083	3.402	.4572	2.017	.3145	1.316
.5000	.0000	.7019	.7226	.3847	.4393	.2346	.2870
	.5000	.7262	1.492	.4532	.9674	.3062	.6703
	1.000	.7399	2.275	.4778	1.469	.3321	1.021
	2.000	.7464	3.867	.4977	2.394	.3535	1.621
1.000	.0000	.8000	1.600	.5336	1.156	.3847	.8786
	.5000	.8000	2.400	.5484	1.693	.4021	1.265
	1.000	.8000	3.200	.5541	2.216	.4092	1.637
	2.000	.8000	4.800	.5617	3.170	.4182	2.267
2.000	.0000	.8768	3.024	.6730	2.478	.5383	2.076
	.5000	.8647	3.915	.6561	3.039	.5193	2.443
	1.000	.8565	4.776	.6453	3.598	.5085	2.835
	2.000	.8494	6.430	.6399	4.599	.5038	3.503

Figure 2. A table as produced by Q-lib (c is numer of servers)

SEQUENTIAL BIFURCATION FOR FACTOR-SCREENING

Bert Bettonvil, Tilburg

Summary: To screen the input variables in simulation problems we propose a modification of sequential bifurcation, a method resembling the binary search technique. Compared to other techniques that can be used in the same field, our method turns out to be very efficient.

1. Introduction

If we are dealing with a simulation problem in which a great many input variables may be varied at will, then before we can study the performance of the simulation outcomes in detail, we have to find out which of the input variables are really important. Especially in those cases in which a simulation run is expensive (time-consuming), this "screening phase" is necessary.

There exist a number of screening techniques, and in this paper we introduce a modification of one of these, namely Sequential Bifurcation (abbriviated to "SB" hereafter), which in its original version was already mentioned in an overview by Jacoby and Harrison in 1962, but was neglected since then. Our modification of SB makes this method very efficient in comparison with the other techniques within the same field.

In this paper we shall restrict ourselves to observations without error. A paper on SB in case of observations with error is in preparation.

2. Model and technique.

Suppose that a simulation outcome, y, may depend on N input variables $x_1, x_2, \ldots, x_N$. We think that just a few a these are really important, but we have no idea of how many, and which of them are. However, we do have some prior

Operations Research Proceedings 1987

information on the effects of $x_1,\dots,x_N$: y can be expressed as a first order linear model

$$y = y(x_1,x_2,\dots,x_N) = \beta_0 + \beta_1 x_1 + \beta_2 x_2 + \dots\dots + \beta_N x_N \qquad (2.1)$$

in which each of the input variables can assume two levels and we know the sign (but not the size) of each of the regression parameters $\beta_1,\dots,\beta_N$. Then we can recode the input variables in such a way that they all assume the levels 0 and 1, and that

$$\beta_i \geq 0 \quad (i=1,\dots,N). \qquad (2.2)$$

In the sequel we restrict ourselves to this recoded model. Finding the important variables is equivalent to finding the positive regression parameters.

For ease of description we assume that $N=2^m$ for some $m \in \mathbb{N}$. This is not a real restriction, for we can add dummy variables to the model with effects known to be zero to achieve this.

We are going to track down the positive β-parameters by accumulating them in groups of size first 2^m, next $2^{m-1},\dots,$finally $2^{m-m}=1$. To avoid long sums of β's, we introduce a special notation (":=" means "is defined as").

DEFINITION 1. The sum of the parameters in the ith group out of 2^j equally sized groups is denoted by $\beta_{i|j}$ $(i=1,2,\dots,2^j;\ j=0,1,\dots,m)$.

DEFINITION 2. $\beta_{0|0} := 0$.

DEFINITION 3. $y_{i|j} := \beta_0 + \beta_{i|j}$.

We start the search for the nonzero β's by performing two simulation runs: $y_{0|0}=y(0,0,\dots,0)$ and $y_{1|0}=y(1,1,\dots,1)$. Because of (2.2) it is clear that

$$y_{0|0} \leq y_{1|0}. \tag{2.3}$$

If $y_{0|0}=y_{1|0}$, the job is done, as this means that $\beta_{1|0}=0$, so $\beta_i=0$ for all $i=1,\dots,N$. Otherwise, at least one of the regression parameters is positive, and we proceed with a next simulation run, in which we take half of the x_i's at value 1, the other half at value 0. Without loss of generality we may take $x_1=\dots=x_{N/2}=0$ and $x_{N/2+1}=\dots=x_N=1$, so we observe $y_{2|1}$. We shall find ourselves in one of the next three situations:

$$y_{2|1} = y_{0|0} \tag{2.4.a}$$

$$y_{2|1} = y_{1|1} \tag{2.4.b}$$

$$y_{0|0} < y_{0|1} < y_{1|0} \tag{2.4.c}$$

because of (2.2). In case (a) all parameters in the second half are zero, so we can restrict ourselves to the first half; in case (b) we know that $\beta_{1|1}=0$ and we must focus on (the parameters contained in) $\beta_{2|1}$; in case (c) both halfs have to be investigated, for both contain at least one important variable. The first half is to be searched by taking the variables in the second quarter at value 1, and the remaining ones at value 0: observe $y_{2|2}$; for an investigation of the second half we take the variables in the last quarter at value 1, those in the other three quarters at value 0: observe $y_{4|2}$. This splitting goes on until we reach groups of size 1, i.e. individual variables.

Example 1. Jacoby and Harrison present an example with $128=2^7$ variables, in which variable numbers 68, 113 and 120 are the only ones with positive effects. In figure 1 the procedure is sketched. Positive β-parameters are underlined. We need 16 observations, whereas Jacoby and Harrison need 29 ones.

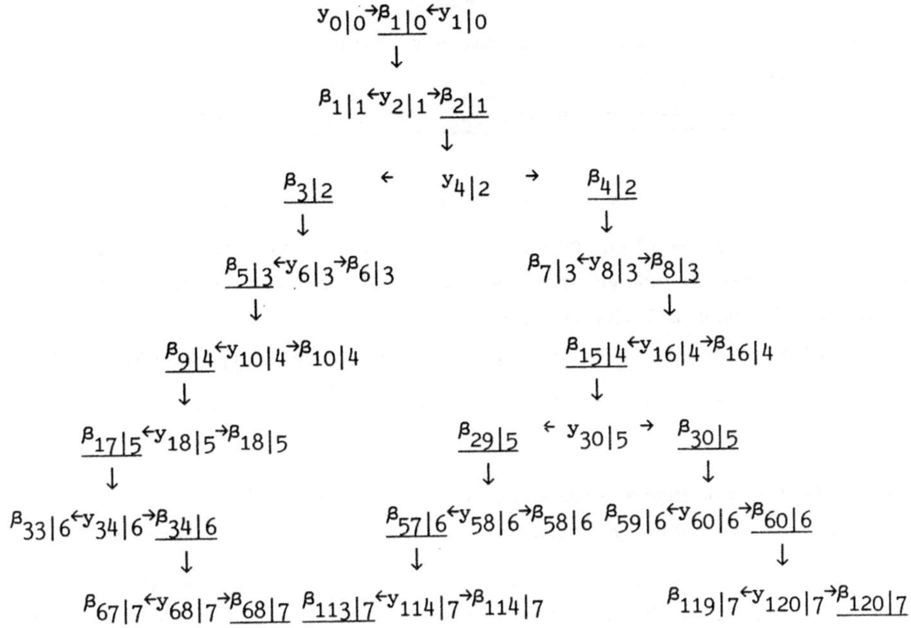

Figure 1. Example of error-free sequential bifurcation.

The basic form of SB was developed, not for use in linear models as (2.1), but for solving a problem posed by Dorfman (1943); the detection of individuals having syphilis. An analysis for syphilis can be made on a sample of blood containing blood from (in Dorfman's case) US soldiers. If none of the members has syphilis, the analysis is negative. Otherwise, the analysis is repeated twice, i.e. for both halfs of the original group, etc. Jacoby and Harrison applied this method to linear models, but overlooked the possibility of reducing the number of observations.

3. Number of runs.

In order to be able to compare our version of SB to techniques, that are applicable to the same kind of problems, we derive the maximum number of observations we need to find k (an unknown number) of positive parameters out of 2^m (with m known) potential parameters. First we give some examples, next we derive an exact upper limit.

Suppose k=0. We observe only $y_{0|0}$ and $y_{1|0}$; the number of observations is 2.

Suppose k=1. We observe $y_{0|0}$ and $y_{1|0}$, and at each level j (j=1,...,m), we have to perform exactly one observation, which will make us find out that at that level exactly one group of parameters has a positive sum, and also, which group does. We need m+2 observations.

Suppose k=2. We observe $y_{0|0}$ and $y_{1|0}$, and $y_{2|1}$ at level 1. If we are lucky, both parameters are contained in $\beta_{1|1}$ or in $\beta_{2|1}$, but if not ($\beta_{1|1}>0$ and $\beta_{2|1}>0$), at level j=2,...,m two observations need to be done. In worst case we need 3+2(m-1)=2m+1 observations; (in best case only m+2).

Now we consider the general case. At level 0 there are always two observations; at level j (j=1,...,m) at most 2^{j-1} observations can ever been made. On the other hand, with k non-zero parameters, at none of the levels we need more than k observations, so for j=1,...,m, the maximum number of observations at level j is the minimum of k and 2^{j-1}. Be ℓ an integer such that $2^{\ell-1}<k\leq 2^{\ell}$, then for $j=1,\ldots,\ell$ the minimum of k and 2^{j-1} is 2^{j-1}; for $j=\ell+1,\ldots,m$ this minimum is k. The maximum number of observations is

$$2+\Sigma_{j=1}^{m}\min\{k;2^{j-1}\} = 2+\Sigma_{j=1}^{\ell}2^{j-1}+\Sigma_{j=\ell+1}^{m}k = 1+2^{\ell}+k(m-\ell). \qquad (3.1)$$

Example 2. In example 1 we had $128=2^7$ variables, of which three were important. According to (3.1), with m=7, k=3, $\ell=2$, the maximum numner of observations is $1+2^2+3(7-2)=20$. Actually we could do with 16 observations because of clustering.

4. Comparison to related techniques.

In this section we compare our version of SB to Two-Stage Group Screening, Multi-Stage Group Screening and SB (Jacoby & Harrison-version), by means of the (worst case) number of runs needed to find k important variables out of $N=2^m$ candidates. For ease of comparison we approximate (3.1) by $k.m=k.{}^2\log N$.

In Two-Stage Group Screening (described in detail by Mauro (1984) and Mauro and Burns (1984)) the variables are accumulated into G groups of size N/G each, and these groups are analyzed by means of a PB-design (Placket and Burman, see Kleijen (1987), p.302) in G+4-(G mod 4) runs. In the second stage the variables in the (say) s important groups are analyzed in sN/G+4-(sN/G mod 4) runs. If we think that k_1 variables are important, then the optimal G is $\sqrt{k_1 N}$, and the total number of observations is about $N(k+k_1)/\sqrt{k_1}$, if k variables are important. If $k_1=k$, we need about $2N\sqrt{k}$ runs, which for large N, small k is far more than $k.{}^2\log N$.

Multi-Stage Group-Screening is a generalization of the former technique, see Li (1962) and Patel (1962). The total number of runs for the analysis of N variables in c stages if k variables are important, is given by Li (1962) to be $ck(N/k)^{1/c}$. The optimal c is ln(N/k), yielding an optimal number of runs of k.e.ln(N/k). Again, this is greater than the number of runs in our version of SB.

In the SB-version of Jacoby and Harrison one starts by observing $y_{0|0}$, $y_{1|1}$ and $y_{2|1}$, and if some $\beta_{i|j-1}$ is positive, both $y_{2i|j}$ and $y_{2i-1|j}$ are observed. The total number of runs is $2^{\ell-1}-1+2k(m-\ell)$, which is about twice the number of runs we need.

Example 3. Suppose we are dealing with N=1024 variables, how many runs do we need to find the k important variables, if k=0,1,...,8. In table 1 the necessary number of runs are tabulated for our version of SB in row "SB", for two-stage group screening in row "G2", for multi-stage group screening in row "GM" and for the Jacoby-Harrison-version of SB in row "JH". In all cases these are worst case numbers, i.e. we assume no clustering.

For G2 as well as GM we assumed k to be guessed right, which is in favour of these methods. For GM we used the formula k.e.ln(N/k), for G2 it is possible to compute the real (worst-case) number of runs.

k	*0*	*1*	*2*	*3*	*4*	*5*	*6*	*7*	*8*
SB	*2*	*12*	*21*	*29*	*37*	*44*	*51*	*58*	*65*
G2	*2*	*68*	*96*	*116*	*136*	*148*	*164*	*172*	*188*
GM	*2*	*18.8*	*33.9*	*47.6*	*60.3*	*72.3*	*83.8*	*94.9*	*105.5*
JH	*3*	*21*	*39*	*55*	*71*	*85*	*99*	*113*	*127*

Table 1. Number of runs for four techniques.

Comparison of SB to other methods in the same field turns out in favour of SB.

References.

Dorfman, Robert 1943. The detection of defective members of large populations, Annals of Mathematical Statistics 14,436-440.

Jacoby, J.E. and S.Harrison 1962. Multi-variable experimentation and simulation models, Naval Research Logistic Quarterly 9, 121-136.

Kleijnen, J.P.C. 1987. Statistical tools for simulation practitioners, Marcel Dekker Inc., New York.

Li, Chou H. 1962. A sequential method for screening experimental variables, American Statistical Association Journal 57, 455-477.

Mauro, Carl A. 1984. On the performance of two-stage group screening experiments, Technometrics 26(3), 255-264.

Mauro, Carl A. and Kevin C. Burns 1984. A comparison of random balance and two-stage group screening designs: a case study, Communications in Statistics, Theory and Methods 13(21), 2625-2647.

Patel, M.S. 1962. Group-screening with more than two stages, Technometrics 4(2), 209-217.

Acknowledgment.

I am indebted to Professors Jack Kleijnen and Peter Sander for their support during my work on this subject.

A DISCRETE SIMULATION-MODEL FOR THE EVALUATION OF TELECOMMUNICATION-NETWORKS

Paul E. Bogerd, The Hague

Abstract

This paper presents a simulation-model and a software-package for the evaluation of small-sized local telecommunication-networks. The model should assist the decision-maker in constructing network-configurations that will offer a high grade of service, even when some of the components in the network are out of order.
The model is based on call-by-call simulation of the traffic, and computes for any configuration the probability of congestion, the mean load on the components, and a number of other responses . The model is incorporated in a highly interactive, user-friendly software-package that runs on a PC, which enables the decision-maker to use the model himself. Both the model and the software are briefly described.
The final sections of this paper report some experiences obtained thus far. It is our belief that the use of a PC in simulation is very well possible, and gives great opportunities in developing user-oriented OR-models.

1. Introduction

In this paper a simulation-model and a software-package for the evaluation of small-sized telecommunication-networks is presented. The model and software, both called ICONS 1) are being developed for the Dutch Ministry of Defence, which plans to set up a number of such networks at different locations. The model is intended to assist the decision-maker in determining the lay-out, the required capacity etc. of each of these networks.
One of the requirements imposed on the model is that it should be useable by the decision-maker himself. This implies that its implementation into a computer-program should be very user-friendly, and that the program should run on a PC.
This paper gives a brief description of the model and of the developed software. In section 2 of the paper, a description of the problem-situation will be given, together with a summary of the requirements imposed on the model. Section 3 presents the simulation model and the principles on which it is based. The software that is developed is described in section 4, followed by a summary of the experience obtained with it in section 5. Finally, some concluding remarks are given in section 6.

1) ICONS is an acronym of Interactive Communication-Network Simulation.

Operations Research Proceedings 1987

2. Purpose and application of ICONS

The networks considered will be used for the transmission of different types of communication. The major part of the traffic-load will consist of voice-communication (ordinary telephone traffic). However, the network will also be used for the transmission of data, and some other types of communication.

The telecommunication-system should meet the following requirements:

- it must deliver a high grade of service, even when, during high traffic load, a number of components (trunks and/or switches) are out of order;
- in over-load situations, high-priority calls supersede low-priority calls (preemption)
- data-communication will be performed using packet-switching, voice-communication will use circuit-switching.

ICONS is built for a situation in which the environment, the network will operate in, largely determines the number of switches. Moreover, the assignment of the subscribers to the switches is also largely determined by environmental factors.

The objective of the decision-maker is to obtain a 'cost effective' network, i.e. a network for which the ratio of 'effectiveness' and 'costs' is high compared to other solutions. ICONS is developed in order to quantify both the cost and the effectiveness of any network-configuration. The effectiveness of any configuration can be expressed in terms of:

- probability of preemption;
- probability of blocking; that is the probability that, due to congestion, an attempt to set up a connection between two subscribers fails;
- average and maximum load of each of the components in the network (switches and trunks);
- average time needed for establishing a connection between two subscribers.

All these criteria are to be evaluated in both 'normal' and 'degraded' situations. ICONS should give insight into the value of each of these criteria.

The decision to use discrete simulation is based on the following observations:

- the number of alternative lay-outs is relatively low;
- the model should have a high degree of flexibility, and should be capable of evaluating different routing-strategies;
- the amount of traffic is relatively low (up to 200 Erlang), thus allowing short running-times;

3. Description of Simulation-model

ICONS is still under development. That is, not all the criteria that are decribed above are already contained in ICONS. In this section the current status of ICONS will be described. However, some of the planned extensions for ICONS will be briefly mentioned.

The simulation-model ICONS does not simulate all the type of traffic that is transmitted through the telecommunications-network. The traffic transmitted using 'packet-switching' is not incorporated in the model. In practice, a small part of the capacity of the network will exclusively be allocated to the handling of packet-switching.

Generation of traffic

The model simulates the traffic and the network on a call-by-call basis. That is, the model contains a 'call-generator' that generates individual calls. In the current version of ICONS, the individual subscribers are not modelled. In ICONS subscribers are put together into 'groups'. These groups are equivalent with groups of users that exist in the environment the network will operate in. In the 'real world' such user-groups will consist of the subscribers that are located at the same physical location.

A user-group can generate several types of traffic. Each traffic-type is characterized by the average length of a single call (ranging from 0 to 'infinity'), and by the number of such calls per hour per subscriber. In the future version of ICONS the 'priority' of the call will also have to be specified. The total amount of traffic generated by a certain user-group is obtained by specifying for each traffic-type the number of generating subcribers.

The receiving user-group is determined by means of a matrix in which for each user-group and each traffic-type the probability of receiving a call is specified.

Routing-algorithm

ICONS uses a static routing-table for the routing of traffic 2). In this table, for each pair of switches, a number of possible routes from one switch to the other is specified. The table is generated by ICONS. The contents of the table remains fixed, irrespective of the actual traffic-load.

When setting up a connection between two switches, the first route in the table will always be the first to be attempted. If one of the switches or trunks on the route are saturated, the route is said to be blocked , and the second route will be chosen, etc. If all the routes are blocked , the attempt to make a call is said to have failed. Whenever a call attempt fails, there will be a retry after a certain amount of time. This is repeated a limited number of times.

One of the major extensions of ICONS will be the inclusion of more sophisticated routing-algorithms.

Output of the model

The simulation model computes statistics on:

- the probability of finding a non-blocking route in the first call-attempt, specified for each pair of user-groups separately;
- the length (number of switches) of the routes that were used;

---

2) See also section 5.

- the number of times a blocking route is encountered;
- the number of times a call-attempt had to be repeated before a non-blocking route was found;
- the average amount of traffic on each of the components of the network (switches, trunks and user-switch-connections).

Moreover, ICONS will trace the used capacity of these components. With this trace, it is possible to obtain a graphical representation of the traffic on each of these components as a function of time.

Future versions of ICONS will generate statistics on:

- the probability of preemption;
- required time to set-up a connection.

4. Description of software.

The requirement that the simulation model should be useable by non-OR-specialists, or even non-experienced computer-users imposed strong requirements on the software for the model. It should be highly interactive, "fool-proof", and run on a (IBM-compatible) PC. This has led to the development of the software-package ICONS.

ICONS is developed using TURBO-PASCAL and its toolboxes. The simulation-model itself is implemented using the simulation-tool "MUST", consisting of a set of PASCAL procedures. With MUST, the development of a object-oriented simulation-model was possible.

At the moment, the entire software-package consits of approximately 500K of code.

ICONS consists of a number of separate modules, each of which will be briefly described in the remaining part of this section.

Database Module

The characteristics of different types of switches, trunks and traffic are described in three files; the components that actually form the network are defined in another three files, making use of the former files.

The Database module enables the user to create and modify these files. Input is checked as much as possible for inconsistencies, and help-information is available for every item that is to be defined.

Check Module

The database management system does not guarantee a total absence of inconsistencies in the database. For instance, it is possible to define 'disconnected' networks. The Check module interprets the information in the database, and checks it for possible inconsistencies. Moreover, the costs of the defined network are computed using some straightforward formulas. Finally, the Check module produces a summary-listing of the input, generates the static routing table, and produces inputfiles for the simulation module.

Simulation Module

This module consists of the actual simulation-program itself. The Simulation-module is interactive: at any point in time the simulation can be interrupted, and insight into the actual status of the communication-system can be obtained. Moreover, a trace can be generated with which it is possible to follow all the events and the behaviour of the system.

Although such a trace gives no (statistical) reliable information on the performance of the system, it will improve the credibility of the model. It is believed that when the user can follow the actions taken during the simulation (and he agrees with them!), it will be more likely that he agrees with the overall results of the simulations.

<u>Graphical Modules</u>

There are two graphical modules available. With the first one, it is possible to obtain a graphical representation of the network. With the second one, it is possible to display some of the results of the simulation. In particular, it is possible to display the load on each of the components in the network as a function of time.

See also section 5.

<u>Master Module</u>

The master module displays some menu's that allow the user to execute the different modules, and to define a number of parameters, such as desired trace-information, etc.

5. Example of use of ICONS.

In practice, ICONS will be used in some iterative procedure:

- firstly, a preliminary layout of a network is determined and evaluated. Most likely, the traffic load can not be very accurately predicted. Hence it is reasonable to perform a sensitivity-analysis by running ICONS more than once, using different traffic-loads.

  In this first evaluation it will become clear which components of the network are likely to induce a significant reduction in the grade of service in the case they might be ill-functioning.
- Secondly, some network-configurations with 'failing' components are evaluated. Whenever the grade of service drops below a certain limit, one should consider modifying the original network in order to improve its reliability. Of course, one should take into account the probability that the different components will fail. The modified network then has again to be evaluated, etc.

The remainder of this section will give an example of the use of ICONS. Consider the network as shown in figure 1. With this network, ICONS is run using 3 different traffic-loads. As an example figure 2 gives the traffic-intensity on switch B as a function of time.

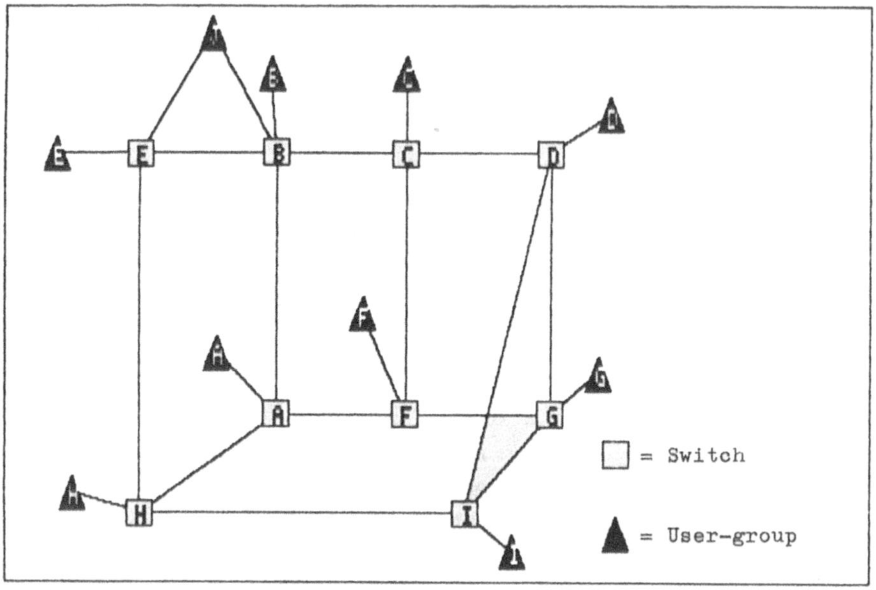

Figure 1. Network

The output of the simulation shows a relatively high traffic-load on switch B. Hence, it is likely that the failure of switch B will result in a significant reduction in grade of service.

Running ICONS again with switch B 'turned off' confirms this assumption. In this case also user-group B is removed from the simulation, since, group B cannot communicate with any of the other groups. From the output of the second set of ICONS-runs, it could be concluded that the addition of a trunk between switch E and switch F was likely to improve the grade of service. Again, another set of ICONS-runs confirms this assumption.

The results of the 9 runs are summarized in figure 3. In this figure, the blocking-probability is shown for a number of traffic-loads, andthree different networks: the original network, the network with switch E 'turned off', and the latter network with trunk E-F added. Notice that the third set of runs yields the best results. However, one should take into account that in this case the traffic-load is reduced due to the omission of user-group B.

The running-time for these 9 simulations was approximately 0.5 hour.

6. Final remarks

The model and software described in this paper are still under development. Nevertheless, their current status allows for an evaluation of performing discrete simulation on a PC.

The main conclusion is that, given the high performance of the current PC's, even time-consuming models such as simulation models can very well be implemented on a PC. Moreover, the possibility of building user-oriented, interactive software will increase the credibility of OR-models, and extends the possible fields of their application.

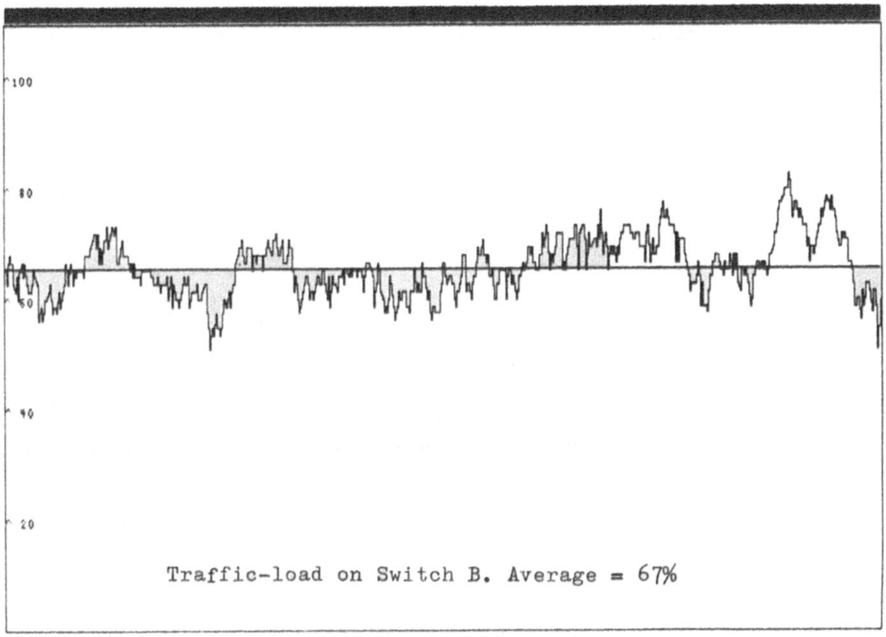

Figure 2. Traffic-load on Switch B as function of time

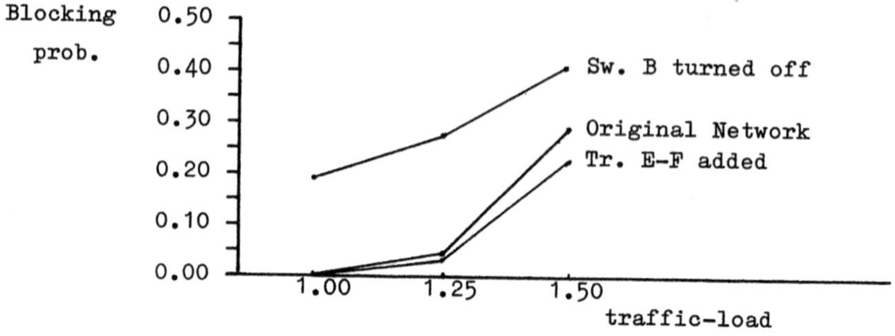

Figure 3. Summary of results of 9 simulation-runs

VERA - Ein Simulationsmodell der Verfügbarkeit von Informationssystemen

Jörn Buhr, Frankfurt

Zusammenfassung: VERA steht für Verfügbarkeit elektronischer Rechnerunterstützung für Anwender. Diese Verfügbarkeit ist abhängig von der Topologie des Systems, d.h. von Direkt- und Alternativwegen durch das System, vom Störverhalten der Rechner und Leitungen, von Recoverystruktur und daraus resultierenden Recoveryzeiten und von Gewichtungen (sofern Durchschnittsverfügbarkeiten gefragt sind).
Ausfallzeiten der Systemkomponenten (Rechner und Leitungen) werden auf der Basis statistisch gewonnener Häufigkeiten und Verteilungen simuliert.
Topologie (incl. Wege durch das System), Störverhalten, Recoveryzeiten und -struktur sowie Gewichtungen können im Sinne eines What-If verändert werden.
Als Ergebnis werden drei Verfügbarkeiten (Einzelkomponente, Durchschnitt, simultan) sowie ein Ausfallzeiten-Überschreitungsrisiko (z.B. wie groß ist das Risiko, daß eine Anwendung in Frankfurt zwischen 9 und 18 Uhr mehr als eine Stunde ausfällt?) errechnet.
Das Modell beruht auf einem dreistufigen System aus Hauptrechnern, Vorschaltrechnern und Leitungsrechnern, die durch ein Leitungssystem untereinander verbunden sind.
Das Programm wurde in APL geschrieben.

1. Topologie eines Informationssystems

Das Informationssystem der Deutschen Lufthansa AG hat eine Netztopologie, die durch 3 Stufen bestimmt ist: Arbeitsrechner (Back-End Computer = BEx), Vorschaltrechner (Front-End Computer = FEx) und Leitungsrechner (Remote Concentrator = RCx).

Abb. 1 und Abb. 2 zeigen Momentaufnahmen des Systems, zwischen denen etwa 1 Jahr liegt.

Die RC sind die anwendernächsten Rechner (Kennzeichnung RCx oder 3 Buchstaben, z.B. HAM für Hamburg). Ihnen nachgelagert (zum Nutzer hin) sind etwa 16.700 Terminals mit zugehörigen Leitungen; diese Komponenten sind nicht mehr im Modell enthalten, weil eine so starke Aufgliederung der Verfügbarkeit gegenwärtig noch nicht gefragt ist und die eigentliche Verfügbarkeitsproblematik auch in den oberen drei genannten Netzwerkebenen liegt.

Operations Research Proceedings 1987

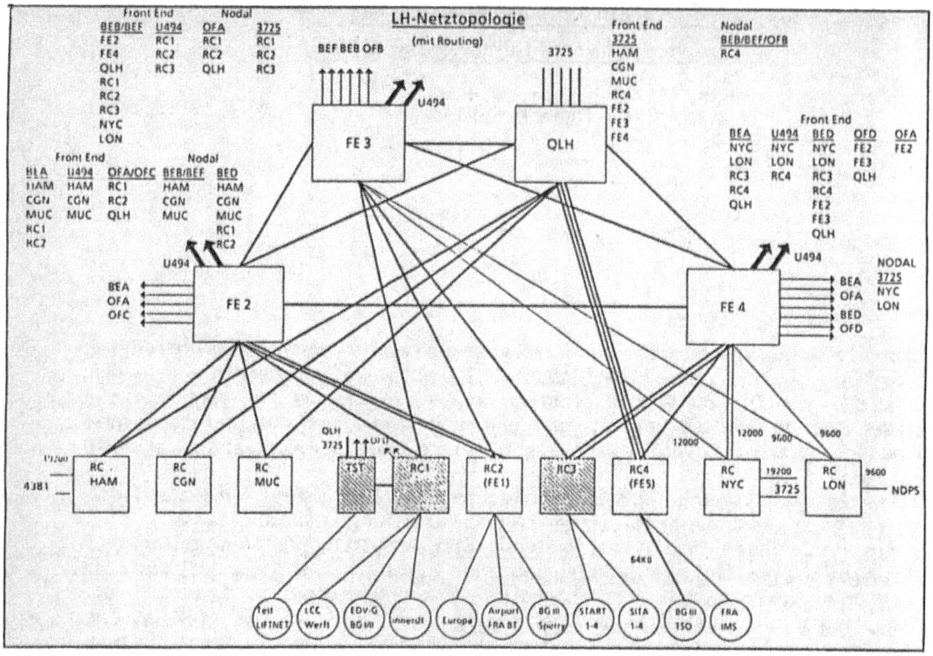
LH-Netztopologie
(mit Routing)
FE 3
QLH
FE 2
FE 4
U494
3725
NDPS

Abb. 1

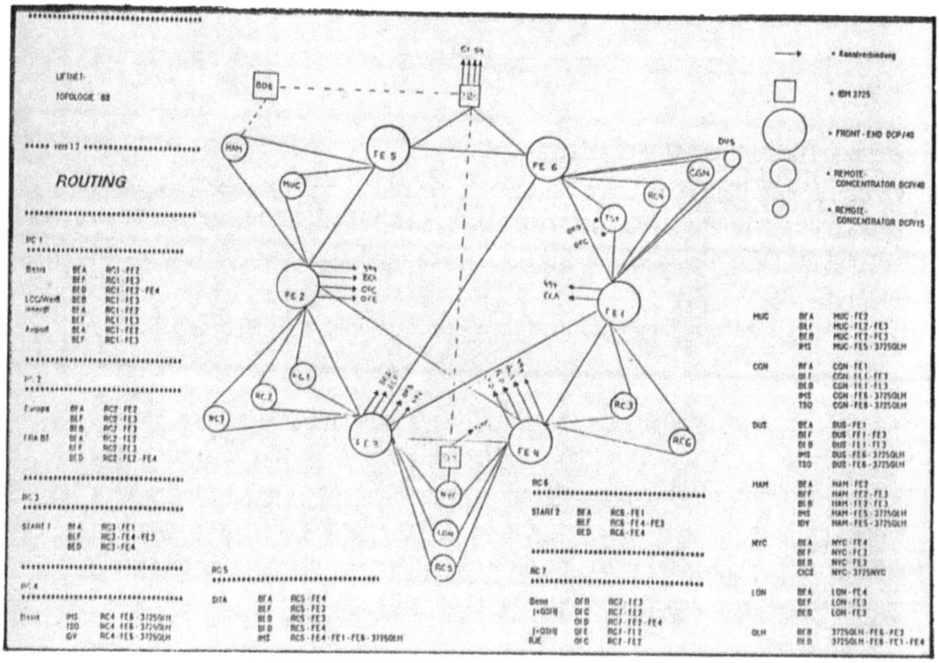
LIFTNET-
TOPOLOGIE '88
ROUTING
IBM 3725

Abb. 2

2. Störungen im System

Störungen im System werden pro Komponente (BEx, FEx, RCx; Leitungen) nach Häufigkeit und Länge erfaßt. Für die Bestimmung der Häufigkeit dient die Anzahl der Störungen pro Jahr bei einer Systemkomponente; für die Ermittlung der Länge einer Störung wird die Durchschnittslänge aller Störungen eines Jahres ermittelt und eine Exponentialverteilung unterstellt.

Die so statistisch ermittelten Werte dienen als Parameter für die Simulation, die mit geeigneten Zufallsgeneratoren arbeitet und einen anfänglich störungsfreien Ausfall-Vektor (der die 3-Minuten-Intervalle eines Jahres enthält) mit Störungen belegt, d.h. der ursprüngliche Einser-Vektor erhält an den Stellen, wo diese Komponente gestört wird, eine Null (Anzahl Störungen: poisson-verteilt; Länge der Störungen: exponentialverteilt, wobei die Länge ein Ein-oder Vielfaches von 3 Minuten ist).
Störungen wurden nicht nach Ursachen oder Arten unterschieden; "geplante" Störungen für Maintenance, Umbau usw. wurden nicht mitgezählt bei der Störungs-Parameter-Ermittlung.

3. Recovery-Zeiten im System

Neben den reinen Störzeiten der einzelnen Systemkomponenten sind die Zeiten für Recovery von Komponenten und deren abhängigen Komponenten zu berücksichtigen.
Hierbei sind nicht nur Recovery-Grund-Ausfallzeiten für die einzelnen Komponenten den Ausfall-Vektoren zuzuschlagen, sondern - je nach möglichem Weg durch das System vom RCx zur BEx - auch noch unterschiedliche Recovery-Zusatz-Ausfallzeiten (die z.B. von der Lastverteilung im System abhängig sind).

4. Routing im System

Jede von einem Terminal angestoßene Anwendung durchläuft die drei Stufen des Systems bis zum Arbeitsrechner und zurück. Dabei kann, wenn der direkte Weg durch eine Störung gesperrt ist, ein Umweg genommen werden. Dieses Routing muß für jeden RC definiert sein (direkte Wege und Alternativwege) und alle für die Anwendung erforderlichen Rechner und Leitungen angeben, verknüpft durch "$\wedge$" oder "$\vee$".

Beispiel:

```
RES

DIE BENOETIGTE RECHNERUNTERSTUETZUNG IST WIE FOLGT DEFINIERT:
RESHAM←HAM∧FE2∧BEA∧(FE2BEA∧(HAMFE2∨(QLH∧HAMQLH∧(QLHFE2∨(FE3∧QLHFE3∧FE3FE2)∨(FE4∧
       QLHFE4∧FE4FE2)∨(FE3∧FE4∧((QLHFE3∧FE3FE4∧FE4FE2)∨(QLHFE4∧FE4FE3∧FE3FE2)))))
       ))
RESCGN←CGN∧FE2∧BEA∧(FE2BEA∧(CGNFE2∨(QLH∧CGNQLH∧(QLHFE2∨(FE3∧QLHFE3∧FE3FE2)∨(FE4∧
       QLHFE4∧FE4FE2)∨(FE3∧FE4∧((QLHFE3∧FE3FE4∧FE4FE2)∨(QLHFE4∧FE4FE3∧FE3FE2)))))
       ))
RESMUC←MUC∧FE2∧BEA∧(FE2BEA∧(MUCFE2∨(QLH∧MUCQLH∧(QLHFE2∨(FE3∧QLHFE3∧FE3FE2)∨(FE4∧
       QLHFE4∧FE4FE2)∨(FE3∧FE4∧((QLHFE3∧FE3FE4∧FE4FE2)∨(QLHFE4∧FE4FE3∧FE3FE2)))))
       ))
RESRC1←RC1∧FE2∧BEA∧(FE2BEA∧(RC1FE2∨(QLH∧RC1QLH∧(QLHFE2∨(FE3∧QLHFE3∧FE3FE2)∨(FE4∧
       QLHFE4∧FE4FE2)∨(FE3∧FE4∧((QLHFE3∧FE3FE4∧FE4FE2)∨(QLHFE4∧FE4FE3∧FE3FE2)))))
       ))
RESRC2←RC2∧FE2∧BEA∧(FE2BEA∧(RC2FE2∨(QLH∧RC2QLH∧(QLHFE2∨(FE3∧QLHFE3∧FE3FE2)∨(FE4∧
       QLHFE4∧FE4FE2)∨(FE3∧FE4∧((QLHFE3∧FE3FE4∧FE4FE2)∨(QLHFE4∧FE4FE3∧FE3FE2)))))
       ))
RESNYC←NYC∧FE4∧BEA∧(FE4BEA∧(NYCFE4∨(FE3∧NYCFE3∧(FE3FE4∨(QLH∧FE3QLH∧QLHFE4)∨(FE2∧
       FE3FE2∧FE2FE4)∨(FE2∧QLH∧((FE3FE2∧FE2QLH∧QLHFE4)∨(FE3QLH∧QLHFE2∧FE2FE4)))))
       ))
RESLON←NYC∧FE4∧BEA∧(FE4BEA∧(LONFE4∨(FE3∧LONFE3∧(FE3FE4∨(QLH∧FE3QLH∧QLHFE4)∨(FE2∧
       FE3FE2∧FE2FE4)∨(FE2∧QLH∧((FE3FE2∧FE2QLH∧QLHFE4)∨(FE3QLH∧QLHFE2∧FE2FE4)))))
       ))
RESRC3←NYC∧FE4∧BEA∧(FE4BEA∧(RC3FE4∨(FE3∧RC3FE3∧(FE3FE4∨(QLH∧FE3QLH∧QLHFE4)∨(FE2∧
       FE3FE2∧FE2FE4)∨(FE2∧QLH∧((FE3FE2∧FE2QLH∧QLHFE4)∨(FE3QLH∧QLHFE2∧FE2FE4)))))
       ))
```

Abb. 3

5. Modellierung

Bei der Modellierung des Systems zeigte sich, daß das Kernproblem, nämlich ob der Weg für eine Anwendung durch das System (direkte und alternative Wege) gestört ist, sich zwanglos als APL-Befehl schreiben ließ.

Die in Abb. 3 dargestellte benötigte Rechnerunterstützung für eine Aufgabe -in diesem Fall Reservierung - stellt eine direkt ausführbare APL-Instruktion dar, die so oder - nach Wunsch - abgeändert, Bestandteil des Programms VERA ist.

Die rechte Seite des Ausdrucks sind die (unter 2. beschriebenen) entsprechend verknüpften Störvektoren der Einzelkomponenten. Die linke Seite des Ausdrucks ist jeweils der Ergebnis-Vektor, der angibt, daß da, wo eine 0 steht, die Rechnerunterstützung für die Anwendung nicht zur Verfügung steht - auch nicht über alle denkbaren Alternativwege durch das System.

Neben der Lösung des o.a. Kernproblems mußte für die verschiedenen Bestimmungsgrößen der Verfügbarkeit der Rechnerunterstützung für eine Anwendung (Netztopologie des Systems, Störungen im System je Einzelkomponente, Recoveryverhalten, Gewichtung der Endbenutzerstellen, Routing pro Anwendung) ein dialogfähiges Programm erstellt werden, welches genügend flexibel und komfortabel ist.

Schließlich sollte die Entwicklung in Form von Prototyping erfolgen.
Für alle diese Erfordernisse war APL (unter Time Sharing) gut geeignet.

6. Risiko-Analyse

Während des Prototyping zeigte sich, daß reine Verfügbarkeitszahlen (Langzeit-Durchschnittswerte) für die Fachabteilung nicht ausreichend sind. Gewünscht wurde auch eine Antwort auf die Frage: Wie groß ist die Wahrscheinlichkeit, daß eine Anwendung innerhalb eines bestimmten Zeitraums (z.B. ein Tag) nicht länger als eine bestimmte Zeitspanne (z.B. 30 Minuten) ausfüllt?
Eine entsprechende Formel wurde eingebaut. In sie gehen die vorher berechnete Verfügbarkeit und Durchschnittslänge eines Ausfalls sowie der gewünschte Zeitraum ein. Die Lösung wird iterativ ermittelt.

7. Resultate

```
         AUSFALL-ZEITEN-UEBERSCHREITUNGS-RISIKO

DURCHSCHNITTLICHE VERFUEGBARKEIT      : 99.12 PROZENT
DURCHSCHNITTL. DAUER EINES AUSFALLS   : 19.67 MINUTEN
ZEITRAUM (IN STUNDEN)                 : 24 STUNDEN

GESAMT-AUSFALLZEIT        UEBERSCHREITUNGS-RISIKO
  PROZENT  MINUTEN        PROZENT
     1.0      14.4        58.6
     2.0      28.8        33.9
     3.0      43.2        19.3
     4.0      57.6        10.9
     5.0      72.0         6.1
     6.0      86.4         3.4
     7.0     100.8         1.8
     8.0     115.2         1.0
     9.0     129.6          .5
    10.0     144.0          .3

         AUSFALL-ZEITEN-UEBERSCHREITUNGS-RISIKO

DURCHSCHNITTLICHE VERFUEGBARKEIT      : 99.12 PROZENT
DURCHSCHNITTL. DAUER EINES AUSFALLS   : 19.67 MINUTEN
ZEITRAUM (IN STUNDEN)                 : 9 STUNDEN

GESAMT-AUSFALLZEIT        UEBERSCHREITUNGS-RISIKO
  PROZENT  MINUTEN        PROZENT
     1.0       5.4        78.4
     2.0      10.8        61.5
     3.0      16.2        48.2
     4.0      21.6        37.8
     5.0      27.0        29.6
     6.0      32.4        23.1
     7.0      37.8        18.1
     8.0      43.2        14.2
     9.0      48.6        11.1
    10.0      54.0         8.7
```

Abb. 4

```
□ VERFUEGBARKEIT VON RESHAM:  99.12 PROZENT         □
-------------------------------------------------------------
| NICHT-VERFUEGBARKEITEN VON RESHAM   PRO TAG               |
| IN MINUTEN                                                |
-------------------------------------------------------------
            INSGESAMT

    MAXIMUM        114
    MINIMUM        0
    DURCHSCHNITT   12.67
    STD.ABWEICH.   20.2
    SPANNE         114
    ANZAHL WERTE   365

            DAVON RCHAM SELBST

    MAXIMUM        30
    MINIMUM        0
    DURCHSCHNITT   2.28
    STD.ABWEICH.   5.76
    SPANNE         30
    ANZAHL WERTE   365

-------------------------------------------------------------
| AUSFALL-LAENGEN  BEI RESHAM IN MINUTEN                    |
-------------------------------------------------------------
            INSGESAMT

    MAXIMUM        93
    MINIMUM        6
    DURCHSCHNITT   19.67
    STD.ABWEICH.   15.94
    SPANNE         87
    ANZAHL WERTE   235

            DAVON RCHAM SELBST

    MAXIMUM        30
    MINIMUM        6
    DURCHSCHNITT   11.91
    STD.ABWEICH.   6.56
    SPANNE         24
    ANZAHL WERTE   70
```

Abb. 5

Abb. 4 und 5 zeigen die wichtigsten Resultate des Programms VERA. Weitere Detailanalysen, z.B. zeitraumbezogen (Jahr, Monat, Woche) sind möglich. Die Aufnahme historischer Stördaten wird erwogen.

8. Verschiedene Verfügbarkeiten

Die unter 7. dargestellten Ergebnisse von VERA beschreiben die Verfügbarkeit von einem RC; eine weitere Detaillierung (örtlich) ist gegenwärtig nicht gefragt.

Für die Berechnung einer Durchschnitts-Verfügbarkeit für alle am System angeschlossenen Anwender wird im allgemeinen eine Gewichtung der an der Anwendung beteiligten RC erforderlich sein.

VERA bietet an:

- eine Gewichtung nach der Anzahl der Endgeräte, die an einem RC hängen
- eine Gewichtung nach der Anzahl der Transaktionen, die über ein RC abgewickelt werden
- eine frei wählbare Gewichtung
- gar keine Gewichtung.

Schließlich stellt VERA noch eine Simultan-Verfügbarkeit des Systems pro Anwendung bereit, welche ausdrückt, daß das System an jeder Stelle für einen Anwender verfügbar ist; diese Verfügbarkeit ist ex definitione in der Regel die niedrigste. Im APL-Programm läßt sie sich fast ohne Mehraufwand erzeugen.

Literatur:

Feller, W.: An Introduction to Probability Theory and Its Applications, Vol. 2, New York - London - Sydney, 1966.

SIMULATION: ANIMIERTE METHODE UND ANWENDUNGEN BEI CIBA-GEIGY

U. Fincke, Basel
W. Vaessen, Basel

Durch neue Entwicklungen auf den Gebieten von Hard- und Software ist eine etablierte Bekannte aus der Palette der Operations Research Methoden wieder stark in den Aufschwung geraten. Das Stichwort zu dieser Entwicklung der Ereignis-orientierten Simulationstechnik ist VISM - Visual Interactive Simulation Modelling. Wir praesentieren eine Uebersicht der Grundprinzipien dieses Verfahrens in Zusammenhang mit komplexen, schlechtstrukturierten Problemen und gehen auf die neuesten Entwicklungen ein.

Schliesslich wollen wir an einigen Beispielen zeigen, wie in unserer Operations Research Gruppe diese Technik angewendet wird, um Management-Entscheidungen von groesserer Tragweite vorzubereiten. Dabei wollen wir Projekte aus den Bereichen Waren-Distribution, Factory-Layout und Produktions-Planung skizzieren, die Hard- und Software, die wir jeweils eingesetzt haben, beschreiben, sowie die Relation zwischen OR-Aufwand und erzieltem Nutzen diskutieren.

ON THE USE OF STOCHASTIC PROCESSES IN MODELING RELIABILITY PROBLEMS

A. Birolini, Zürich

Abstract

Stochastic processes are powerful tools for the investigation of reliability and availability of repairable equipment and systems. Because of the involved models and in order to be mathematically tractable, these processes are generally confined to the class of regenerative stochastic processes. This contribution uses these processes in solving some reliability problems encountered in pratical applications. Investigations deal with different kinds of reliabilities and availabilities for one item, series, parallel, and series/parallel structures.

1. Introduction

For complex equipment and systems, reliability analysis is generally performed at two different levels. At subassembly level, the designer performs failure rate and failure mode analyses to check fulfilment of reliability requirements, and to detect and eliminate reliability weaknesses as early as possible in the design phase. At equipment and system level, the reliability engineer also investigates time behaviour, taking into account reliability, maintainability, and logistical aspects. Depending upon the system complexity, upon the assumed distribution functions for failure-free and repair times, and with thought toward maintenance policy, investigations are performed either analytically, making use of stochastic processes, or numerically with the help of Monte Carlo simulations.

This contribution introduces briefly the stochastic processes used in reliability analyses and investigates some problems encountered in practical applications. It is a part of [3]. The models investigated here assume that systems have only one repair crew and that no further failure can occur at system down. Repair and failure rates are generalized step-by-step, up to the case in which the involved process is regenerative with only one (or a few) regeneration state(s).

Operations Research Proceedings 1987

2. Stochastic processes used in modeling reliability problems

Stochastic processes used in the modeling of reliability problems include renewal and alternating renewal processes, Markov processes with a finite state space, semi-Markov processes, regenerative stochastic processes with only one (or a few) regeneration state(s), and some kinds of non-regenerative stochastic processes. The reliability models covered by each of these processes are summarized in Table 1.

Stochastic processes	Can be used in modeling	Background	Degree of difficulty
Renewal processes	Spare parts reservation in the case of arbitrary failure rates and negligible replacement or repair times	Renewal theory	Medium
Alternating renewal processes	One-item renewable structures with arbitrary failure and repair rates	Renewal theory	Medium
Markov processes (finite state space, time-homogeneous)	Systems of arbitrary structure whose elements have constant failure and repair rates	Differential equations	Low
Semi-Markov processes (at least embedded)	Some systems whose elements have constant failure rates and arbitrary repair rates	Integral equations	Medium
Regenerative processes with only one (or a few) regeneration state(s)	Systems of arbitrary structure whose elements have constant failure rates and arbitrary repair rates; some reduntant structures whose elements have quite general failure rates and arbitrary repair rates	Integral equations	High
Non regenerative processes (e.g. superimposed renewal or alternating renewal processes)	Systems of arbitrary structure whose elements have arbitrary failure and repair rates	Sophisticated methods, partial differential equations	High to very high

Table 1. Stochastic processes used in reliability and availability analyses

The general theory of renewal processes, alternating renewal processes, Markov processes, and semi-Markov processes can be found in the literature, e.g. in [6,7,11,12,14,15]. A comprehensive introduction with emphasis on reliability theory is in [3,4].

All equations necessary for the investigation of reliability models based on Markov and semi-Markov processes are given in Table 2; in particular those equations necessary for the computation of the reliability function $R_{Si}(t)$, the mean time-to-system-failure $MTTF_{Si}$, the point-availability $PA_{Si}(t)$, the interval-reliability $IR_{Si}(t,t+\theta)$ and the stationary value of the availability PA_S and or the interval-reliability $IR_S(\theta)$.

For Markov processes the transition rate ρ_{ij} can be obtained directly from the failure and repair rates of the involved elements, taking care of the system relia-

	Reliability function	Point-availability	Interval-reliability
Semi-Markov processes	$R_{Si}(t)=1-Q_i(t)+\sum_{Z_j\in M}\int_0^t dQ_{ij}(x)R_{Sj}(t-x)$ $MTTF_{Si}=T_i+\sum_{Z_j\in M} p_{ij}MTTF_{Sj},\quad Z_i\in M$ $Q_{ij}(x)=\Pr\{\tau_{ij}\leq x\cap\tau_{ik}>\tau_{ij}$ for all $k\neq j\}$ $=p_{ij}F_{ij}(x)$ $p_{ij}=\Pr\{\tau_{ik}>\tau_{ij}$ for all $k\neq j\}$, $p_{ii}=0$ $F_{ij}(x)=\Pr\{\tau_{ij}\leq x\mid\tau_{ik}>\tau_{ij}$ for all $k\neq j\}$	$PA_{Si}(t)=\sum_{Z_j\in M}P_{ij}(t),\ i=0,1,\dots,m$ $PA_S=\sum_{Z_j\in M}T_j/T_{jj}$ $P_{ij}(t)=\delta_{ij}(1-Q_i(t))+\sum_{k=0}^{m}\int_0^t dQ_{ik}(x)\cdot P_{kj}(t-x)$ $i,j=0,1,\dots,m;\ \delta_{ii}=1,\ \delta_{ij}=0$ for $j\neq i$ $Q_i(x)=\sum_{j=0}^{m}Q_{ij}(x);\quad T_i=\int_0^\infty(1-Q_i(x))dx$	computation has to be problem oriented; for constant failure rates and in the stationary case on has $IR_S(\theta)=\sum_{Z_j\in M}\frac{T_j}{T_{jj}}R_{Sj}(\theta)$ $T_{jj}=\frac{1}{p_j}\sum_{k=0}^{m}p_kT_k$ p_j from $p_j=\sum_{k=0}^{m}p_kp_{kj}$ with $p_0+p_1+\dots+p_m=1$
Time homogeneous Markov processes (method of integral equations)	$R_{Si}(t)=e^{-\rho_i t}+\sum_{Z_j\in M}\int_0^t\rho_{ij}e^{-\rho_i x}R_{Sj}(t-x)dx$ $MTTF_{Si}=\frac{1}{\rho_i}+\sum_{Z_j\in M}\frac{\rho_{ij}}{\rho_i}MTTF_{Sj},\quad Z_i\in M$ ρ_{ij}= transition rate (see definitions) $\rho_{ii}=0;\ \rho_i=\sum_{j=0}^{m}\rho_{ij}$	$PA_{Si}(t)=\sum_{Z_j\in M}P_{ij}(t),\ i=0,1,\dots,m$ $PA_S=\sum_{Z_j\in M}p_j$ $P_{ij}(t)=\delta_{ij}e^{-\rho_i t}+\sum_{k=0}^{m}\int_0^t\rho_{ik}e^{-\rho_i x}\cdot P_{kj}(t-x)dx$ $i,j=0,1,\dots,m;\ \delta_{ii}=1,\ \delta_{ij}=0$ for $j\neq i$	$IR_{Si}(t,t+\theta)=\sum_{Z_j\in M}P_{ij}(t)R_{Sj}(\theta),$ $i=0,1,\dots,m$ $IR_S(\theta)=\sum_{Z_j\in M}p_jR_{Sj}(\theta)$ p_j from $\rho_jp_j=\sum_{k=0}^{m}p_k\rho_{kj}$ with $p_0+p_1+\dots+p_m=1$
Time homogeneous Markov processes (method of differential equations)	$R_{Si}(t)=\sum_{Z_j\in M}P'_{ij}(t)$ $MTTF_{Si}=\frac{1}{\rho_i}+\sum_{Z_j\in M}\frac{\rho_{ij}}{\rho_i}MTTF_{Sj},\quad Z_i\in M$ $P'_{ij}(t)=P'_j(t)$ with $P'_j(t)$ from $\dot{P}'_j(t)=-\rho_jP'_j(t)+\sum_{Z_k\in M}P'_k(t)\rho_{kj}$ with $P'_i(0)=1,\ P'_j(0)=0$ for $j\neq i$,	$PA_{Si}(t)=\sum_{Z_j\in M}P_{ij}(t),\ i=0,1,\dots,m$ $PA_S=\sum_{Z_j\in M}p_j$ $P_{ij}(t)=P_j(t)$ with $P_j(t)$ from $\dot{P}_j(t)=-\rho_jP_j(t)+\sum_{k=0}^{m}P_k(t)\rho_{kj}$ with $P_i(0)=1,\ P_j(0)=0$ for $j\neq i$	$IR_{Si}(t,t+\theta)=\sum_{Z_j\in M}P_{ij}(t)R_{Sj}(\theta),$ $i=0,1,\dots,m$ $IR_S(\theta)=\sum_{Z_j\in M}p_jR_{Sj}(\theta)$ p_j from $p_j\rho_j=\sum_{k=0}^{m}p_k\rho_{kj}$ with $p_0+p_1+\dots+p_m=1$

Definitions:

- $R_{Si}(t)$ = Pr{system up in $(0,t)$ \| Z_i is entered at $t=0$}, $Z_i\in M$
- $MTTF_{Si}$ = E{system failure free time \| Z_i is entered at $t=0$} $=\int_0^\infty R_{Si}(t)dt=\tilde{R}_{Si}(0)$
- $PA_{Si}(t)$ = Pr{system up at t \| Z_i is entered at $t=0$}, $i=0,1,2,\dots,m$
- PA_S = Pr{system up at t in stationary case or for $t\to\infty$} $=\lim_{s\to 0}s\tilde{PA}_{Si}(s)$, $i=0,1,2,\dots,m$
- AA_S = $\lim_{t\to\infty}\frac{1}{t}$ E{system up time in $(0,t)$} $=PA_S$
- $IR_{Si}(t,t+\theta)$ = Pr{system up in $(t,t+\theta)$ \| Z_i is entered at $t=0$}, $i=0,1,2,\dots,m$
- $IR_S(\theta)$ = Pr{system up in $(t,t+\theta)$ in stationary case or for $t\to\infty$}
- $M,\bar{M}$ = set of the up states, set of the down states
- $P_{ij}(t)$ = Pr{system in state Z_j at t \| Z_i is entered at $t=0$}, $i,j=0,1,2,\dots,m$
- $Q_{ij}(x)$ = Pr{sojourn time in $Z_i\leq x\cap$ next transition in Z_j \| Z_i is entered at $x=0$}
- ρ_{ij} = $\lim_{\delta t\to 0}\frac{1}{\delta t}$ Pr{transition from Z_i to Z_j in $(t,t+\delta t)$ \| system in Z_i at t}

S stays for system; $\tilde{u}(s)=\int_0^\infty u(t)e^{-st}dt$ = Laplace transform of $u(t)$

Table 2. Main relations for the reliability and availability computation bility block diagram. The cases of a 1-out-of-2 and of a majoritary (2-out-of-3) redundancy are given in Fig. 1.

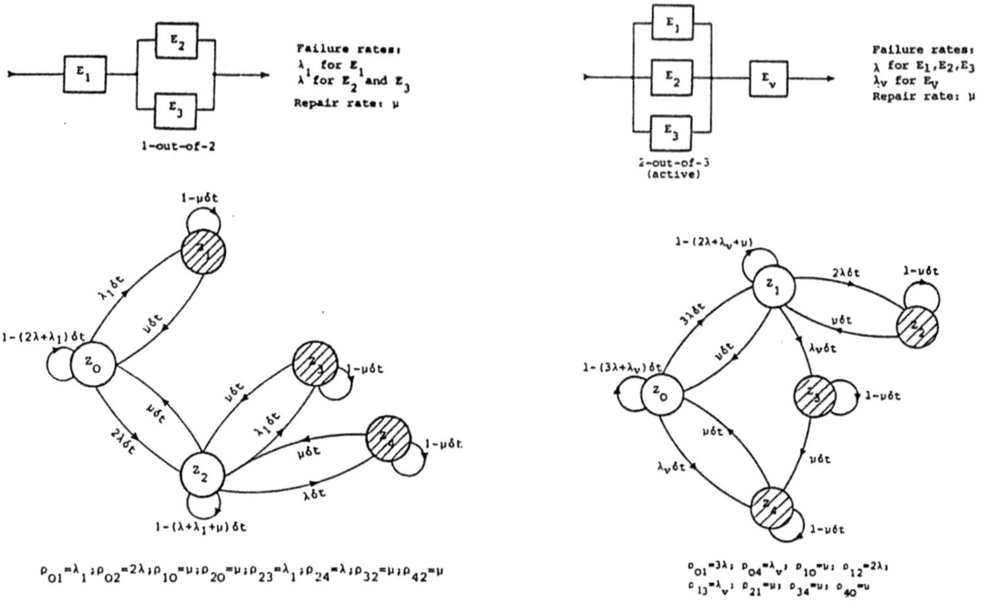

Fig. 1. Reliability block diagram, diagram of the transition probabilities in $(t,t+\delta t)$, and transition rates (ρ_{ij}), for a 1-out-of-2 active redundancy with switch in series and majority redundancy 2-out-of-3 with voter (constant failure and repair rates, only one repair crew, no further failure at system down, t arbitrary, $\delta t \to 0$)

The transition diagrams of Fig. 1 assume that at system failure (notched states in Fig. 1) no further failure can occur. Leaving out this assumption would lead to more complicated state spaces [3,4] without an appreciable change (in practical application) at the reliability or availability values.

3. Applications to reliability problems

3.1 One-item repairable structures

One-item repairable structures are generally investigated by assuming that their stochastic behaviour can be described by an alternating renewal process. This assumption has to be verified by complex item, because it implies that after each repair the item is like new. Some important results for such an item in stationary state are given in Table 3.

Quantity	Expression: Arbitrary failure and repair rates	Expression: Const. failure & repair rates (λ,μ)	Remarks, assumptions
1. Pr{item up at t = 0} (p)	$\frac{MTTF}{MTTF+MTTR}$	$\frac{\mu}{\lambda+\mu}$	$MTTF = E\{\tau_i\},\ i\geq 1$ $MTTR = E\{\tau_i'\},\ i\geq 1$
2. Distribution function of the up time starting at $t=0$ $(F_A(x) = Pr\{\tau_0 \leq x\})$	$\frac{1}{MTTF}\int_0^x (1-F(y))dy$	$1-e^{-\lambda t}$	The same holds for the forward recurrence-time in the up state $(\tau_{Ru}(t))$
3. Distribution function of the down time starting at $t=0$ $(G_A(x) = Pr\{\tau_0' \leq x\})$	$\frac{1}{MTTR}\int_0^x (1-G(y))dy$	$1-e^{-\mu t}$	The same holds for the forward recurrence-time in the down state $(\tau_{Rd}(t))$
4. Renewal density $h_{du}(t)$, $h_{ud}(t)$	$\frac{1}{MTTF+MTTR}$	$\frac{\lambda\mu}{\lambda+\mu}$	$h_{du}(t) = ph_{duu}(t)+(1-p)h_{dud}(t)$ $h_{ud}(t) = ph_{udu}(t)+(1-p)h_{udd}(t)$ $p = MTTF/(MTTF+MTTR)$ as in point 1
5. Point-availability	$\frac{MTTF}{MTTF+MTTR}$	$\frac{\mu}{\lambda+\mu}$	PA = Pr{item up at t}, $t\geq 0$
6. Average-availability	$\frac{MTTF}{MTTF+MTTR}$	$\frac{\mu}{\lambda+\mu}$	$AA = \frac{1}{t}E\{\text{up time in } (0,t)\},\ t>0$
7. Interval-reliability	$\frac{1}{MTTF+MTTR}\int_\theta^\infty (1-F(y))dy$	$\frac{\mu}{\lambda+\mu}e^{-\lambda\theta}$	$IR(\theta) = Pr\{\text{item up in } (t,t+\theta)\},\ t,\theta\geq 0$

Table 3. Main results for onte-item repairable structures in a stationary state

3.2 Systems without redundancy

From a reliability point of view, systems without redundancy are series structures. Assuming that the system consists of n independent elements with failure rates λ_{0i} (i=1,2,...,n) and arbitrarily distributed repair times with mean $MTTR_i$, and that no further failure can occur at system down, one obtains for the reliability function:

$$R_{S0}(t) = e^{-\lambda_0 t} \qquad \text{with} \qquad \lambda_0 = \sum_{i=1}^{n} \lambda_{0i}. \tag{1}$$

For the stationary value of the point-availability it follows

$$PA_S = \frac{1}{1+\sum_{i=1}^{n} \lambda_{0i} MTTR_i}, \tag{2}$$

and for that of the interval reliability

$$IR_S(t,t+\theta) = PA_S(t)\, e^{-\lambda_0 \theta}. \tag{3}$$

Investigations with arbitrary failure rates are still continuing.

3.3 One-out-of-two redundancies

One-out-of-two redundancies have been largely investigated in the literature. Table 4 gives the results for some cases for which a regenerative process still exist (see [3] for more in the case of warm redundancy).

			1-out-of-2 standby redundancy	1-out-of-2 warm redundancy	1-out-of-2 active redundancy
Elements E_1, E_2	Distribution of the failure-free times	in the operating state	$F(t)$	$1-e^{-\lambda t}$	$1-e^{-\lambda t}$
		in the reserve state	——	$1-e^{-\lambda_r t}$	$1-e^{-\lambda t}$
	Distribution of the repair times	in the operating state	$G(t)$	$G(t)$	$G(t)$
		in the reserve state	——	$G(t)$	$G(t)$
	Mean time	to-failure	$MTTF=\int_0^\infty (1-F(t))dt$	$1/\lambda$ bzw. $1/\lambda_r$	$1/\lambda$
		to-repair	$MTTR=\int_0^\infty (1-G(t))dt$	MTTR	MTTR
1-out-of-2 redundancy	Mean time-to-system-failure for the case in which the system is new at $t=0$ ($MTTF_{SO}$)		$MTTF+\dfrac{MTTF}{1-\int_0^\infty f(t)G(t)dt}$	$\frac{1}{\lambda}+\frac{1}{(\lambda+\lambda_r)(1-\tilde{g}(\lambda))}$ $\approx\frac{1}{\lambda}(1+\frac{1}{(\lambda+\lambda_r)MTTR})$ ♦	$\frac{1}{\lambda}+\frac{1}{2\lambda(1-\tilde{g}(\lambda))}$ $\approx\frac{1}{\lambda}(1+\frac{1}{2\lambda MTTR})$ ♦
	Stationary-state point-availability and average-availability ($PA_S = AA_S$)		$\dfrac{MTTF}{\int_0^\infty t\,d(F(t)G(t))}$	$\dfrac{\lambda+\lambda_r(1-\tilde{g}(\lambda))}{\lambda(\lambda+\lambda_r)MTTR+\lambda\tilde{g}(\lambda)}$	$\dfrac{2-\tilde{g}(\lambda)}{2\lambda MTTR+\tilde{g}(\lambda)}$
	Stationary-state interval-reliability ($IR_S(\theta)$)		$\approx R_{SO}(0)$ ◇	$\approx R_{SO}(0)$ ◇	$\approx R_{SO}(0)$ ◇

◇ $R_{SO}(t) = \Pr\{\text{system up in } (0,t) \mid \text{state } Z_0 \text{ is entered at } t=0\}$; ♦ for $\lambda\, MTTR \ll 1$ follows $\tilde{g}(\lambda) \approx 1-\lambda\, MTTR$

Table 4. Main results for repairable 1-out-of-2 redundancies (only one repair crew)

3.4 k-out-of-n redundancy

In practical applications it is generally assumed that all elements of a k-out-of-n redundancy are identical with constant failure rates λ (respectively λ in operating state and λ_r in reserve state) and constant repair rate μ. The stochastic process governing the system is then a birth and death process. Assuming only one repair crew and no further failure at system down, the results depend on λ, λ_r, μ and n-k only. Table 5 gives these results for n-k=1 and n-k=2. The generalization of the repair rate leads to a regenerative stochastic process with only two regeneration states [3,4]. The generalization of the failure and repair rates leads to non-regenerative stochastic processes.

		Mean time-to-system-failure for the case in which the system enters state Z_0 at t=0 ($MTTF_{S0}$)	Stationary-state point-availability and average-availability ($PA_S = AA_S$)	Stationary-state interval-reliability ($IR_S(\theta)$)
n - k = 1	general case	$\frac{\nu_0+\nu_1+\mu}{\nu_0\nu_1}$	$\frac{\nu_0\mu+\mu^2}{\nu_0\nu_1+\nu_0\mu+\mu^2}$	$\frac{\mu^2 R_{S0}(\theta)+\nu_0\mu R_{S1}(\theta)}{\nu_0\nu_1+\nu_0\mu+\mu^2} = R_{S0}(\theta)$
	n = 2 k = 1	$\frac{2\lambda+\lambda_r+\mu}{\lambda(\lambda+\lambda_r)}$	$\frac{\mu(\lambda+\lambda_r+\mu)}{(\lambda+\lambda_r)(\lambda+\mu)+\mu^2}$	$\frac{\mu^2 R_{S0}(\theta)+\mu(\lambda+\lambda_r)R_{S1}(\theta)}{(\lambda+\lambda_r)(\lambda+\mu)+\mu^2} = R_{S0}(\theta)$
	n = 3 k = 2	$\frac{4\lambda+\lambda_r+\mu}{2\lambda(2\lambda+\lambda_r)}$	$\frac{\mu(2\lambda+\lambda_r+\mu)}{(2\lambda+\lambda_r)(2\lambda+\mu)+\mu^2}$	$\frac{\mu^2 R_{S0}(\theta)+\mu(2\lambda+\lambda_r)R_{S1}(\theta)}{(2\lambda+\lambda_r)(2\lambda+\mu)+\mu^2} = R_{S0}(\theta)$
n - k = 2	general case	$\frac{\nu_2(\nu_0+\nu_1+\mu)+\mu(\nu_0+\mu)+\nu_0\nu_1}{\nu_0\nu_1\nu_2}$	$\frac{\nu_0\nu_1\mu+\nu_0\mu^2+\mu^3}{\nu_0\nu_1\nu_2+\nu_0\nu_1\mu+\nu_0\mu^2+\mu^3}$	$\frac{\mu^3 R_{S0}(\theta)+\mu^2\nu_0 R_{S1}(\theta)+\mu\nu_0\nu_1 R_{S2}(\theta)}{\nu_0\nu_1\nu_2+\nu_0\nu_1\mu+\nu_0\mu^2+\mu^3} = R_{S0}(\theta)$
	n = 3 k = 1	$\frac{1}{\lambda}+\frac{\lambda(2\lambda+3\lambda_r+\mu)+\mu(\lambda+2\lambda_r+\mu)}{\lambda(\lambda+\lambda_r)(\lambda+2\lambda_r)}$	$\frac{\mu[(\lambda+2\lambda_r)(\lambda+\lambda_r)+\mu(\lambda+2\lambda_r)+\mu^2]}{(\lambda+2\lambda_r)[\lambda(\lambda+\lambda_r)+\mu(\lambda+\lambda_r)+\mu^2]+\mu^3}$	$\frac{\mu^3 R_{S0}(\theta)+\mu^2(\lambda+2\lambda_r)R_{S1}(\theta)+\mu(\lambda+2\lambda_r)(\lambda+\lambda_r)R_{S2}(\theta)}{(\lambda+2\lambda_r)[\lambda(\lambda+\lambda_r)+\mu(\lambda+\lambda_r)+\mu^2]+\mu^3} = R_{S0}(\theta)$
	n = 5 k = 3	$\frac{1}{3\lambda}+\frac{3\lambda(6\lambda+3\lambda_r+\mu)+\mu(3\lambda+2\lambda_r+\mu)}{(3\lambda+2\lambda_r)(3\lambda+\lambda_r)3\lambda}$	$\frac{\mu(3\lambda+2\lambda_r)(3\lambda+\lambda_r+\mu)+\mu^3}{(3\lambda+2\lambda_r)(3\lambda+\lambda_r)(3\lambda+\mu)+\mu^2(3\lambda+2\lambda_r)+\mu^3}$	$\frac{\mu^3 R_{S0}(\theta)+\mu^2(3\lambda+2\lambda_r)R_{S1}(\theta)+\mu(3\lambda+2\lambda_r)(3\lambda+\lambda_r)R_{S2}(\theta)}{(3\lambda+2\lambda_r)(3\lambda+\lambda_r)(3\lambda+\mu)+\mu^2(3\lambda+2\lambda_r)+\mu^3} = R_{S0}(\theta)$
n - k arbitrary		$MTBF_{Sj} = \int_0^\infty R_{Sj}(t)\,dt$	$PA_S = AA_S = \sum_{j=0}^{n-k} P_j$ $P_j = \pi_j / \sum_{i=0}^{n-k+1} \pi_i$ with $\pi_0 = 1$ and $\pi_i = \frac{\nu_0\nu_1\ldots\nu_{i-1}}{\mu^i}$	$IR_S(\theta) = \sum_{j=0}^{n-k} P_j R_{Sj}(\theta)$

$R_{Sj}(t)$, according to equation (264); $\nu_i = k\lambda+(n-k-i)\lambda_r$, $i = 0,1,\ldots,n-k$; λ, λ_r = failure rates ($\lambda_r = \lambda \rightarrow$ active, $\lambda_r \equiv 0 \rightarrow$ stanby); μ = repair rate

Table 5. **Main results for repairable k-out-of-n redundancies (constant failure and repair rates, one repair crew, no further failure at system down)**

3.5 Series/parallel structures

As an example of a series/parallel structure let us consider a majority redundancy with n=1, i.e. a 2-out-of-3 redundancy in series with a voter (Fig. 1). Let λ be the failure rate of the three parallel elements and λ_v that of the voter. The repair times are distributed according to G(x) with density g(x) and mean MTTR. The system has only one repair crew, no repair priority on the voter and no further failure at system down. Because of the assumptions made the process has 5 states (Z_0 to Z_4) of which only states Z_0, Z_1 and Z_4 are regenerative (use Fig. 1 only to visualize the states). The corresponding semi-Markov transition probabilities are (Tab.2):

$$Q_{01}(x) = \frac{3\lambda}{3\lambda+\lambda_v}(1-e^{-(3\lambda+\lambda_v)x})$$

$$Q_{10}(x) = \int_0^x g(y)e^{-(2\lambda+\lambda_v)y}dy$$

$$Q_{12}(x) = \int_0^x 2\lambda e^{-(2\lambda+\lambda_v)y}(1-G(y))dy$$

$$Q_{121}(x) = \int_0^x \frac{2\lambda}{2\lambda+\lambda_v}(1-e^{-(2\lambda+\lambda_v)y})g(y)dy$$

$$Q_{13}(x) = \frac{\lambda_v}{2\lambda}Q_{12}(x)$$

$$Q_{134}(x) = \frac{\lambda_v}{2\lambda}Q_{121}(x)$$

$$Q_{04}(x) = \frac{\lambda_v}{3\lambda}Q_{01}(x)$$

$$Q_{40}(x) = G(x). \tag{4}$$

$Q_{12}(x)$ and $Q_{13}(x)$ will be used to compute the reliability function. $Q_{121}(x)$ and $Q_{134}(x)$ take care of a transition throughout the non-regenerative states Z_2 and Z_3, respectively. They will be used to compute the point-availability.

For the reliability functions $R_{S0}(t)$ and $R_{S1}(t)$ one obtains (Tab.2)

$$R_{S0}(t) = e^{-(3\lambda+\lambda_v)t}+\int_0^t 3\lambda e^{-(3\lambda+\lambda_v)x}R_{S1}(t-x)dx$$

$$R_{S1}(t) = e^{-(2\lambda+\lambda_v)t}(1-G(t))+\int_0^t g(x)e^{-(2\lambda+\lambda_v)x}R_{S0}(t-x)dx. \tag{5}$$

The Laplace transform (Table 2) of $R_{S0}(t)$ is then given by

$$\tilde{R}_{S0}(s) = \frac{s+5\lambda+\lambda_v-3\lambda\tilde{g}(s+2\lambda+\lambda_v)}{(s+2\lambda+\lambda_v)[s+\lambda_v+3\lambda(1-\tilde{g}(s+2\lambda+\lambda_v))]} \tag{6}$$

and the mean time-to-system failure $MTTF_{S0}$ by

$$MTTF_{S0} = \frac{5\lambda+\lambda_v-3\lambda\tilde{g}(2\lambda+\lambda_v)}{(2\lambda+\lambda_v)[\lambda_v+3\lambda(1-\tilde{g}(2\lambda+\lambda_v))]}. \tag{7}$$

The point-availabilities $PA_{Si}(t)$ can be computed from Table 2 with $M = \{Z_0, Z_1,\}$ and $P_{ij}(t)$ from the following system of integral equations:

$$P_{00}(t) = e^{-(3\lambda+\lambda_v)t} + \int_0^t 3\lambda e^{-(3\lambda+\lambda_v)x} P_{10}(t-x)dx + \int_0^t \lambda_v e^{-(3\lambda+\lambda_v)x} P_{40}(t-x)dx$$

$$P_{10}(t) = \int_0^t g(x)e^{-(2\lambda+\lambda_v)x} P_{00}(t-x)dx + \int_0^t \frac{2\lambda}{2\lambda+\lambda_v} (1-e^{-(2\lambda+\lambda_v)x})g(x)\, P_{10}(t-x)dx$$
$$+\int_0^t \frac{\lambda_v}{2\lambda+\lambda_v} (1-e^{-(2\lambda+\lambda_v)x})g(x)\, P_{40}(t-x)dx$$

$$P_{40}(t) = \int_0^t g(x)P_{00}(t-x).$$

$$P_{01}(t) = \int_0^t 3\, e^{-(3\lambda+\lambda_v)x} P_{11}(t-x)dx + \int_0^t \lambda_v e^{-(3\lambda+\lambda_v)x} P_{41}(t-x)dx$$

$$P_{11}(t) = e^{-(2\lambda+\lambda_v)t}(1-G(t)) + \int_0^t g(x)e^{-(2\lambda+\lambda_v)x} P_{01}(t-x)dx$$
$$+\int_0^t \frac{1}{2\lambda+\lambda_v} (1-e^{-(2\lambda+\lambda_v)x})g(x)\,(2\lambda P_{11}(t-x)+\lambda_v P_{41}(t-x))dx$$

$$P_{41}(t) = \int_0^t g(x)P_{01}(t-x)dx. \tag{8}$$

The interval-reliability $IR_{S0}(t,t+\theta)$ can be computed by

$$IR_{S0}(t,t+\theta) = P_{00}(t)R_{S0}(\theta)+P_{01}(t)R_{S1}(\theta), \tag{9}$$

with $P_{00}(t)$ and $P_{01}(t)$ from (8) and $R_{S0}(\theta)$ and $R_{S1}(\theta)$ from (5). The approximation, which assumes that state Z_1 is regenerative at each time point, is good if $MTTR \ll 1/\lambda$ holds.

Asymptotic behaviour exists, independent of initial conditions at t=0, and leads to

$$PA_S = \frac{2\lambda+\lambda_v+\lambda(1-\tilde{g}(2\lambda+\lambda_v))}{(2\lambda+\lambda_v)(1+(3\lambda+\lambda_v)MTTR)+\lambda(\lambda_v MTTR-2)(1-\tilde{g}(2\lambda+\lambda_v))} \tag{10}$$

$$IR_S(\theta) \approx \frac{[2\lambda+\lambda_v)-2\lambda(1-\tilde{g}(2\lambda+\lambda_v))]R_{S0}(\theta)}{(2\lambda+\lambda_v)(1+(3\lambda+\lambda_v)MTTR)+\lambda(\lambda_v MTTR-2)(1-\tilde{g}(2\lambda+\lambda_v))}. \tag{11}$$

Equation (11) only considers the first term of (9). This approximation holds for $MTTR \ll 1/\lambda$.

Generalization of repair and failure rates leads to a non-regenerative stochastic process.

4. Conclusions

Stochastic processes are powerful tools for the investigation of reliability problems; in particular of the time behaviour of repairable equipment and systems.

If all elements are independent and have constant failure and repair rates, time-homogeneous Markov processes with a finite state space can be used. Markov processes also arise if the repair and/or failure-free times have an Erlang distribution (supplementary states). If repair rates are general, but the system has only one repair crew, the involved process contains an embedded semi-Markov process. A further generalization leads to processes with only one (or a few) regeneration state(s), or even to non-regenerative stochastic processes.

References

[1] H. Ascher et al., Repairable systems reliability, New York: Marcel Dekker, 1984.

[2] R.E. Barlow et al., Mathematical theory of reliabilty, New York: Wiley, 1965; Statistical theory of reliability and life testing, New York: Holt, Rinehart & Winston, 1975.

[3] A. Birolini, On the use of stochastic processes in modeling reliability problems, Lecture Notes in Economics and Math. Systems, vol 252, Berlin/Heidelberg/New York/Tokyo: Springer-Verlag 1985

[4] A. Birolini, Qualität und Zuverlässigkeit technischer Systeme: Springer-Verlag, 1985 (engl. ed. in prep.).

[5] D.R. Cox, The analysis of nonMarkovian stochastic processes by the inclusion of supplementary variables, Proc. Cambridge Phil.Soc., vol. 51, pp.433-441, 1955; Renewal theory, London: Methuen, 1962.

[6] W. Feller, An introduction to probability theory and its applications, New York: Wiley, vol I 1957, vol II 1966.

[7] K.W. Gaede, Zuverlässigkeit Mathematischer Modelle, München: Carl Hanser Verlag, 1977.

[8] B.V. Gnedenko et al., Mathematical methods of reliability theory, New York: Academic, 1969.

[9] U. Höfle-Isphording, Zuverlässigkeitsrechnung, Berlin/Heidelberg/New York: Springer-Verlag, 1978.

[10] S. Osaki et al. (eds.), Stochastic models in reliability theory, Lecture Notes in Economics and Math.Systems, no.235, Berlin/Heidelberg/New York/Tokyo: Springer-Verlag, 1984.

[11] E. Parzen, Stochastic processes. San Francisco: Holden-Day, 1962.

[12] W.L. Smith, Regenerative stochastic processes, in Proc. Int. Congr. Math., vol.3, pp.304-305, 1954; Regenerative stochastic processes, Proc.Royal Soc., London, Ser.A, vol.232, pp.6-31, 1955; Remarks on the paper: Regenerative stochastic processes, Proc. Royal Soc., London, Ser.A, vol.256, pp.496-501, 1960.

[13] A.D, Solovyev, Standby with rapid renewal, Eng. Cybernetics, no.1, pp.49-64, 1970; Asymptotic behaviour of the time of first occurrence of a rare event, Eng. Cybernetics, no.6, pp. 1038-1048, 1971.

[14] S.K. Srinivasan, Stochastic point processes and their applications, London: Griffin, 1974.

[15] H. Störmer, Semi-Markoff-Prozesse mit endlich vielen Zuständen, Lecture Notes in Operations Research and Math.Systems, no.34, Berlin/Heidelberg/New York: Springer-Verlag, 1970.

Analytical solution of the truncated hyperexponential queues with both balking and reneging

Abou-El-Ata, M.O.   Al-Seedy, R.O.

The truncated interarrival hyperexponential queue: $H_2/M/C/N$ (in case of two branches) has been treated numerically by Gupta [1] and analytically in some special cases of both C=1,2 and N=1,2,3 by Al-Seedy and el [2].

In this paper it is aimed to give the analytical solution of the truncated interarrival hyperexponential queue: $H_2/M/C/N$ in the general case of both values C and N and adding the two concepts of balking and reneging. The discipline considered is FIFO one.

REFERENCES

[1] Gupta, S.K., "Queues with hyper-Poisson input and exponential service time distribution with state dependent arrival and service rates". Op. Res. 15, (1967).

[2] Al-Seedy, R.O., and Abou-El-Ata, M.O., "The truncated queues and their measures of effectiveness". M. Sc., Tanta Univ., 1983.

Sojourn Times in Feedback Queues

J.L. van den Berg, O.J. Boxma
Amsterdam

In many modern computer-communication systems, a job may be processed in several phases, or a job may generate new tasks. Such phenomena can be modeled by service systems with feedback. In the queueing literature, attention has been mainly devoted to single-service queues with so-called Bernoulli feedback: when a customer (task) completes his service, he departs from the system with probability $1-p$ and is fed back with probability p. In the present study a more general feedback mechanism is allowed: when a customer completes his i-th service, he departs from the system with probability $1-p(i)$ and is fed back with probability $p(i)$. We mainly restrict ourselves to the case of a Poisson external arrival process and identically, negative exponentially, distributed service times at each service. The resulting queueing model has the property that the joint queue-length distribution of type-i customers, $i=1,2,...$, is of product-form type. This property is exploited to analyse the sojourn-time process.

Our main result is a formula for the joint distribution of the successive sojourn times of a customer in the system. As a by-result, it is shown that the sojourn time distributions of all individual cycles are identically, negative exponentially, distributed. Also, the correlation between the sojourn times of the j-th and k-th cycle of a customer is calculated; furthermore, the distribution of the total sojourn time is derived.

Various choices of the feedback probabilities lead to interesting special cases. In some detail we study the case of so-called deterministic feedback: $p(i)=1$, $i=1,2,...,N-1$; $p(N)=0$. In this case, in which each customer requires exactly N services, the total sojourn time distribution has a particularly simple form. In the limit $N\to\infty$, with mean service time shrinking to zero as $1/N$, the sojourn time distribution tends to the sojourn time distribution in the M/D/1 queue with processor sharing. Similarly, Bernoulli feedback can be shown to lead to the M/M/1 queue with processor sharing.

When service times are not negative exponentially distributed, the joint queue-length distribution no longer is of product-form type. A set of equations for the mean sojourn times at each cycle can be derived. This set is solved for the case of deterministic feedback, with the service times at each cycle having the same first and second moments. It is proved that the mean sojourn times are equal from the second cycle onwards.

REFERENCES

[1] VAN DEN BERG, J.L., BOXMA, O.J., GROENENDIJK, W.P. (1987) Sojourn times in the M/G/1 queue with deterministic feedback, Report OS-R8708, Centre for Mathematics and Computer Science, Amsterdam.

[2] VAN DEN BERG, J.L., BOXMA, O.J. (1987) Sojourn times in feedback queues, Report OS-R8710, Centre for Mathematics and Computer Science, Amsterdam.

Exchangeable items in repair systems: Delay times

Hans Daduna, Hamburg

We consider a repair station where arriving customers deliver failed items to the station and then join an external customer queue in order of their arrival. The items are assumed to be exchangeable, i.e., customers accept any repaired item which is given to them. An item leaving the repair station is given to the customer at the head of the customer queue who then immediately departs from the queue.

A consequence of this organisation is that customers usually do not obtain the item they delivered to the station unless the station itself is a single server under "First-come-first-served" regime. If the station is modeled as a Jackson network of queues the system orientated performance measures, like throughput or utilization, are not altered by introducing the external customer queue (instead of giving back to each customer his own item). But it turns out that the most important customer orientated performance measure "sojourn time distribution" is changed compared with the "individual passage time distribution".

We compute the sojourn time distribution for a customer in the customer queue for the simplest non-trivial network: two exponential parallel queues with different service intensities.

This work was stimulated by the paper of BERG and POSNER; they investigated the case when the repair station is modeled by a M/M/c/∞-FCFS queueing systems.

Reference:

BERG, M.; POSNER, M.J.M.: Customer Delays in M/M/c Repair Systems with Spares. Naval Res. Logistics Quarterly 32, 1985, 287-299.

OPPORTUNITY-BASED PREVENTIVE MAINTENANCE

Rommert Dekker - Amsterdam, The Netherlands

By preference, preventive maintenance is carried out on units during periods when they are not required for operation. Such periods can be created artificially (annual shutdowns, which are costly) or can be due to external events, in which case they are called maintenance opportunities. It is difficult to make efficient use of these opportunities as they occur randomly and are of restricted length. At each of these opportunities a maintenance engineer is faced with the problem which maintenance packages, if any, he should carry out and with what priority.

In this presentation we investigate a mathematical model developed for this problem. As the size of the problem is quite large (the dimension is equal to the number of packages, which may be 30 or more), a complete analytical optimization was impossible. Therefore, we followed a hierarchical approach, first solving the problem for each package separately and then combining the results into a procedure for the total problem. This procedure is based on policy improvement in Markov decision chains and was checked with simulation.

Preventive Replacements at Opportunities

Matthijs Dijkstra, Rotterdam

In the theory of preventive maintenance a basic model deals with the optimal replacement time of a single component with a stochastic lifetime. An extension of this model is the 'Opportunity Preventive Maintenace' (O.P.M.) model. This model describes a component that can only be replaced at stochastic points in time (the opportunities for preventive maintenance).

A maintenance policy is to perform the replacement only if the component age has exceeded a prespecified threshold value. We investigated the value of policies of this kind and determined different sets of modelparameters, according to component lifetime distributions and inter opportunity time distributions. The result is a determination of quantities of interest for the value of preventive maintenance.

When the T.B.O. distribution is exponential the problem can be solved analytically. A computer program has been written for determining the quantities of interest in this special case. Whan the T.B.O. distribution is not exponential the problem can be tackled with simulation or discretisation methods. Because of working with other computer programs we became aware that not only the mean but also the variance of the T.B.O. distribution has its consequences for the value of preventive maintenance.

ÜBER DIE STATIONÄRE VERTEILUNG VON MARKOV-KETTEN VOM M/G-TYP

Egbert Falkenberg, Darmstadt

Eine Markov-Kette ist vom M/G-Typ, falls die Übergangsmatrix P der Markov-Kette die Gestalt

$$P = \begin{pmatrix} B_0 & B_1 & B_2 & B_2 & \cdots \\ C & A_1 & A_2 & A_3 & \cdots \\ & A_0 & A_1 & A_2 & \cdots \\ & & A_0 & A_1 & \cdots \\ & 0 & & \cdot & \cdots \end{pmatrix}$$

hat. Dabei sind A_i m×m Matrizen, B_i sind k×m Matrizen, B_0 ist eine k×k Matrix und C ist eine m×k Matrix. Sei $\vec{\pi} = (\vec{\pi}_n)_{n\in\mathbb{N}_0}$ mit den k- bzw. m-dimensionalen Teilvektoren $\vec{\pi}_0$ bzw. $\vec{\pi}_n$, $n \in \mathbb{N}$, die stationäre Verteilung der Markov-Kette.

Üblicherweise wird zur Approximation von $\vec{\pi}$ die Matrix P ersetzt durch eine endliche Matrix $\tilde{P}$. Dabei fehlen jedoch brauchbare Fehlerabschätzungen.

Hier wird eine Methode vorgestellt, die es erlaubt, für beliebige $n \in \mathbb{N}$ die ersten Teilvektoren $\vec{\pi}_0,\vec{\pi}_1,\ldots,\vec{\pi}_n$ mit Hilfe eines endlichen linearen Gleichungssystems exakt zu berechnen. Für $\vec{\pi}_n$ mit großem n werden dann notwendige und hinreichende Bedingungen für eine asymptotische Entwicklung der Form

$$\vec{\tau}_n = \delta^n \vec{b} + O(\gamma^n)\vec{e} \tag{*}$$

mit $1 > \delta > \gamma > 0$, $\vec{b} \in \mathbb{R}_+^m$ und $\vec{e} = (1,\ldots,1)' \in \mathbb{R}^m$ hergeleitet. Der Vektor $\vec{b}$ und die Konstanten δ können dabei explizit bestimmt werden. Besonders einfache Ergebnisse gelten für den Sonderfall, daß die $(A_i)_{i\in\mathbb{N}_0}$ aus einem Ph/G/1-Bedienungsmodell stammen. Andererseits folgt aus den notwendigen Bedingungen für (*) , daß die Vermutung von Takahashi (1981) über das asymptotische Verhalten der stationären Verteilung der Anzahl der Kunden in einem G/G/c-Bedienungsmodell ohne zusätzliche Voraussetzungen nicht gilt.

Literatur

Takahashi, Y.: Asymptotic exponentiality of the tail of the waiting-time distribution in a Ph/Ph/c queue. Adv. Appl. Prob. 13, 619-630 (1981)

A DISCRETE-TIME QUEUE WITH STATE-DEPENDENT ARRIVALS

Manfred Kramer, Konstanz

Summary: We study the queueing system Geometric/G/1/N with state-dependent arrival rates in discrete time. We develop a stable algorithm for computing the stationary waiting time distribution with respect to an entering customer.

0. Introduction

Queues in discrete time have gained widespread attention during the past years especially in modeling computer and communication systems. One reason for this fact is certainly that a discrete time basis is often more appropriate when events can occur only at regularly spaced points in time. As an example we mention the operation mode of a terminal concentrator in a computer network. In this context customers represent message frames from several input sources waiting for transmission on a common output channel. Messages comprise usually a fixed number of characters that can be transmitted only at predetermined instants /1/.

A basic model for studying such processes is the Geometric/G/1/N queue with state-dependent arrivals. For the stationary probabilities of the corresponding model in continuous time with Poisson input and phase-type service an efficient algorithm is well known /2/. The discrete-time queue with independent arrivals occurring in batches and phase-type service has been investigated in /3/. Another variant allowing variable input and service processes can be found in /1/.

In this paper we study the stationary waiting time distribution in the basic model mentioned above. We develop an efficient and numerically stable algorithm for computing this distribution by simple recurrences without solving a system of linear equations.

1. The Queue Length Process

Let us imagine a time axis divided into contiguous time slots numbered 1,2,... and consider a queueing system with the following arrival and service process:

(i) A customer arrives just prior to a slot mark with probability λ_n, when there are n customers in the system, $n=0,\dots,N-1$.

(ii) Customers are served by a single server. The service of a customer starts or ends only just after a slot mark and consists of k phases with probability f_k, $k=1,\dots,K$. A phase is completed with probability μ at the end of a time slot.

The last assumption implies that the duration of one phase (k phases) is determined by the first (k^{th}) success in a Bernoulli trial process and is therefore subject to a geometric (Pascal) distribution with parameter μ (and k). Thus the probability that a service lasts for j slots is given by

$$F_j = \sum_{k=1}^{K} f_k \binom{j-1}{j-k} \mu^k (1-\mu)^{j-k} \quad , \quad j=1,2,\dots \tag{1}$$

Define P_0 to be the average length of an idle period. For the subsequent busy period define $P_{n,k}$ ($p_{n,k}$) to be the average number of slots with n customers in the system and k phases of the current service to be completed just before (at) the next slot mark. Since a newcomer may enter the system just prior to a slot mark it follows

$$p_{n+1,k} = \lambda_n P_{n,k} + (1-\lambda_{n+1}) P_{n+1,k} \quad , \quad k=1,\dots,K \tag{2}$$

for $n=0,\dots,N-1$, where $P_{0,k} = 0$. Immediately after a slot mark a phase may expire and after the last phase of a service the customer departs. Thus $p_n = \mu p_{n+1,1}$ is the mean number of customers in a busy cycle that leave n customers behind on departure. A new service is then initiated requiring k slots with probability f_k. These events are summarized in the following equations

$$P_{n+1,k} = \mu p_{n+1,k+1} + (1-\mu) p_{n+1,k} + (p_{n+1} + p_0 \delta_{n,0}) f_k, \quad k=1,\dots,K \tag{3}$$

for $n=0,\dots,N-1$.

After a complete busy cycle as many customers have left a queue of n customers behind on departure as customers have met the same queue length level on entrance. Yet the average number of entering customers equals the product of the expected sojourn time the process spends at queue length level n and of the arrival rate λ_n. Hence it follows $p_0 = \lambda_0 P_0$ and

$$p_n = \lambda_n \sum_{k=1}^{K} P_{n,k} \tag{4}$$

for n=1,...,N-1.

2. The Embedded Queue Length

The queue length averages $p_{n+1,k}$ are related to the averages p_n at departure instants by

$$p_{n+1,k} = \sum_{m=0}^{n+1} p_m q_{m,n+1,k}\ , \quad k=1,\ldots,K\ , \tag{5}$$

where $q_{m,n+1,k}$ is the probability for n+1 customers in the system on condition that there have been m customers just after the last departure epoch and k phases of the current service are yet to be completed. Clearly $q_{0,n+1,k} = q_{1,n+1,k}$ holds. Inserting (5) in (2), (3) and comparing the coefficients of p_m we get for n=0,...,N-1 after some algebra

$$\begin{aligned} q_{m,n+1,k} &= \mu\{\lambda_n q_{m,n,k+1} + (1-\lambda_{n+1}) q_{m,n+1,k+1}\} \\ &\quad + (1-\mu)\{\lambda_n q_{m,n,k} + (1-\lambda_{n+1}) q_{m,n+1,k}\} \\ &\quad + \{\lambda_n \delta_{m,n} + (1-\lambda_{n+1})\delta_{m,n+1}\} f_k,\quad m=1,\ldots,n+1,\ k=1,\ldots,K, \end{aligned} \tag{6}$$

where $q_{m,0,k} = 0$ and $q_{m,n+1,K+1} = 0$. If $0<\lambda_{n+1}<1$ and $0<\mu<1$ this equation can be solved for $q_{m,n+1,k}$ to yield a recurrence relation involving nonnegative coefficients only. It can be evaluated straightforwardly for k=K,...,1, m=1,...,N-1 and n=m-1,...,N-1 in that order.

Let $q_{m,n}$ ($Q_{m,n}$) be the probability for n (at least n) customers in the system just after a departure on condition that there have been m customers after the last one. We have obviously $q_{m,n} = \mu q_{m,n+1,1}$ and the recurrences

$$Q_{m,n} = q_{m,n} + Q_{m,n+1} \tag{7}$$

with $Q_{n+1,n} = 1$. Using $p_n = \mu p_{n+1,1}$ in (5) we arrive at the following reduced system

$$p_n = \sum_{m=0}^{n+1} p_m q_{m,n} \tag{8}$$

for n=0,...,N-1.

We now prove inductively that the solution of this system can be obtained by computing p_n for n=0,...,N-1 successively from

$$p_n q_{n,n-1} = \sum_{m=0}^{n-1} p_m Q_{m,n} \tag{9}$$

starting with $p_0 = 1$. Substituting $q_{n,n-1} = 1-Q_{n,n}$ equation (9) can be rewritten as

$$p_n = \sum_{m=0}^{n} p_m Q_{m,n} . \tag{10}$$

This is trivially true for n=1 since $Q_{0,0} = 1$. Substituting (7) for $Q_{m,n}$ and (8) for p_n on the left-hand side yields another instance of the hypothesis (9) with n replaced by n+1. This completes the proof by induction.

3. Queue Length and Waiting Time

The queueing process is regenerative with respect to the successive busy cycles. Thus the stationary probability that the queueing process is in a given state when a new customer arrives can be expressed as the average number of these occurrences in a busy cycle divided by the average number of arrivals in a busy cycle /4,p.285/. For the latter average it follows from (4) that

$$c = \sum_{n=0}^{N-1} p_n \tag{11}$$

Therefore, the stationary probability π_n ($\pi_{n,k}$) that a customer enters a system with n customers (and a service in progress requiring k phases till completion) is

$$\pi_n = \frac{1}{c} p_n \tag{12}$$

for n=0,...,N-1 and

$$\pi_{n,k} = \frac{1}{c} \lambda_n P_{n,k} \quad , \; k=1,\ldots,K \tag{13}$$

for n=1,...,N-1 respectively. π_n is also the stationary probability that a departing customer leaves n customers behind.

The waiting time of a customer is composed of the sum of all time slots required to serve the customer himself and all the customers met in the system. Let w_k be the stationary probability that his waiting time covers k phases. Because a customer sees the state probabilities π_0 and $\pi_{n,k}$ upon arrival we obtain the convolution formula

$$w_k = \pi_0 f_k + \sum_{n=1}^{N-1} \sum_{j=1}^{k} \pi_{n,j} f_{k-j}^{(n)} \quad , \ k=1,\dots,NK-1, \tag{14}$$

where $f_k^{(n)}$ denotes the probability that n independent services need k phases. Thus a customer waits for j slots with probability

$$W_j = \sum_{k=1}^{NK-1} w_k \binom{j-1}{j-k} \mu^k (1-\mu)^{j-k} \quad , \ j=1,2,\dots \tag{15}$$

until his service is finished.

4. Concluding Remarks

The recurrence relations derived above are easy to implement on a computer. Since only nonnegative terms are involved the resulting algorithm is highly stable. It can be shown that its complexity is $O(N^2K^2)$.

References

/1/ J.P. CLAUDE, D. SERET, Files d'attente et processus discrets in: N. Abu El Ata (ed.), *Modelling Techniques and Tools for Performance Analysis*, 101-112 North-Holland, Amsterdam-New York-Oxford 1986

/2/ R.A. MARIE, J.M. PELLAUMAIL, Steady-State Probabilities for a Queue with a General Service Distribution and State-Dependent Arrivals *IEEE Trans. on Software Engineering* 9 (1983), 109-113

/3/ M.F. NEUTS, Computational Uses of the Method of Phases in the Theory of Queues *Comp. and Maths. with Appls.* 1 (1975), 151-166

/4/ H.C. TIJMS, *Stochastic Modeling and Analysis* J. Wiley, New York 1986

APPROXIMATIONS TO THE LIFETIME DISTRIBUTION OF K-OUT-OF-N SYSTEMS WITH COLD STANDBY

D.P. Kroese
W.C.M. Kallenberg
Enschede

Consider the following k-out-of-n system with cold standby. Let S be a system that consists of n identical components and k component positions. In order that the system works, each position has to be occupied by a working component. The components are however subject to failure and they cannot be repaired. We are now interested in the lifetime distribution of the entire system for large n and fixed k.

Let M_n be the lifetime of the k-out-of-n system. For k=1, M_n is simply the sum of n-k+1 i.i.d. random variables. In this case the Central Limit Theorem applies and we can even expand the system lifetime distribution by the well known Edgeworth expansions . For k>1 this is no longer the case. We can however consider $N_i(t)$, the number of failures during [0,t] at position i. Let $N^*(t)=N_1(t)+\cdots+N_k(t)$, then obviously $\{M_n \le t\} \Leftrightarrow \{N^*(t) \ge n-k+1\}$. For each component position i an expansion of $P(N_i(t)\le x)$ can be found. It can also be shown that the expansion of $P(N^*(t)\le x)$ equals the convolution of the expansions of $P(N_i(t)\le x)$, $i=1,\ldots,k$. With the relationship between $N^*(t)$ and M_n stated above, an expansion of $P(M_n\le t)$ can be constructed. Our main result is given in the next theorem.

Theorem

Let M_n be the lifetime of a k-out-of-n system. Let F be the lifetime distribution of the components . Assume that F is non-lattice and that F(0)=0. If the expectation μ, the variance σ^2 and the third central moment μ_3 of F finitely exist, then

$$\sup_{x\in\mathbb{R}} \left| P\left(\frac{M_n-(n-k+1)\mu k^{-1}}{\sqrt{n-k+1}\,\sigma k^{-1}} \le x \right) - \Phi(x) - \frac{\varphi(x)}{\sqrt{n-k+1}} \left\{ (1-x^2)\frac{\mu_3}{6\sigma^3} + \frac{1}{2}(k-1)\left(\frac{\sigma}{\mu} - \frac{\mu}{\sigma}\right)\right\} \right| = o(n^{-1/2}) \quad \text{as } n\to\infty.$$

This leads for example to the following Edgeworth approximation.

$$P(M_n \leq x) \approx \Phi(y_n) + \frac{\varphi(y_n)}{\sqrt{n-k+1}} \left\{ (1-y_n^2)\frac{\mu_3}{6\sigma^3} + \frac{1}{2}(k-1)\left(\frac{\sigma}{\mu} - \frac{\mu}{\sigma}\right) \right\},$$

where $y_n = \dfrac{x-(n-k+1)\mu k^{-1}}{\sqrt{n-k+1}\ \sigma k^{-1}}$.

By using this approximation we can improve considerably on the known normal approximation of the system lifetime distribution. Numerical work shows excellent agreement with the theoretical error structure of the approximations.

Approximations for production/inventory models with general inputs

Jan-Kees van Ommeren, Amsterdam

In this paper we consider a single product production/inventory model with the following characteristics. The production facility can be either on with producing rate 1, or off. The produced goods are stored in a buffer with capacity M. The inventory level is controlled by a (m,M)-switch-over policy with $0<m<M$. Under this policy the production facility is turned on when the inventory level drops below some specified constant m and is turned off when the buffer is full.

The customers arrive according to a renewal process. Their demand sizes are independent and have a general distribution function. We assume that the expected demand per unit time is less than 1.

Sometimes the demand size of an arriving customer exceeds the present inventory. We distinguish two classes of systems which react differently on these demands. The first class consists of systems which do not deliver subsequently and so the excess demand is lost. The systems which belong to the second class allow backlogging, or equivalently, allow a negative inventory level.

Interesting performance measures for the systems where no backlogging is allowed are among others the fraction of customers whose demands are not completely satisfied and the fraction of demand that is not delivered. For systems with backlogging such performance measures are the fraction of customers whose demand cannot be met from stock on hand and the expected inventory level just after an arrival of a customer given that this level is negative.

Under some mild assumption on the distribution function of the demand sizes we derive exponential expansions (as $m\to\infty$ and $M-m\to\infty$) for the above mentioned performance measures. The derivation is based upon a comparison of the production/inventory model with an unbounded single server queue which has the same arrival process and where the input sizes are equally distributed as the demand sizes in the original system.

COMPARISON OF THE THROUGHPUT IN A TANDEM SERIES OF QUEUES WITH FINITE BUFFERS

Ad Ridder, Leiden

We consider a tandem series of queues with single servers and finite buffers such as a production or communication line. A blocking protocol tells us what will happen if an intermediate buffer is full. In "communication blocking" a completed job of the preceding server is restarted by that server again. In "production blocking" after service completion the job blocks the server until a place comes free in the buffer.

We assume a Poisson arrival stream to the system and exponential service requirements at the servers. Then we show that the throughput of the system with production blocking is larger than with communication blocking, a result which intuitively follows from the description of the two blocking protocols. The method we develop for that purpose, is based on comparison of the sample paths of the processes during regeneration cycles.

SHORT-TERM RELIABILITY AND AVAILABILITY OF PRODUCTION SYSTEMS

Aat Schornagel - Amsterdam, The Netherlands

Availability of production systems is defined as the ratio production/demand. Owing to unexpected events such as unit breakdowns, delayed preventive maintenance times and start-up procedures, short-term (or periodic) availability is subject to random changes around its long-term average. System availability depends on system configuration, unit reliability characteristics and capacity, maintenance and operational strategies, location and capacity of buffer stocks etc.

Our review of the reliability literature indicates that short-term availability has never been studied. In our presentation we indicate how information on short-term availability may contribute to (economic) optimisation in the design and during operation of production systems and processes.

We propose a unified approach to compute the probability distribution of short-term system availability. First, all forthcoming probability distributions are approximated by phase-type distributions. This enables application of well known techniques from the theory of Markov processes. When the state space becomes too large, we aggregate states by combining units in subsystems to "super" units. Comparison with simulation results shows a satisfactory accuracy performance of this approximative method. Furthermore, we address the sensitivity of the availability distribution to input parameters.

Some Aggregation Methods for Closed Queuing Networks *)

Rudo J. Wijbrands, Eindhoven

Zusammenfassung: Dieser Beitrag beschreibt ein neues Aggregationsverfahren für die Analyse von geschlossenen Warteschlangennetzen. Das Verfahren ist basiert auf der Mittelwertanalyse. Das Verfahren wird erklärt an Hand des klassischen Beispiels eines Rechensystems. Über Erfahrungen bezüglich der Genauigkeit wird berichtet.

Summary: In this paper we study some aggregation methods for the analysis of closed queuing networks. The methods are based on the mean value analysis. The methods are explained by looking at a classical example in computer performance evaluation, a CPU-disk model. Some numerical results are added to illustrate the accuracy of the proposed methods.

1. Introduction

One of the most popular algorithms for solving product form closed queuing networks is the mean value analysis (MVA) algorithm (refer Reiser & Lavenberg [3]). It has gained its great popularity thanks to its intuitively appealing formulation: a set of recursive relations entirely in terms of throughputs, sojourn times and queue lengths.
The application of the algorithm is restricted by its computational complexity. For a closed queuing network system with M single server stations and R types of clients to be distinguished, the complexity of algorithm is of the order:

$$O(\,M\;R\prod_{r=1}^{R}(\,K_r+1\,)\,) \tag{1}$$

where K_r denotes the number of clients of type r in the system. Up to now, no algorithms have been developed which reach a major break through with respect to this computational complexity. Most larger problems can be dealt only approximately, e.g. by applying the Schweitzer approach [4].

In this paper we study a problem in which the complexity of the analysis is that high that we have to deal with approximation methods. The model we study is not suitable for the above suggested Schweitzer type of approach.
We consider a computer system consisting of a number of processors and disk units. Refer Figure 1. Jobs which are offered to the system in first instance are stored in a memory queue; the number of jobs which have simultaneous access to the system is only limited. If the number of jobs in the system decreases below a certain multiprogramming level limit, then a new job is released from the memory queue and admitted to the system. The newly admitted job is assigned to one of the processors. The job is alternately served by this CPU and one of the disk units, until it is completed and leaves the system. It can never obtain service from any other CPU, however.
The method of analysis we propose for this model, is a traditional one. In a decomposition step we study

*) The investigations were supported in part by the Foundation for Computing Science in the Netherlands, SION, with financial aid from the Netherlands Organization for the Advancement of Pure Research (ZWO).

a closed queuing network model consisting of the CPU's and the disk units. This model is solved by mean value analysis. Next, in a composition step the CPU-disk model is aggregated to a single station with a queue dependent service rate. The service rates for this station are equivalent with the throughputs obtained for the CPU-disk model. The composite model can be analyzed as a birth and death process. This paper is restricted to the decomposition step of this approach.
As the routing in the CPU-disk model is partly fixed, a job always obtains service from the same processor, the number of client types to be distinguished in this model is at least equal to the number of CPU's. This may have serious consequences for the computational complexity (1). We note however that such a distinction between jobs in the memory queue is not necessary. This observation suggests a natural way of aggregating and thus reducing the complexity; jobs which appear to be identical if they could run on the same processor are to be aggregated. It is the aim of this paper to investigate the possibilities of such an aggregation.

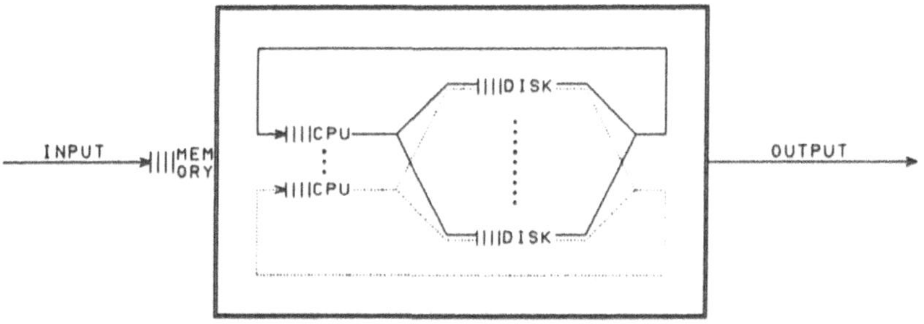

Figure 1: A queuing model of a single queue access multiple processor computer system

The remainder of this paper consists of the following sections. In Section 2 the principles of the mean value analysis are briefly summarized. In Section 3 we discuss four aggregation ideas. Numerical results are discussed in Section 4. Some concluding remarks are added in Section 5.

We note that an extended version of this paper, [5], is available upon request.

2. The mean value analysis algorithm

To introduce our notations, we will briefly discuss the principles of the mean value analysis algorithm for closed single server station queuing networks. We will use the following notation:

- c, d, i — station indices for respectively a CPU, a disk, and an arbitrary station
- r, s — indices denoting the type of client, $r, s = 1, 2, \ldots, R$
- $\underline{k}$ — system population in vector notation,

  $$\underline{k} \in \left\{ (0,0,\ldots,0), (0,0,\ldots,0,1), \ldots\ldots, (K_1, K_2, \ldots, K_R) \right\}$$

- $\underline{e}_r$ — vector denoting a population of a single client of type r
- $S_{i,r}[\underline{k}]$ — mean sojourn time at station i for type r clients, given population $\underline{k}$
- $N_{i,r}[\underline{k}]$ — mean number of clients at station i of type r, given population $\underline{k}$
- $\Omega_r[\underline{k}]$ — mean cycle time for clients of type r, given a system population $\underline{k}$; a cycle consists of a predefined number of visits to each of the network stations

$\Lambda_r[\underline{k}]$ mean throughput rate for clients of type r and given the system population $\underline{k}$

$w_{i,r}$ mean workload for clients of type r at station i

$f_{i,r}$ mean number of visits per cycle to station i for a particular type r client

We assume the processors to spread their attention equally over all clients in their queue; the processor sharing (PS) service discipline. In the mean value scheme this yields the following relation for the sojourn time of a single visit to a CPU:

$$S_{c,r}[\underline{k}] = (\, 1 + \sum_{s=1}^{R} N_{c,s}[\underline{k}-\underline{e}_r]\,)w_{c,r} \tag{2}$$

This relation is based on the arrival theorem for closed queuing networks. This theorem states that in product form queuing networks an client arriving at a certain queue will observe the equilibrium situation as if this arriving client observed the queue while not being a part of the system population. The arrival theorem was first mentioned in Lavenberg & Reiser [1].
The disk units are assumed to handle requests in first-come-first-served (FCFS) order. The mean value relation we use to compute the sojourn times in the next sections, is the following:

$$S_{d,r}[\underline{k}] = \sum_{s=1}^{R} N_{d,s}[\underline{k}-\underline{e}_r]w_{d,s} + w_{d,r} \tag{3}$$

We note that in order to obtain an exact mean value scheme, the workloads for each of the different client types in this relation ought to have identical exponentially distributions.

The cycle time for a client of type r is simply a summation over the sojourn times:

$$\Omega_r[\underline{k}] = \sum_c f_{c,r}S_{c,r}[\underline{k}] + \sum_d f_{d,r}S_{d,r}[\underline{k}] \tag{4}$$

The throughput of the system is defined as the average number of clients which finish a service cycle per unit of time. Noting that there are k_r clients of type r in the system and that relation (4) yields in fact the average sojourn time for such a cycle of service, we can apply Little's formula ([2]) to the system as a whole, and thus find the system throughput:

$$\Lambda_r[\underline{k}] = \frac{k_r}{\Omega_r[\underline{k}]} \tag{5}$$

The mean value scheme is completed by applying Little's formula to each of the separate queues in the network:

$$N_{i,r}[\underline{k}] = f_{i,r}\Lambda_r[\underline{k}]S_{i,r}[\underline{k}] \tag{6}$$

The relations (2) and (3) define the recursion in the mean value scheme. The recursion goes back until the zero population vector, at which level it is known that all queues are empty.

3. Aggregation

Now let us assume we consider a system as described in the introduction of this paper; a system in which each of the clients may visit only one of the CPU's. Furthermore, let us assume that in accordance with the workload characteristics we decide to aggregate a number of client types r to an aggregated type q. This could, but not necessarily, be a logical aggregation, e.g. aggregating all batch jobs or all compiler jobs. To denote this aggregation, we introduce the following notation:

Θ_q Set of types r which are aggregated in type q

The workload characteristics of the aggregated type are largely determined by the relative throughput of the component parts. In case the system and workload characteristics are balanced, the aggregation is more or less trivial:

$$K_q = \sum_{r \in \Theta_q} K_r \tag{7}$$

$$f_{i,q} = \sum_{r \in \Theta_q} \frac{K_r}{K_q} f_{i,r} \tag{8}$$

$$w_{i,q} = \sum_{r \in \Theta_q} \frac{f_{i,r} K_r}{\sum_{s \in \Theta_q} f_{i,s} K_{i,s}} w_{i,r} \tag{9}$$

The problem which remains is of a reduced complexity and can be solved again with the mean value scheme (2)-(6).

As an alternative for the direct aggregation of the workload characteristics, (7)-(10), we suggest an adjustment of the mean value scheme. Again we aggregate the types of clients $r \in \Theta_q$ to a new type q. However in each of the recursion steps of the mean value scheme, we act as if we are dealing with the original R type of clients problem. Redefining $\underline{k}$ as the aggregated population vector $(k_1, k_2, \ldots, k_q, \ldots, k_Q)$, this leads to the following set of equations:

$$S_{i,r}[\underline{k}] = \begin{cases} (1 + \sum_{s=1}^{R} N_{i,s}[\underline{k} - \underline{e}_q])w_{i,r} & ,i \text{ a } PS \text{ queue} \\ \sum_{s=1}^{R} N_{i,s}[\underline{k} - \underline{e}_q]w_{i,s} + w_{i,r} & ,i \text{ a } FCFS \text{ queue} \end{cases} \tag{11}$$

$$\Lambda_r[\underline{k}] = \frac{\frac{K_r}{K_q} k_q}{\sum_i f_{i,r} S_{i,r}[\underline{k}]} \tag{12}$$

$$N_{i,r}[\underline{k}] = f_{i,r} \Lambda_{i,r}[\underline{k}] S_{i,r}[\underline{k}] \tag{13}$$

$$\Omega_q[\underline{k}] = \frac{k_q}{\sum_{r \in \Theta_q} \Lambda_r[\underline{k}]} \tag{14}$$

Note that if we aggregate two identical types of clients, both the aggregation (7)-(10) and the scheme (11)-(14) will yield the exact results.

Both aggregation algorithms neglect the fact that the aggregation increases the variability of in particular the queue length distributions of the CPUs. An increasing variability usually causes the system throughput to drop. We may look at this problem in terms of the arrival theorem. The arrival theorem states that a client arriving at a station will observe a system queue as if he was looking at the equilibrium of the system with one client of the arriving type removed from the population. A removal of a client from the system population has two effects on the observed equilibrium. The direct effect is that the queue lengths for the clients of the type from which one client is removed reduces; summing over all queues, we find one client less. A second, indirect effect is that some probably minor shifts in the lengths of the queues of other types may occur, but conserving their total lengths. If we use the set of equations (11)-(14) and we study an example including two CPU's, the removal of an aggregated client can be seen as the removal of half a CPU 1 client and half a CPU 2 client. Looking at the sojourn time at the disk unit we see that, as both types of clients are approximately equal, this removal of two times half a client has about the same effect on the total queue length at this station as the removal of a client of the arriving type; the arriving client observes two chains, each containing half a client less than they should contain according to the system population. At the CPU queue, however, we observe only half of the direct effect, as only one of the aggregated types visits this CPU. This will yield an overestimation of the queue length and therewith the sojourn time at the CPU. This observation leads to the following suggestion for an improved approximation of the sojourn time at stations at which only one type $r \in \Theta_q$ arrives:

$$S_{c,r}[\underline{k}] = (1 + \sum_{s \neq r} N_{c,s}[\underline{k}-\underline{e}_q] + N_{c,r}[\underline{k}-\frac{K_q}{K_r}\underline{e}_q])w_{c,r} \tag{15}$$

$$S_{d,r}[\underline{k}] = \sum_{s \neq r} N_{d,s}[\underline{k}-\underline{e}_q]w_{d,s} + N_{d,r}[\underline{k}-\frac{K_q}{K_r}\underline{e}_q]w_{d,r} + w_{d,r} \tag{16}$$

where

$$\frac{K_q}{K_r}\underline{e}_q = (0,0,...0,\frac{K_q}{K_r},0,...,0)$$

a population vector containing type r clients only. As this population vector is likely to contain a non-integer number of clients, an interpolation might be required. If we denote by $\lceil x \rceil$ the smallest integer equal to or larger than x, and by $\lfloor x \rfloor$ the largest integer equal to or smaller than x, then this interpolation leads to the following expression for the average queue lengths:

$$N_{i,r}[\underline{k}-\frac{K_q}{K_r}\underline{e}_q] = \begin{cases} 0 & ,k_q < \frac{K_q}{K_r} \\ N_{i,r}[\underline{k}-\frac{K_q}{K_r}\underline{e}_q] & ,k_q \geq \frac{K_q}{K_r}, \frac{K_q}{K_r} \text{ integer} \\ \left(\lceil\frac{K_q}{K_r}\rceil - \frac{K_q}{K_r}\right)N_{i,r}[\underline{k}-\lfloor\frac{K_q}{K_r}\rfloor\underline{e}_q] + & \\ \left(\frac{K_q}{K_r} - \lfloor\frac{K_q}{K_r}\rfloor\right)N_{i,r}[\underline{k}-\lceil\frac{K_q}{K_r}\rceil\underline{e}_q] & ,\text{elsewhere} \end{cases} \tag{17}$$

The use of equations (15)-(17) is based on the idea that the direct influence of the removal of a client is most important, whereas the indirect effect is almost independent from the type of client removed.

A logical question to ask next is why not use equations (15)-(17) for all the stations. The advantage would be that the total number of type r clients found in the queues is indeed one less than the number of type r clients present. The total number of clients of the aggregated type q which is found upon arrival, however, would be less than $k_q - 1$. This effect overrides the beneficial effect of a more correct queue r approximation. Again, however, we may think of an artificial way to overcome these problems, by suggesting the following relations:

$$S_{c,r}[\underline{k}] = (1 + N_{c,r}[\underline{k}-\frac{K_q}{K_r}\underline{e}_q] + \sum_{s \notin \Theta_q} N_{c,s}[\underline{k}-\underline{e}_q] + \sum_{\substack{s \in \Theta_q \\ s \neq r}} \frac{k_q}{k_q - 1} N_{c,s}[\underline{k}-\underline{e}_q])w_{c,r} \tag{18}$$

$$S_{d,r}[\underline{k}] = w_{d,r} + N_{d,r}[\underline{k}-\frac{K_q}{K_r}\underline{e}_q]w_{d,r}$$

$$+\sum_{s \notin \Theta_q} N_{d,s}[\underline{k}-\underline{e}_q]w_{d,s} + \sum_{\substack{r' \in \Theta_q \\ s \neq r}} \frac{k_q}{k_q-1} N_{d,s}[\underline{k}-\underline{e}_q]w_{d,s} \qquad (19)$$

The effect of using these relations is that the sum of all queue lengths used to compute the sojourn times for a client of type r, yields exactly the total population considered in the recursion step minus this type r client.

4. Some numerical experiments

To allow a numerical evaluation of the aggregation methods discussed in the preceding section, we have created a test-bed of CPU-disk models. We distinguish between the examples according to four problem characteristics: the number of client types, the number of CPU's, the bottleneck situation and the differences in the workload of the clients to be aggregated. For these characteristics we have defined the following problem categories:

- Ia) One type of client per CPU. After aggregating, a problem remains with a single client type.
- Ib) Two types of clients per CPU. In that case the aggregation defines two types of clients.
- IIa) Two CPU's.
- IIb) Three CPU's.
- IIIa) The CPU is the bottleneck (the bottleneck stations have an utilization between 75% and 90%). The number of disk units is that large that there is no queuing at these devices. In the model the disk units are replaced by so called infinite-server stations.
- IIIb) The CPU is the bottleneck. The average utilization of the disk units is about half the utilization of the bottleneck station.
- IIIc) The CPU is the bottleneck. The average disk utilization is about 75% of the bottleneck station.
- IIId) The disk units define the system bottleneck. The CPU utilization is about 75% of the utilization of this bottleneck.
- IVa) All CPU's identical. No differences in workload.
- IVb) All CPU's identical. Each CPU has a private system disk such that the disk workload of the types of clients to be distinguished differs.
- IVc) The CPU's differ in processor speed by a factor 2 or 3. Each CPU has a private system disk. The types to be aggregated have approximately the same service requirements per visit, but differ in the number of visits required to finish their work.
- IVd) The CPU's differ in speed by a factor 2 or 3. The multiprogramming levels are proportional with the CPU speed. The number of visits before a job is finished is identical for each of the CPU's. Each CPU has access to a private system disk.

Note that for the problems resulting under IVa and IVb, the direct aggregation of the workload via (8)-(10) is identical to the mean value aggregation through (11)-(14).

For each combination of problem categories, we have defined one example, such that our test-bed comprises of 64 examples. Some results involving the accuracy of the cycle time approximation are summarized below. More detailed results are reported in [5].

The following aggregation methods were compared:

1: Direct aggregation of workload; equation (8)-(10)

2: Aggregation in the mean value scheme; equation (11)-(14)

3: Idem, but with adjustment (15)-(17)

4: Idem, but with adjustment (17)-(19)

	Average error (%)				**Maximum error (%)**			
EXAMPLE\METHOD	1	2	3	4	1	2	3	4
IA)	7.15	6.68	0.80	0.80	13.80	13.05	1.57	1.74
IB)	7.79	7.21	1.62	1.61	20.45	14.96	2.87	2.95
IIA)	5.67	5.38	1.12	1.13	11.15	9.79	2.40	2.43
IIB)	9.48	8.69	1.57	1.56	20.45	14.96	2.87	2.95
IIIA)	9.02	8.50	1.16	1.16	14.96	14.96	2.74	2.74
IIIB)	10.63	9.60	1.67	1.78	20.45	13.90	2.87	2.95
IIIC)	7.47	6.99	1.48	1.55	14.72	11.79	2.64	2.73
IIID)	3.19	3.04	1.07	0.88	7.55	6.96	1.62	1.47
IVA)	7.03	7.03	1.39	1.42	14.96	14.96	2.87	2.95
IVB)	6.92	6.92	1.35	1.33	13.79	13.79	2.69	2.76
IVC)	7.67	7.26	1.33	1.32	15.30	13.90	2.68	2.73
IVD)	8.69	6.92	1.32	1.30	20.45	13.32	2.53	2.60
ALL	7.58	7.03	1.35	1.34	20.45	14.96	2.87	2.95

Table 1: Cycle time errors in CPU-disk model with aggregation over the processors;

In Table 1 the average and maximum errors found in comparing the exact and approximate aggregated cycle times are given. The differences in the approximations through (15)-(17) and (17)-(19) are only moderate. This is largely explained by the fact that the test-bed contained no examples in which the disk workload differed among the types of clients to be aggregated.

The maximum errors observed are twice the average for method 2, 3 and 4, and the triple for method 1. So the mean value aggregations seem to guarantee a reasonable accuracy.

A potential improvement of these methods is found in the observation that all the errors are in fact overestimations of the cycle times, refer [5].

	Average error (%)				**Maximum error (%)**			
EXAMPLE\METHOD	1	2	3	4	1	2	3	4
IIA)	1.33	0.14	0.14	0.70	6.81	0.66	0.66	1.75
IIB)	1.04	0.04	0.04	0.40	3.56	0.33	0.33	1.30
IIIA)	1.14	0.16	0.16	1.02	3.55	0.66	0.66	1.75
IIIB)	0.93	0.09	0.09	0.73	1.85	0.32	0.32	1.29
IIIC)	1.99	0.05	0.05	0.27	6.81	0.19	0.19	0.48
IIID)	0.57	0.04	0.04	0.05	1.58	0.12	0.12	0.28
IVA)	1.33	0.13	0.13	0.55	3.55	0.66	0.66	1.59
IVB)	1.19	0.06	0.06	0.54	3.12	0.19	0.19	1.75
IVC)	0.71	0.07	0.07	0.47	1.93	0.35	0.35	1.25
IVD)	1.40	0.07	0.07	0.52	6.81	0.32	0.32	1.57
ALL	1.16	0.08	0.08	0.52	6.81	0.66	0.66	1.75

Table 2: Cycle time errors in CPU-disk model with aggregation per processor

Although with this paper we intended to study a special type of aggregation, aggregation over the processors, we have also performed a small exercise in aggregating the clients per processor instead. The results of this exercise, which comprises of the 32 examples convening under problem characteristic Ib, are summarized in Table 2. The results suggest that this type of aggregation is much more robust. Especially aggregation method 2 behaves very well under these conditions. Note that, as each of the composing types in the aggregation have the same routing pattern, aggregation methods 2 and 3 are identical.

5. Conclusions

We have discussed and numerically evaluated some aggregation methods based on the mean value scheme. Each of these methods yields the performance characteristics for the whole range of aggregated population vectors from zero to maximum. So they can be applied in a decomposition approach to analyze larger systems.

We have shown that in aggregating types of clients which partly visit different stations, the direct workload aggregation is insufficient. An aggregation in the mean value scheme gives much better results with just a small loss in computational efficiency as compared to the direct workload aggregation.

Three variants of the mean value scheme based aggregation method were discussed. The first method just differed from the ordinary mean value scheme in the recursion used. In some situations it can be proved to be identical to the direct workload aggregation. This method proved to be especially accurate in the situation where the types of clients to be aggregated have the same routing pattern.

Two variants were discused which were especially accurate in case these routing patterns differed. methods suggested may be applied to tackle other large problems as well.

References

1. LAVENBERG, S.S. AND REISER, M., "Stationary State Probabilities at Arrival Instants for Closed Queueing Networks with Multiple Types of Clients," *J.Appl.Probab.* **17** (1980), 1048-1061.
2. LITTLE, J.D.C., "A Proof for the Queuing Formula: $L=\lambda W$," *Oper.Res.* **9** (1961), 383-387.
3. REISER, M. AND LAVENBERG, S.S., "Mean-Value Analysis of Closed Multichain Queuing Networks," *J.ACM* **27** (1980), 313-322.
4. SCHWEITZER, P.J., SEIDMANN, A., AND SHALEV-OREN, S., "The Correction Terms in Approximate Mean Value Analysis," *Oper.Res.Lett.* **4** (1986), 197-200.
5. WIJBRANDS, R.J., "On an Aggregation Method for CPU-Disk Models," *Memorandum COSOR 86-05*, Dep. of Math. and Comput.Sci., Eindhoven U. of Techn., Eindhoven (1986).

COMPUTER PROGRAM FOR DETERMINING AN OPTIMUM SOLUTION IN LONG-TERM FOREST EXPLOITATION PROCESS

Janez Barle, Janez Grad, Ljubljana

In the paper we describe a possible usage of dynamic programming, based upon Bellman's principle, in order to obtain the optimum solution of a predefined long-term forest exploitation process. In this process measures and activities are introduced at each step within the iterative solution process, and their effects on the intermediate results and the final solution are stated in the form of mathematical relations. The whole forest area which is taken into consideration is divided into a number of homogeneous parts - segments. For each of these segments, five special functions are imposed by means of which we direct (guide) the process within a number of time intervals, either years or decades. These functions are: (F1) the maximum possible tree growth capacity, (F2) exploitation capacity, (F3) the quality of the existing tree specimens, (F4) level of administration - care for improved growing conditions, and (F5) stage in segment development, based upon the age (oldness) of the tree specimens in it. The functions help to optimize exploitation endeavours. At each step of the iterative solution process four possible activities can be imposed: (A1) no activity at all, (A2) exchange of the existing tree specimen with a new one, (A3) rarefying, and (A4) restoration. The exploitation policy is defined by a sequence of chosen activities during the iterative process. The stated functions, activities and time intervals are part of the mathematical model and dynamic programing process respectively. The problem and the corresponding computer program can be extended by incorporating some additional functions and activities. Program prototype in FORTRAN was written and tested on DEC-10 computer at the University of Ljubljana, Yugoslavia. In the paper we also comment the problems which are associated with dynamic programming applied to such type of problem.

Minimierung der maximalen erwarteten Verspätung in EO- Netzplänen

Matthias Bücker, Karlsruhe

Zusammanfassung: Aufbauend auf dem $1|prec|f_{max}$- Problem wird ein ähnliches Modell für Anordnungsbeziehungen von Typ der sogenannten GERT- Netzpläne vorgestellt. Diese Änderung bewirkt, daß das daraus resultierende Problem NP- schwer wird. Betrachtet man jedoch eine Einschränkung der GERT- Netzpläne, die sogenannten EO- Netzpläne, so läßt sich der Algorithmus für das $1|prec|f_{max}$- Problem auf das daraus entstehende Problem übertragen.

Summary: Based on the $1|prec|f_{max}$- problem a similar model will be introduced using precedence constraints of the type of so- called GERT- networks. This modification causes the resulting problem to be NP- hard. Investigating instead of GERT- networks a restriction of these to the so- called EO- networks shows that the algorithm for the $1|prec|f_{max}$- problem can be tranfered to the new problem.

1. Einleitung

Im Jahre 1973 hat E.L. Lawler einen Algorithmus angegeben, mit dem das $1|prec|f_{max}$- Problem in polynomialer Zeit gelöst werden kann (s. LAWLER[73]). Die dort vorausgesetzten Anordnungsbeziehungen implizieren jedoch, daß alle Vorgänge ausgeführt werden. Diese Voraussetzung wird bei der Betrachtung von Anordnungsbeziehungen vom Typ der GERT- Netzpläne fallengelassen; die Vorgänge werden in einem solchen Modell nur noch mit gewissen Wahrscheinlichkeiten ausgeführt, woraus sich in natürlicher Weise eine Zusammenfassung mehrere Vorgänge zu sogenannten Operationen ergibt. Diese Verallgemeinerung führt dazu, daß das Problem der Minimierung der maximalen (erwarteten) Verspätung der Operationen NP- schwer wird, auch wenn alle anderen Voraussetzungen (inbesondere, daß nur eine Maschine zur Abarbeitung der Operationen (Jobs) zur Verfügung steht) gleich bleiben. Beschränkt man sich jedoch auf einen bestimmten Typ von GERT- Netzplänen, die sogenannten EO- Netzplänen, und setzt hier Zyklenfreiheit und Zulässigkeit[1] voraus (was wir im folgenden immer tun wol-

---

1) Zum Konzept der GERT- bzw. EO- Netzpläne und der in diesen Netzplänen verwendeten Knotentypen, sowie des Begriffs "zulässig" sei auf NEUMANN/ STEINHARDT[79] verwiesen.

Operations Research Proceedings 1987

len), so kann der von Lawler angegebene Algorithmus für die Zielfunktion "Minimierung der maximalen erwarteten Verspätung der Operationen" übertragen werden. Dabei ist es in diesen speziellen Netzplänen möglich die erwarteten Verspätungen der Operationen mit Hilfe linearer Gleichungssysteme zu bestimmen.

2. Begriffe und Definitionen für Scheduling- Probleme in GERT- Netzplänen

Da wir uns hier mit Scheduling- Problemen in zyklenfreien, zulässigen GERT- Netzplänen beschäftigen wollen, sollen hier die dafür benötigten Begriffe eingeführt werden. Die Knotenmenge eines GERT- Netzplans bezeichnen wir mit V, die Menge der Pfeile (Vorgänge) mit E.

Zusätzlich zu den sonst üblichen zwei Pfeilgewichten p_{ij} und $F_{ij}(t)$), die die Ausführungswahrscheinlichkeit bzw. die Verteilungsfunktion der Dauer des Pfeils (Vorgangs) $\langle i,j\rangle \in E$ angeben (unter der Bedingung, daß Knoten $i \in V$ aktiviert bzw. $\langle i,j\rangle \in E$ ausgeführt wird), führen wir noch eine 3. Komponente $f_{ij} \in \mathbb{R}$ ein, die den geforderten Fertigstellungstermin für Vorgang $\langle i,j\rangle \in E$ angibt (unter der Bedingung, daß $\langle i,j\rangle \in E$ ausgeführt wird).

Wird in einem GERT- Netzplan nun ein stochastischer Knoten aktiviert, so wird genau ein wegführender Pfeil (Vorgang) ausgeführt, während bei der Aktivierung eines deterministischen Knotens alle wegführenden Pfeile ausgeführt werden. Wir führen daher den folgenden Begriff ein (s. RUBACH[84]) ($\mathcal{S}(i) \subseteq V$: Menge aller Nachfolger von Knoten $i \in V$):

<u>DEF.:</u> Gegeben sei ein GERT- Netzplan N.

i) Sei $i \in V$ ein stochastischer Knoten, dann heißt der durch $\{i\} \cup \mathcal{S}(i)$ induzierte Unternetzplan von N eine stochastische Operation.

ii) Sei $i \in V$ ein deterministischer Knoten und $j \in \mathcal{S}(i)$, dann heißt der durch $\{i, j\}$ induzierte Unternetzplan von N eine deterministische Operation.

<u>BEZ.:</u> i) Mit $\mathcal{O}$ wollen wir die Menge aller Operationen eines GERT- Netzplans bezeichnen.

ii) Mit r_o bezeichnen wir die Quelle einer Operation $o \in \mathcal{O}$ und mit S_o die Menge aller Senken von o.

iii) $P = (o_1,\dots,o_n)$ $(o_i \in \mathcal{O},\ o_i \neq o_j,\ i \neq j,\ i, j \in \{1,\dots,n\})$ wollen wir Prioritätsliste nennen.

(o_i hat höhere Priorität als o_j $:\Longleftrightarrow$ o_i steht in P vor o_j)

Da während jeder Ausführung eines zyklenfreien, zulässigen GERT- Netzplans von jeder Operation $o \in \mathcal{O}$ höchstens ein Vorgang ausgeführt wird, können wir uns im folgenden auf die Betrachtung von Operationen beschränken. Die "Verspätung", die wir nun einer Operation $o \in \mathcal{O}$ zuordnen, stellt wegen der stochastischen Ablaufstruktur eine Zufallsvariable dar, die abhängt von einer Prioritätsliste P (es steht nur eine Maschine zur Bearbeitung der Vorgänge zur Verfügung), sowie vom Abschlußzeitpunkt eines Vorgangs $\langle i,j\rangle \in o$.

<u>BEZ.:</u> Sei N ein GERT- Netzplan und Ω die Menge aller Realisierungen dieses Netzplans. Gegeben sei ferner eine Prioritätsliste P, eine Realisierung $\omega \in \Omega$, sowie eine Operation $o \in \mathcal{O}$.

i) $C_{ij}(P,\omega)$ sei der Abschlußzeitpunkt des Vorgangs $\langle i,j\rangle \in E$ unter der Prioritätsliste P in der Realisierung $\omega \in \Omega$.

ii) $$L_o(P,\omega) := \begin{cases} C_{ij}(P,\omega) - f_{ij}, & \text{Vorgang } \langle i,j\rangle \in o \text{ wird in der Realisation } \omega \text{ ausgeführt} \\ 0, & \text{sonst} \end{cases}$$

nennen wir die Verspätung von Operation o.

Ziel ist es jetzt, eine Prioritätsliste P^* zu finden, die $\max_{o \in \mathcal{O}}\{\mathbb{E}[L_o(P)]\}$ ($\mathbb{E}[\cdot]$: Erwartungswert) minimiert. Probleme dieser Art sollen im folgenden mit $1|GERT|\max\{\mathbb{E}[L_o]\}$ bzw. $1|EO|\max\{\mathbb{E}[L_o]\}$ bezeichnet werden, je nachdem, ob ein zyklenfreier, zulässiger GERT- oder EO- Netzplan den Anordnungsbeziehungen zugrunde liegt.

Daß das $1|GERT|\max\{\mathbb{E}[L_o]\}$- Problem nicht in polynomialer Zeit gelöst werden kann, wird im folgenden Abschnitt gezeigt.

3. Die Komplexität des $1|GERT|\max\{\mathbb{E}[L_o]\}$- Problems

Um zu zeigen, daß das $1|GERT|\max\{\mathbb{E}[L_o]\}$- Problem NP- schwer ist, führen wir das Minimum- Cover- Problem (MC) auf unser Problem zurück ((MC) ist NP- schwer (s. GAREY/ JOHNSON[79])).

Unter (MC) verstehen wir dabei ein Problem der folgenden Art:

Gegeben sei eine endliche Menge S, ein Mengensystem $Q \subseteq \mathbb{P}(S)$ (Potenzmenge von S) und eine Zahl $k \in \mathbb{N}$ mit $k \leq |S|$.

(MC): Gibt es eine Teilmenge $Q' \subseteq Q$ mit $|Q'| \leq k$ und $\bigcup_{A \in Q'} A = S$?

<u>THEOREM:</u> Das $1|GERT|\max\{\mathbb{E}[L_o]\}$- Problem ist NP- schwer.

<u>Bew.:</u> Gegeben sei ein Minimum- Cover- Problem.

Betrachte nun den GERT- Netzplan mit Knotenmenge V und Pfeilmenge E:

$V = \{1, n, n+1\} \cup Q \cup S$

$E = \{\langle 1,i\rangle | i \in Q\} \cup \{\langle i,j\rangle | i \in Q, j \in i\} \cup \{\langle j,n\rangle | j \in S\} \cup \{\langle n,n+1\rangle\}$.

Dabei seien die Knoten aus $\{1\} \cup Q$ deterministische EO- Knoten[1), die Knoten aus S stochastische IO- Knoten, n ein stochastischer UND- Knoten und n+1 ein stochastischer EO- Knoten (vergl. Abb. 1). (Da dieser Netzplan nur Operationen enthält, die aus einem Pfeil mit dessen Anfangs- und Endknoten bestehen, werden die Operationen hier im folgenden durch die Angabe des entsprechenden Pfeiles gekennzeichnet.)

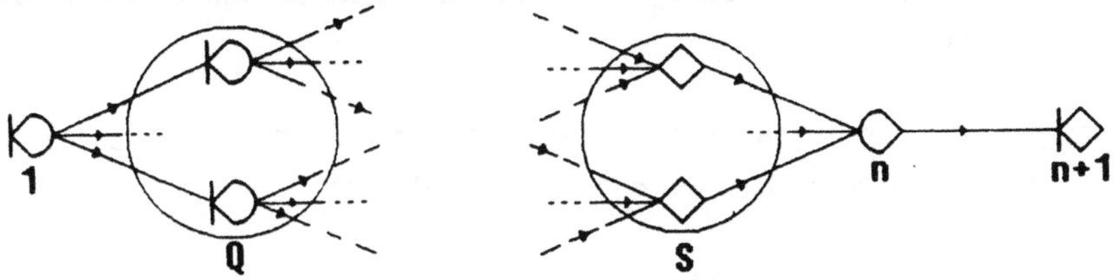

Abb. 1

Die Pfeilgewichte legen wir wie folgt fest:

$p_{ij} = 1 \qquad \forall \langle i,j\rangle \in E$

$F_{ij}(t) = \begin{cases} 0, & \text{für } t < 1 \\ 1, & \text{sonst} \end{cases} \qquad \forall \langle i,j\rangle \in E$

$$f_{ij} = \begin{cases} 2|S| + 1 + \sum_{Q' \in Q} |Q'|, & \text{für } \langle i,j\rangle \neq \langle n,n+1\rangle \\ 2|S| + 1 + k, & \text{für } \langle i,j\rangle = \langle n,n+1\rangle \end{cases}$$

Wie man sofort sieht, lautet die Antwort auf (MC) genau dann "Ja", wenn gilt, daß der Zielfunktionswert des $1|GERT|\max\{\mathbb{E}[L_o]\}$- Problems kleiner oder gleich 0 ist, da für alle $\langle i,j\rangle \neq \langle n,n+1\rangle$ offensichtlich $\mathbb{E}[L_{\langle i,j\rangle}(\cdot)] \leqq 0$ gilt, während $\mathbb{E}[L_{\langle n,n+1\rangle}(\cdot)] \leqq 0$ nur dann gelten kann, wenn höchstens k Knoten aus der Menge Q aktiviert werden, bevor die Operation $\langle n,n+1\rangle$ ausgeführt wird (diese Knoten entsprechen dann gerade der Menge Q'). ■

Damit ist gezeigt, daß das $1|GERT|\max\{\mathbb{E}[L_o]\}$- Problem NP- schwer ist. Deshalb wollen wir uns als nächstes mit dem leichteren Problem $1|EO|\max\{\mathbb{E}[L_o]\}$ beschäftigen.

---

1) Zu den in GERT- Netzplänen verwendeten Knotentypen sei noch einmal auf NEUMANN/STEINHARDT[79] verwiesen.

4. Ein Algorithmus zur Lösung des $1|EO|\max\{\mathbb{E}[L_o]\}$- Problems

Als erstes sollen nun einige Hilfsmittel zusammengestellt werden:

SATZ 1: Gegeben sei ein zyklenfreier, zulässiger EO- Netzplan N, eine Prioritätsliste $P = (o_1,\dots,o_n)$ $(o_i \in \mathcal{O})$ und für ein $i \in \{1,\dots,n\}$ die Prioritätsliste $P' = (o_1,\dots,o_{i-1},o_n,o_i,\dots,o_{n-1})$, dann gilt:

$$\mathbb{E}[L_{o_n}(P')] \leqq \mathbb{E}[L_{o_n}(P)]$$

Bew.: Unter P' wird o_n in keiner Realisation von N später als unter P ausgeführt. ∎

DEF.: Eine Operation $o \in \mathcal{O}$ eines zyklenfreien EO- Netzplans N heißt Schlußoperation, wenn $S_o \subseteqq S$ (= Senkenmenge von N) gilt.

BEZ.: i) $O_S \subseteqq \mathcal{O}$ soll die Menge aller Schlußoperationen bezeichnen.

ii) Mit P_o bezeichnen wir die Menge aller Prioritätslisten, in denen die Operation $o \in \mathcal{O}$ an letzter Stelle steht.

$(P_o = \{P|P = (o_1,\dots,o_{n-1},o),\ o_1,\dots,o_{n-1},o \in \mathcal{O}\})$

LEMMA 1: Gegeben sei ein zyklenfreier, zulässiger EO- Netzplan N, $o \in \mathcal{O}$ und $P, Q \in P_o$, dann gilt:

$$\mathbb{E}[L_o(P)] = \mathbb{E}[L_o(Q)]$$

Bew.: Da o in P und Q an letzter Stelle steht, wird o unter P und unter Q für jede Realisation von N zur gleichen Zeit ausgeführt. ∎

Damit können wir nun zeigen:

SATZ 2: Gegeben sei ein zyklenfreier, zulässiger EO- Netzplan N.

$o \in O_S$ und $P \in P_o$ seien so gewählt, daß $\mathbb{E}[L_o(P)] = \min_{q \in O_S} \{\mathbb{E}[L_q(Q)]\,|\,Q \in P_q\}$ gilt,

dann folgt:

Es existiert eine Prioritätsliste $P^* \in P_o$, die $\max_{o' \in \mathcal{O}} \{\mathbb{E}[L_{o'}(P)]\}$ minimiert.

Bew.: O.B.d.A. können wir davon ausgehen, daß in einer optimalen Prioritätsliste die letzte Operation eine Schlußoperation ist, da es keinen Vorteil bringen kann, eine Schlußoperation vor eine Operation zu stellen, die vor dieser Schlußoperation ausgeführt werden muß.

Gegeben sei nun eine Prioritätsliste $Q \in P_q$ mit $q \in O_S$ in der Form

A	o	B	q

. Betrachte nun die Prioritätsliste $P \in P_o$ der folgenden

Form: | A | B | q | o |. Da o, q $\in O_S$, also keinen anderen Operationen "vorangehen" müssen, gilt für alle Operationen o' aus den Bereichen A und B wegen Satz 1: $\mathbb{E}[L_{o'}(P)] \leqq \mathbb{E}[L_{o'}(Q)]$. Außerdem gilt wegen Satz 1: $\mathbb{E}[L_q(P)] \leqq \mathbb{E}[L_q(Q)]$. Aus den Voraussetzungen folgt: $\mathbb{E}[L_o(P)] \leqq \mathbb{E}[L_o(Q)]$, womit insgesamt $\max_{o' \in \mathcal{O}} \{\mathbb{E}[L_{o'}(P)]\} \leqq \max_{o' \in \mathcal{O}} \{\mathbb{E}[L_{o'}(Q)]\}$ und $P \in P_o$ gilt. ■

Damit können wir jetzt eine optimale Prioritätsliste P^* für einen gegebenen zyklenfreier, zulässiger EO- Netzplan N wie folgt bestimmen:

Schritt 0: Setze $i = |\mathcal{O}|$.

Schritt 1: Bestimme die Menge O_S von N und eine Operation $o \in O_S$ mit

$$\mathbb{E}[L_o(P)] = \min_{q \in O_S} \{\mathbb{E}[L_q(Q)] \,|\, Q \in P_q\} \quad (P \in P_o).$$

Schritt 2: Setze o an die i- te Stelle der Prioritätsliste.

Schritt 3: Entferne o aus N, setze $i = i-1$ und gehe zu Schritt 1, falls $i > 0$, sonst terminiere.

BEM.: Wegen Lemma 1 kann in Schritt 1 eine beliebige Prioritätsliste $P \in P_o$ bzw. $Q \in P_q$ verwandt werden.

Wie die bei der Ausführung des Algorithmus auftauchende Schwierigkeit, $\mathbb{E}[L_o(P)]$ $(P \in P_o)$ zu bestimmen überwunden werden kann, wird im nächsten Abschnitt beschrieben.

5. Die Berechnung von $\mathbb{E}[L_o(P)]$

Gegeben sei ein zyklenfreier, zulässiger EO- Netzplan N mit Knotenmenge $V = \{1,\ldots,v\}$ und Pfeilmenge E. Die Quelle des Netzplans sei der Knoten mit der Nummer 1, $D \subseteqq V$ sei die Menge der deterministischen Knoten in V und $d_{ij} = \mathbb{E}[D_{ij}]$ $(D_{ij} \sim F_{ij})$ $\forall \langle i,j\rangle \in E$, $\mathcal{R}(i) \subseteqq V$ die Menge der von $i \in V$ ereichbaren Knoten und $\overline{\mathcal{R}}(i) \subseteqq V$ die Menge aller Knoten, von denen $i \in V$ erreichbar ist. Dann bestimme für eine gegebene Operation $o \in O_S$:

$$D_o := D \cap \overline{\mathcal{R}}(r_o)$$

$$M_o := \{k \in V \,|\, (\exists i \in D_o\colon i \neq r_o \wedge k \in \mathcal{S}(i) \wedge r_o \notin \mathcal{R}(k)) \vee (r_o \in D_o \wedge k \in \mathcal{S}(r_o) \wedge k \notin S_o)\}$$

$$V_o := \bar{\mathfrak{R}}(r_o) \cup (\bigcup_{k\in M_o} \mathfrak{R}(k))$$

$$E_o := \{\langle i,j\rangle \in E | i,\ j \in V_o\}$$

$$\mathcal{S}_o(i) := \mathcal{S}(i) \cap V_o$$

$$p_{ij}^{(o)} := \begin{cases} p_{ij} & ,\ \text{falls} \sum_{j\in\mathcal{S}_o(i)} p_{ij} \geq 1 \\ p_{ij}[\sum_{j\in\mathcal{S}_o(i)} p_{ij}]^{-1} & ,\ \text{sonst} \end{cases}$$

$$q_j^{(o)} := \sum_{\langle i,j\rangle\in E_o} q_i^{(o)} p_{ij}^{(o)} \quad (q_1^{(o)} := 1)$$

$$t^{(o)} := \sum_{\langle i,j\rangle\in E_o} q_i^{(o)} p_{ij}^{(o)} d_{ij}$$

$$d^{(o)} := \sum_{\langle i,j\rangle\in Eo} p_{ij} d_{ij}$$

$$f^{(o)} := \sum_{\langle i,j\rangle\in Eo} p_{ij} f_{ij}$$

<u>SATZ 3:</u> Gegeben sei ein zyklenfreier, zulässiger EO- Netzplan N sowie eine Operation $o \in O_S$, dann gilt für $P \in P_o$:

$\mathbb{E}[L_o(P)] = q_{r_o} [t^{(o)} + d^{(o)} - f^{(o)}]$, wobei q_{r_o} mit Hilfe des linearen Gleichungssystem $q_j := \sum_{\langle i,j\rangle\in E} q_i p_{ij}$ $(q_1 := 1)$ berechnet werden kann.

<u>Bew.:</u> Wie man sich leicht klarmacht, bildet der durch V_o induzierte Netzplan N_o mit den Pfeilgewichten $p_{ij}^{(o)}$ und $F_{ij}(t)$ $(\langle i,j\rangle \in E_o)$ ebenfalls einen zyklenfreien, zulässigen EO- Netzplan, in dem die Quelle der Operation o r_o mit Wahrscheinlichkeit 1 aktiviert wird.Die Zyklenfreiheit und Zulässigkeit von N_o bedingen, daß die $q_j^{(o)}$ die Aktivierungswahrscheinlichkeiten der Knoten $j \in V_o$ darstellen. (Analoges gilt für das Gleichungssystem $q_j := \sum_{\langle i,j\rangle\in E} q_i p_{ij}$) (s. NEUMANN[86]). Da $P \in P_o$ ist, gilt daß $t^{(o)}$ die erwartete Dauer bis zur Aktivierung der Quelle von o unter P angibt (unter der Bedingung, daß o ausgeführt wird). $d^{(o)}$ bzw. $f^{(o)}$ geben die erwartete Dauer bzw. den erwarteten Fertigstellungstermin der Operation o an (unter der Bedingung, daß o ausgeführt wird). Damit berechnet sich die erwartete Verspätung von o unter der Bedingung, daß o ausgeführt wird gemäß $t^{(o)} + d^{(o)} - f^{(o)}$. Daraus folgt sofort, daß $\mathbb{E}[L_o(P)] = q_{r_o} [t^{(o)} + d^{(o)} - f^{(o)}]$ gilt. ■

Abschließend soll dieses Vorgehen noch an einem Beispiel verdeutlicht werden.

BSP.: Gegeben sei folgender EO- Netzplan, in dem die Operation o durch die schwarzen Knoten gekennzeichnet ist.

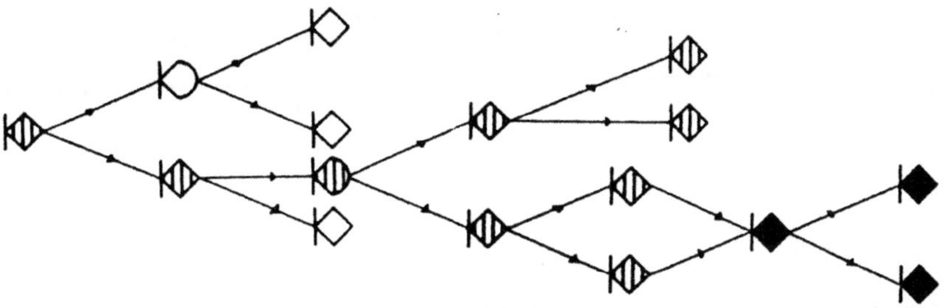

Abb. 2

Die Mengen D_o, M_o, V_o, E_o, $\varphi_o(i)$, sowie die Werte für $p_{ij}^{(o)}$, $q_j^{(o)}$ und $t^{(o)}$ ergeben sich durch Betrachtung des Netzplans der von den schraffierten Knoten und der Quelle von o induziert wird. Dieser Netzplan enthält genau die Vorgänge, die ausgeführt werden können unter der Bedingung, daß o ausgeführt wird. (Die Werte für $d^{(o)}$ und $f^{(o)}$ ergeben sich durch die Betrachtung der Operation o.)

Literatur:

GAREY, M.R., JOHNSON, D.S., *Computers and Interactability*, Freeman, San Francisco, 1979

LAWLER, E.L., "Optimal Sequencing of a Single Machine Subjekt to Precedence Constraints", *Managment Sience*, Vol. 19 (1973), 544- 546

NEUMANN, K., *Stochastic Project Networks: Temporal Analysis, Scheduling, and Cost Optimization*, Technical Report am Institut für Wirtschaftstheorie und Operations Research WIOR- 280, 1986

NEUMANN, K., STEINHARDT, U., *GERT Networks*, Lecture Notes in Economics and Math. Systems, Vol. 172, Springer, Berlin, 1979

RUBACH, T., *Stochastische Reihenfolgeplanung mit Hilfe von GERT- Netzplänen*, Dissertation an der Universität Fridericiana Karlsruhe, 1984

APPROXIMATION VON ERWARTUNGSWERTEN KONVEX-KONKAVER FUNKTIONEN

Karl Frauendorfer
Zürich

Die Berechnung von Erwartungswerten spielt in der stochastischen Optimierung eine wesentliche Rolle. Wir betrachten im folgenden

$$E(\phi) := \int_{\Xi} \int_{\Theta} \phi(\xi,\eta)\, dG(\eta)\, dF(\xi) ,$$

wobei $\xi \in \mathbb{R}^k$ und $\eta \in \mathbb{R}^l$ multivariate Zufallsgrössen repräsentieren mit den dazugehörigen Verteilungsfunktionen $F(\xi)$, $G(\eta)$; ϕ sei konvex in ξ und konkav in η; $\Xi \subset \mathbb{R}^k$ bzw. $\Theta \subset \mathbb{R}^l$ seien k-dimensionale bzw. l-dimensionale Rechtecke, die den Träger von ξ bzw. η enthalten.

Der Approximationsgedanke liegt nun darin, Funktionen $L(\xi,\eta)$ bzw. $U(\xi,\eta)$ zu ermitteln, die in ξ und in den einzelnen Komponenten η_j $(1 \leq j \leq l)$ bzw. in den einzelnen Komponenten ξ_i $(1 \leq i \leq k)$ und in η linear sind und für welche die Beziehung

$$L(\xi,\eta) \leq \phi(\xi,\eta) \leq U(\xi,\eta) \qquad \forall(\xi,\eta) \in \Xi \times \Theta$$

gilt. Mittels Anwendung der Jensen-Ungleichung und der Verallgemeinerung der Edmundson-Madansky Ungleichung lassen sich

$$E(L) := \int_{\Xi} \int_{\Theta} L(\xi,\eta)\, dG(\eta)\, dF(\xi) ,$$

$$E(U) := \int_{\Xi} \int_{\Theta} U(\xi,\eta)\, dG(\eta)\, dF(\xi)$$

berechnen, die eine untere bzw. obere Schranke für $E(\phi)$ darstellen. Durch sukzessives Partitionieren von $\Xi \times \Theta$ werden diese untere und obere Schranken verbessert und konvergieren monoton gegen $E(\phi)$.

DOES IT PAY TO SOLVE INFINITE-STAGE MARKOV DECISION PROBLEMS WHEN FINITE-STAGE SOLUTIONS ARE ASKED FOR?

Gerhard Hübner, Hamburg

It is well-known for Markov decision problems that there are more and faster algorithms for the infinite-stage case than for finite-stage problems. In the latter case (essentially) only standard successive approximation methods are applicable in connection with extrapolating bounds and eliminating of non-optimal actions. But if the infinite-stage solution is known it may be used to approximate the finite-stage value.

It seems to be an unsolved question in which cases it is more efficient first to evaluate the infinite-horizon case and then to use the information thus obtained for the finite-stage problem.

This question is tackled by comparing the direct and indirect MacQueen-type bounds of Hinderer (1976). In the indirect case two types of errors arise: one for approximating the infinite horizon value and one for the final few-stages value.

The main answer consists in a rule of thumb: The indirect method should be effective only if the method used for approximating the infinite-stage value is at least twice as fast as normal value iteration. According to numerical results of the last few years this may hardly be reached by methods such as policy iteration, policy value iteration, Gauß-Seidel iteration, or linear programming. On the other hand the indirect method may be advantageous if the infinite-stage value

- is obtained by an analytical solution,
- exploits the special structure of the problem,
- is used for a couple of scrap values, or
- is needed for other purposes, too.

Hinderer, K.(1976) Estimates for finite-stage dynamic programs. J.Math.Anal.Appl. 55, 207-238.

EIN MONOTONES STOPPROBLEM BEI PARTIELLER INFORMATION

Uwe Jensen, Stuttgart-Hohenheim

Betrachtet wird ein Modell, in dem mögliche Gewinne, beschrieben durch einen stochastischen Prozeß (X_t), $t \geq 0$, nur bis zu einem zufälligen Zeitpunkt ζ realisiert werden können. Nach dem Zeitpunkt ζ verfallen die Gewinne und es wird eine Endauszahlung $r(\zeta)$ fällig, die vom Zeitpunkt ζ abhängen und negativ sein kann. Gesucht ist eine Stopzeit, die den erwarteten Gewinn maximiert, also ein Ausgleich zwischen einer möglichst langen Beobachtungsdauer verbunden mit höheren Gewinnen und dem unvorhersehbaren Eintreten des Zeitpunktes ζ. Dieses Modell, welches in der Literatur unter speziellen Annahmen behandelt wurde, hat verschiedene Anwendungsaspekte; z.B. als Modell in der Zuverlässigkeitstheorie (Nach welcher störungsfreien Betriebszeit sollen Gewinne realisiert werden, bevor bei einer Störung zur Zeit ζ die Gewinne verfallen und Kosten $-r(\zeta)$ anfallen?) oder im zeitdiskreten Fall als Modell für Spiele wie "DOUBLE OR QUIT" und "ALLES ODER NICHTS".

Gesucht ist insbesondere eine Lösung des Stopproblems bei partieller Information. Stehen beispielsweise Informationen über den Gewinn X_t nur mit zeitlicher Verzögerung oder nur in Form des Erwartungswertes EX_t zur Verfügung, dann ist im Stopproblem zur Konkurrenz nur eine Teilmenge aller Stopzeiten zugelassen. Zur Charakterisierung optimaler Stopzeiten wird eine Semimartingaldarstellung benutzt, die auch beim Übergang zu bedingten Erwartungswerten stabil bleibt. Unter geeigneten Struktur- und Monotoniebedingungen können optimale Stopzeiten angegeben werden. An einem Beispiel wird gezeigt, wie sich unterschiedliche Informationsstände auf die optimale Stopzeit und den zugehörigen erwarteten Gewinn auswirken.

STATE-ACTION FREQUENCIES IN MULTI-OBJECTIVE AND CONSTRAINED MDP'S

L.C.M. Kallenberg, Leiden

It is well known that discounted and undiscounted Markov-decision problems (MDP's) can be solved by linear programming (LP). The interpretation of the variables used in the LP formulation is not so well known. This interpretation as state-action frequencies can be used for the solution of some multi-objective and constrained MDP's. We will give two examples.

Consider a discrete finite MDP with state space E, action sets A(i), $i \in E$, immediate rewards r_{ia} and transition probabilities p_{iaj}, $j \in E$, for $i \in E$ and $a \in A(i)$.
Let $(\beta_1, \beta_2, \ldots, \beta_N)$ be a given initial distribution. For any policy π, define $\phi(\beta,\pi)$ by

(1) $$\phi(\beta,\pi) := \liminf_{T\to\infty} \frac{1}{T} \Sigma_{t=1}^{T} \Sigma_{j,a} \mathbb{P}_{\beta,\pi} (X_t = j,\ Y_t = a) r_{ja} ,$$

where X_t, Y_t are the state and action at time t.
It is conceivable that the use of this expected average reward criterion may yield an optimal policy which is unacceptable to a risk-sensitive decision maker. Such a decision maker may prefer a criterion which includes a penalty for the "variability" in the stream of rewards. An important question is: which criterion is conceptually meaningful and mathematically tractable? We show that

(2) $$v(\beta,\pi) := \Sigma_{i,a} \{r_{ia} - \phi(\beta,\pi)\}^2 x_{ia}(\pi) ,$$

where $x_{ia}(\pi)$ is the frequency of the pair (i,a) given policy π, is a satisfactory choice for the variability. Consider the optimization problem

(3) $$\max_{\pi} \{\phi(\beta,\pi) - \lambda\, v(\beta,\pi)\} ,\ \lambda \in [0,\infty) .$$

It turns out that an optimal policy for (3) can be derived from the following convex quadratic program

(4) $$\max_{\pi} \{\Sigma_{i,a}(r_{ia} - \lambda r_{ia}^2) x_{ia} + \lambda(\Sigma_i \Sigma_a r_{ia} x_{ia})^2\}$$

subject to the constraints of the LP corresponding to criterion (1).

The second problem is to find a policy that maximizes(1) subject to the sample-path constraint

(5) $$\mathbb{P}_{\beta,\pi}\{\limsup_{T\to\infty} \frac{1}{T} \Sigma_{t=1}^{T} c(X_t,Y_t) \leq \alpha\} = 1 ,$$

where $c(\cdot,\cdot)$ is some cost function which has to be bounded in the average by α. Such a problem may occur in controlled queueing models. In the unichain case it can be shown, using the state-action frequencies, that an optimal policy can be found by the solution of the usual LP with the additional constraint

(6) $$\Sigma_{i,a} c(i,a)x_{ia} \leq \alpha .$$

AGGREGATION - DISAGGREGATION ALGORITHMS FOR DISCRETE STOCHASTIC SYSTEMS

Grzegorz Reyman and Jan van der Wal, Eindhoven

Abstract

In this paper an aggregation - disaggregation method is formulated for a finite horizon Markov decision process with two-dimensional state and action spaces. This second dimension of the state and the action contains a similar type of information in which aggregation is both natural and simple. The quality of the approach is illustrated by an example.

1. Introduction

In practice Markov decision models typically give rise to very large state spaces. Then straightforward computation of an optimal strategy is out of the question, but sometimes decomposition or aggregation and disaggregation might work very well. See e.g. Courtois [1], Whitt [5], Mendelssohn [2] and Schweitzer et al. [3].

In this paper we consider a finite horizon Markov decision model with two dimensional state and action spaces, which due to one of the two dimensions are large. Aggregation in this dimension in the action space leeds to the same aggregation in the state space. Typical examples are models with a limited resource which is used for the actions and for which the state contains the remaining amount of resource, or models where in a certain time a total production level has to be achieved. The action contains the production for the next period and the state information about what already has been produced.

A simple example of the latter is the following. In the next 20 hours we have to produce 4288 items on a machine. The production speed of the machine can be varied between 0 and 320 per hour. The machine may however breakdown and the probability of a breakdown increases if the production speed is increased. Repair times can be reduced at additional costs. If at the end of the planning period not all 4288 items are ready, penalty costs are incurred. Assuming the machine breaks down only at the end of an hour, we obtain a fairly standard Markov decision problem.

Let the action set be

$$\{ (r,0), (p,0), (p,64), (p,128), \ldots, (p,320) \}$$

where r stands for repair and p for produce, then the state space will be

$$\{ (d,0), (d,64), (d,128), \ldots, (d,4288), (u,0), (u,64), (u,128), \ldots, (u,4288) \}$$

with d denoting that the machine is down and u that it is up. In this case, however, the outcome of the optimization may be poor. On the other hand, if the action set is taken to be

$$\{ (r,0), (p,0), (p,1), \ldots, (p,320) \} ,$$

the state space will be

Operations Research Proceedings 1987

$$\{ (d,0), (d,1),..., (d,4288), (u,0), (u,1),..., (u,4288) \} ,$$

the computed solution will be optimal but at the cost of an enormous increase in computing time.
It is clear that one has to find a balance between accuracy and computing time, in particular if we realize ourselves that the model used is not a perfect description of reality. In problems of this type an aggregation-disaggregation approach may work well (cf. Veugen et al. [4]).
The remainder of this paper is organized as follows. In Section 2 the model is introduced, Section 3 contains four aggregation-disaggregation algorithms and in Section 4 the performance of the presented algorithms is compared for the 20 hour production planning problem formulated above.

2. Model

We consider a finite horizon Markov decision problem with T periods : $0,1,...,T-1$. The state and action spaces of the problem are two dimensional and denoted by $I \times X$ and $A \times U$ respectively. The sets I and A are small compared to $X = \{0,1,...,N\}$ and $U = \{0,1,...,M\}$. The aggregation will be performed on X and U.
Think of x in the state space pair (i,x) as the used amount of resource or the number of items already produced, and of the u in the action pair (a,u) as the amount of resource used for the execution of a or the number of items to be produced in the next period.
This interpretation of x and u suggests the following simple transition structure. If in the state (i,x) the action (a,u) is taken the system makes a transition to the state $(j,x+u)$ with a probability $p_{ij}^a(u)$, with $\sum_j p_{ij}^a(u) = 1$. So only states (j,y) , with $y=x+u$ can be reached. As a result of the action (a,u) in (i,x) there is an immediate reward $r(i,x,a,u)$.
Further a terminal reward $V_0(i,x)$ is obtained if, as a result of the last action, the system is in the state (i,x) at the time N, eg. to cover used resource or shortage.
We denote by $V_T(i,x,\pi)$ the expected reward for the initial state (i,x) when the strategy π is used. Then an optimal Markov strategy can be obtained by the following dynamic programming recursion :

For $n=0,1,...,T-1$ compute for all $(i,x) \in I \times X$

$$V_{n+1}(i,x) = \max_{a,u} \{ r(i,x,a,u) + \sum_j p_{ij}^a(u) V_n(j,x+u) \} \tag{1}$$

The Markov strategy $\pi^* = (f_{T-1}, f_{T-2},..., f_0)$, with $f_n : I \times X \to A \times U$ and $f_n(i,x)$ maximizing the right hand side in (1), is optimal.
Often at time n only a subset of the states need to be considered. For example in the production problem the initial state is (machine works, 0 produced) thus at time 1 only states (i,x) with $0 \le x \le 320$ are possible. Also the set of possible actions in a state may be a subset of $A \times U$. For simplicity of notation we will have the same state and action space for all n .
From the form of the transition probabilities it follows that aggregation in the set U leads to an aggregation in the set X. For example considering only multiplies of k for u means that x will be a multiple of k as well, given it is a multiple of k initially. This is the simplest form of aggregation : restriction to multiples. Choose k and define $\hat{X} = \{0,k,2k,...,\alpha k\}$ and $\hat{U} = \{0,k,2k,...,\beta k\}$ with α (β) the largest integer with $\alpha k \le N$ ($\beta k \le M$) respectively . Solve the recursive scheme (1) with $I \times X$ replaced by $I \times \hat{X}$ and $(a,u) \in A \times \hat{U}$. Sometimes a value for k can be chosen for which the solution is close to optimal and the computing time remains acceptable.
In the next section we present four aggregation-disaggregation algorithms for the case the above simple approach fails.
When describing these algorithms the following notation will be useful,

$$X_l = \{0 , 2^l , 2 \cdot 2^l , 3 \cdot 2^l , ... , \alpha \cdot 2^l \}$$

with α the largest integer for which $\alpha \cdot 2^l \le N$. Similarly we define U_l.

3. Aggregation - Disaggregation algorithms

In this section we subsequently consider four aggregation-disaggregation algorithms.

3.1. Algorithm A

The simplest form of aggregation-disaggregation (AD) is AD in the action space only. For each u and for each state (i, x) a separate AD routine is executed to obtain $V^A_{n+1}(i,x)$. For $n = 0, 1, \ldots, T-1$ and $(i, x) \in I \times X$ the value $V^A_{n+1}(i,x)$ is computed as follows.
Consider a sequence of aggregations in the action space, say $U_L, U_{L-1}, \ldots, U_1, U_0$. First compute in the state (i,x) for each a the best $u \in U_L$,i.e., determine

$$\max_{u \in U_L} \{ r(i,x,a,u) + \textstyle\sum_j p^a_{ij}(u) V^A_n(j,x+u) \} . \tag{2}$$

Denote this maximizing u by $u^A_L(i,x,a,n)$. Next, compare this u with two neighboring values in U_{L-1} . So, using the notation

$$U_l(u) := \{u-2^l, u, u+2^l\} \cap U_l \quad , \; u \in U_{l+1} , \tag{3}$$

this next step can be written as

$$\max_{u \in U_{L-1}(u^A_L(i,x,a,n))} \{ r(i,x,a,u) + \textstyle\sum_j p^a_{ij}(u) V^A_n(j,x+u)\} .$$

Call the maximizer $u^A_{L-1}(i,x,a,n)$ and repeat the procedure for $l = L-2, L-3, \ldots, 1$ by computing

$$\max_{u \in U_l(u^A_{l+1}(i,x,a,n))} \{ r(i,x,a,u) + \textstyle\sum_j p^a_{ij}(u) V^A_n(j,x+u)\} .$$

And finally we obtain

$$V^A_{n+1}(i,x) = \max_a \max_{u \in U_0(u^A_1(i,x,a,n))} \{ r(i,x,a,u) + \textstyle\sum_j p^a_{ij}(u) V^A_n(j,x+u)\} . \tag{4}$$

Denote this maximizing pair by $(a,u)^A(i,x,n)$. Once we have computed the $V^A_T(i,x)$ for all, or merely the initial, (i,x) we have a supposedly optimal value and the various $(a,u)^A(i,x,n)$ together constitute the supposedly optimal strategy.
Two remarks have to be made. First note that the u^A_l need not to be unique. So we need a tie-break rule, for instance pick the smallest one. Secondly, and far more important, we see that this algorithm will not necessarily find the optimal value and optimal strategy, since the procedure described above may only find a local optimum in some sense. It is not guaranteed that the obtained strategy is nearly optimal. However, in most realistic examples the behavior of the various cost and probability functions is more or less continuous. So it is likely that if we take a modest value of L to begin with, the algorithm will work well. An attempt to safeguard against bad solutions is to extend the sets $U_l(u^A_{l+1}(i,x,a,n))$ with other values of u around values which were not too bad in the previous step. We will come back to this idea in algorithm D.

3.2. Algorithm B

This second AD algorithm can be seen as a repeated version of algorithm A for various aggregation levels in the state space. In the first step the problem is considered to have state space $I \times X_L$ and action space $A \times U_L$. In the second step the problem has state space $I \times X_{L-1}$ and action space $A \times U_{L-1}$ and is solved by algorithm A starting with action space $A \times U_L$. In the final, L-th, step algorithm A is executed with state space $I \times X$ and $A \times U$ as action space . Hopefully not all steps have to be executed. We intend to stop as soon as the computed optimal values $V^{B,l-1}_T$ and $V^{B,l}_T$ for the given initial state (i_0, x_0) are sufficiently close. (It is assumed that $x_0 \in X_l$ for all l).
Formally B runs as follows :
Step 1

Calculate for $n=0,1,...,T-1$ and all $(i,x)\in I\times X_L$

$$V_{n+1}^{B,1}(i,x) = \max_a \max_{u\in U_L} \{ r(i,x,a,u)+\sum_j p_{ij}^a(u) V_n^{B,1}(j,x+u) \} \tag{5}$$

until we get $V_T^{B,1}(i_0,x_0)$.

Step l, $l=2,...$ (algorithm A with state space $I\times X_{L-l+1}$ and action space $A\times U_{L-l+1}$)

For $n=0,1,...,T-1$ and $(i,x)\in I\times X_{L-l+1}$ compute $V_{n+1}^{B,l}(i,x)$ as follows.
For each a first determine

$$\max_{u\in U_L} \{ r(i,x,a,u)+\sum_j p_{ij}^a(u) V_n^{B,l}(j,x+u) \} \tag{6}$$

and denote the minimizer by $U_L^{B,l}(i,x,a,n)$.
Next compute for $m=L-1,...,L-l+2$

$$\max_{u\in U_m(u_{m+1}^{B,l}(i,x,a,n))} \{ r(i,x,a,u)+\sum_j p_{ij}^a(u) V_n^{B,l}(j,x+u) \} . \tag{7}$$

And finally

$$V_{n+1}^{B,l}(i,x) = \max_a \max_{u\in U_{L-l+1}(u_{L-l+2}^{B,l}(i,x,a,n))} \{ r(i,x,a,u)+\sum_j p_{ij}^a(u) V_n^{B,l}(j,x+u) \} . \tag{8}$$

Stop
if $V_T^{B,l-1}(i_0,x_0)$ and $V_T^{B,l}(i_0,x_0)$ are sufficiently close, where (i_0,x_0) is the initial state.

3.3. Algorithm C

C can be seen as a refinement of B. Instead of merely a repetition of A's, as B is, we also use the information about which values of u were good for a, that has been obtained in the previous step.
Formally C runs as follows.
Step 1

Calculate for $n=0,1,...,T-1$, for all $(i,x)\in I\times X_L$ for each a

$$\max_{u\in U_L} \{ r(i,x,a,u)+\sum_j p_{ij}^a(u) V_n^{C,1}(j,x+u) \} \tag{9}$$

and denote the maximizer by $u_L^C(i,x,a,n)$.

Compute

$$V_{n+1}^{C,1}(i,x) = \max_a \max_{u\in U_L} \{ r(i,x,a,u)+\sum_j p_{ij}^a(u) V_n^{C,1}(j,x+u) \} \tag{10}$$

Step 1 , 1=2,3,...

First construct for each $n=0,1,...,T-1$ and each triple $(i,x,a)\in I\times X_{L-l+1}\times A$ the set $U_{L-l+1}^C(i,x,a,n)$ of possible values for u in the maximization. If $x\in X_{L-l+2}$ then

$$U_{L-l+1}^C(i,x,a,n) = U_{L-l+2}(u_{L-l+2}^C(i,x,a,n)) \tag{11}$$

If, however, $x \in X_{L-l+2}$ then $u_{L-l+2}(i,x,a,n)$ has not been determined in the previous iteration. Therefore the information is used that has been obtained for the neighbors of (i,x) : $(i,x-2^{L-l+1})$ and $(i,x+2^{L-L+1})$. So for $x \in X_{L-l+2}$

$$U^C_{L-l+1}(i,x,a,n) \tag{12}$$

$$:= U_{L-l+1}(u_{L-l+2_C}(i,x-2_{L-l+1},a,n)) \cup U_{L-l+1}(u_{L-l+2}(i,x+2_{L-l+1},a,n))$$

Next compute for each a

$$\max_{u \in U^C_{L-l+1}(i,x,a,n))} \{ r(i,x,a,u) + \sum_j p^a_{ij}(u) V^{C,l}_n(j,x+u) \} \tag{13}$$

and denote the maximizing u by $u_{L-l+1}(i,x,a,n)$.

Finally compute

$$V^{C,l}_{n+1}(i,x) = \max_a \max_{u \in U^C_{L-l+1}(i,x,a,n)} \{ r(i,x,a,u) + \sum_j p^a_{ij}(u) V^{C,l}_n(j,x+u) \} . \tag{14}$$

Stop if $V^{C,l-1}_T(i_0,x_0)$ and $V^{C,l}_T(i_0,x_0)$ are sufficiently close.

None of these three algorithms is guaranteed to produce an optimal or even nearly optimal solution. As argued before the structure of the problem is such that one might hope for a reasonably good solution provided the level of aggregation L is not too large to begin with. The Figure 1 shows the major problem for the method.

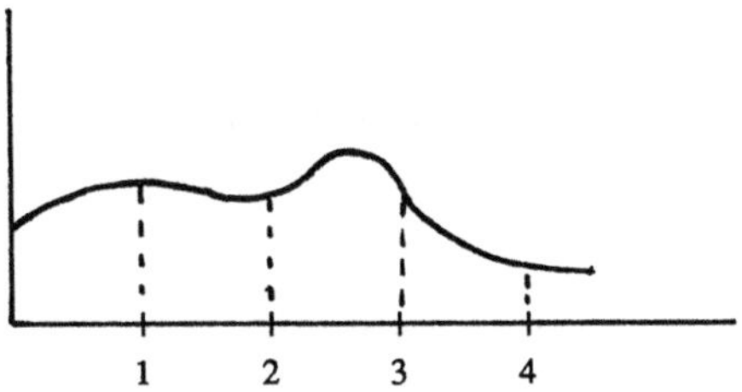

Figure 1

If at a certain stage the points indicated as 0,1,...,4 are calculated, the optimal one seems to be 1. From then on only the interval (0,2) will be considered whereas the optimal value lies between 2 and 3 !
To diminish the risk of making mistakes of this kind we suggest to increase the sets wherein we search for the best values of u. In algorithm D below this is done by taking into consideration also neighborhoods of u's which were only nearly optimal in the previous step. We formulate D as an extension of algorithm C.

3.4. Algorithm D

Algorithm D is executed in exactly the same way as algorithm C with the exception that the sets U^C defined in (11) and (12) are constructed differently.

Step 1

Calculate for $n=0,1,...,T-1$, for all $(i,x) \in I \times X_L$ for each a

$$\max_{u \in U_L} \{ r(i,x,a,u) + \sum_j p_{ij}^a(u) V_n^{D,1}(j,x+u) \} \tag{15}$$

and construct the set $U_L^{D,2}(i,x,a,n)$ of values for u which nearly maximize (15). Nearly means that they come within ε_1 of the optimum.

Compute $V_{n+1}^{D,1}$ as in (10).

Step l , l=2,3,...

First construct the sets $U_{L-l+1}^{D,1}(i,x,a,n)$ of values of u that have to be considered.

For $x \in X_{L-l+2}$ define

$$U_{L-l+1}^{D,1}(i,x,a,n) = \bigcup_{u \in U_{L-l+1}^{D,2}(i,x,a,n)} U_{L-l+1}(u) \tag{16}$$

For $x \in X_{L-l+2}$ define

$$U_{L-l+1}^{D,1}(i,x,a,n) = U_{L-l+1}^{D,1}(i,x-2^{L-l+1},a,n) \cup U_{L-l+1}^{D,1}(i,x+2^{L-l+1},a,n). \tag{17}$$

Next compute for each a

$$\max_{u \in U_{L-l+1}^{D,1}(i,x,a,n))} \{ r(i,x,a,u) + \sum_j p_{ij}^a(u) V_n^{D,l}(j,x+u) \} . \tag{18}$$

and construct the sets $U_{L-l+1}^{D,2}(i,x,a,n)$ of actions which maximize (18) within ε_l . $U^{D,2}$ will be a subset of $U^{D,1}$.

Finally compute $V_{n+1}^{D,l}$ analogously to (14)

$$V_{n+1}^{D,l}(i,x) = \max_a \max_{u \in U_{L-l+1}^{D,1}(i,x,a,n)} \{ r(i,x,a,u) + \sum_j p_{ij}^a(u) V_n^{D,l}(j,x+u) \} . \tag{19}$$

4. Numerical results

The four algorithms have been tested on the production problem mentioned in the introduction : producing 4288 items in 20 hours. The detailed description of the problem is as follows.

- The state set $I = \{0,1\}$ where 0 indicates that the machine is down and 1 that it is available for production
- If the machine is down there are 3 actions, 0 : do nothing, 1 : normal repair, 2 : fast repair
- In state 1 there is only one action, 0 : produce, but the production rate can be varied between 0 and 320 per hour.
- If the machine is down, the costs of the normal repair are 15, the costs of the fast repair 60. The probability that at the end of the hour the repair turns out to be successful is 1/5 for a normal repair and 1/2 for a fast repair.

- The failure probability $p_{10}^{0}(u)$ depending on the production rate is given in Figure 2

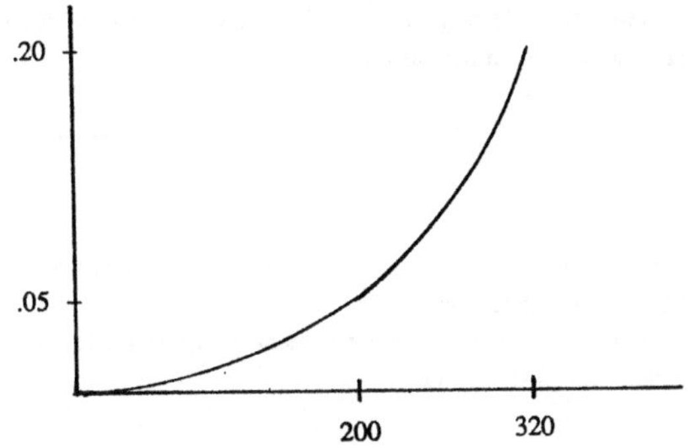

$$p_{10}^{0}(u) = \frac{u}{8000-20u}$$

- The shortage costs at the end of the planning period are 2 per item. If at the end of the 20 hours the machine is down there is a terminal cost of 20.

When presenting the results, two things are important : the accuracy of the final result and the computing time needed to get this result.
To obtain the exact solution we solved the problem with X and $U=\{0,1,2,...,4288\}$. In the aggregation we used grid sizes 64,32,16,8,4,2 and 1 if necessary.
The number of states in X_l varies as follows

l	0	1	2	3	4	5	6
Number of states in X_l	4289	2145	1073	537	269	135	68

The exact result :

Minimal cost 480.08

Computing time 3142.4
(All computing times are given in seconds on a VAX 11/750).
The results for the algorithms A,B,C and D are as follows.

grid	Minimal costs			
	A	B	C	D
64	500.03	500.03	500.03	500.03
32	486.54	486.54	486.54	486.54
16	481.22	481.22	481.22	481.22
8	480.37	480.37	480.37	480.37
4	480.17			
2	480.09			
1	480.08			
total computing time	552.5	63.5	18.9	31.2

In algorithms B,C and D we stopped when the difference between two successive iterations was less than 1%. In D the sets of nearly optimal u values consisted of these actions for which the relative difference from optimality was within 1%.
The complete, exact, solution of the problem with state space $I \times X_L$ and action space $A \times U_L$ for $L = 6,5,4$ and 3 (grids

64,32,16 and 8) gave exactly the same result as A,B,C and D in a computation time 80.6.
As we see from the results, the structure of the problem is such that the fact that the algorithms check only a subset of all u values in the grid does not lead to any loss of optimality. Clearly, one may come up with cases which the problem, signalled in Figure 1, of looking in the wrong area, does give an increase in costs.

5. Conclusions

We have presented four aggregation-disaggregation algorithms for Markov decision processes in which aggregation in the action space leads to the same level of aggregation in the state space.
The algorithms have been tested on a simple production problem. Although more numerical experience is wanted we are convinced that in particular algorithm D is a very accurate and fast heuristic.

6. References

[1] Courtois, P.-J. (1977), Decomposability: Queueing and Computer System Application, Academic Press, New York.

[2] Mendelssohn, R. (1982), An iterative aggregation procedure for Markov decision processes, Operations Reseach 30, 62-73.

[3] Schweitzer, P.J., Puterman, M. and K.W. Kindle (1981), Iterative aggregation - disaggregation procedures for solving discounted semi-Markovian reward processes, Working paper series, No. 8123, Graduate School of Management, University of Rochester.

[4] Veugen, L.M.M., Van der Wal, J. and J. Wessels (1985), Aggregation and disaggregation in Markov dicision models for inventory control, EJOR 20, 248-254.

[5] Whitt, W. (1978,1979), Approximations of dynamic programs I and II, MOR 3, 231-243 and 4, 179-185.

On the Number of Value Determination Steps in Policy Value Iteration

Jörg Stein, Hamburg

We consider discounted Markovian decision processes with infinite horizon and finite state and action spaces. While in successive approximation in every step a maximizing policy is determined, policy value iteration algorithms perform after each maximization a few value determination steps with the policy just found. We call the number of fixed policy iterations the "order" of value determination.

Since we need more time to perform a maximization step than to do one fixed policy iteration, in general policy value iteration takes less total computation time than successive approximation. If the currently used policy is very different from the optimal one, it seems to be inefficient to determine the value with high accuracy by taking a high order. Otherwise if the actual policy is already the optimal one, which in general will be the case quite a few iterations before the stopping criterion is satisfied, value determination of high order is very efficient.

We present an algorithm which chooses the order after every maximization step separately. If a maximization step yields a large improvement in relation to preceeding value determination steps, we conject that the new policy may be suboptimal too, and take a low order. Otherwise we perform a large number of fixed policy iterations.

In several numerical tests this algorithm turned out to be more efficient than fixed order algorithms. Moreover an optimal fixed order is unknown as well.

The curse of non-stationarity in applying stochastic models

Jacob Wijngaard, Eindhoven

Multi-dimensionality has been indicated by Bellmann as the most important factor making the use of dynamic programming models difficult: the curse of dimensionality.

Applying stochastic models in production control has its own curse: non-stationarity. There are many, easy to analyze, stationary stochastic models available that describe the behaviour of real situations reasonably well, except that the situations just change too fast: changing markets, new products, new technologies, ...
It is important to formulate models such that one gets as much stationarity as possible.
A one-stage mult-product production-inventory situation with N identical products will be considered to illustrate this point.
Set-up times are long and it is assumed therefore that average lot-sizes have to be larger than customer order sizes.
It is shown that the behaviour of each of the individual inventories may be approximated by

$$I_i(t) := I(t)/N + y_i$$

where $I(t)$ is the total inventory and y_i is independently normally distributed. So, the $I_i(t) - I(t)/N$ converge much faster to stationarity than the $I_i(t)$. It is possible to use stationary models for the behaviour of the $I_i(t) - I_i(t)/N$ even if the behaviour of the $I_i(t)$ is not stationary.
The result leads to a hierarchical approach of production control: simple one-dimensional non-stationary models can be used at the aggregate level (total inventory) while multi-dimensional stationary models can be used at item level.

OPTIMAL CAPACITY EXPANSION IN A CHEMICAL PLANT

Yvo M.I. Dirickx
V. Oomes
ENSCHEDE

Confronted with the happy circumstance of linearly growing demand for its homogeneous product, a particular chemical company had to decide for the next decade or so when and what kind of capacity expansion was necessary to minimise expected operating costs while meeting demand.
The technical process was analysed for different equipment configurations (the operating efficiency was analysed) using simulation. The results of the simulation study were used as input for a continuous time dynamic programming model that was set up to determine an optimal capacity expansion programme. Numerical results will be presented.

THE FORMULATION OF THE NASH BARGAINING PROBLEM AS A HIERARCHICAL CONTROL PROBLEM

Harri Ehtamo, Jukka Ruusunen and Raimo P. Hämäläinen
Espoo, Finland

Abstract : It is shown that the Nash bargaining problem can be solved as a hierarchical decision problem, where at the top level of the hierarchy a fixed point equation is solved to obtain the bargaining solution. As a practical application of the approach the computation of a cooperative energy management policy for a power pool consisting of N interconnected power systems is considered.

1. Introduction

Many problems that emerge in society deal with complex systems which involve multiple parties as decision makers (DMs). Often, the goals of different parties are more or less conflicting. If the DMs pursue their own interests, a noncooperative solution is the result. On the other hand, if the DMs are able to act cooperatively, they seek a solution that considers the interests of all the DMs.

In this paper our interest is in dynamic N–player bargaining problems. In static bargaining games Nash's bargaining scheme provides a way according to which a cooperative policy for the DMs can be found; see e.g. Roth (1979) for details. As will be shown, it also provides a way to study cooperation in dynamic, multiagent decision making, because the bargaining problem can be solved as a two–level hierarchical decision problem. This is computationally attractive, since the original problem, which is nonlinear in the DMs' objectives, is now imbedded into a family of problems which are linear in the DMs' objectives. At the lower level of the hierarchy a weighted sum of the DMs' objectives is maximized; at the top level a fixed point equation is solved to obtain the bargaining solution.

As a practical application of our approach we shall consider the computation of a cooperative policy for N interconnected power plants. In this example the lower level problem can be further decomposed into N subsystems corresponding to each individual power plant. These subsystems are coordinated by a second–level coordinator in the resulting three–level hierarchy. The second–level coordinator observes the reactions of the individual DMs to given power prices, and then adjusts the price function such that power balance is achieved.

Operations Research Proceedings 1987

2. The Nash bargaining solution

According to Nash [Nash (1950)] a solution to the bargaining problem is the unique outcome which maximizes the geometric average of the gains which the DMs realize reaching an agreement instead of settling for the status quo outcome.

Let Ω_i be the decision set for DMi, $1 \leq i \leq N$, set $\Omega = \Omega_1 \times \ldots \times \Omega_N$, and let J_i, $J_i : \Omega \to R$, be the objective functional for DMi. Set

$$F_i(u) = J_i(u) - D_i, \quad 1 \leq i \leq N, \tag{1}$$

$$F(u) = \prod_{i=1}^{N} F_i(u), \quad u \in \Omega, \tag{2}$$

where $D_i \in R$ is a status quo outcome for DMi. Then we say that $u^* \in \Omega$ is a Nash bargaining solution if it is a solution to the following problem, called BP:

$$\max_{u \in \Omega} F(u) \tag{3}$$

$$\text{subject to } F_i(u) \geq 0, \quad 1 \leq i \leq N. \tag{4}$$

The following theorem can be proved [for the proofs of the theorems presented here see Ehtamo *et al.* (1987)]:

Theorem 1: Let $u^* \in \Omega$ be such that

$$\sum_{i=1}^{N} \mu_i F_i(u) \leq \sum_{i=1}^{N} \mu_i F_i(u^*) \quad \text{for all } u \in \Omega, \tag{5}$$

where

$$\mu_i = \prod_{j \neq i} F_j(u^*), \; \mu_i > 0, \; 1 \leq i \leq N. \tag{6}$$

Then u^* solves (3), (4).

Remark 1: Note that $\mu_i > 0$ for all i implies $F_i(u^*) > 0$ for all i. Namely, from (5), (6) it follows that $F_k(u^*) > 0$ for some k (provided there is $u \in \Omega$ such that $F_i(u) \geq 0$ for all i). Since $F_i(u^*)\mu_i = F(u^*) = F_k(u^*)\mu_k > 0$ for all i, it follows that $F_i(u^*) > 0$ for all i.

Define

$$S = \{(F_1(u), \ldots, F_N(u)) \mid u \in \Omega\} \subset R^N, \tag{7}$$

let $\Lambda^{\leq} \subset R^N$ be the convex cone $\Lambda^{\leq} = \{\lambda \in R^N \mid \lambda_i \leq 0, \; \forall i\}$, and set $Y = S + \Lambda^{\leq}$. Suppose that BP has a solution $u^* \in \Omega$ (observe that BP always has a solution $u^* \in \Omega$ if $\{y \in Y \mid y_i \geq 0, \; \forall i\}$ is compact, since the geometric average is a continuous function). Then the corresponding optimal outcome $\left(J_1(u^*),\ldots,J_N(u^*)\right)$ is Pareto–optimal. Moreover, if Y is a convex set, then the optimal outcome $\left(J_1^*,\ldots,J_N^*\right)$ is unique, and we have

Theorem 2: Let Y be a convex set, and let $u^* \in \Omega$ solve BP with $F_i(u^*) > 0$ for all i. Then (5) holds, where μ_i is as in (6).

Remark 2: It can be shown that if Ω is a convex set in a vector space U, and $F_i : \Omega \to R$, $1 \le i \le N$, are concave functionals on Ω, then Y is convex.

Assume that for each

$$\mu \in \Lambda^> = \{\lambda \in R^N \mid \lambda_i > 0,\ \forall i\} \tag{8}$$

the problem

$$\max_{u\in\Omega} \sum_{i=1}^{N} \mu_i F_i(u) \tag{9}$$

has a solution $\omega(\mu) \in \Omega$. Set

$$T_i(\mu) = \prod_{j\neq i} F_j(\omega(\mu)),\ 1 \le i \le N, \tag{10}$$

$$T(\mu) = [T_1(\mu),\ldots,T_N(\mu)]'. \tag{11}$$

Then $T : \Lambda^> \to R^N$. If T has a fixed point $\mu^* \in \Lambda^>$, i.e.

$$T(\mu^*) = \mu^*,\ \mu^* \in \Lambda^>, \tag{12}$$

then the corresponding $u^* = \omega(\mu^*)$ solves BP. A convenient method for solving BP's is thus the following algorithm

Step 1: Set $k = 0$ and choose a starting vector $\mu^0 \in \Lambda^>$.

Step 2: For $\mu^k \in \Lambda^>$ compute $\omega(\mu^k)$ and $T(\mu^k)$.

Step 3: If $\| T(\mu^k) - \mu^k \| < \epsilon$, set $\mu^* = \mu^k$. Else, choose the next trial point $\mu^{k+1} \in \Lambda^>$, set $k := k+1$, and go to Step 2.

3. Hierarchical control of interconnected power systems

Consider a power pool consisting of N interconnected power systems controlled by individual decision makers. The problem is to find a management policy such that all the systems achieve lower total production costs over the planning period compared to the situation, where no energy exchange is possible.

The dynamics of the individual power systems are described by the state equations

$$\dot{x}_i = A_i x_i + B_i u_i, \quad x_i(0) = x_{i0},$$
$$y_i = C_i x_i, \tag{13}$$

$1 \le i \le N$, where $x_i(t) \in R^{n_i}$ describes extraction steam flow from the boiler and pressure levels in the boiler and the headers, $u_i(t) \in R$ is the fuel flow to the boiler, and $y_i(t) \in R$ is the electric power generated by the turbine–generator. All the variables are deviations from the corresponding equilibrium levels. The cost of production for DMi is assumed to be a quadratic function of the fuel flow

$$J_i = \int_0^T (\frac{1}{2} r_i u_i^2(t) + s_i u_i) dt, \tag{14}$$

where T is the planning period, and $r_i, s_i > 0$ are constants.

The deviation of local demand from the equilibrium level in system i at time t is denoted by $d_i(t)$. The status quo outcome $(D_1, \ldots, D_N)$ corresponds to the situation, where no energy exchange takes place. It is calculated by setting $y_i(t) = d_i(t)$ for all i, and computing the corresponding $D_i = J_i$ from (13) and (14).

If there is an interconnection between the systems, it is possible to exchange energy. In this case electric power supply in system i at time t is $y_i(t) + P_i(t)$, where $P_i(t) \in R$ is the electric power transferred to system i at time t (if $P_i(t) < 0$, electric power is transferred from system i to the network). Power balance in system i requires that power supply must cover local demand, i.e.

$$y_i(t) + P_i(t) = d_i(t), \quad t \in [0, T], \quad 1 \le i \le N. \tag{15}$$

On the other hand, power balance in the network requires

$$\sum_{i=1}^{N} P_i(t) = 0 \quad \text{for} \quad t \in [0, T]. \tag{16}$$

Set $u = (u_1, \ldots, u_N)$, $F_i(u) = D_i - J_i(u)$, where u belongs to an appropriate function space U. The cooperative management problem can then be stated as follows: Find $u \in U$ such that the functional (2) is maximized subject to the constraints (4), (13), (15) and (16).

Remark 3: In fact $u \in \Omega = \{u \in U \mid u_i(t) \ge -u_i^e, \ t \in [0, T], \ 1 \le i \le N\}$, where the constant $u_i^e > 0$ is the equilibrium level of the fuel flow for system i, $1 \le i \le N$. However, it will be assumed that the problem has interior solutions; hence the control constraints are neglected.

We use the algorithm presented in Section 2 to find a cooperative management policy. Let $\mu \in \Lambda^>$, and consider

$$\min_{u \in U} \sum_{i=1}^{N} \mu_i J_i(u) \tag{17}$$

subject to (13), (15) and (16).

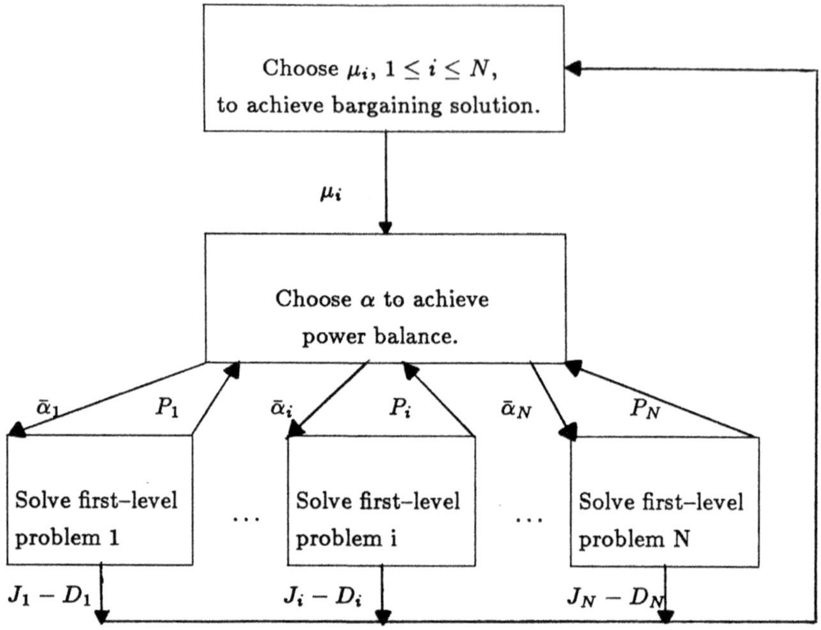

Fig. 1. The three–level coordination structure.

Define

$$\tilde{J} = \sum_{i=1}^{N} \mu_i \tilde{J}_i = \sum_{i=1}^{N} \mu_i \left\{ \int_0^T \left[\frac{1}{2} r_i u_i^2 + s_i u_i + \bar{\lambda}_i' (A_i x_i + B_i u_i - \dot{x}_i) \right.\right.$$

$$\left.\left. + \bar{q}_i (d_i - y_i - P_i) + \bar{\alpha}_i P_i \right] dt \right\}, \tag{18}$$

where $\bar{\lambda}_i = \mu_i^{-1} \lambda_i$, $\bar{q}_i = \mu_i^{-1} q_i$, $\bar{\alpha}_i = \mu_i^{-1} \alpha$, $\lambda_1, \ldots, \lambda_N$ are the costate variables, and $q_1, \ldots, q_N$, α are the Lagrange multipliers related to the constraints (15) and (16). A hierarchical approach [see e.g. Singh (1980)] can be used to solve problem (17). At the first level α is treated as a known function and each $\tilde{J}_i$ is minimized with respect to (u_i, P_i, x_i). At the second level the function α is manipulated until power balance, Eq. (16), is achieved. This can be done by choosing the price function α such that the value of the dual function $\varphi(\alpha)$ associated with $\tilde{J}$ is maximized. Solving the first–level problems together with the second–level problem then gives the solution to problem (17) for $\mu \in \Lambda^{>}$.

To calculate the bargaining solution we add the third level to the hierarchy. The three–level hierarchy is illustrated in Figure 1. The weight vector $\mu^* \in \Lambda^{>}$ satisfying (12) is calculated at the top level according to the algorithm of Section 2.

As a numerical example consider a power pool consisting of three ($N = 3$) identical power systems. The systems differ only with respect to local demands. The numerical values used in the computations are

$$A_i = \begin{pmatrix} -0.033 & 0.000 & 0.000 \\ 0.067 & -0.060 & 0.060 \\ 0.000 & 0.030 & -0.043 \end{pmatrix}, \quad B_i = \begin{pmatrix} 0.033 \\ 0.000 \\ 0.000 \end{pmatrix},$$
$$C_i = \begin{pmatrix} 0.000 & 0.000 & 0.280 \end{pmatrix},$$
$$r_i = 6.5, \; s_i = 13.0, \; i = 1,2,3. \tag{19}$$

Local demands are

$$d_i(t) = E_i sin(\omega_i t + \delta_i), \; i = 1,2,3,$$
$$\begin{array}{lll} E_1 = 5.0 & E_2 = 6.0 & E_3 = 7.0 \\ \omega_1 = 2\pi/T & \omega_2 = 2\pi/T & \omega_3 = 2\pi/T \\ \delta_1 = 0.0 & \delta_2 = -\pi/4 & \delta_3 = -\pi/2. \end{array} \tag{20}$$

$x_i(0)$ is determined by $d_i(0)$, $i = 1,2,3$. At the top–level the weight vectors were generated by

$$\mu^{k+1} = \varsigma\mu^k + (1-\varsigma)T(\mu^k),$$

where the damping factor $\varsigma \in (0,1)$.

Figure 2 illustrates the fuel flows corresponding to the status quo and the bargaining solution. In the bargaining solution the controls are smaller than in the status quo. In Figure 3 the cooperative production rates together with the status quo production rates are illustrated. It can be seen that the production rate corresponding to the cooperative solution is shifted in time in Systems 1 and 3 compared to the status quo situation. Values for final weights and costs are

	μ_i^*	J_i^*	D_i
System 1	0.4037	185 790	216 730
System 2	0.3217	259 870	300 680
System 3	0.2710	355 480	399 870

which shows that cooperation does, indeed, reduce costs in all power systems in the planning period. According to Theorem 1 the solution obtained is a solution for the bargaining problem associated with (13)–(16), (19), (20).

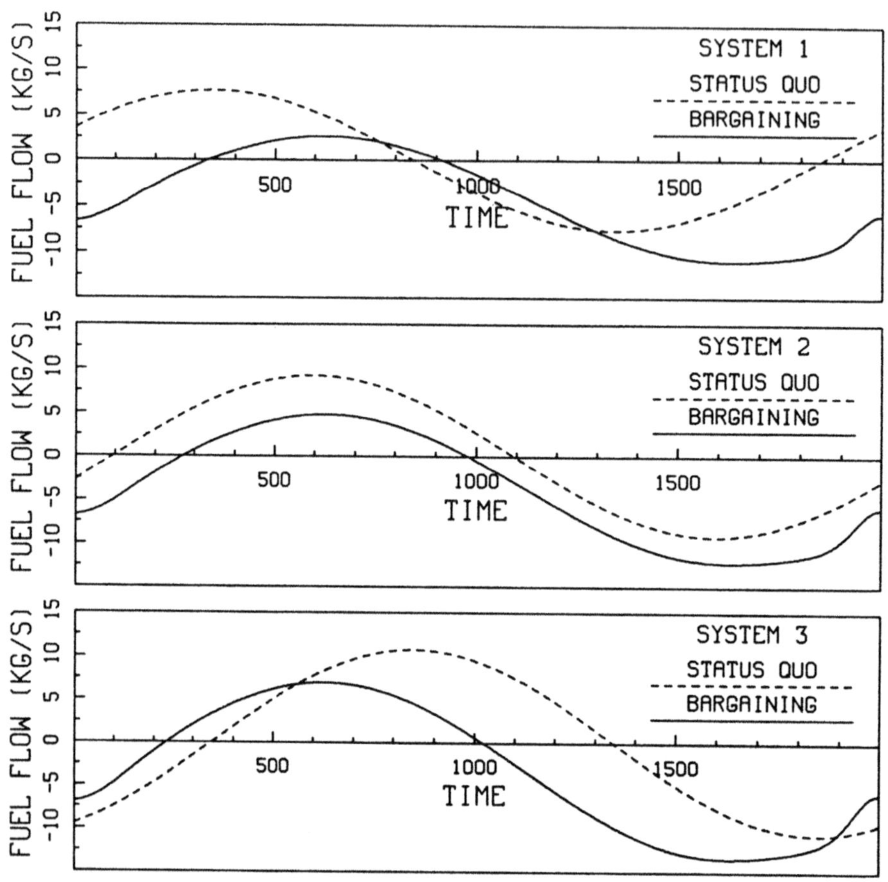

Fig. 2. Fuel flows corresponding to status quo and bargaining.

4. Conclusion

In this paper we have proposed a new way of solving bargaining problems by transforming the original product–form problem into a set of linear ones. This means that the bargaining problem can be solved as a two–level hierachical decision problem. In many practical situations the lower level of the hierarchy forms a large–scale system which, in turn, can be hierarchically controlled. This is the case in the energy management example presented in the paper.

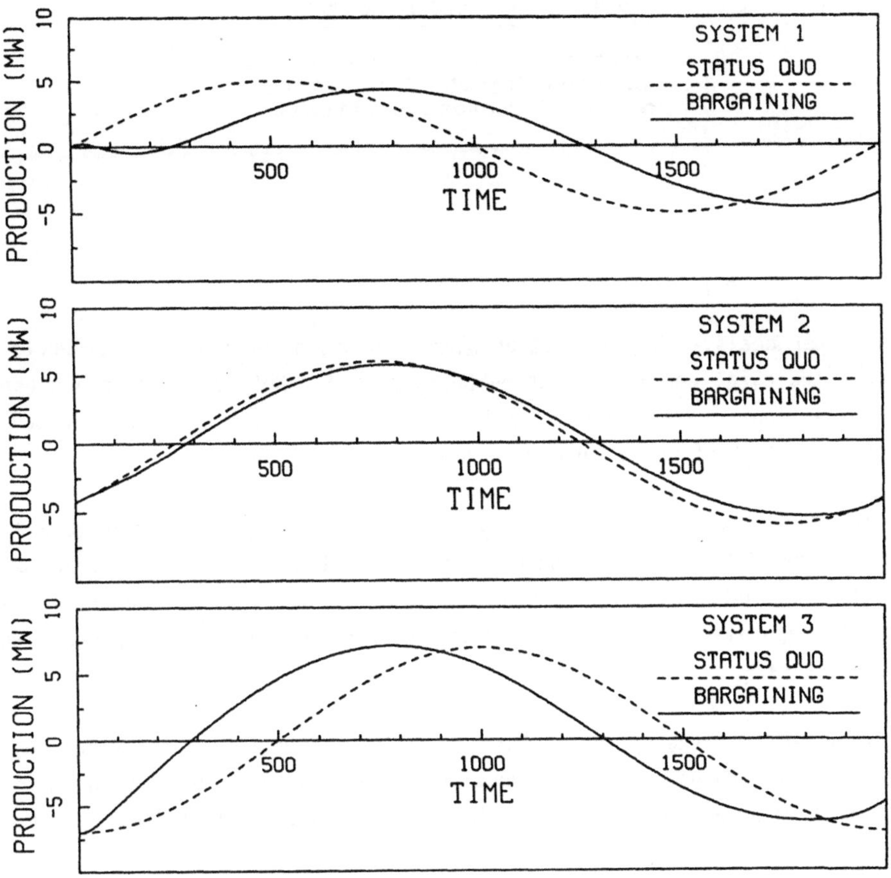

Fig. 3. Production rates corresponding to status quo and bargaining.

5. References

Ehtamo, H., J. Ruusunen, and R. P. Hämäläinen (1987), "On the Computation of the Nash Bargaining Solution with an Energy Management Example," Helsinki Univ. Tech., Syst. Anal. Lab., Rep. A20.

Nash, J. F. (1950), " The Bargaining Problem", Econometrica, Vol. 28, pp. 155–162.

Roth, A. E. (1979), "Axiomatic Models of Bargaining," Springer–Verlag, Berlin Heidelberg.

Singh, M. G. (1980), "Dynamical Hierarchical Control," North Holland, Amsterdam.

OPTIMAL PRODUCTION - MIX

Richard F. Hartl, Vienna
Johannes Krauth, Stuttgart
Joachim Warschat, Stuttgart

This paper deals with a production plant, on which two different products can be produced. The plant consists of three Subsystems S_i. Before or after a phase of separate processing in Subsystems S_1 and S_2 the two products have to be processed in subsystem S_3. Each of these subsystems has a limited capacity.

In the first part, we assume empty stocks at the beginning, and at a fixed time, T, in the future certain quantities X_i of the two products have to be delivered to the customers. Facing linear holding costs, concex production costs, and stringent capacity constraints the problem is to decide when to produce which product at what rate.

It is shown that the optimal solution consists of up to *six* different regimes and that the time paths of the production rates need not be monotonic. These results, which can be obtained analytically, are also illustrated in several numerical examples.

We also analyze a variation of the model with non-convex production costs.

Finally the case is considered where the terminal demand at time T is replaced by a continuous and seasonally fluctuating demand rate. It is demonstrated that the optimal production rates show an interesting and non-trivial behavior. In particular, it may happen that on intervals where the demand for one product increases the optimal production rate decreases. This is also demonstrated by computer plots for some numerical examples.

TECHNOLOGICAL PROGRESS IN A DYNAMIC MODEL OF THE FIRM

Onno van Hilten, Maastricht

1. Introduction

In most dynamic models of the firm the inputs of the production process, capital and labour, are homogeneous. That means that each unit of such an input has exactly the same characteristics. For instance, the characteristics are not dependent on the date of purchase. But it may happen that, as time progresses, capital goods become available which give a higher output per unit of input. This is called technological progress (which will be abbreviated as TP). Taking this into account leads to the use of vintage models, in which capital goods of different vintage are not treated as being identical. Examples of this approach are for instance models by Malcomson [3] and Nickell [4]. Those models concentrate on the investment behaviour of firms. The purpose of this paper is to study a model which incorporates TP and in which investment policy and financial policy are determined simultaneously.

2. Technological progress

There are a lot of different kinds of TP but only one kind will be incorporated here. The following definitions are needed:

$Q(t,v)$ = output produced with capital goods of vintage v at time t
$K(t,v)$ = stock of capital goods of vintage v at time t
$L(t,v)$ = level of employment on capital goods of vintage v at time t

$$K(t,v)/Q(t,v) = c_1, \text{ for all t and v} \tag{1}$$

This last definiton does not mean that all capital goods are identical. The way in which they differ shows up in the amount of labour per unit of output:

$$L(t,v)/Q(t,v) = c_2(v), \text{ for all t} \tag{2}$$

The TP will now be represented by the fact that c_2 is not a constant, but a decreasing function of time. This TP is labour-augmenting: labour productivity of newer vintages is higher since c_2 is decreasing: you get more working force per worker. The TP will also be embodied: it only applies to newly bought capital goods: c_2 is a function only of v, not of t. The amount of labour per unit of out-

Operations Research Proceedings 1987

put for vingate v is constant over the entire lifetime of that vintage. Finally, the TP is not induced by the activities of the firm, it is exogeneously given. I assume exponential depreciation at a constant rate and furthermore I assume that a vintage is never scrapped. This is of course a debatable assumption. I will say something about that later on. I now can sum up all vintages which leads to the following differential equations for the capital and labour stock:

$$K(t) = \int_{-\infty}^{t} K(t,v)dv \longrightarrow \dot{K}(t) = I(t)-aK(t) \quad (3)$$

$$L(t) = \int_{-\infty}^{t} L(t,v)dv \longrightarrow \dot{L}(t) = c_2(t)/c_1.I(t)-aL(t) \quad (4)$$

Since capital productivity is the same for all capital goods, I get this very simple relation between capital and output:

$$Q(t) = K(t)/c_1 \quad (5)$$

Note that there are no substitution possibilities between capital and labour. The model is therefore called a clay-clay vintage model. I define a concave sales function S(Q) and I will use the term operating income for the expression:

$O[K(t),L(t)] = S[Q(t)] - w(t).L(t) - a.K(t)$

3. The model

$$\max_{I,D} \int_0^z e^{-it} D(t)dt + e^{-iz} X(z) \quad (6)$$

$$\dot{X}(t) = (1-f)(O[K(t),L(t)]-r.Y(t))-D(t) \quad (7)$$

$$\dot{K}(t) = I(t)-a.K(t) \quad (8)$$

$$\dot{L}(t) = \frac{c_2(t)}{c_1} \cdot I(t)-a.L(t) \quad (9)$$

$$K(t) = X(t)+Y(t) \quad (10)$$

$$0 \leq Y(t) \leq bX(t) \quad (11)$$

$$D(t) \geq 0 \quad (12)$$

$$I(t) \geq 0 \quad (13)$$

$X(0)$, $K(0)$ and $L(0)$ given (14)

where I(t) stands for gross investment, D(t) for dividends, X(t) for equity, Y(t) for debt, r for the interest rate, f for the corporation profit tax rate and i for the time preference rate of the shareholders. The objective of the firm is to maximize the shareholders value of the firm (eq. 7). There are three state variables: capital, labour and equity: debt can be eliminated, using (10). The firm has two control variables, gross investment and dividend pay-out. The amount of debt is limited (eq (11)) and investments are irreversible (eq. (13)).
The Maximum-principle was used to derive necessary conditions for optimality and the indirect method of Hesteness and Russak was used to adjoin the pure state constraints (see for instance [1]; mathematical derivations can be obtained from the author upon request). The costate variable of the state equation for the labour stock will play an important role. In general one cannot interpret the costate variable as a shadow price when one uses the Hestenes-Russak approach, but since L does not appear in the pure state constraints, it is allowed here.

The shadow price of labour (SP_L) measures the contribution to the objective function of an extra unit of labour. In that sense we can interpret the following formula (SP_E is the shadow price of equity):

$$\frac{d}{dt} SP_L = (1-f).w(t).SP_E(t)+a.SP_L(t) \tag{15}$$

$$\Rightarrow SP_L(t) = -\int_t^z SP_E(\tau).(1-f).w(\tau)e^{-a(\tau-t)}d\tau \tag{16}$$

$\exp[-a(\tau-t)]$ is what is left at time τ of a unit of labour recruited at time t. So $(1-f).w(\tau).\exp[-a(\tau-t)]$ is the after tax wage cost at time τ 'belonging' to a unit of labour recruited at time t. Paying out these wages results in a reduction of equity. In terms of the objective function this means a negative contribution of that amount multiplied by the shadow price of equity. To arrive at the shadow price of labour at time t, we have to sum up all these contributions. To avoid confusion I should say here that this shadow price of equity is <u>not</u> equal to the costate variable of equity (it follows from the optimality conditions that $SP_E(t)=\exp[-i.t]$+(the Lagrange multiplier of the constraint $D\geq 0$)). The minus sign indicates that the shadow price of labour is in fact a cost.

4. Optimal policies when there is no technological progress

Using the four inequality constraints (see eqs. (11)-(13)) one can distinguish sixteen different paths. Each combination of binding and non-binding constraints determines a path. For simplicity I will assume that the cost of equity is cheaper than the cost of debt ($i<[1-f].r$) and I will only consider paths <u>without debt</u>:

path 1: I>0, D>0
path 3: I=0, D>0
path 5: I>0, D=0

The path connecting procedure of Van Loon [2] was used to make strings of paths. Such a string then denotes an optimal policy for the whole planning period. Without TP and assuming that the wage rate is constant, the only possible strings are 3-1 and 5-1.

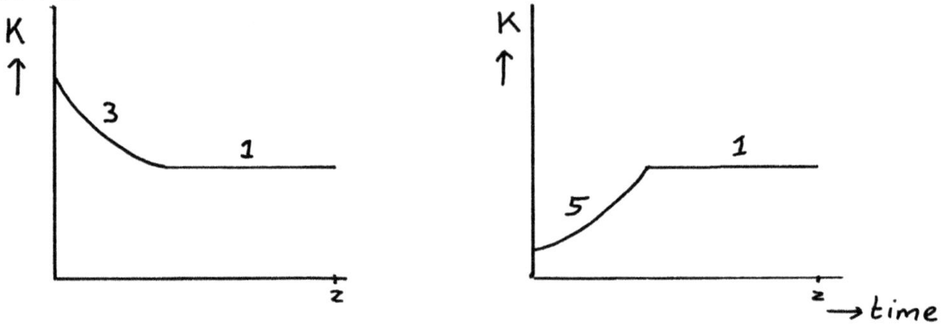

On path one marginal revenue equals marginal cost:

$$\frac{dS}{dQ} = C_X = c_1\{w(t).\frac{c_2(t)}{c_1} + a + \frac{i}{1-f}\} \tag{17}$$

This marginal cost C_X is now a constant (c_2 is now a constant!) and due to the convavity of S, this leads to a constant capital stock. The marginal cost is measured before tax and consists of wages, depreciation and cost of capital. String 5-1 occurs when marginal revenue exceeds marginal cost at t=0, 3-1 occurs when marginal revenue is less than marginal cost at t=0.

5. Optimal policies when there is technological progress

With TP the situation is different. Now we have on path one the following equality:

$$\frac{dS}{dQ} = C_X(t) + \frac{\dot{c}_2(t).e^{it}.SP_L(t)}{1-f} \tag{18}$$

Since the TP is labour-augmenting, it is not reasonable to assume that wages are constant. Now C_X is no longer a constant. Moreover we now have an extra cost term. Note that the derivative of c_2 is negative and SP_L is negative, so this extra term is positive. Given the interpretation of SP_L, one can see that $\exp[i.t].SP_L(t)/(1-f)$ is the before tax current value of the labour cost associated with a unit of labour recruited at time t. The derivative of c_2 can be seen as the '<u>speed of TP</u>' and so the extra cost term in (18) can be seen as the <u>extra</u> labour cost you incur by attracting a unit labour <u>now</u> instead of a moment later. You do not take advantage of the future effects of TP, and the size of these future effects is measured by the rate f TP. Total marginal cost, the right hand side of (18) is called the user cost of capital (UCC).

$$UCC(t) = c_1\{w(t)\cdot\frac{c_2(t)}{c_1}+a+\frac{i}{1-f}\}+\frac{\dot{c}_2(t).e^{it}.SP_L(t)}{1-f} \qquad (19)$$

One of the great problems in this model is that now marginal cost is not given exogeneously as in the case without TP. The user cost of capital at time t depends on the shadow price of labour on time t and that shadow price depends on the shadow price of the equity over the remaining part of the planning period. But anyhow, with TP path one is, just as before, the only possible final path and it remains valid that marginal revenue equals marginal cost on path one. But now this does not necessarily results in a constand capital good stock. For instance, if this user cost of capital is rising, dS/dQ must be rising and so Q and K are decreasing due to the concavity of S.

To be more precise, I make some further assumptions.

$$c_2(t) = c_2(0).e^{-kt} \quad ,k>0 \qquad (20)$$

$$w(t) = w(0).e^{lt} \quad ,l>0 \qquad (21)$$

Note that the wage rate is rising. At first sight it might seem logical to conclude that if the rise of the wage rate equals the TP,TP does not matter, at least not on the final path. But this is not true. The reason is that the TP only applies to the <u>newest</u> capital goods, while the rise of the wage rate applies to <u>all</u> workers. The user cost of capital and the capital stock are constant on the final path one only if i+a-k=0 and 2k=l. Furthermore it can be seen that if the rise of the wage rate is smaller than the TP, the capital good stock is rising on the final path one. But is not necessarily so that the capital goods stock is decreasing if the rise of the wage rate is greater than the TP. It was possible to derive these results because on path one the shadow price of equity is very simple (SP_E=exp[-i.t], so then in fact SP_L and UCC only depend on known parameters:

$$SP_L(t)= -\int^{z} e^{-i\tau}.(1-f).w(\tau).e^{-a(\tau-t)}d\tau \qquad (22)$$

Just as in the case without PT I want to make strings of paths. Althoug UCC is not exogeneously given, it is possible to derive that the user cost of capital is a decreasing function of time during the whole planning interval if i+a-k<=0 and 2k>1. The fact that UCC is a decreasing function can be used in the path connecting procedure. Now I not only get 3-1 and 5-1 as optimal strings, but also 3-5-1 and 1-5-1. Let us try to understand why string 3-5-1 could be optimal and under which conditions. As we have seen before on path 1 we have dS/dQ=UCC and on path 3 dS/dQ is rising, since the firm does not invest. The analogy of the situation without TP would be this: if at t=0 marginal revenue is less than marginal cost, stay on path three until dS/dQ=UCC and then stay on path one. Let us assume that at t=0 marginal revenue is less than marginal cost[1]. What could be the reason that 3-5-1 is the optimal string instead of 3-1 and when is this so? My hypothesis is as follows: it may happen that on path one the firm cannot grow quickly enough to maintain dS/dQ=UCC.

On the final path one we have a simple formula for the user cost of capital, since SP_E=exp[-i.t] (see (19)). All parameters in this formula are exogeneous, so the rate of decrease of UCC is fixed. To maintain dS/dQ=UCC the firm will have to grow at a certain speed. But the speed of growth is determined by investment and investments are bounded by the availability of cash. Suppose now that if the firm follows policy 3-1 (figure one), it is not able to maintain dS/dQ=UCC on path one up to time z. A way to avoid this is to start growing earlier, on path five. Then you get the solution as depicted in figure two. Remember that on path five all the cash-flow is spent on investments while on path one dividends are paid out. It is not possible for the firm to jump on path one earlier, because on path one dS/dQ=UCC must be satified. Therefore path five is needed.

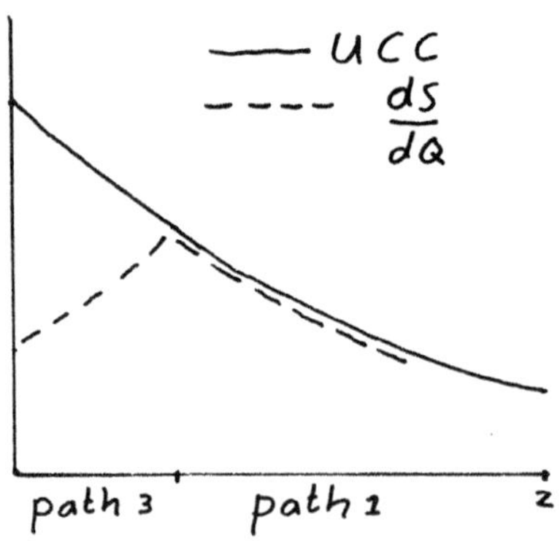

Figure one: string 3-1

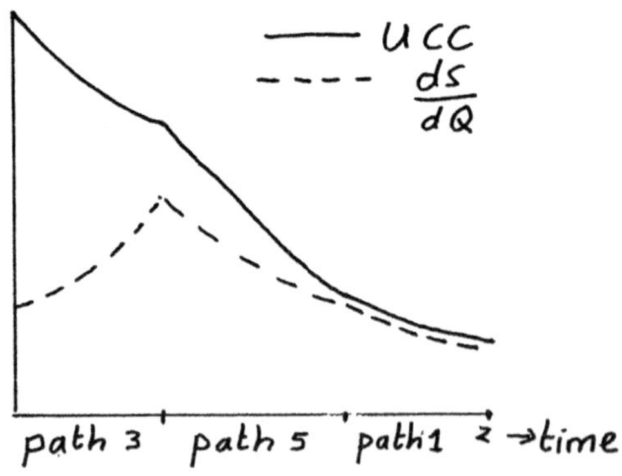

Figure two: string 3-5-1

This hypothesis looks reasonable, but I have not been able to prove it yet. Due to the fact that the user cost of capital at t=0 depends on the optimal string (see footnote one), it is very difficult to pin down the inital conditions under which a certain optimal string occurs. Related to this problem is the question under what conditions on the shape of S the hypothesis becomes relevant. But the answer to this question is rather complex since the number of relevant parameters is quite large and since it also depends on the endogeneous variables capital stock and time. Furthermore, the construction and interpretation of strings is even more difficult when UCC is rising and I can't say anything if UCC oscillates.

6. Conclusion

I will end up with some conclusions. The biggest problem is that the user cost of capital formula which plays such an important role, contains the costate variables of equity and labour. The reason for this seems to be the fact that investment and finance are treated simultaneously: Malcomson and Nickell, who only treat investment do find an expression for UCC without auxilliary variables. A consequence is that it is not possible to find a feedback control. The only way out of this problem seems to me to find an explicit solution for all state, adjoint and Lagrange multipliers, but that requires a great deal of luck and I don't think we are that lucky here.

As I mentioned in the beginning, I assume in this model that capital goods are never scrapped. It would be very interesting to introduce as a new control variable the service life of capital goods. The solution becomes more complicated then and I did not succeed to apply the path connecting procedure.

In spite of the problems, there are some interesting results when we compare this model with TP to the model without TP. The final level of the firm is constant only under very specific circumstances and the road towards that final level is much more complex.

1) If 3-1 is the optimal string, UCC at t=0 only depends on exogeneous parameters; but if 3-5-1 is the optimal string, UCC at t=0 depends on endogeneous parameters, since on path five SP_E equals exp[-i.t]+(the Lagrange multiplier of D>=0). So it is a bit strange to assume that marginal revenue is less then marginal cost and then to find the optimal string, because marginal cost at t=0 depends on the optimal string. But it can be shown that the UCC (t=0) 'belonging to' 3-5-1 is larger than the UCC (t=0) 'belonging to' 3-1, which we assume to be larger than marginal revenue at t=0.

References

[1] FEICHTINGER, G., R.F. HARTL, 1986, Optimale Kontrolle ökonomischer Prozesse, De Gruyter, Berlin
[2] LOON, P.J.J.M. VAN, 1983, A Dynamic theory of the firm: production, finance and investment, Springer-Verlag, Berlin
[3] MALCOMSON, J.M., 1975, Replacement and the rental value of capital equipment subject to obsolescence, Journal of Economic Theory, vo. 10, pp. 24-41
[4] NICKELL, S.J., 1975, A closer look at replacement investment, Journal of Economic Theory, vol. 10, pp. 54-88

A DYNAMIC INVESTMENT RULE FOR AN IRREVERSIBLE PROJECT

Werner Jammernegg, Graz

We deal with the question whether to invest in an already selected project, e.g. a productive asset like a machine. Assuming that the investment projects of a firm are economically independent and indivisible, the answer of the previous question reduces to the application of the net-present-value-rule. But this is only correct if the considered project is a short-lived one. If the investment project is at least irreversible in the short run, the question when to invest becomes important too. For this type of project the optimal investment decision must be investigated in a dynamic framework.

We derive in this paper the optimal investment policy for an irreversible project in a discrete time framework over a finite planning horizon which is bounded by the project's economic life. The project is assumed to be traded on the capital markets and its market values are independent over time with known distributions. The opportunity to invest in the project in a planning period is characterized by a known probability. The optimality criterion is the discounted utility of the market value less the capital cost of the project.
If this utility is non decreasing we show that the optimal investment decisions are also non decreasing functions of the market value: the project is undertaken if its utility minus capital cost is greater than a positive quantity, the net-present-value-premium.

We investigate the change of the investment behaviour when varying the project's planning horizon and the probability that an investment opportunity occurs.
Finally we consider the effect of a variation of the distribution of the market value with respect to first and second order stochastic dominance, respectively. We present results using an algorithm for a linear utility function and Erlang-distributed market values.

The firm's dynamic investment policy

Peter M.Kort
Tilburg

In this paper we develop dynamic models of the firm with convex (increasing marginal costs) and concave (decreasing marginal costs) adjustment cost functions . The conclusion is that in the convex model investment appears to have a continuous course during the time and in the concave case investment expenditures take place in a "jump pattern".
We further introduce the net present value and the internal rate of return within dynamic models of the firm (with convex adjustment costs , concave adjustment costs , activity analysis , tax parameters and a growing economic environment).
The above models all have in common that they assume a certain future . We can extend the analysis by the introduction of uncertainties and the use of stochastic control theory . We can prove , that , - depending on the firm's capital good stock , its amount of cash , the uncertainty of the firm's investment and the attitude of the shareholder towards risk - the firm makes a choice between three destinations of its revenue : increase the amount of cash , invest the money or pay out dividend .

THE CONTROL OF ENVIRONMENTAL POLLUTION OF A FIRM

Mikulas Luptacik, Vienna

The subject of this survey paper is the behaviour of the profit maximizing entrepreneuer under environmental constraints. In the first part the following question is analyzed: how does the control of pollution affect the optimal investment-employment decisions of a firm? The relevant question connected with the strategies of environmental policy is that of their instruments. For this purpose two versions of the model are considered (FEICHTINGER-LUPTACIK, 1987): model A with emission certificates and model B with environmental standards. The nature of the optimal solution for both a static and a dynamic model is analyzed and an economic interpretation is given. Using the theory of optimal control it can be shown for model A that along the optimal path the increasing expenditure in pollution control is accompanied by an increasing stock of employees. In model B the firm does not abate until the pollution constraint becomes active. In model C which combines the two previous models the firm will abate before the level of pollution hits the given constraint. For this model the question is answered: how large the emission charge must be set such that in a stationary state the emission don't exceed the environmental standard.

The sensitivity analysis of equilibrium values with respect to the changes in the parameters of the model establishes the negative effects of a higher discount rate and positive effects of a higher output price on the hiring rate, employment and on the level of abatement equipment. The important conclusion is that higher charges for pollution do not necessarily result in higher equilibrium values of expenditure for pollution control.

The second part of the paper is devoted to the question: how the firm should combine the different production techniques and the abatement activity in order to maximize the shareholder's value of the firm? (HARTL-LUPTACIK, 1986). When will the firm introduce the new - from the point of view of environmental quality progressive - technology? For this purpose a model of dynamic activity analysis with two production techniques using labour and capital as inputs and yielding marketable goods as well as pollution as two sorts of output is considered. The second activity is more efficient but it causes more pollution per unit of capital. The third one is an abatement activity using labour and capital as inputs. Regarding the instruments of environmental policy the concept of emission certificates is used. The optimal control problem is solved using Green's theorem approach in the linear case and phase-plane techniques in the non-linear version. If the initial stock of equity of the firm is relatively small, the firm uses the more efficient technology (with higher pollution) first in order to accumulate equity and switches to the "cleaner" technology later.

In the case of a large discount rate in the long run only the dirty activity is used.

On the other hand, if the decision maker is sufficiently far sighted in the long run only the clean activity is used. Under reasonable assumptions no cleaning by third activity takes place.

The sensitivity analysis from the environmental point of view shows that income tax and emission charge should be large enough in order to reduce the level of pollution. On the other hand the income tax and the emission charge should not be larger than certain critical levels because otherwise the firm hasn't enough equity to use the clean technology.

References:

FEICHTINGER, G. - LUPTACIK, M. (1987): "Optimal production and abatement policies of a firm", European Journal of Operational Research 29, North-Holland, 274-285.

HARTL, R. - LUPTACIK, M. (1986): "Environmental constraints and choice of technology: A dynamic activity analysis", Forschungsbericht Nr. 82 des Instituts für Ökonometrie, OR und Systemtheorie der TU Wien.

An Application of Control Theory to the Economic Analysis of the Firm's Management in Centrally Planned Economy

Krystyna Strzała, Göttingen

Zusammenfassung: Verschiedene Aspekte des Entscheidungsfindungsprozesses in Zentralverwaltungswirtschaften (u.a. Motivation des Managements, Zielvorstellungen der Unternehmen) werden diskutiert. Ein linear-quadratisches Entscheidungsmodell eines Bauunternehmens in Polen wird dargestellt und mit Hilfe des Algorithmus - beruhend auf dem Diskrete-Minimum-Prinzip von R.S.Pindyck - gelöst. Zur Ermittlung der optimalen Planungsstrategien werden verschiedene Simulationen durchgeführt und die Ergebnisse präsentiert.

Summary: Aspects of decision making process at the firm level in centrally planned economy (manager's motivation, the 'desired state' of a firm) are discussed. To solve a quadratic-linear tracking model an Algorithm based on the Discrete-Minimum Principle proposed by R.S. Pindyck is used. The results of optimal control experiments aiming at the optimization of the firm's position made on a small econometric model for building enterprise in Poland are presented and discussed.

Introduction

Thirty years of constant attempts to manage the centrally planned economy neglecting market forces have failed. Recent reforms in Poland as well as in other socialist countries have tended to increase the role of the market variables in stimulating economic processes.

Due to the reforms which have been implemented in Poland since the middle of the 70s, autonomy of firms has grown. Strictly planned directives for production have almost been eliminated and the role of other monetary instruments has been increased. These should have created to opportunity to optimize the functioning of the firms, but analyzing the firm's behavior in the period 1975-1980, especially in the housing market in Poland, some questions arise:

- were the firms really interested in economic optimization while still playing the dominant role on the 'seller's market' and continuously faced with shortage in input?
- what kind of objective function is to be chosen in optimization of the firm's behavior?
- what is the 'desired state' of the enterprise?

In order to answer these questions some of the features of the firm's management in the centrally planned economy are to be discussed.

1 Features of the decision-making process at the firm level

As a consequence of the methods and the institutional solutions by which the Polish economy was controlled, the feature of it is excess demand on consumption goods markets as well as

Operations Research Proceedings 1987

in the production sphere. Excess demand is followed by shortage, which coincides with the 'seller's market' [7]. Planning in the form of planning bargaining between the central planner superior authority and state enterprises, creates such phenomena as 'plan stretching', cost and price manipulation, all of which led to a situation in which the functioning of firms is basically resource-constrained and simultaneously the level of the firms' revenues do not depend on their's resource utilization [8].

If a firm operates at a loss, that does not lead to real banktruptcy i.e. to a ceasing operation. The firm receives additional credit, or its tax is reduced, or it is granted a subsidy, or the selling price is raised. In one of these ways, it survives financial difficulties.

The firm as a buyer tries to acquire as much input as possible to keep shortage from hinderring production. This leads to such well-known phenomena as hoarding of raw materials, labor and fixed equipment i.e. to the so-called 'freezing of production funds'.

On the other hand, the firm as a seller faces an almost insatiable demand. In the case of a 'seller's market', negotiable prices and excess demand, it is almost always able to spill the costs over onto the buyers. As a result, the costs of production are high and regularly included in prices.

Due to the reforms which have been carried out since the middle of the 70s, the form of planning has changed from the physical norm method to the financial method. The large number of physical planning indices determined by the higher planning organs has been replaced by financial factors. Planning bargaining has been replaced by redistribution bargaining. The firm's managerial staff is bargaining with higher planning authorities in order to ensure that the latter create better financial circumstances for it.

The normal degree of that process is different in each production sphere, depending mainly on the 'importance' of the outputs. It also varies within the production branch, depending on whether the firm is powerful or weak.

In the building industry in Poland, all these negative phenomena have been highly concentrated. The building enterprises have been working at a loss for a long time, mainly because of the existing excess demand and high social pressure. These factors have made, this one of the highest priority industries in the economy as a whole. Building enterprices were more readily granted allowances and benefits and were therefore not interested in economizing on production.

The described before circumstances in the planning process in Poland and especially those connected with the building industry, tempted firms' managers to react with requests, complaining and bargaining to acquire planned conditions which were easy to fullfil. The efficiency of the plan did not come up for discussion.

The recent reforms introduced into the Polish economy aim at increasing the profitability criterion in controlling the functioning of firms. In the planning process, the profitability criterion is considered in many different kinds of decisions as granted credits or allowances for new investments. Because of that, decision-makers in firms have to adapt themselves to the new circumstances and start to analyze not only the feasibility but the efficiency of production as well. On the other hand, in the behavior of firms' managers, the tendencies toward independence which lead them in the direction of a profit calculus can be observed. These tendencies are connected with the new possibilities created throw reforms, i.e. with material incentives linked to profit (the manager's bonus, the profit share of the staff which complements wages, and the dependence of the firm's welfare funds on profit).

In these new circumstances, it seems to be reasonable to try to apply an econometric decision model in order to describe the functioning of the firm and as a tool of an economic analysis and practical decision making in the preparation of the firm's plan.

2 The major problems in constructing an econometric decision model

Many problems in economics could be formulated as dynamic models in which control or policy variables are used to move a system over time from a less desirable to a more desirable position. The problem of the firm in the market economy is classically specified in such a way: "... Here the controls are pricing and production levels, the dynamic system is a model of production and sales, and the desired position is high levels of profits." [5, p.1]. Needless to say, that this is not the case with a firm in a centrally planned economy.

Firm's behavior can be guided by many different motives, such as increasing the volume of production, acquiring and increasing profits, or, last but not least, achieving a 'good marks' from the superior administrative authority. The goals specified in firms' production plan, after which they should strive, are relatively detailed and often not consistent (e.g. increasing the volume of production - lowering of the production costs). Because of that, the objective preference function to be optimized has to fullfil certain conditions:

1 targets have to be measurable
2 it should integrate all these detailed goals laid down by the higher administrative authority
3 it has to be consistent with the objective preference function of the economy as a whole

Taking into account that the characteristic features of firms in the building industry in Poland are such as: 'freezing of production funds', capacity under-utillization, high production costs, the rush work at the end of a month or a quarter, it is difficult to conclude not only what the desired state of the firm should be, but what kind of objective function is to be chosen as well.

If we assume that the production plan made by the firm's managers is made up correctly, we can assume that the desired state of the firm is that described by the production plan. On the other hand, firm and firms managers are judged on the basis of the production plan they created and then fullfilled. Also we know that the planned levels are often under-estimates, we take the levels specified in the production plan, for the time being, as the desired levels of the firm [5].

Taking into account all the characteristic features of the situation of the firm in the socialist economy, it goes without saying that to develop an objective preference function which could fullfil all constraints is a complicated task.

There are four measures, which are basically implemented for judging the firm's position: volume of production, gross output, overall costs, and net output. All these measures have some disadvantages.

Measuring the results of the firm by the volume of production or size of gross output could bring absurd responses, because costs of raw materials and capital costs would be totally ignored under these objectives [9].

Implementing the overall costs could lead to economizing on costs but would not encourage firms to increase output or implement technological progress.

In the case of the 'seller's market', maximizing net output could result in activities not consistent with the profit of the economy as a whole (e.g. manipulating of the production structure).

Another way to solve this problem, is to construct the weighted function, in which all or some objectives mentioned above could be included, i.e. quadratic tracking functional [10].

In a case of the linear-quadratic econometric model, an explicitly specified preference function is optimized subject to constraints derived from an econometric equation system. These models are the most often used, especially for deriving optimal stabilization policies (see e.g. [5], [4]). In spite of all the well known disadvantages of the quadratic-linear approach, some of the advantages of these models should not be forgotten [3]. Firstly, the quadratic objective function, in the case of the linear equality constraints, leads to linear decision rules which are easy to handle. Secondly, these models are rather well suited for simulation studies.

In the case, in which the desired values are equal to the planned values, the quadratic-linear tracking model seems to be proper to checking the consistency of the different versions of the production plan as well as to comparing the effects caused by changes in controlling rules (administrative authorities).

3 Econometric decision model for the building enterprise

Several econometric models describing different aspects of the functioning of the building enterprise in Poland, have been constructed (see [11]). In this paper, I will present the results of the simulation runs made on one of them. These simulation exercises aimed at checking the consistency of the firm's production plan and at the optimization of the firm's position. The models were constructed in 1982 by using quarterly data for the 'Gdanskie Przedsiebiorstwo Budowlane', covering the period 1975-1980. The model presented below has 10 stochastic equations. The endogenous variables are:

FA fixed assets in GPB (GPB-Gdanskie Przedsiebiorstwo Budowlane)

E employment

GVF gross value of finished building objects

NO net output

WB wage bill

AW average wages

PL productivity of labor, defined as PL = GO/E

PFA productivity of fixed assets, defined as PFA = GO/FA

OC overall costs

GO gross output

The exogenous and lagged endogenous variables are:

IO investment outlays

PBP productivity subject to building production

TSB technological structure of building production

MA coefficient of material absorbtion

VA value of material stocks

Z1 dummy variable, Z1 = 1 in 1.q.1979 and 1980,
= 0 in others quarters

Z2 dummy variable, Z2 = 1 in 1.q.1976, 1979, and 1980
= 0 in the others

Z1 and Z2 reflect the unexpected changes in the rules by which the building firms were controlled. The model contains six lagged endogenous variables, i.e. FA1, E1, NO1, WB1, AW1, and GO1 .

The structural form of the model has been estimated by using 2SLS method. The following results have been obtained (the brackets beneath each estimated coefficient contain the t-statistic. Also shown are the values of the corrected coefficient of determinancy R2, standard error SE, and Durbin-Watson (DW) or h-statistic for autocorrelation).

$$FA = \underset{(0{,}6683)}{7{,}765} + \underset{(21{,}90)}{0{,}9673}\ FA1 + \underset{(4{,}245)}{0{,}000692}\ IO$$

R2 = 0,9616 SE = 3,71 DW = 2,296 h = - 0,6746

$$E = \underset{(1{,}745)}{2706} - \underset{(-6{,}269)}{4{,}173}\ FA + \underset{(2{,}928)}{0{,}005311}\ IO - \underset{(-2{,}148)}{3{,}116}\ TSB$$

R2 = 0,8513 SE = 44,64 DW = 1,759

$$GVF = \underset{(-1{,}052)}{-149{,}2} + \underset{(15{,}72)}{0{,}1003}\ E + \underset{(23{,}52)}{2{,}246}\ PBP$$

R2 = 0,9646 SE = 2,90 DW = 1,32

$$NO = \underset{(7{,}291)}{100{,}7} + \underset{(10{,}31)}{0{,}6354}\ GVF - \underset{(-1{,}757)}{0{,}8051}\ NO1 - \underset{(-3{,}14)}{0{,}09852}\ MA - \underset{(-13{,}01)}{0{,}2024}\ VA$$

R2 = 0,9513 SE = 4,23 DW = 1,99 h = 0,1891

$$WB = \underset{(1{,}192)}{5{,}531} + \underset{(3{,}816)}{0{,}7438}\ WB1 + \underset{(2{,}112)}{0{,}0514}\ TSB - \underset{(-3{,}576)}{2{,}879}\ Z2$$

R2 = 0,6776 SE = 1,09 DW = 1,792 h = 0,9402

$$AW = -1444 + 151{,}3\ WB + 0{,}4902\ AW1 + 17{,}83\ TSB$$
$$(-2{,}03)\ (4{,}762)\ (4{,}685)\ (2{,}744)$$

$R2 = 0{,}8623$ $SE = 194{,}96$ $DW = 2{,}028$ $h = 0{,}084$

$$PL = -17{,}55 + 0{,}0706\ FA + 0{,}066198\ AW + 1{,}605\ PBP$$
$$(-1{,}902)\ (1{,}836)\ (3{,}198)\ (15{,}39)$$

$R2 = 0{,}9538$ $SE = 3{,}09$ $DW = 1{,}231$

$$PFA = 368{,}1 + 0{,}2977\ E - 142{,}8\ Z1$$
$$(1{,}905)\ (2{,}483)\ (-3{,}123)$$

$R2 = 0{,}6738$ $SE = 53{,}25$ $DW = 1{,}753$

$$OC = -116{,}6 + 0{,}6168\ GVF + 0{,}2894\ OC1 + 0{,}6216\ TSB + 0{,}2899\ VA$$
$$(-2{,}893)\ (3{,}598)\ (2{,}80)\ (2{,}198)\ (8{,}039)$$

$R2 = 0{,}8017$ $SE = 11{,}77$ $DW = 1{,}965$ $h = -0{,}2079$

$$GO = -88{,}83 + 0{,}07575\ E + 0{,}07549\ GO1 + 2{,}593\ PBP$$
$$(-2{,}878)\ (6{,}719)\ (1{,}855)\ (15{,}47)$$

$R2 = 0{,}9198$ $SE = 5{,}02$ $DW = 1{,}668$ $h = 0{,}7094$

From the econometric point of view, the results are satisfactory. The signs of estimated coefficient are as expected, except the one before NO 1 (lagged net output) in equation describing NO (net output). This problem occurs because of the way the net output is calculated (see [11]).

In presented model, there are no variables with lags greater than one period. Owing to this, the suitably rewritten reduced form of the model is identical to the state-variable form of the model (see [5],[10]).

As the state-variables, all non-lagged endogenous variables are defined. Among the exogenous variables, the set of control variables, including the investment outlays (IO) and technological structure of the building production (TSB), is distinguished.
In order to solve this optimal control model, an algorithm based on the Discrete-Minimum Principle proposed by R.S.Pindyck has been used.

4 Results of the optimal control experiments

After testing the model for stability, controllability and forecasting properties, six simulation runs have been executed (see [11]. All simulations have been run for six periods, beginning with the third quarter of 1980 and ending with the fourth quarter of 1981. For the state variables three variants of the desired levels (related to the three versions of the production plan), and for the control variable TSB two levels (version A - minimal level of TSB and B - planned levels), have been chosen.

In each run the objective function (i.e. the Q and R matrices) has been changed to illustrate the different aspects of the optimal solution as well as the dynamic properties of the model. It has been attempted to determine the weight matrices Q and R in such a way as to reflect the subjectivity of the properties of the economic targets.

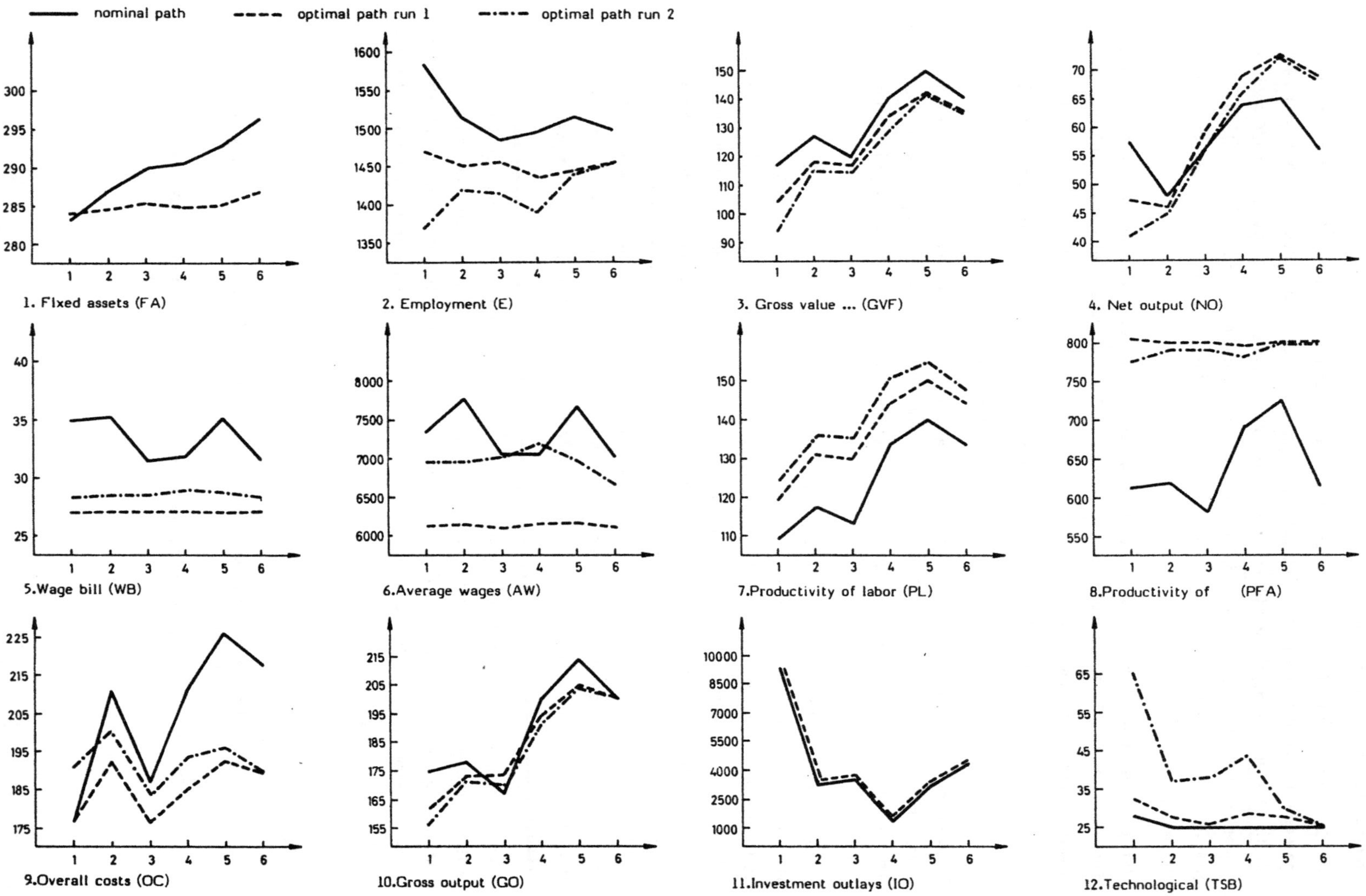
nominal path
optimal path run 1
optimal path run 2
1. Fixed assets (FA)
2. Employment (E)
3. Gross value ... (GVF)
4. Net output (NO)
5.Wage bill (WB)
6.Average wages (AW)
7.Productivity of labor (PL)
8.Productivity of (PFA)
9.Overall costs (OC)
10.Gross output (GO)
11.Investment outlays (IO)
12.Technological (TSB)

The assumption are as follows:

Run 1: The nominal (desired) values for the state variables and for IO (investment outlays) are equal to the planned levels. The priorities for the targets variables i.e. GVF (gross value of finished building objects) and OC (overall costs), have been laid down by the firm's decision-makers as 1:2, for the control variables IO (investment outlays) and TSB (technological structure of building production) as 1:2 either. For the other state and control variables, priorities are equal 0.

Run 2: The nominal values for the state and control variables are the same as in Run 1. The priorities for the targets GVF and OC are as in Run 1, and for WB (wage bill) and AW (average wages) are as 1:1 .

For the scarcity of place only chosen results are presented in graphic form (diagrams 1-12).

Taking into account the results some general comments can be formulated:

Firstly, comparing the nominal and optimal values, it can be seen that the values are closer to each other with growing 't'. This is the consequence of the stability properties of the model. The optimal values are relatively close to the desired values, except WB (wage bill) and AW (average wages) in Run 1.

Secondly, the results show that ther exist an influence of the weights on the optimal values of the targets variables. Comparing the optimal values for WB (wage bill) and AW (average wages) in these two runs, it can be seen, that as a result of assuming the non-zero weights for WB and AW, the optimal values for these variables appear closer to the nominal values in the second run than in the first one. Also, average optimal value for the wage bill and average wages have grown, these levels are still lower than the desired levels. Non-zero weights for WB and AW led to the lower levels of the NO (net output), GVF (gross value of the finished building objects) and OC (overall costs) at the beginning of the planning period (in comparison with Run 1). In the last quarters the differences nearly disappeared (see diagrams 3, 4, 10 respectively).

Thirdly, bearing in mind, that the objective preference function accumulates penalties only over a finite time period - namely six quarters, it is not surprising that the optimal trajectory for some variables behave strangely during the last two quarters of the planning period. This could occur, because whatever happens after the planning period is over, will not enter into the cost function (see diagrams 9, 12).

References

[1] AOKI, M. **Optimal Control and System Theory in Dynamic Economic Analysis.** Amsterdam 1976.

[2] FRIEDMAN, B.M. **Economic Stabilization Policy. Methods in Optimization.** Amsterdam 1975.

[3] GRUBER, J. **Introduction: Towards Observed Preferences in Econometric Decision.** In: J.Gruber (Ed.), Econometric Decision Models (Lecture Notes in Economics and Mathematical Systems, 208), Berlin-Heidelberg-NewYork 1983, pp.1-9.

[4] JOHANSEN, L. The Report of the Commitee on Policy Optimization-UK. **Journal of Economic Dynamic and Control,** Vol.1(1979), pp.101-109.

[5] KENDRICK, D. **Stochastic Control for Economic Models.** New York 1981.

[6] KORNAI, J. **Economics of Shortage.** Vol.1 and Vol.2, Amsterdam 1980.

[7] KORNAI, J. **Growth, Shortage and Efficiency.** Oxford 1982.

[8] KORNAI, J. **Niedobor w Gospodarce.** Warszawa 1985.

[9] MORISHIMA, M. **The Economic Theory of Modern Society.** Cambridge 1976.

[10] PINDYCK, R.S. **Optimal Planning for Economic Stabilization.** Amsterdam 1973.

[11] STRZAŁA, K. **Proba Wykorzystania Modelu Ekonometrycznego w Sterowaniu Optymalnym.** (dissertation) Sopot 1982.

[12] TAPIERO, Ch.S. **Managerial Planning: An Optimum and Stochastic Control Approach.** New York 1977.

RESOURCE EXTRACTION: IMPERFECT VS. PERFECT SUBSTITUTES

Franz Wirl, Wien

Summary: The recent literature about exhaustible resource extraction emphasized the analysis of perfect substitutes ("backstop"), but overlooked the important case of imperfect substitutes. However, the hypothesis of imperfect substitutes, exhaustible and renewable, seems more appropriate for such important commodity markets like the market of primary energy carriers. This paper investigates whether properties of resource extraction programmes, which are typical for the perfect substitute hypothesis, e.g. strict sequential exploitation of the reservoirs, discontinuous and instant supply by backstop technologies etc., carry over to the case of limited and imperfect substitution.

1. Introduction

A number of 'real world problems' have been overlooked in the literature of optimal resource depletion, despite the large recent interest in such problems. In particular, most stories on exhaustible resource extraction and the availability of substitutes are told within the perfect substitute paradigm; see e.g. Nordhaus (1973) who coined the term backstop, Dasgupta/Heal (1979), Heal (1979), Kemp/-Long (1980), Stiglitz/Dasgupta (1981), (1982), (1983). This paper extends this conventional paradigm and considers imperfect substitutes and it investigates those properties which are considered as robust within the standard framework.

Section 2 introduces the supply relations. Section 3 summarizes the (partly well known) results of perfect substitute extraction. Section 4 introduces imperfect, but exhaustible substitutes, which draws on Wirl (1986), (1987). Section 5 introduces a renewable backstop technology, which is an imperfect substitute to the resource. This interaction between the two industries is modelled as a dynamic Stackelberg game between a resource extracting monopolist and a competitive synfuel industry.

Operations Research Proceedings 1987

2. Supply

Consider firms which extract an identical product from two different reservoirs (for a treatment of common pool problems see Khalatbari (1976)). Considerable simplifications will be introduced in the following to derive and to state the major properties with a minimum of technical assumptions. Assume, that the extraction from the first pool is costless but that extraction from the second pool involves constant marginal costs c of production. The production facilities may be technically independent so that the outputs x(t), y(t) can be independently adjusted for each facility at any instant of time t. Typical examples for this description of the production technology are: extraction of oil of similar quality from different and separated fields, mining of coal surface and in situ.

In other instances the two resource reservoirs are located in one physical pool and the production of both resources is determined by choosing the extraction rate for one resource, the marker. This description applies to the important example of (joint) production of oil and associated gas.

3. Some Propositions on the extraction of perfect substitutes.

Suppose - within this section - that the consumers are indifferent between consuming the good x or y if both goods are equally priced and would otherwise only consume the cheaper product. Hence, the inverse demand function f will depend on aggregate supply only:

$$p=f(x+y) \qquad (1)$$

Furthermore assume, that the choke price $\bar{p}=f(0)>c$ is finite and that the (monopolist's) revenue function is concave.

Proposition 1: The aggregate supply of both resource based commodities declines continuously to zero at the ultimate depletion date. The extraction moves sequentially and discontinuously from the exploitation of the first resource pool to the second and more costly extractable reservoir. This result applies to competitive as well as to monopolistic markets.

This result is well established, e.g. Herfindahl (1967) proved the sequential exploitation property, see also Kemp (1980).

If the two (perfect) substitutes are the output of joint extraction effort as in the case of oil and associated gas, $y(t)=\theta x(t)$ then the producer may decide whether to sell, $z(t)$ sales with $z(t)\leq y(t)$, all, $z(t)=y(t)$ or dispose some amounts of the second resource: $z(t)<y(t)$. The resources are named 'oil' and 'gas' to motivate the analysis and the results.

Proposition 2: A noncompetitive producer supplies aggregate sales $x(t)+z(t)$ continuously, which decline monotonically over time. The optimal policy starts flaring all gas: $z=0$, but switches at a particular date to the opposite policy marketing all gas: $z=\theta x$.

This proposition is proven in Wirl (1988).

4. Resource extraction properties of imperfect substitutes

This Section supposes, that the two commodities are resource based but imperfect substitutes from the perspective of the consumers. More precisely assume that the demand is governed by a system of inverse demand functions[1]:

$$p^1=f^1(x,y) \quad (2)$$

$$p^2=f^2(x,y) \quad (3)$$

which satisfies the following conditions: f is twice continuously differentiable with negative first derivatives such that the determinant of the Jacobi matrix is positive. These conditions are derived from the hypothesis of positive cross price elasticities and dominance of the (negative) own price elasticities. Furthermore, suppose that the backstop (or choke) prices p^1,p^2 are finite.

Additionally assume that the profit function[2] $\pi(x,y):=p^1x+(p^2-c)y$, which is implied from the demand relations (7) and (8) is concave. Furthermore assume, that the marginal profit (and hence the marginal

[1] Extraction and sales may differ in the case of joint production of substitutable resources. Hence sales, $z(t)$, rather extraction $y(t)=\theta x(t)$ must then enter the inverse demand relations.

[2] The same modification as in Footnote 1 must be applied to cover the case of joint extraction, i.e. y must be replaced by z.

revenues due to linear costs) from each commodity are negatively affected by a simultaneous increase in the sales of the competing substitute, i.e. $\pi_{xy}<0$. This condition states, that marginal profits, i.e. the profit from selling an additional of x deteriorates if the sales of competing product y are simultaneously increased.

Proposition 3: The extraction of imperfectly substitutable resources characterized by different extraction costs and separate extraction facilities is in general neither strictly sequential nor monotonically. This applies to both market regimes: monopoly and competition.

This result deviates substantially from those models which predict a discontinuous transition from the use of one resource to the other. The proof of **Proposition 3** requires some counterexamples; Wirl (1986) provides simple examples based on linear demand relations which establish the above proposition. The following proposition characterizes the case of joint production.

Proposition 4: The marketing of joint products from an extraction facility with possible disposal of the associated, 'second' product y, involves three distinct strategies over time: complete disposal, z=0, of the by-product, partial disposal, $0<z<y=\theta x$, and full utilization $z=y=\theta x$. The extraction of the first resource, the marker, declines monotonically. The sale strategy z(t) proceeds continuous across the above mentioned three phases.

This result again contradicts the rule of thumb derived within the conventional paradigm of perfect substitutes; for a formal proof see Wirl (1987). The properties of the above proposition comply with market tendencies; e.g. Opec decreased the ratio of flared gas during the last twenty, see Opec Secretariat (1984) . This leads to increased availability of natural gas and may in part explain the recent oil price deterioration. Moreover, the contractual binding between oil and gas prices[1] is a serious obstacle to increase the market share of gas, as suggested in **Proposition 4**. Hence, a downward revision of gas to oil price ratios seems likely (everything else remaining unchanged; e. g. increased environmental concern may distort this inference).

[1] The gas price is more or less strictly pegged to liquid fuel prices

5. Imperfect Substitutes as Backstop Technologies

The papers of Kemp/Long (1982), Heal(1979) and Salant(1979), but also Hoel(1983) and of some other authors consider a market where a resource exploiting monopoly faces competition from a competitive industry which can supply a perfect substitute at cost c. The solution is well known and shown in Figure 1. The broken (resource x') and dotted (backstop, renewable commodity y') lines indicate this solution.

Proposition 5: **The monopoly chooses a price strategy which rises along the formula that the percentage growth of the marginal revenues must be equal to the rate of discount until the price increases reach the costs of the backstop technology. Then the monopolist will marginally undercut the costs of its competitors until the resource stock is depleted. The competitive industry enters instantenuously at the date of resource depletion and supplies then the entire market for the remainder.**

Now suppose that the know how about an imperfect but renewable substitute is already available in the public domain, e.g. fast breeder reactors could substitute for liquid fuels, and that no economies of scale exist so that a competitive industry will supply this commodity. Then, the resource extracting cartel will attempt to maximize the present value of the profit stream:

$$\pi^1(x,y)=p^1(x,y)x \tag{4}$$

π^1 concave with negative second derivatives:

$$\max_{x,S} \int_0^R e^{-rt}\, \pi^1(t)\, dt \tag{5}$$

$$x(t)\geq 0 \tag{6}$$

$$\dot{X}(t)=x(t)\ ,\ X(0)=0,\ X(T)=R \tag{7}$$

subject to the laws of demand, eq. (2), the differential equation (7) which accounts for the available resources, and the non-negativity constraint (6).

However, the competitive industry will simultaneously supply the substitute if the the production costs can be recovered at the current market conditions. And if this is possible, the competitive industry will produce exactly that amount when the market price is equal to the costs: $p^2=c$. Therefore,

$$y(t) = \begin{cases} 0 & \text{if: } p^2(x(t),0) < 0 \\ \text{from: } p^2(x(t),y(t))=c & \text{if: } p^2(x(t),0) \geq 0 \end{cases} \quad (8)$$

the supply of synfuels depends on the output of the resource extracting monopolist: $y=\alpha(x)$, $\alpha'=-p^2_x/p^2_y<0$ for $y>0$ due to the implicit function theorem.

Hence an intertemporal Stackelberg equilibrium $(x^s(t),y^s(t))$ must maximize (5) and satisfy (8) but will account for the impact of his supply strategy on the competitors supply. Proceeding in a conventional and current value formulation the following conditions are derived[1]:

$$H=\pi^1+\mu x \quad (9)$$

$$H_x=\pi^1_x+\pi^1_y y'+\mu=0 \quad (10)$$

$$\dot{\mu}=r\mu \quad (11)$$

The optimal depletion strategy is described by the well known condition that the extended marginal revenues (= marginal profits) equal the resource shadow price and must rise at the rate of interest. More precisely the (negative) impact of additional supply on the competing industry's output must be added to the (conventional) marginal revenues. Suppose, that $\pi^1(x,y(x))$ is concave in x, then the following proposition is immediately verified:

<u>Proposition 5</u>: The monopolist extraction is determined by the usual condition that (modified) marginal revenues grow at the rate of interest. This leads to a continuous market penetration of synfuels until ultimately the resource stock is depleted.

Differentiation of (10) and (11) for the periods of competition and solving for x, y yields: $\dot{x}<0$, $\dot{y}>0$. Moreover, it is routine to

[1] The arguments, in particular the index of time t is suppressed in the sequel. Furthermore, low case indices denote partial derivatives.

establish that the synfuel industry penetrates smoothly and continuously the market until it covers the entire market. Suppose, that the backstop technology enters discontinuously at date τ, i.e. $y(\tau_-)=0$, but $y(\tau_+)>0$. However, the Hamiltonian must remain continuous across this date: $H(\tau_-)=H(\tau_+)$, hence:

$$\pi^1(x_-,0)+\mu x_-=\pi^1(x_+,y_+)+\mu x_+ \tag{12}$$

which implies that resource extraction is discontinuosly reduced: $x_+<x_-$. However from a concave revenue schedule π^1 follows:

$$\pi^1(x_-,0)-\pi^1(x_+,y_+)\geq-\pi^1{}_x(x_-,0)(x_+-x_-)-\pi^1{}_y(x_-,0)y_+ \tag{13}$$

and inserting (12) and some rearrangements imply:

$$[\pi^1{}_x(x_-,0)+\mu](x_+-x_-)\geq-\pi^1{}_y(x_-,0)y_+ \tag{14}$$

The inequality (14) is, however, contradictory, because the term within the squared parenthesis vanishes and the right handside is positive, due the assumptions about the demand (x and y are substitutes).

The Figure 1 compares the solutions under the different assumptions of a perfect versus an imperfect substitute and the latter case corresponds to the bold lines.

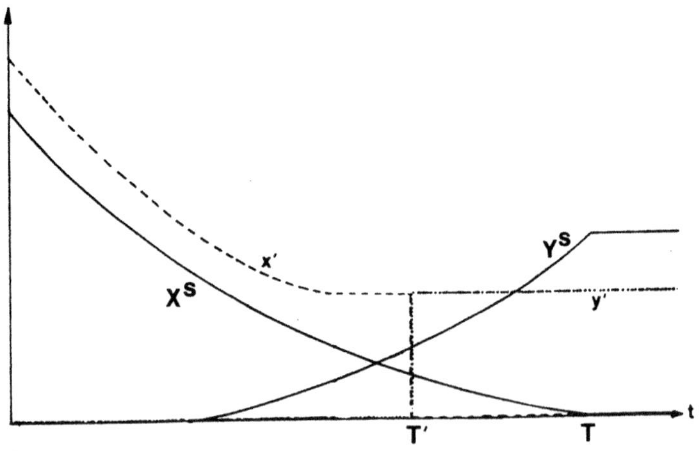

Figure 1

References:

Dasgupta Partha/Heal Geoffrey M. (1979): Economic Theory and Exhaustible Resources, Cambridge Economic Handbooks.

Heal Geoffrey M. (1979): Uncertainty and the Optimal Supply Policy for an Exhaustible Resource, in R.S. Pindyck(ed.): *Advances in the Economics of Energy and Resources*, Vol 2, 119-147,JAI Press

Herfindahl O.C. (1967):Depletion and Economic Theory, in M.M.Gaffney (ed.) *Extractive Resources and Taxation*, University of Wisconsin Press, 63-69.

Hoel Michael (1983):Monopoly Resource extractions under the presence of predetermined substitute production, *Journal of Economic Theory*, **30**, 201-212.

Kemp Murray .C./Long Ngo van (1980): Exhaustible Resources, Optimality and Trade, Amsterdam, North Holland.

Kemp Murray C./Long Ngo van (1982):On the Development of a Substitute for an Exhaustible Resource, in W.Eichhorn (ed.): *Economic Theory of Natural Resources*, Physica, Würzburg-Wien.

Khalatbari F. (1976): Market Imperfections and Optimal Rate of Depletion of an Exhaustible Resource, *Economica*, **44**, 409-414.

Nordhaus William D. (1973):The Allocation of Energy Resources, *Brookings Papers on Economic Activity*, Vol 3.

Opec Secretariat (1984): OPEC Annual Statistical Bulletin 1983.

Salant Stephen W. (1979): Staving off the Backstop:Dynamic Limit Pricing with a Kinked Demand Curve in R.S. Pindyck(ed.): *Advances in the Economics of Energy and Resources*, Vol 2, 187-204, JAI Press.

Stiglitz Joseph E./Dasgupta Partha (1981): Market structure and Resource Extraction under Uncertainty, *Scandinavian Journal of Economics*, **83**, 318-333.

Stiglitz Joseph E./Dasgupta Partha (1982): Market structure and Resource Depletion: A Contribution to the Theory of Intertemporal Monopolistic Competition, *Journal of Economic Theory*, **28**, 128-164.

Stiglitz Joseph E./Dasgupta Partha (1983): Strategic Considerations in Invention and Innovation: the Case of Natural Resources, *Econometrica*, **51**, 1439-1448.

Wirl Franz (1986), Resource Extraction of Imperfect Substitutes, Inst. für Energiewirtschaft, T. U. Wien.

Wirl Franz (1988), Joint Production of Substitutable, Exhaustible Resources or: Is Flaring Gas Rational, to appear in *Journal of Economic Dynamics and Control*.

Probabilistic Analysis of the Simplex Method

Karl Heinz Borgwardt
Augsburg

1. Introduction

Since the invention of the Simplex Method in 1947 by George B. Dantzig it had been an open problem for mathematicians why this algorithm works so efficiently in practice. This question became even more interesting when in the beginning of the seventies Klee, Minty, Jeroslow and other authors recognized that the usual variants of the Simplex Method are nonpolynomial in the sense of worst-case complexity theory. That means that there is no polynomial in the dimension of linear programming problems which acts as an upper bound for the respective numbers of pivot steps required for solving all the possible problems.

Obviously it was a great advantage for the Simplex Method that people had not known this negative and bad fact from the beginning. So the users got the impression that this is a very reliable, fast and efficient tool for solving their problems. They did not care very much about the reasons for that behaviour. On the other hand an immediate nonpolynomiality proof would very likely have led to a very cautious and detached judgement on the quality of the method as we have learned from many other cases (especially in combinatorial optimization). The persuasive power of polynomiality proofs can be observed very clearly at the example of the ellipsoid method (which is not efficient in practice at all) and at the boom and the public attention for the algorithm of Karmarkar.

Since the beginning of the seventies a great effort has been made to clarify the "normal" quality of the Simplex Method and in 1977 — ten years ago, the first breakthrough was achieved by the author. In the following time several authors tried to improve the results, find sharper bounds in different directions studying different stochastic models. This paper is to give a brief survey of these approaches and explain the advantages and drawbacks of the main results in this field of research.

Throughout the paper we will deal with linear programming problems of the following type. (Note that different types can easily be transformed into that form.)

$$\begin{array}{lll} \text{Maximize} & v^T x & \text{(a linear objective)} \\ \text{where} & a_1^T x \le b^1 & \\ & \vdots & \text{resp. } Ax \le b \text{ with } A = \begin{pmatrix} a_1^T \\ \vdots \\ a_m^T \end{pmatrix} \\ & a_m^T x \le b^m & \end{array}$$

with $v, x, a_1, \ldots, a_m \in \mathbb{R}^n$, $\quad b \in \mathbb{R}^m$ and $m \ge n$.

Operations Research Proceedings 1987

We call this an (m, n)-problem to characterize the two dimensions of the problem.

The restrictions $Ax \leq b$ define a convex polyhedron. In the vertices (edges) of that polyhedron at least n $(n-1)$ restrictions $a_i^T x \leq b^i$ are active. And we know that if there are vertices and if there are optimal points, there is a vertex among the optimal points. Finally, we know that every nonoptimal vertex is incident to an edge which improves the objective. These facts are exploited in the Simplex Method which works as follows.

Phase I: Find a vertex of $X = \{x \in \mathbb{R}^n \mid Ax \leq b\}$.
If this is impossible, STOP.

Phase II: Construct a sequence of vertices $x_0, x_1, \ldots, x_s$ starting with the vertex x_0 (found in Phase I) such that successive vertices are adjacent and that $v^T x$ is strictly increasing.
Stop as soon as an optimal vertex is reached or as soon as it becomes clear that optimal points do not exist.

The procedure for Phase I is closely related to that for Phase II. Only a slight modification of the problem is necessary. So it makes sense to concentrate on Phase II for the moment. The number s characterizes the number of vertex exchanges which have to be performed. Under nondegeneracy assumptions (which are valid in our analysis) this number coincides with the number of pivot steps for solving the Phase II-task.

Our definition of Phase II is not yet complete, because we should declare a rule for determining the successor vertex if more than one improving adjacent vertices are available. Such a rule characterizes the **variant** of the algorithm.

Throughout the paper we use the following nondegeneracy assumptions.

Every submatrix of (A, b) and of $\binom{A}{v}$ is of full rank.

This guarantees that each vertex corresponds to a set of exactly n active restrictions, that an optimal point is unique (if existing) and that every pivot step leads to a new vertex.

Carrying out a probabilistic analysis of the Simplex Method requires the fixation of a variant and the decision for a certain stochastic model simulating the real-world-distribution of linear programming problems.

It should be mentioned that the number of steps highly depends

- on the chosen stochastic model,
- on the variant which is used.

Studying the average behaviour means calculating the expected number of steps under the given stochastic model. In the following we will demonstrate and explain the two main stochastic models, the main ideas for evaluation of the expected values and the main results, which showed the polynomiality of the average behaviour.

2. The Sign-Invariance-Model

We begin with the most simple stochastic model which could be analyzed, the so-called Sign-Invariance-Model (MAY & SMITH (1982)). Let $A = \begin{pmatrix} a_1^T \\ \vdots \\ a_m^T \end{pmatrix}$ and $b = \begin{pmatrix} b^1 \\ \vdots \\ b^m \end{pmatrix}$ be given. To determine the direction of the inequalities ($a_i^T x \leq b^i$ or $a_i^T x \geq b^i$) we perform m Bernoulli-experiments (we flip a coin m times) with probability $\frac{1}{2}$. So the m directions are independent, $\geq$ and $\leq$ are equally likely.

Consequently, a given data set (A, b) induces 2^m different linear programming problems. By multiplication with (-1) it is possible to transform the restrictions into the standard form $Ax \leq b$. Averaging over the probability space now means to solve all these 2^m problems, to count the numbers of required pivot steps and to divide the sum by 2^m. The feasibility regions of the individual problems will be called cells. Of course, some of the cells will be empty. Then the algorithm has the task to recognize infeasibility. A first analysis for the average Phase II-complexity was done by HAIMOVICH (1983) and ADLER. They investigated and used a parametric Simplex variant which works as follows.

Let v be the objective of the problem and let u be an auxiliary objective such that $\begin{pmatrix} A \\ v \\ u \end{pmatrix}$ has only full-rank submatrices. We call a point $y \in X$ (the feasible region) (u, v)-cooptimal if $u^T z \leq u^T y$ for all $z \in X$ with $v^T z = v^T y$.

Now we can find a Simplex-Path connecting all (u, v) cooptimal points according to the parametric algorithm of GASS & SAATY (1955). That means that we successively calculate the solution-vertices of the problem

$$\begin{array}{ll} \text{Maximize} & (u + \rho v)^T x \\ \text{subject to} & x \in X \\ \multicolumn{2}{l}{\text{with } \rho \text{ running from } -\infty \text{ to } +\infty.} \end{array}$$

We can assume that Phase I is already done and that a start vertex x_0 has been found such that x_0 is a maximal vertex with respect to a given auxiliary objective u. Constructing a vertex-sequence according to the algorithm described above touches only (u, v)-cooptimal vertices. So the number of cooptimal vertices can be used as an upper bound for the number of pivot steps.

Haimovich comes to the following conclusion.

Theorem (Haimovich 1983). *The expectation value of the numbers of pivot steps used on cooptimal paths in the sign invariance model and under the condition that a cooptimal path exists, is not greater than* $n\frac{m-n+2}{m+1}$.

Proof. There are 2^m possible cells and $\binom{m}{n}$ intersection points of n restriction hyperplanes. We call the intersection sets of $n - 1$ restriction hyperplanes **lines**.

Every such line itself is intersected by $m - n + 1$ restriction-hyperplanes. So we obtain $m - n + 2$ segments on each line. Each segment belongs to a different cell and every segment is cooptimal in exactly one cell. Now we know that nondegeneracy guarantees that

1) Every cell has at most one $v^T x$-maximal vertex and at most one (u, v)-cooptimal path.
2) Each intersection point of n hyperplanes and every line segment is optimal in exactly one cell.
3) We have $\binom{m}{n} + \binom{m}{n-1}$ cells with (u, v)-cooptimal paths.
3) There are $\binom{m}{n-1}(m - n + 2)$ line segments.

So he concludes $E_{m,n}$ (number of line segments on a cooptimal path/cooptimal path exists) $= n\frac{m-n+2}{m+1}$.

Here an analysis of Phase I is not yet done. We have a fictive start vertex based on the auxiliary objective u.

That necessary generalization was independently done in papers of TODD (1983), ADLER & MEGIDDO (1983), ADLER, KARP & SHAMIR (1983). Their methods were based on work of SMALE (1982) who had analyzed a similar model and algorithm but had not achieved polynomiality. We refer to the paper of Adler, Karp and Shamir, because its notation is closest to our geometrical concept. Again, we want to solve the problem

$$\begin{array}{ll} \text{Maximize} & v^T x \\ \text{subject to} & a_1^T x \leq b^1, \ldots, a_m^T x \leq b^m. \end{array}$$

But now we use a Lexicographic Constraint-By-Constraint algorithm for doing the task of Phase I and Phase II together. For this purpose we use relaxed feasibility regions $X^{(k)} = \{x \in \mathbb{R}^n \mid a_1^T x \leq b^1, \ldots, a_k^T x \leq b^k\}$ for $n \leq k \leq m$. Note that $X^{(m)} = X$. Our algorithm runs through $m - n + 1$ stages.

Stage 0: Determine the unique vertex $\bar{x} \in X^{(n)}$ and choose u as $u = \varepsilon^1 a_1 + \varepsilon^2 a_2 + \ldots + \varepsilon^n a_n$ with $\varepsilon > 0$ sufficiently small.
Go to Stage 1.

Stage k: $(1 \leq k \leq m - n)$
Start at $\bar{x}$, the maximal vertex for $u^T x$ on $X^{(k-1+n)}$. If $\bar{x} \in X^{(k+n)}$, then go to Stage $k+1$. Else use the parametric algorithm to maximize $(-a_{k+n}^T x)$ until $a_{k+n}^T x \leq b^{k+n}$ is achieved. The last traversed edge contains a point x with $a_{k+n}^T x = b^k$. All path-points are $(u, -a_k)$-cooptimal, hence x maximizes $u^T x$ on $X^{(k+n)}$. If it is impossible to achieve $a_{k+n}^T x \leq b^{k+n}$ then the original problem is infeasible. STOP. Else set $\bar{x} = x$ and go to Stage $k + 1$.

Stage $m-n+1$: Start at $\bar{x}$ which is $u^T x$-maximal on $X^{(m)} = X$. Apply the parametric algorithm and find the optimal point or find out that there is no solution.

Adler, Karp and Shamir sum up over all 2^{n+k} instances in Stage $k+1$ and over all the basic cones. A fixed vertex is cooptimal in exactly $n+1$ of 2^{n+k} instances, because it is incident to $2n$ segments. Either it is optimal or two of its incident segments are cooptimal. The authors come to an expected value of at most $2^{n+1}(n+1)$ for arbitrary choice of u. But the lexicographic choice of u in Stage 0 enables them to prove that

$$E_{m,n}(s^{CBC}) \leq 2(n+1)^2.$$

Theorem (ADLER, KARP & SHAMIR (1983); ADLER, MEGIDDO (1983); TODD (1983)). *For solving the complete (m,n)-problem, the expected number of pivot steps is not greater than $2(n+1)^2$ under the sign invariance model.*

3. Drawbacks of the Sign-Invariance-Model

We already mentioned that not every problem instance has a feasible cell. We know that m restrictions (nondegenerate) partition the n-dimensional space into $\binom{m}{0}+\binom{m}{1}+\ldots+\binom{m}{n}$ cells. On the other hand the number of generated problems is 2^m. So for $m \gg n$ the quotient

$$\frac{\text{number of feasible problems}}{\text{number of generated problems}} \quad \text{tends to 0 very fast.}$$

Only **conditioning** on feasible problems can avoid averaging over a lot of infeasible (and much less difficult) problems. Also, the expected number of vertices is low. We have

$$E_{m,n}\text{ (vertices per generated problem)} = \frac{2^n\binom{m}{n}}{2^m}$$

$$E_{m,n}\text{ (vertices per feasible problem)} = \frac{2^n\binom{m}{n}}{\binom{m}{0}+\ldots+\binom{m}{n}}$$

which tends to 2^n for $m \to \infty$, n fixed and is similar to the vertex number of an n-dimensional cube (a very simple polyhedron for linear programming).

The most important aspect is the average redundancy rate. If we introduce the m-th restriction, the number of cells increases from

$$\binom{m-1}{0}+\ldots+\binom{m-1}{n} \text{ to } \binom{m}{0}+\ldots+\binom{m}{n}.$$

That means that some cells have been reproduced and some have been divided by the last restriction. We can calculate the number of divided cells. It is

$$\binom{m}{0}-\binom{m-1}{0}+\binom{m}{1}-\binom{m-1}{1}+\ldots+\binom{m}{n}-\binom{m-1}{n}=\binom{m-1}{0}+\ldots+\binom{m-1}{n-1}.$$

Consequently the new restriction is relevant only in $2[\binom{m-1}{0}+\ldots+\binom{m-1}{n-1}]$ of 2^m generated and $\binom{m}{0}+\ldots+\binom{m}{n}$ feasible cells. Hence both the pure and the conditional redundancy rate tends to 1 very quickly.

Also the expected number of nonredundant constraints is low. It is calculated by multiplication of m with the nonredundancy rate. Whereas the pure expected number again tends to 0, the conditional expected number of nonredundant constraints behaves like

$$\frac{m2\{\binom{m-1}{0}+\ldots+\binom{m-1}{n-1}\}}{\binom{m}{0}+\ldots+\binom{m}{n}} \rightarrow 2n \text{ for } m\rightarrow\infty \text{ and } n \text{ fixed.}$$

This is again typical for a cube in $\mathbb{R}^n$. These results show that the Sign-Invariance-Model hardly generates problems with more than $2n$ nonredundant constraints. So the expected values for s and $s^{\mathrm{CB\check{C}}}$ do not give reliable explanations for the behaviour of the Simplex Method in the asymptotic case ($m \gg n$). Instead of proving the quality of the Simplex Method, they reflect the special properties of the chosen stochastic model.

4. The Rotation-Symmetry-Model

The above mentioned danger can be avoided by using a different model, the rotation-symmetry-model. The author analyzed that model in several papers (BORGWARDT (1977, 1982a, 1982b etc.)). We deal with problems of the type

$$\begin{array}{ll} \text{Maximize} & v^T x \\ \text{subject to} & a_1^T x \leq b^1 \\ & \vdots \\ & a_m^T x \leq b^m \\ \multicolumn{2}{l}{\text{where } b^i > 0 \text{ for } i=1,\ldots,m.} \end{array}$$

Here we have feasibility in any case because $0 \in X$. So we are sure to solve "hard" problems. Our stochastic assumptions are as follows:

> Let $a_1,\ldots,a_m$, v, $b^1,\ldots,b^m$ be independent random variables, let $a_1,\ldots,a_m$, v be distributed identically and symmetrically under rotations of $\mathbb{R}^n$ and let the components of b be distributed identically over $(0,\infty)$.

Without loss of generality we can simplify these assumptions by "normalization". That means that we define X by $a_1^T x \leq 1,\ldots,a_m^T x \leq 1$ and that we demand $a_1,\ldots,a_m$, v to be distributed independently, identically and symmetrically under rotations. It should be noted that under this model the following nondegeneracy assumption is valid with probability 1: Every submatrix of $\binom{A}{v^T}$ and of $(A\ \mathbb{1})$ has full rank.

The variant is again a parametric algorithm, the so-called "shadow vertex algorithm". Here we start again at a vertex maximizing an auxiliary objective $u^T x$ and construct a sequence of (u,v)-cooptimal vertices in order to reach the $v^T x$-maximum. All these cooptimal vertices will be mapped into vertices of a two dimensional polyhedron, if we project X orthogonally onto the plane spanned by u and v. That means that they become vertices of "the shadow" of X, which explains the term "shadow vertex". Since every (u,v)-cooptimal vertex is also a shadow vertex, we know that S, the number of shadow vertices, is an upper bound for s.

Now we observe the following one-to-one correspondence. Every vertex of X is the intersection point of n restriction hyperplanes. Let us denote these restrictions $a_{\Delta^1}^T x = 1, \ldots, a_{\Delta^n}^T x = 1$ and let x_Δ be the intersection point. Hence we have a one-to-one correspondence between the vertex x_Δ and an index set $\Delta = \{\Delta^1, \ldots, \Delta^n\} \subset \{1, \ldots, m\}$ with $\Delta^1 < \Delta^2 < \ldots < \Delta^n$. On the other hand Δ corresponds uniquely to the n-element set $\{a_{\Delta^1}, \ldots, a_{\Delta^n}\}$. For our analysis of the average number of shadow vertices the following geometrical facts are extremely useful. Let us denote by Y the convex hull of $(0, a_1, \ldots, a_m)$.

Lemma. *1. x_Δ is a vertex of X if and only if the convex hull of $\{a_{\Delta^1}, \ldots, a_{\Delta^n}\}$ lies completely on the boundary of Y.*

2. Let x_Δ be a vertex of X. Then it is even a shadow vertex if and only if the convex hull of $\{a_{\Delta^1}, \ldots, a_{\Delta^n}\}$ is intersected by span(u,v).

The following integral expression results from that lemma.

$E_{m,n}(S) = \binom{m}{n} \int\limits_{\mathbb{R}^n} \cdots \int\limits_{\mathbb{R}^n} P$ (CH$(a_1, \ldots, a_n)$ is a boundary simplex of Y and CH$(a_1, \ldots, a_n)$ is intersected by span(u,v)) $dF(a_1) \ldots dF(a_n)$.

Here F denotes the distribution function under consideration. This integral expression can be simplified because of rotational symmetry. So the boundary event and intersection event become independent for fixed $a_1, \ldots, a_n$. The first depends only on the distance of the hyperplane through $a_1, \ldots, a_n$ to 0 and the second is proportional to the spherical measure of the convex cone spaned by $a_1, \ldots, a_n$. Nevertheless it took several years to evaluate this integral sufficiently well. In the following I want to list some of the main results for our Phase I-analysis.

Asymptotic Results for $m \to \infty$ and fixed n

Theorem (Borgwardt 1977–1984). *There is a function $\varepsilon(m,n)$ asymptotically tending to 0 such that for*

Gaussian distribution on $\mathbb{R}^n$:

$$E_{m,n}(S) \leq \sqrt{\ln m}\, n^{\frac{3}{2}}\, 2^{\frac{3}{2}}\, \frac{20}{9}(1+\varepsilon(m,n))$$

$$E_{m,n}(S) \geq \sqrt{\ln m}\, n^{\frac{3}{2}}\, 2^{\frac{3}{2}}\, \frac{3}{5}(1-\varepsilon(m,n)).$$

Uniform distribution on unit ball:

$$E_{m,n}(S) \leq m^{\frac{1}{n+1}}\, n^2\, 2\Big(1+\frac{1}{\sqrt{2}}\Big)(1+\varepsilon(m,n))$$

$$E_{m,n}(S) \geq m^{\frac{1}{n+1}}\, n^2\, 2\frac{1}{3}(1-\varepsilon(m,n)).$$

Uniform distribution on unit sphere:

$$E_{m,n}(S) \leq m^{\frac{1}{n-1}}\, n^2\, 2\Big(1+\frac{1}{\sqrt{2}}\Big)(1+\varepsilon(m,n))$$

$$E_{m,n}(S) \geq m^{\frac{1}{n-1}}\, n^2\, 2\frac{1}{7}(1-\varepsilon(m,n)).$$

General distributions with bounded support:

$$E_{m,n}(S) \leq m^{\frac{1}{n-1}}\, n^2\, \sqrt{2\pi}\,(1+\varepsilon_F(m,n))$$

(ε depends on F here).

The difference in the order of growth for different distributions results from different redundancy rates. For instance uniform distribution on unit sphere leads to redundancy rate 0, whereas Gaussian distribution induces a redundancy rate rather close to 1. Note that:
The restriction $a_i^T x \leq 1$ is redundant if and only if $a_i \in \text{Int}\, Y = \text{Int CH}(0, a_1, \ldots, a_m)$. This can never occur for a distribution on the unit sphere.
Now we could try to find the minimal possible asymptotic growth.

Theorem (Borgwardt 1979–1984). *All results for $m \to \infty$ and fixed n.*

1) *For all distributions where $P(\|a\| \geq r)^{-1}$ is of polynomial order for $r \to \infty$ we know that $E_{m,n}(S)$ is bounded by a constant $C(n)$ not depending on m. (Example $C(n) = n^{\frac{5}{2}}$ can be realized).*
2) *For all distributions with bounded support $E_{m,n}(S) \to \infty$.*
3) *For every $\delta > 0$ there is a distribution with bounded support such that $E_{m,n}(S) = 0(m^{\delta})$ as a function of m.*

Note the similarity of 1) with the result for the Sign-Invariance-Model. It shows that such a small growth can even be achieved by analyzing only feasible problems when the redundancy rate is kept small.

This is a great advantage of our model. We can vary the redundancy rate from almost 1 to 0 and observe the behaviour, whereas the Sign-Invariance-Model has a fixed redundancy rate very close to 1. In 1981 we achieved a polynomial upper bound.

Theorem (Borgwardt 1981). *For all distributions satisfying our conditions independence, identity, rotational symmetry we have* $E_{m,n}(S) \le e\pi(\frac{\pi}{2} + \frac{1}{e})n^3 m^{\frac{1}{n-1}}$.

5. The Inclusion of Phase I

As in the analysis of Phase II under the Sign-Invariance-Model we had some trouble with the auxiliary objective function $u^T x$. These complications disappear completely by our construction of a probabilistically analyzable method for inclusion of Phase I.

For demonstrating that algorithm we introduce the following problems with additional restrictions

$$\begin{array}{lll} \text{Problem} & I_k & (k = 2, \dots, n) \\ \text{Maximize} & v^T x & \\ \text{subject to} & Ax \le b & \text{and } x^{k+1} = \dots = x^n = 0. \end{array}$$

That means we fix components $k+1, \dots, n$ for a while to 0. Note that I_n is our original problem.

Initialization:

Set $k = 2$. Find a vertex of I_2. Solve I_2 by means of the parametric algorithm.
If this is impossible STOP.
Else store the solution vertex $(x^1, x^2, 0, \dots, 0)^T$.

Typical Step:

$(k = 3, \dots, n)$.
Use the solution point $(x^1, \dots, x^{k-1}, 0, \dots, 0)^T$ for I_{k-1}. It is located on an edge of the I_k-polyhedron.

1) Find an adjacent vertex $(\overline{x}^1, \dots, \overline{x}^k, 0, \dots, 0)^T$ of the I_k-polyhedron to that edge.
2) Solve I_k by means of the parametric algorithm by setting $u = (0, \dots, 0, 1, 0, \dots, 0)^T$ (1 in kth-component).
   If the problem has unbounded objective STOP.
   Else store the maximal vertex.
3) If $k = n$ STOP. Else set $k = k + 1$. Go to 1).

Here there is no doubt about the choice of u. The algorithm is by far not the most efficient one but it has the tremendous advantage that we can do a probabilistic analysis by $n-1$-times application of the shadow-vertex algorithm.

So we get our main result.

Theorem (Borgwardt 1981). *For every distribution satisfying our three conditions our complete method does not require more than* $m^{\frac{1}{n-1}} n^4 \cdot C$ *steps on the average.*

Based on that method we could also prove that problems with sign constraints, i. e.,

$$\begin{array}{ll} \text{Maximize} & v^T x \\ \text{subject to} & Ax \leq b \;\; x \geq 0 \text{ and } b > 0 \end{array}$$

can be solved in $m^{\frac{1}{n-1}} n^4 C$ steps on the average.

But still our analysis had one drawback. For our algorithm it is necessary to know a feasible point (to start with), here we have the origin. If we knew a different feasible point, it would work too, we only had to transform the coordinate system. But how should we handle problems of the form

$$\begin{array}{ll} \text{Maximize} & v^T x \\ \text{subject to} & a_1^T x \leq b^1, \ldots, a_m^T x \leq b^m \text{ (general form)} \\ & \text{with arbitrary } b^i? \\ & \text{We call that problem } P_n. \end{array}$$

If there is a negative b^i we do not know a feasible point.
Now we reformulate our restrictions as

$$a_i^T x \leq b^i \iff a_i^T x \leq 1 - \tilde{b}^i \text{ with } \tilde{b}^i = 1 - b^i.$$

We embed our polyhedron in $\mathbb{R}^{n+1}$ by postulating

$$(a_i^1, a_i^2, a_i^3, \ldots, a_i^n, \tilde{b}^i) \begin{pmatrix} x^1 \\ \vdots \\ x^n \\ x^{n+1} \end{pmatrix} \leq 1 \iff a_i^T x + \tilde{b}_i x^{n+1} \leq 1.$$

In level $x^{n+1} = 0$ we have a polyhedron and a problem which satisfies our requirements about the distribution of the restrictions, because the a_i are distributed independently, identically and symmetrically under rotations. That means our analysis would hold for the solution of those problems $\tilde{P}_n$. But we must be aware of the fact that we have to find the $v^T x$-optimum in level $x^{n+1} = 1$, where $a_i^T x + \tilde{b}_i x^{n+1} \leq 1$ means that $a_i^T x + \tilde{b}_i \leq 1 \iff a_i^T x \leq 1 - \tilde{b}_i$. Hence we rediscover our problem in level $x^{n+1} = 1$.

Now we proceed as follows.

1) Solve the problem $\tilde{P}_n$ (by means of our complete method)

$$\begin{array}{ll} \text{Maximize} & v^T x + 0x^{n+1} \\ \text{subject to} & a_i^T x + \tilde{b}_i x^{n+1} \leq 1 \text{ and } x^{n+1} = 0 \; (i = 1, \ldots, m). \end{array}$$

If this problem has no solution STOP (unboundedness of the objective is guaranteed also in P_n).
Else store the solution vertex and go to 2).

2) Apply the shadow vertex algorithm in order to

$$\begin{array}{ll} \text{Maximize} & x^{n+1} \\ \text{subject to} & a_i^T x + \tilde{b}_i x^{n+1} \leq 1 \ (i = 1, \ldots, m). \end{array}$$

Every point on the path is (v, e_{n+1})-optimal. If the maximal value of x^{n+1} is less than 1 then our problem is infeasible $\rightarrow$ STOP.

Else: As soon as level $x^{n+1} \geq 1$ is reached, find a point of level $x^{n+1} = 1$ on the last edge. Now $(x^1, \ldots, x^n)^T$ is the optimal point for P_n. STOP.

We ask for the average complexity of that method. It is clear that the solution of $\tilde{P}_n$ requires not more than $m^{\frac{1}{n-1}} n^4 C$ on the average.

We need to know the effort for Step 2. And we can prove the following

Theorem (Borgwardt 1987). *Problems of the kind*

$$\begin{array}{ll} \textit{Maximize} & v^T x \\ \textit{subject to} & a_1^T x \leq b^1, \ldots, a_m^T x \leq b^m \end{array}$$

where $v, a_1, \ldots, a_m$ are distributed independently, identically and symmetrically under rotations, where the b^i's are independent real variables, also independent of the a_i's and uniformly distributed over an interval $[-q, +q] \subset \mathbb{R}$, then the method described above does not require more than $m^{\frac{1}{n-1}} n^4 C_2$ pivot steps on the average.

So we have an algorithm — by far not the most efficient one — which admits a probabilistic analysis, which solves all LP-problems and which is polynomial in the average number of steps. And the analysis is done under the "hard" rotational symmetry model.

References

[1] Adler, I., Karp, R. & Shamir, R. [1983b]: *A Simplex Variant Solving an $m \times d$ Linear Program in $0(\min(m^2, d^2))$ Expected Number of Pivot Steps*, University of California, Computer Science Division, Berkeley, December 1983.

[2] Adler, I. & Meggido, N. [1983]: *A Simplex Algorithm where the Average Number of Steps is Bounded Between two Quadratic Functions of the Smaller Dimension*, Department of Industrial Engineering and Operations Research, University of California, Berkeley, California, December 1983.

[3] Borgwardt, K. H. [1977]: *Untersuchungen zur Asymptotik der mittleren Schrittzahl von Simplexverfahren in der linearen Optimierung*, Dissertation Universität Kaiserslautern.

[3] Borgwardt, K. H. [1982a]: *Some Distribution-Independent Results About the Asymptotic Order of the Average Number of Pivot Steps of the Simplex Method*, Mathematics of Operations Research 7 (1982), 441–462.

[5] Borgwardt, K. H. [1982b]: *The Average Number of Pivot Steps Required by the Simplex-Method is Polynomial*, Zeitschrift für Operations Research 26 (1982), 157–177.

[6] Borgwardt, K. H. [1987]: *The Simplex Method — A Probabilistic Analysis*, Springer Verlag Berlin, Heidelberg, New York, 1987.

[7] Haimovich, M. [1983]: *The Simplex Algorithm is Very Good! — On The Expected Number of Pivot Steps and Related Properties of Random Linear Programs*, 415 Uris Hall, Columbia University, New York, April 1983.

[8] May, J. H. & Smith, R. L. [1982]: *Random Polytopes: Their Definition, Generation and Aggregate Properties*, Mathematical Programming 24 (1982), 39–54.

[9] Shamir, R. [1987]: *The Efficiency of the Simplex Method: A Survey*, Management Science 33 (1987), 301–334.

[10] Smale, S. [1983]: *On the Average Speed of the Simplex Method*, Mathematical Programming 27 (1983), 241–262.

[11] Todd, M. J. [1983]: *Polynomial Expected Behavior of a Pivoting Algorithm for Linear Complementarity and Linear Programming Problems*, Mathematical Programming 35 (1986), 173–192.

A METHOD OF REFERENCE POINT APPROXIMATION IN VECTOR OPTIMIZATION

Johannes Jahn, Erlangen

Zusammenfassung: In dieser Arbeit wird ein interaktives Verfahren zur numerischen Lösung von Vektoroptimierungsproblemen vorgestellt. Diese Methode verwendet beliebig wählbare Referenzpunkte. Numerische Ergebnisse werden für einige lineare Vektoroptimierungsprobleme angegeben, die aus der Literatur bekannt sind.

Abstract: In this paper we present an interactive method for the numerical solution of vector optimization problems. This method works with reference points which may be chosen arbitrarily. Numerical results are given for some linear vector optimization problems known from the literature.

1 Introduction to the Algorithmic Conception

Interactive methods in vector optimization use certain information of the decision maker for the computation of best compromise solutions (e.g., see the book of HWANG-MASUD [3]). This required information is more or less hard to obtain. If an engineer or an economist treats a concrete vector optimization problem, sometimes he has certain goals or aspiration levels which may be given as a reference point. Then the decision maker wants to realize his goals in a best way. In this case a so-called reference point method is a useful solution technique for the considered vector optimization problem.

To be more specific, we have the following assumption:

$$\left.\begin{array}{l}\text{Let } S\subset\mathbb{R}^n \text{ be a given nonempty constraint set,}\\ \text{and let } f:S\to\mathbb{R}^m \text{ be a given vector function.}\end{array}\right\} \quad (1)$$

Under this assumption we consider the vector optimization problem

$$\min_{x\in S} f(x) \qquad (2)$$

for which minimal solutions are defined as follows:

Definition 1:

Let the assumption (1) be satisfied. An element $\bar{x}\in S$ is called a *minimal*

Operations Research Proceedings 1987

solution (or an *Edgeworth-Pareto optimal point* or a *functional efficient point*) of the vector optimization problem (2), if there is no $x \in S$ with

$$f_i(x) \leq f_i(\bar{x}) \text{ for all } i \in \{1,\ldots,m\}$$

and

$$f(x) \neq f(\bar{x}).$$

The set of all minimal solutions of the problem (2) is denoted by M.

For the following we assume that the vector optimization problem (2) is well posed or in other words,

let the set M of minimal solutions of the problem (2) be nonempty. (3)

If we assume - as we indicated it before - that the decision maker may give a reference point $y \in \mathbb{R}^m$ which should be realized as good as possible, then it is evident to solve the following approximation problem:

$$\min_{x \in M} \|y-f(x)\| \tag{4}$$

where $\|\cdot\|$ denotes for example a weighted Chebyshev norm in $\mathbb{R}^m$. A solution of the approximation problem (4) is a minimal solution of the original problem which is as close as possible to the aspiration levels of the decision maker. In principle any norm may be chosen in (4), but the weighted Chebyshev norm is more useful since it can be better interpreted in the case of real-world problems.

Notice that for nonlinear vector optimization problems the approximation problem (4) is not always solvable. In this case it may be helpful to replace "min" by "inf" in problem (4) or to determine so-called ε-minimal points (e.g., see [8] and [10]).

The approximation problem (4) was already investigated in [6]. For the case m=2 an algorithm for the solution of the approximation problem (4) was given in [6, p. 13-15] which cannot be applied to vector optimization problems with more than two goals.

Based on the preceding remarks we formulate the following *method of reference point approximation*:

Step 0: The decision maker chooses a weighted Chebyshev norm $\|\cdot\|$.

Step 1: The decision maker chooses a reference point $y^{(1)} \in \mathbb{R}^m$. Set $i:=1$.

Step 2: Compute a solution of the approximation problem

$$\min_{x \in M} \|y^{(i)}-f(x)\|.$$

Step 3: This solution is presented to the decision maker who terminates the iteration process or proceeds as follows: The decision maker chooses another reference point $y^{(i+1)}$ and continues the procedure with $i:=i+1$ in step 2.

This algorithm can be extended in several ways, and therefore it describes a whole class of methods. For instance, it may be desirable to change also the weights of the Chebyshev norm during the iteration process. Moreover, at each iteration additional information, e.g. concerning trade-offs, may be given to the decision maker such that the selection of another reference point becomes simpler.

From a practical point of view with this method of reference point approximation one needs only some few parameters which are often simple to obtain. But on the other hand the subproblem in the second step is very hard to solve.

There are several reference point approaches (e.g., compare the investigations in [13], [2], [12] and [9]) which are closely related to the above presented method. For these methods the constraint set M of the subproblem is replaced by the set S and therefore the subproblem is simpler to solve. But on the other hand one needs additional assumptions on the reference point and the vector optimization problem.

2 Theoretical Results

In this section we turn our attention to the approximation problem (4) which arises as a subproblem in the method of reference point approximation. In general, this approximation problem is hard to solve (if it is solvable at all), since in the nonlinear case the set M cannot be described by finitely many inequality or equality constraints. With the following theorem we see in the case of a nonlinear vector optimization problem that the problem (4) is in fact a complicated semi infinite optimization problem.

Theorem 2:

Let the assumptions (1) and (3) be satisfied, and let the approximation problem (4) be solvable for some $y \in \mathbb{R}^m$. An element $\bar{x} \in M$ is a solution of the problem (4) if and only if $\bar{x}$ is a solution of the following optimization problem:

$$\left.\begin{array}{l} \min \|y-f(x)\| \\ \text{subject to the constraints} \\ x \in S \\ \sum_{i=1}^{m} f_i(x) = \min \sum_{i=1}^{m} f_i(x') \\ \qquad \text{subject to the constraints} \\ \qquad x' \in S \\ \qquad f_i(x') \leq f_i(x) \text{ for all } i \in \{1,\ldots,m\}. \end{array}\right\} \quad (5)$$

Proof:

For the proof of this theorem we show that $x \in M$ holds if and only if x satisfies the constraints of the problem (5).

(a) For an arbitrary $x \in M$ we obtain with Definition 1 that $x \in S$ and

$$\begin{array}{l} \sum_{i=1}^{m} f_i(x) = \min \sum_{i=1}^{m} f_i(x') \\ \qquad \text{subject to the constraints} \\ \qquad x' \in S \\ \qquad f_i(x') \leq f_i(x) \text{ for all } i \in \{1,\ldots,m\}. \end{array}$$

(b) Now take any $x \notin M$. If $x \notin S$ then the constraints of the problem (5) are not satisfied. If $x \in S$ then there exists some $x' \in S$ with

$$f_i(x') \leq f_i(x) \text{ for all } i \in \{1,\ldots,m\}$$

and

$$f(x') \neq f(x).$$

Consequently we have

$$\sum_{i=1}^{m} f_i(x) > \sum_{i=1}^{m} f_i(x'),$$

and therefore the constraints of the problem (5) are also not satisfied. □

Notice that the optimization problem (5) is equivalent to the following problem:

$$\begin{array}{l} \min \|y-f(x)\| \\ \text{subject to the constraints} \\ x \in S \end{array}$$

$$\sum_{i=1}^{m} f_i(x') \geq \sum_{i=1}^{m} f_i(x) \text{ for all } x' \in S \text{ with } f_i(x') \leq f_i(x) \text{ for all } i \in \{1,\dots,m\}.$$

In general, this semi infinite optimization problem is numerically hard to solve because the parameter set of the given inequality constraints is a subset of $\mathbb{R}^n$.

The approximation problem (4) can be simplified essentially, if the reference point is a lower bound of the image set $f(S)$. This result is summarized in the following theorem.

Theorem 3:

Let the assumptions (1) and (3) be satisfied, and let the approximation problem (4) be solvable for some $y \in \mathbb{R}^m$ with

$$y_i \leq f_i(x) \text{ for all } x \in S \text{ and all } i \in \{1,\dots,m\}.$$

If $\bar{x} \in S$ is an image-unique solution of the problem

$$\min_{x \in S} \|y - f(x)\| , \qquad (6)$$

i.e.

$$\|y - f(\bar{x})\| < \|y - f(x)\| \text{ for all } x \in S \text{ with } f(x) \neq f(\bar{x}),$$

then $\bar{x}$ is a solution of the problem (4).

Proof:

It is well known that every image-unique solution $\bar{x} \in S$ of the problem (6) belongs also to M (e.g., compare [7, Thm. 5.15, a)]). Because of the inequality

$$\|y - f(\bar{x})\| \leq \|y - f(x)\| \text{ for all } x \in M \subset S$$

the element $\bar{x}$ is then also a solution of the approximation problem (4).

□

In Theorem 3 one can drop the assumption on the image-uniqueness of the solution, if we weaken the minimality notion.

Definition 4:

Let the assumption (1) be satisfied. An element $\bar{x} \in S$ is called a *weakly minimal solution* of the vector optimization problem (2), if there is no $x \in S$ with

$$f_i(x) < f_i(\bar{x}) \text{ for all } i \in \{1,\dots,m\}.$$

Let M_w denote the set of weakly minimal solutions.

It is evident that $M \subset M_w$. If we replace the approximation problem (4) by the problem

$$\min_{x \in M_w} \|y-f(x)\| \tag{7}$$

then we obtain the following result:

Theorem 5:

Let the assumption (1) be satisfied, let the set M_w of weakly minimal solutions be nonempty, and let the approximation problem (7) be solvable for some $y \in \mathbb{R}^m$ with

$y_i \le f_i(x)$ for all $x \in S$ and all $i \in \{1,...,m\}$.

Then every solution of the problem (6) is also a solution of the problem (7).

Proof:

If $\bar{x} \in S$ is a solution of problem (6), then we have $\bar{x} \in M_w$ (e.g., compare [7, Thm. 5.25]). Since $M_w \subset S$ we conclude that $\bar{x}$ is also a solution of the problem (7). □

The two preceding theorems show that under certain assumptions a solution of the approximation problems (4) and (7) can be obtained by solving simpler problems.

3 Linear Vector Optimization Problems

Although the described method of reference point approximation has several interesting properties, the subproblem in the second step is very hard to solve numerically if one treats a nonlinear vector optimization problem. In the case of a linear vector optimization problem with a bounded constraint set this approximation problem can be solved at least for smaller problems. In the next subsection we describe how to solve this subproblem. Numerical results for some standard problems are presented in the last subsection.

3.1 Determination of a Solution of the Subproblem

In this subsection we consider the problem (2) as a linear vector optimization problem of the type

$$\left.\begin{array}{l} \min Cx \\ \text{subject to the constraints} \\ Ax \leq b \\ x \in \mathbb{R}^n . \end{array}\right\} \quad (8)$$

Here let C denote a real (m,n) matrix, let A denote a real (q,n) matrix, let b be an element of $\mathbb{R}^q$, and $\leq$ should be understood in a componentwise sense. For the following we assume that the constraint set

$$S := \{x \in \mathbb{R}^n \mid Ax \leq b\}$$

is nonempty and bounded. Then the constraints describe a bounded convex polytope in $\mathbb{R}^n$. If we choose in $\mathbb{R}^m$ the weighted Chebyshev norm $\|\cdot\|$ as

$$\|y\| := \max_{1 \leq i \leq m} w_i |y_i| \quad \text{for all } y \in \mathbb{R}^m$$

with appropriate weights $w_1, \ldots, w_m > 0$, then for any fixed reference point $y \in \mathbb{R}^m$ we consider the Chebyshev approximation problem (4) which can be written as

$$\min_{x \in M} \max_{1 \leq i \leq m} w_i |y_i - (Cx)_i| \ . \tag{9}$$

It is well known in linear vector optimization that the set M consists of certain facets and edges of S which can be described by certain vertices. With the aid of a modified simplex method (e.g., see [15], [14], [4] and [12]) it is possible to compute these vertices which are minimal solutions. In other words: One can compute a partition of the set M of minimal solutions with

$$M = M_1 \cup M_2 \cup \ldots \cup M_\ell$$

where $\ell \in \mathbb{N}$ and $\emptyset \neq M_j \subset M$ for all $j \in \{1, \ldots, \ell\}$ such that: For every M_j ($j=1, \ldots, \ell$) there exist s_j vertices $x^{(j_1)}, \ldots, x^{(j_{s_j})} \in M$ with

$$M_j = \{ x \in S \mid x = \sum_{k=1}^{s_j} \lambda_k x^{(j_k)} \text{ with } \lambda_k \geq 0 \text{ and } \sum_{k=1}^{s_j} \lambda_k = 1 \} \tag{10}$$

(see Fig. 1). For instance, with the method of ISERMANN (see [4] and [5]) these vertices $x^{(j_k)}$ and facets and edges M_j can be computed.

Instead of solving the problem (9) we solve the ℓ approximation problems

$$\min_{x \in M_j} \max_{1 \leq i \leq m} w_i |y_i - (Cx)_i| \tag{11}$$

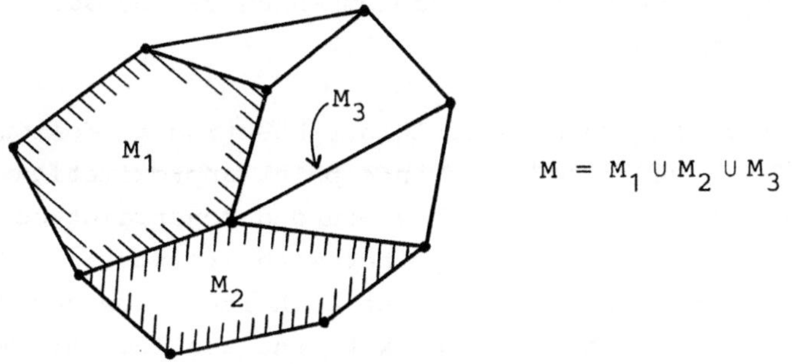

Fig. 1. Partition of the set M

for j=1,...,ℓ and among the solutions of these ℓ problems we choose a solution with the smallest minimal value. With the aid of the equation (10) the Chebyshev approximation problem (11) can be written as

$$\min \max_{1\le i\le m} w_i \left| y_i - \sum_{k=1}^{s_j} \lambda_k (Cx^{(j_k)})_i \right|$$

subject to the constraints

$$\sum_{k=1}^{s_j} \lambda_k = 1$$

$$\lambda_1,\ldots,\lambda_{s_j} \ge 0 .$$

This problem is equivalent to the problem

$$\min \lambda_o$$

subject to the constraints

$$\left.\begin{aligned} -\frac{1}{w_i}\lambda_o - \sum_{k=1}^{s_j} \lambda_k (Cx^{(j_k)})_i &\le -y_i \\ -\frac{1}{w_i}\lambda_o + \sum_{k=1}^{s_j} \lambda_k (Cx^{(j_k)})_i &\le y_i \end{aligned}\right\} \text{ for all } i\in\{1,\ldots,m\}$$

$$\sum_{k=1}^{s_j} \lambda_k = 1$$

$$\lambda_o\in\mathbb{R};\ \lambda_1,\ldots,\lambda_{s_j} \ge 0 .$$

This is a linear optimization problem which can be solved with the aid of the simplex method.

We summarize the results of this subsection as follows: For the application of the method of reference point approximation to a linear vector optimization problem with a bounded constraint set one determines firstly the facets and edges together with its vertices describing the set of minimal solutions and then at each iteration one solves ℓ linear programs and chooses the solution with the smallest minimal value in order to get a solution of the subproblem.

3.2 Numerical Results

In the following we apply the method of reference point approximation to three linear vector optimization problems which are well known from the literature. The weights w_i of the Chebyshev norm are set to 1. The information on the vertices describing the set M of minimal solutions is taken from [5]. In the corresponding tables we list the obtained results for different reference points.

Example 6 ([14, p.465], [4, p.719], [5, p.11]):

$$\min \begin{pmatrix} -4 & -1 & -2 \\ -1 & -3 & 1 \\ 1 & -1 & -4 \end{pmatrix} x$$

subject to the constraints

$$\begin{pmatrix} 1 & 1 & 1 \\ 2 & 2 & 1 \\ 1 & -1 & 0 \end{pmatrix} x \le \begin{pmatrix} 3 \\ 4 \\ 0 \end{pmatrix}$$

$$x \ge 0 .$$

Number of vertices which are minimal solutions: 5

Number of facets or edges consisting only of minimal solutions: 2

reference point	best approximation to the reference point	distance
$y^{(1)} = (-10, -8, -15)$	(0, 1.0833, 1.8333)	6.5833
$y^{(2)} = (-8, -6, -13)$	(0, 1.0833, 1.8333)	4.5833
$y^{(3)} = (-7, -5, -12)$	(0, 1.0833, 1.8333)	3.5833
$y^{(4)} = (-7, -5, -11)$	(0, 1.1667, 1.6667)	3.1667
$y^{(5)} = (-7, -5, -10)$	(0, 1.25, 1.5)	2.75

Example 7 ([5, p.15], [12, p.244]):

$$\min \begin{pmatrix} -1 & -3 & 2 & 0 & -1 \\ -3 & 1 & 0 & -3 & -1 \\ -1 & 0 & -2 & 0 & -3 \end{pmatrix} x$$

subject to the constraints

$$\begin{pmatrix} 2 & 4 & 0 & 0 & 3 \\ 0 & 0 & 2 & 5 & 4 \\ 5 & 0 & 0 & 0 & 0 \\ 0 & 0 & 0 & 2 & 0 \\ 5 & 5 & 2 & 0 & 0 \end{pmatrix} x \le \begin{pmatrix} 27 \\ 35 \\ 26 \\ 24 \\ 36 \end{pmatrix}$$

$x \ge 0$.

Number of vertices which are minimal solutions: 11

Number of facets or edges consisting only of minimal solutions: 4

reference point	best approximation to the reference point	distance
$y^{(1)}$ = (-15, -40, -20)	(5.2, 0.0526, 0, 4.9368, 2.5789)	7.0632
$y^{(2)}$ = (-13, -39, -19)	(5.2, 0, 0, 4.9273, 2.5909)	6.0273
$y^{(3)}$ = (-12, -38, -19)	(5.2, 0, 0, 4.7455, 2.8182)	5.3455
$y^{(4)}$ = (-12, -37, -19)	(5.2, 0, 0, 4.5636, 3.0455)	4.6636
$y^{(5)}$ = (-12, -37, -18)	(5.2, 0, 0, 4.7455, 2.8182)	4.3455
$y^{(6)}$ = (-12, -37, -17)	(5.2, 0.0526, 0, 4.9368, 2.5789)	4.0632
$y^{(7)}$ = (-11, -35, -16.5)	(5.2, 0.1053, 0, 4.6737, 2.9079)	2.5763

Example 8 ([11] (cited in [5]), [5, p.18]):

$$\min \begin{pmatrix} 0 & 0 & 1 & 0 & 0 & 0 \\ 0 & 0 & 0 & 1 & 0 & 0 \\ 0 & 0 & 0 & 0 & 1 & 0 \\ 0 & 0 & 0 & 0 & 0 & 1 \end{pmatrix} x$$

subject to the constraints

$$\begin{pmatrix} 2 & 4 & 0 & 0 & 0 & 0 \\ 4 & 3 & 0 & 0 & 0 & 0 \\ 1 & 1 & 0 & 0 & 0 & 0 \\ 0 & -1 & 0 & 0 & -1 & 1 \\ -1 & -1 & 0 & 0 & 0 & 0 \end{pmatrix} x \le \begin{pmatrix} 24 \\ 28 \\ 8 \\ -5.75 \\ -7 \end{pmatrix}$$

$$(1, 0, 1, -1, 0, 0)\, x = 6$$

$x \ge 0$.

Number of vertices which are minimal solutions: 3

Number of facets or edges consisting only of minimal solutions: 2

reference point	best approximation to the reference point	distance
$y^{(1)} = (-5,-5,-2,\ -2)$	(5.3929,2.1429,0.6071,0,3.6071,0)	5.6071
$y^{(2)} = (-2,-3,\ 0,\ -2)$	(4.9643,2.7143,1.0357,0,3.0357,0)	3.0357
$y^{(3)} = (\ 1,-1,\ 0.5,\ 0)$	(3.8333,4.0833,2.1667,0,1.6667,0)	1.1667

4 Conclusion

Although the presented method of reference point approximation is simple to apply for the interactive solution of vector optimization problems, the arising subproblems are numerically difficult to solve. In the third section we pointed out how to solve these subproblems successfully in the case of a linear vector optimization problem. But we should also note that even for linear problems of medium size it is very difficult to compute a partition of the set of minimal solutions, since the number of vertices which are minimal solutions may increase dramatically (e.g., compare [1]).

5 References

[1] J.P. Dauer and Y.-H. Liu, "Solving Multiple Objective Linear Programs in Objective Space", to appear.

[2] M. Grauer, "Reference Point Optimization - The Nonlinear Case", in: P. Hansen (ed.), *Essays and Surveys on Multiple Criteria Decision Making* (Springer, Berlin, 1983), p. 126-135.

[3] C.-L. Hwang and A.S.M. Masud, *Multiple Objective Decision Making - Methods and Applications* (Springer, Berlin, 1979).

[4] H. Isermann, "The Enumeration of the Set of All Efficient Solutions for a Linear Multiple Objective Program", *Operational Research Quarterly* 28 (1977) 711-725.

[5] H. Isermann, "Users Manual for the EFFACET Computer Package for Solving Multiple Objective Linear Programming Problems" (manuscript, University of Bielefeld, 1984).

[6] J. Jahn, "Some Characterizations of the Optimal Solutions of a Vector Optimization Problem", *OR Spektrum* 7 (1985) 7-17.

[7] J. Jahn, *Mathematical Vector Optimization in Partially Ordered Linear Spaces* (Peter Lang, Frankfurt, 1986).

[8] P. Loridan, "ε-Solutions in Vector Minimization Problems", *J. Optim. Theory Appl.* 42 (1984) 265-276.

[9] A.M.J. Skulimowski, "A Sufficient Condition for Θ-Minimality of Distance Minimization Procedures", to appear.

[10] T. Staib, "On Two Generalizations of Pareto Minimality", *J. Optim. Theory Appl.* (to appear).

[11] R.E. Steuer, Repertoire of Multiple Objective Linear Programming Test Problems (Working Paper in Business Administration, University of Kentucky, Lexington, 1978).

[12] R.E. Steuer, *Multiple Criteria Optimization: Theory, Computation, and Application* (John Wiley, New York, 1986).

[13] A.P. Wierzbicki, "The Use of Reference Objectives in Multiobjective Optimization", in: G. Fandel and T. Gal (eds.), *Multiple Criteria Decision Making - Theory and Application* (Springer, Berlin, 1980), p. 468-486.

[14] P.L. Yu and M. Zeleny, "The Set of All Nondominated Solutions in Linear Cases and a Multicriteria Simplex Method", *J. Math. Anal. Appl.* 49 (1975) 430-468.

[15] M. Zeleny, *Linear Multiobjective Programming* (Springer, Berlin, 1974).

Ein Verfahren zur Lösung des Kompensationsmodells der stochastischen linearen Programmierung

Jürgen Böttcher, Hamburg

Das betrachtete Kompensationsmodell lautet

$$\min_{x\in X} c'x + \sum_{k=1}^{r} p^k Q(x,h^k) \quad \text{mit} \quad Q(x,h^k) = \min_{y\geq 0} \{q'y \mid Wy = h^k - Tx\}$$

und $X := \{x \mid Ax = b, x \geq 0\}$. Hierbei ist A eine $m_1 \times n_1$-Matrix und W eine $m_2 \times n_2$-Matrix. Die weiteren Dimensionen ergeben sich entsprechend. Weiterhin sei h^k die Ausprägung einer diskreten Zufallsvariablen mit $P(h^k) = p^k > 0$, $k = 1,\ldots,r$.

Ein derartiges Problem entsteht z.B. bei Produktionsmodellen mit stochastischer Nachfrage, wenn die Produktionsentscheidung vor Kenntnis der Nachfrage getroffen werden muß und Kompensationsmöglichkeiten zum Ausgleich der Differenz zwischen Produktion und Nachfrage vorhanden sind.

In diesem Modell ist eine (nicht separable) stückweise lineare Zielfunktion über einem konvexen Polyeder zu minimieren. Das Modell ist auch als großes LP mit dualer block-angularer Struktur formulierbar. Die erste Formulierung führt zu Verfahren der nichtlinearen (nicht differenzierbaren) Programmierung, die zweite Formulierung führt zu Dekompositionsverfahren der linearen Programmierung.

Im Vortrag wird ein Verfahren vorgestellt, das beide Ansätze kombiniert. Das Problem wird zerlegt in die r parametrischen linearen Programme $Q(x,h^k)$ und ein Masterproblem, das alle für die aktuelle Lösung aktiven Nebenbedingungen enthält. Bei Ausführung des Verfahrens zeigt das Masterproblem die Optimalität der Lösung, oder es liefert eine Abstiegsrichtung s. Dann ergibt die Minimierung der stückweise linearen Funktion $c's\lambda + \Sigma_{k=1}^{r} p^k Q(x + s\lambda, h^k)$ eine bessere Lösung.

Probleme mit großem r (typisch für Kompensationsmodelle ist $r \gg 100$) können unter Verwendung effizienter Datenstrukturen gelöst werden:

- Betrachte nur dual zulässige Basen für die parametrischen linearen Programme und speichere diese in einer Graphenstruktur.
- Ordne alle h^k mit dgl. optimalen Basis in einem binären Baum nach Dominanz und Indifferenz gemäß einer durch die Basisspalten erzeugten unvollständigen Kegelordnung.

THE MAXIMAL DISTANCE IN A POLYHEDRON

Stane Indihar, Maribor

The question how to find two points in a bounded polyhedron X for which the euclidean distance is maximal leads to the following non-linear programming (NLP) problem

$$\max\left\{\|x - y\|^2 = x^Tx + y^Ty - 2x^Ty \;\middle|\; x \in X,\ y \in X\right\}. \qquad (P_d)$$

The problem (P_d) is a special case of the NLP-problem

$$\max\left\{f(x,y) \;\middle|\; x \in X,\ y \in Y\right\}, \qquad (P)$$

$$f(x,y) = a^Tx + b^Ty + x^TCy + x^TCy + x^TEx + y^TFy\ ,$$

$$X = \left\{x \in R^m \;\middle|\; Ax \leq p,\ x \geq 0\right\}\ ,\quad Y = \left\{y \in R^n \;\middle|\; By \leq q,\ y \geq 0\right\},$$

where E, F are positive semidefinite matrices and X, Y are bounded polyhedra. For a fixed $y \in Y$ $(x \in X)$ the function $f(x,y)$ is convex, therefore an extreme point $\bar{x} \in X$ $(\bar{y} \in Y)$ exists for which $f(x,y)$ reaches the maximal value on X (on Y). It follows that a pair $(\tilde{x},\tilde{y})$ of extreme points exists which is an optimal solution of (P).

Introducing the vectors and matrices

$$z = \begin{bmatrix} x \\ y \end{bmatrix},\ c = \begin{bmatrix} a \\ b \end{bmatrix},\ r = \begin{bmatrix} p \\ q \end{bmatrix},\ D = \begin{bmatrix} A & 0 \\ 0 & B \end{bmatrix},\ G = \begin{bmatrix} 0 & C \\ C^T & 0 \end{bmatrix},\ H = \begin{bmatrix} E & 0 \\ 0 & F \end{bmatrix},$$

the problem (P) can be transformed in a general quadratic programming problem

$$\max\left\{c^Tz + z^T(\tfrac{1}{2}G + H)z \;\middle|\; Dz \leq r,\ z \geq 0\right\}.$$

The method for solving problems (P) and (P_d) is based on a (finite) algorithm [1] for maximazing a convex quadratic function and on a (finite) algorithm [2] for minimizing a concave function over a bounded polyhedron.

[1] Czochralska, I., The method of bilinear programming for nonconvex quadratic programming, Zastosowania Matematyki 17 (1982) 515 - 525.

[2] Thoai, N.V., Anwendung des Erweiterungsprinzip zur Lösung konkaver Optimierungsaufgaben, Mathem. Operationsforsch. Statist. 12 (1981) 45 - 51.

NUMERISCHE SENSITIVITÄTSANALYSE EINES ERNÄHRUNGSPROBLEMS UNTER BERÜCKSICHTIGUNG DER SCHWANKUNG ALLER EINGABEDATEN

Christian Jansson, Kaiserslautern

Bei der Beurteilung der Lösung von linearen Optimierungsaufgaben ist nach Ansicht vieler Autoren eine Sensitivitätsanalyse oder postoptimale Analyse ein wichtiger Bestandteil. In (1) und (2) sind Algorithmen beschrieben, die für das lineare Optimierungsproblem

$$\operatorname*{Max}_{x \in X} c^t x \ , \quad X := \{x \in \mathbb{R}^n \mid Ax = b,\ x \geq 0\}$$

mit den Eingabedaten P = (A,b,c) eine Sensitivitätsanalyse mit Berücksichtigung der Schwankung aller Koeffizienten von P ermöglichen. (A ist eine m×n-Matrix, $b \in \mathbb{R}^m$ und $c \in \mathbb{R}^n$.) Dabei werden die Schwankungsbereiche der Eingabedaten durch Angabe oberer und unterer Schranken dargestellt.

Die Algorithmen berechnen unter sehr allgemeinen Voraussetzungen obere und untere Schranken für die Komponenten aller auftretenden optimalen Ecken und für die optimalen Zielfunktionswerte der Schar linearer Optimierungsprobleme, deren Eingabedaten innerhalb dieser Schwankungsbereiche liegen.

Es wird über eine vollständige numerische Sensitivitätsanalyse eines Ernährungsproblems mit Hilfe dieser Algorithmen berichtet. Diese Sensitivitätsanalyse wird mit anderen bekannten Methoden der postoptimalen Analyse verglichen.

Literatur:

(1) Jansson, C.: Zur Linearen Optimierung mit unscharfen Daten. Dissertation, Kaiserslautern (1985).

(2) Jansson, C.: Zur Linearen Optimierung mit unscharfen Daten. Operations Research Proceedings 1986, Springer-Verlag (1987), 610-617.

TRANSFORMATION OF NONSMOOTH AND NONCONVEX PROGRAMMING PROBLEMS

Ivan Meško, Maribor

Some applicable nonlinear optimization models can be expressed in the form of mixed integer linear programming problems [1]. Using these expressions we are not in need of computer programs for solving original problems even if for original problems suitable algorithms exist. The work of time is the transmission and control of data and not the solving of mixed integer linear programming problems [2].

Let $f_{ij}: R^n \to R$ be linear functions for $i \in E_j \cup L_j$, $j=1,\dots,k$. Then

$$B = \bigcup_{j=1}^{k} \{x \in R^n \mid f_{ij}(x) = b_{ij} \text{ for } i \in E_j,\ f_{ij}(x) \le b_{ij} \text{ for } i \in L_j\}$$

is a polyhedron which can be nonconvex. If B is bounded it can be expressed in the form

$$B = \{x \in R^n \mid u_j = 0 \text{ or } 1,\ d(u_j - 1) \le f_{ij}(x) - b_{ij} \le d(1-u_j) \text{ for } i \in E_j,$$
$$f_{ij}(x) - b_{ij} \le d(1-u_j) \text{ for } i \in L_j,\ j=1,\dots,k,\ \sum_{j=1}^{k} u_j \ge 1\}$$

where d is a suitable large enough constant.

Consider the function

$$f(x) = \begin{cases} f_1(x) & \text{for } g(x) \le 0 \\ f_2(x) & \text{for } g(x) \ge 0 \end{cases}$$

where $f_1(x)$, $f_2(x)$ and $g(x)$ are real functions for $x \in R^n$ and $g(x)$ let be bounded. If $f_1(x) = f_2(x)$ for $g(x) = 0$, then $f(x)$ is single valued for $g(x) = 0$. The function $f(x)$ can be expressed in the form

$$f(x) = h(x,u) = uf_1(x) + (1-u)f_2(x)$$

subject to zero-one variable u and $-du \le g(x) \le d(1-u)$, where d is a large enough constant.

For the important case $f(x) = |g(x)|$ we obtain

$$|g(x)| = y + z$$

subject to suitable constant d, zero-one variable u and

$$g(x) = z - y \qquad 0 \le y \le du \qquad 0 \le z \le d(1-u) \qquad -du \le g(x) \le du$$

References

[1] I. Meško: Piecewise Linear Approximations in R^n. In: H. Isermann (Ed.), Operations Research Proceedings 1986, Springer 1987, 618-625.

[2] U.H. Suhl: Solving Large-Scale Mixed-Integer Programs with Fixed Charge Variables. Mathematical Programming 32 (1985), 165-182.

LINEARE STOCHASTISCHE OPTIMIERUNG MIT VAGEN DATEN

Heinrich ROMMELFANGER
Jochen WOLF

Frankfurt

Bei dem Bemühen reale Entscheidungsprobleme in lineare Optimierungsmodelle abzubilden, hat man zumeist Schwierigkeiten mit der Festlegung der Koeffizienten und der Restriktionsgrenzen. Den überwiegenden Teil dieser Daten kann der Entscheidungsträger nur näherungsweise angeben, während die Modellannahmen die Zuordnung eindeutiger Werte fordern.

Ein schon mittlerweile als klassisch zu bezeichnender Weg, dieser Ungenauigkeit der Daten zu begegnen, besteht darin, Umweltzustände so auszuwählen, daß ihre Eintrittswahrscheinlichkeiten bekannt sind und alle zustandsspezifischen Daten eindeutig festgelegt werden können. Als Abbild des Realproblems erhält man dann ein lineares stochastisches Optimierungsmodell.

Ein neuerer Weg, die Ungenauigkeit der Daten in das Modell zu integrieren, ist die Darstellung vager Daten in Form unscharfer Mengen. Die Realität wird so in ein Fuzzy-LP-Modell abgebildet.

Die oft gestellte Frage, ob der stochastische Ansatz oder das Fuzzy-Modell die bessere Methode zum Lösen des Realproblems ist, wird in dieser Arbeit beantwortet mit "sowohl Wahrscheinlichkeiten als auch Fuzzy sets". Wie gezeigt werden wird, beschreiben diese beiden Instrumentarien unterschiedliche Aspekte der Ungenauigkeit.

Es wird vorgeschlagen, in Abhängigkeit des Informationsstandes Umweltzustände zu unterscheiden und die dann noch vorhandenen zustandsspezifischen Ungenauigkeiten in Form von unscharfen Mengen auszudrücken.

Für die sich so ergebenden Optimierungssysteme werden Wege zur Bestimmung einer optimalen Lösung aufgezeigt, die sich sowohl an den Verfahren der stochastischen Optimierung als auch an der Fuzzy-Entscheidungstheorie und der Fuzzy-Optimierung orientieren.

MATHEMATICAL PROGRAMMING IN PRACTICE

J. Telgen, Utrecht

Mathematical programming probably is the operations research technique most used in practice. Also in operations research curricula it usually takes up most of the teaching time and effort.

Nevertheless I experience a wide and widening gap between on the one hand operations research in practice and on the other hand mathematical programming as it is taught.

Mathematical programming in practice (as a tool of operations research) is quite different from the mathematical programming taught. Students learn the steps of the simplex method, but when (and if!) they use it, they take a software program and type "GO" or "OPTIMIZE".

They will see only input (the LP-model) and output (solution report) and use the algorithm in much the same way as we use the square root algorithm in our pocket calculator.

Apart from the above point on the importance of algorithmic details still too much emphasis is put on the method instead of the problem. I know of no mathematical programming (or operations research) curriculum where problem recognition and problem formulation are predominant. And exactly those two items are what is important in practice. Are we working for the right client; what is this client's problem; what can he do with a solution; why does he optimize/satisfice; many questions like these are the ones to be solved in practice. And they are hardly addressed in our teaching.

The situation described above is different for very large companies and research laboratories where large scale LP problems are solved on a daily basis. For such organisations the current mathematical programming education is probably adequate. But it is not for all other purposes!

This statement is illustrated by three examples on "Scheduling meetings" (Interfaces vol 15, no 4), "Take-over of a laundry firm", and a "Cutting stock problem for a wash leather firm".

Berge's Strong Perfect Graph Conjecture for 4-Chromatic Graphs

S.M. Baas , Enschede

1. Introduction

Berge's well known conjecture (SPGC) can be formulated as follows: "A graph G is perfect if and only if it contains neither an odd hole nor an odd anti-hole". In this formulation an odd hole is a chordless circuit on an odd number of nodes and an odd anti-hole the complement of an odd hole. In the past the conjecture has been verified for several classes of graphs, among which planar and 3-chromatic graphs.
In this summary we present an outline for a proof that SPGC is valid for graphs in which a maximum clique has four nodes. As a direct consequence such graphs will be 4-chromatic, i.e., the nodes can be coloured with four different colours such that any two adjacent nodes have different colours. Then a formulation stating that SPGC is true for 4-chromatic graphs is allowed.
The proof is based on the notion of *preperfectness;* a graph G is said to be preperfect if it contains neither odd holes nor odd anti-holes and if in addition it is minimal imperfect (= critical perfect), i.e., G is imperfect but all its induced proper subgraphs are perfect. Minimal imperfectness implies several useful properties such as: each minimum independent set has a node in common with each maximum clique except one (clique = complete subgraph n-clique = clique on n nodes, independent set = complement of a clique).
Actually a verification of SPGC would mean that there exist no preperfect graphs at all.

2. An outline for a proof

The proof is constructive: it produces an independent set $\mathcal{I}$ that has a node in common with all maximum cliques, in violation of the condition that in a mimimal imperfect graph one maximum clique should be excluded from $\mathcal{I}$.
Using another well-known property of minimal imperfect graphs we can choose an arbitrary node $u_0 \in N$ (the node-set of G), remove it from G and next partition $N \setminus \{u_0\}$ into four disjoint independent sets $S_i, i = 1,2,3,4$. Now consider the bipartite subgraph $G_{1,2} = (S_{1,2}, E_{1,2})$ induced in G by $G_{1,2} = S_1 \cup S_2$. In line with the one property quoted above concerning minimal imperfect graphs, each maximum clique in G has an edge in common with $G_{1,2}$, except two cliques, one (say K_{c1}) which has a node, v_{c1}, only in S_1 and one (K_{c2}) which has a node, v_{c2} only in S_2. K_1 and K_2 are said to be *critical cliques* for $G_{1,2}$ and v_{c1}, v_{c2} *critical nodes* in $G_{1,2}$.
Again by minimal imperfectness we have $u_0 \in K_1 \cap K_2$. Moreover each node $u \notin S_{1,2}$ is contained in at least one non-critical maximum clique and is therefore adjacent to at least one edge in $G_{1,2}$.

DEFINITION 1:
Let $G_b = (N_b, E_b)$ be a bipartite subgraph of G. Then the *triangular neighbourhood* of G_b is:

$$\Delta(G_b = \{u \in N \setminus N_b | u \text{ is adjacent to an edge in } E_b\}.$$

DEFINITION 2:
Let G_b be a bipartite subgraph of G and $u \in \Delta(G_b)$. Then a node $v \in N_b$ is a *contact-node for u* if $uv \in E$. The set of all contact-nodes for u with G_b is the *node-contact* $NC(u;G_b)$. An edge $e \in E_b$ is a *contact-edge for u* if $u \in \Delta(e)$. The set of all contact-edges for u with G_b is the *edge-contact* $EC(u;G_b)$. A contact-node $v \in NC(u;G_b)$ is *isolated* if it is not contained in a contact-edge.

In terms of these definitions it can be concluded that each node not in $S_{1,2}$ is contained in $\Delta(G_{1,2})$ or, equivalently, has edge-contact with $G_{1,2}$. To avoid the use of a tiresome lengthy terminology a bipartite subgraph induced by two disjoint maximum independent sets is further called a *skeleton* in G. Apart from $G_{1,2}$ we could also use the skeleton $G_{3,4}=S_3 \cup S_4$, as a basis for our analysis. Obviously properties derived with respect to $G_{1,2}$ equally well apply to $G_{3,4}$. Now given the absence of odd holes and odd anti-holes as required by preperfectness, various interesting results can be derived, specifying the coherence between (parts of) $G_{1,2}$ and the triangular neighbourhood. The proof that SPGC is true for 4-chromatic graphs makes ample use of such structural results. It is constructive in the sense that an independent set $\mathcal{I}$ that has a node in common with each maximum clique is gradually built up, starting with u_0. The underlying process consists of three different phases, as will be explained hereafter. The selection of u_0 as a first element in $\mathcal{I}$ automatically implies that the two cliques K_{c1} and K_{c2} which are critical for $G_{1,2}$ are *saturated* by $\mathcal{I}$, i.e. these two cliques already have a node in common with $\mathcal{I}$. Now suppose that at some stage $u \in \Delta(G_{1,2})$ has been added to $\mathcal{I}$. Let $e \in E_{1,2}$ be included in $\mathcal{I}$. Consequently, if K is a maximum clique that contains e as an edge but is not saturated yet by $\mathcal{I}$, a node $u' \in K \cap \Delta(G_{1,2})$ must be selected to be added to $\mathcal{I}$ as a next element. These observations give rise to the following definition.

DEFINITION 3:
An *edge-contact path*, or simply *ec-path* $P_{ec}=u_p e_p e_{p+1} ....e_{q-1} u_q$, generated on a skeleton $G_s=(N_s,E_s)$ in G is a sequence of alternately nodes $u_i \in \Delta(G_s), i=p,p+1,...,q$ and edges $e_i \in E_s, i=p,p+1,...,q-1$, such that:
(i) $\{u_p,u_{p+1},...,u_q\}$ is an independent set
(ii) $e_p \in EC(u_p;G_s), e_{q-1} \in EC(u_q;G_s)$
(iii) for $i=p+1,p+2,...,q-1: e_{i-1},e_i \in EC(u_i;G_s)$
An ec-path is elementary if or $p \leq i < j \leq q$:

$$EC(u_i;G_s) \cap EC(u_j;G_s) \neq \emptyset \Rightarrow j=i+1.$$

Now the problem considered in phase 1 of the constructive part of the proof mainly concerns the existence of ec-paths all of which must have u_0 as initial node ($p=0$ in the definition). The complication that may occur in this respect essentially bears upon condition (i) in definition 3: the independence of consecutive nodes $u_0,u_1,u_2,....$
It can be shown that, by using suitable selection criteria, one can proceed with the construction of an arbitrary ec-path until a situation is met in which both candidates, the one in S_3 and the one in S_4 are adjacent to u_0. This situation is elucidated in the following figure.

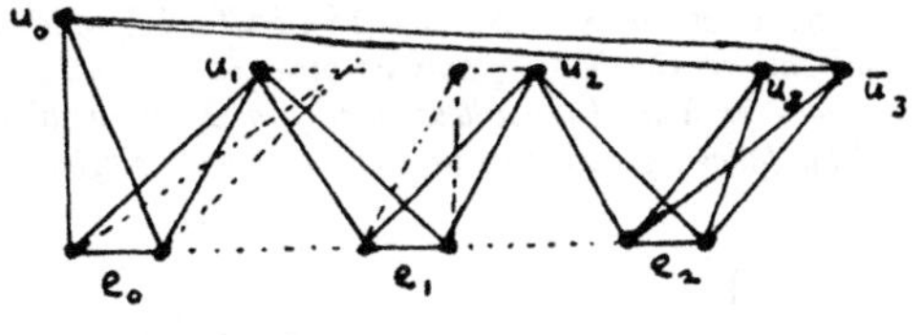

However the occurrence of such complications can be excluded by prepefectness:

THEOREM 1:
Let G be a preperfect graph and $p_{ec}=u_0e_0u_1......e_{r-1}u_r$ an elementary ec-path generated on a skeleton G_s in G. Furthermore let $E_r \in EC(u_r;G_s), e_r \neq e_{r-1}$ form a 4-clique with an edge $\bar{e}=u\bar{u}$. If both $u_0u \in E$ and $u_0\bar{u} \in E$, then there exists a 5-clique in G containing $u_0,u,\bar{u}$ together with an edge $e \in E_s$. □

Hence in a preperfect graph in which the maximum clique-size $\omega(G)=4$, a situation as shown in the above figure cannot occur. Consequently the construction of ec-paths from u_0 can be continued without further complications, by adding new, mutually independent nodes $u \in S_{3,4}$ and associated contact-edges in G_s, until no new nodes and contact-edges are available anymore.

Due to the aforementioned selection criteria the nodes in different ec-paths are also independent. Consequently the so-called *reachable set* for u_0, $\mathfrak{R}(u_0) \subset S_{3,4}$, i.e., the set of all nodes which are reachable from u_0 along an ec-path, forms an independent set that saturates all maximum cliques which contain an edge in one of the corresponding ec-paths. This completes phase 1, with $\mathfrak{I}_1 = \mathfrak{R}(u_0)$ as a first contribution to $\mathfrak{I}$.

Next suppsoe that everywhere in $G_{1,2}$ the distances between consecutive contact-nodes associated with nodes in $\mathfrak{I}_1$ are even: all paths connecting two such contact-nodes have even length, i.e., they contain an even number of edges. Consider such a path and number its nodes $1,2,\ldots,2t+1$. Obviously all odd nodes belong to S_1 and the even nodes to S_2, or just the reverse. Moreover in a connected part in $S_{1,2}$ consisting of paths of which only the terminal nodes are contact-nodes for nodes in $\mathfrak{I}_1$, the numbering can be performed consistently in the sense that the terminal nodes are all contained in S_1 or all in S_2.

Now in this case an independent set $\mathfrak{I}_3$ that saturates all the remaining maximum cliques is easily constructed by selecting the even nodes in each connected part, as illustrated in the following figure.

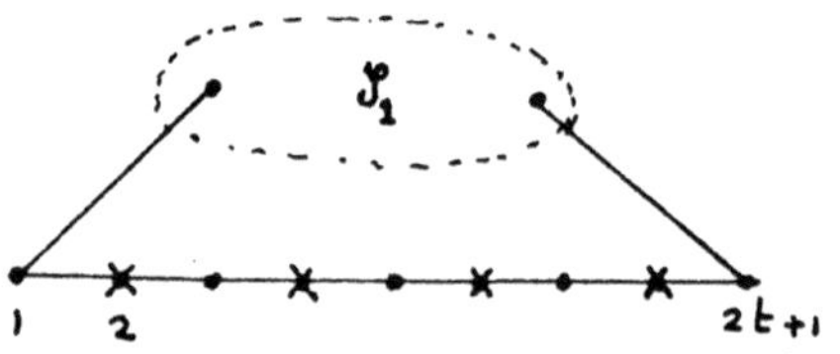

Clearly each remaining maximum clique, since it contains an edge in $\mathfrak{I}_{1,2}$, has precisely one node in common with $\mathfrak{I}_3$. In fact the construction of $\mathfrak{I}_3$ must be regarded as the third and concluding phase in the whole process. it remains to be shown what has to be done if not all the paths connecting consecutive contact-nodes in $S_{1,2}$ (relative to $\mathfrak{I}_1$) have even length. This brings us to phase 2, in which similar techniques are used as in phase 1.

THEOREM 2:

Let G be a preperfect graph, G_s a skeleton in G and $S \subset \Delta(G_s)$ an independent set in which any two nodes can be connected along an ec-path on G_s.

If for u, $\bar{u} \in S$ there exist nodes $v \in NC(u;G_s), \bar{v} \in NC(\bar{u};G_s)$ which are terminal nodes for an odd path in G_s that contains no other contact-nodes relative to S, then there exists an even hole H (chordless circuit) in G_s, and a subset $T \subseteq S$ that generates one of the following two structures on H (called "singleton" respectively "odd multifoil").

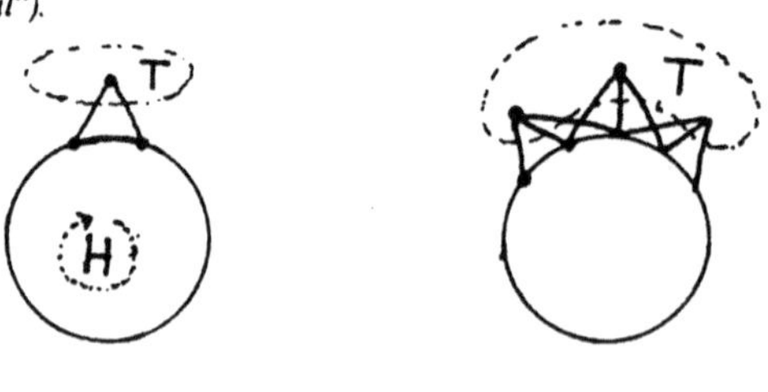

The next theorem in some sense gives a "complementary" result.

THEOREM 3:
Let $T \subseteq S$ generate a singleton or an odd multifoil on an even hole H in a skeleton G_s within a preperfect graph G. If each edge in H is contained in at least one maximum clique then there exists an independent set $S' \subset \Delta(G_s)$ in which any two nodes can be connected along an ec-path on G_s, such that a subset $T' \subseteq S'$ generates either a singleton or a multifoil on the part of H not covered by T. Moreover each path in G_s connecting a contact-node $v \in NC(u;G_s), u \in T$, with a contact-node $v' \in NC(u';G_s), u' \in T'$ and contains no other contact-nodes, has even length.

On the basis of suitable selection criteria similar to those which were used in phase 1, one can construct S' with the help of ec-paths such that also independence of u_0 is guaranteed. By a repeated application of this process, as long as required by the presence of odd paths, a sequence of independent sets $S'_i \subset S_{3,4}, i=1,2,...,s$ can be produced until finally all remaining paths in $S_{1,2}$ are even. With $\mathfrak{I}_2 = \cup_{i=1}^{s} S'_i$ phase 2 is completed and the process can turn to phase 3, described before.
With $\mathfrak{I} = \mathfrak{I}_1 \cup \mathfrak{I}_2 \cup \mathfrak{I}_3$ the construction of an independent set that saturates all maximum cliques in G has succeeded. This completes the outline of the proof.

REFERENCES

BERGE, C., *Graphs and Hypergraphs*, North-Holland Publishing Company, New York, Oxford (1979).
LOVASZ., L., Normal Hypergraphs and the Perfect Graph Conjecture, *Discrete Mathematics 2* (1979). 253-267.
TUCKER, A.C., The Strong Perfect Graph Conjecture for Planar Graphs, *Can. J. Math. 25* (1973), 103-114.

Scheduling Periodically Recurring Events and Processes

Peter Brucker, Osnabrück

Summary: Periodically recurring events (e.g. periodic arrival of trains at some station) are to be scheduled in order to get optimal solutions with respect to one of the following objectives:
(i) minimizing the maximum time between adjacent events (e.g. minimizing the maximum waiting time of passengers), (ii) minimizing the average waiting time, (iii) maximizing the minimal distance between adjacent events (e.g. maximizing the safety distance between trains). Such situations may be modelled by regular or irregular polygons with vertices on a circle line. The problem is, to move the polygons relative to each other such that an objective function depending on the distances between vertices is optimized.

For the case of two irregular polygons an efficient solution is given which is based on a parametric approach. This solution technique may be also applied for approximating a set of points in $\mathbb{R}^n$ by points of a rectangular grid. For solving the n-polygon problem an enumerative procedure is presented in which special structured linear programs are to be solved as subproblems. These linear programs are closely related to network flow problems.

Finally we consider problems in which the events are replaced by nonoverlapping processes (i.e. the vertices on the circle line are replaced by intervalls). The problem is to move the different pattern of intervalls on the circle line relative to each other such that the maximum overlapping is minimal.

A two commodity flow formulation for the vehicle routing problem

Martin Desrochers
Antoon Kolen
Abilio Lucena

The formulation used for the vehicle routing problem is given by

$$\text{minimize } (1/Q_0)\sum_{i,j=0}^{n} c_{ij}(x_{ij}^p + x_{ij}^q)$$

$$\text{s.t. } \sum_{j=0}^{n} x_{ij}^p - \sum_{j=0}^{n} x_{ji}^p = \begin{cases} \sum_{k=1}^{n} d_k & i=0 \\ -d_i & i=1,\dots,n \end{cases}$$

$$\sum_{j=0}^{n} x_{ij}^q - \sum_{j=0}^{n} x_{ji}^q = \begin{cases} -\sum_{k=1}^{n} d_k & i=0 \\ d_i & i=1,\dots,n \end{cases}$$

$$\sum_{j=0}^{n} (x_{ij}^p + x_{ij}^q) = Q_0 \quad i=1,\dots,n$$

$$\sum_{j=0}^{n} (x_{oj}^p + x_{oj}^q) = mQ_0$$

$$x_{ij}^p + x_{ij}^q \in \{0, Q_0\},\ x_{ij}^0, x_{ij}^q \geqslant 0,\ i,j = 0,1,\dots,n,$$

where m is the number of vehicles of capacity Q_0 located at the depot 0, d_i is the demand of client i, $i=1,\dots,n$ and c_{ij} is the transportation cost from i to j, $i,j=0,1,\dots,n$. The p-commodity is used to deliver d_i, to client i from the depot, the q-commodity is used to pick-up d_i from client i and deliver it to the depot. It is easy to show that the graph $G=(\{0,1,\dots,n\},A)$ with $(i,j)\in A \Leftrightarrow x_{ij}^p + x_{ij}^q = Q_0$ decomposes into m cycles with total demand in each cycle less than or equal to Q_0, i.e., this is a valid formulation for the vehicle routing problem.

The LP-relaxation of the formulation is used as a starting point for a cutting plane algorithm, the result of which is used as a lower bound in a branch and bound procedure.

HIERARCHISIERUNG VON RESTRIKTIONEN UND DER BALAS - ALGORITHMUS

Roland Dillmann
Wuppertal

Der Enumerationsprozeß beim Balas - Algorithmus läßt sich verkürzen, wenn die Einhaltung strategisch wichtiger Restriktionen bevorzugt gesichert wird. Dies leistet der Übergang von

$$\begin{array}{lcl} c^t x \to \min & & c^t x \to \min \\ Ax \leq b & \text{zu} & DAx \leq Db \\ x_i \in \langle 0, 1 \rangle & & x_i \in \langle 0, 1 \rangle \end{array}$$

mit $D = \mathrm{Diag}(g_1, \ldots, g_m)$. Dabei werden die Gewichte g_i so gewählt, daß die Einhaltung der bevorzugten Restriktion die Variablenauswahl beim Vorwärtsschritt entscheidend bestimmt. Heuristisch gewonnene Informationen und Schnitte können erfolgreich einbezogen werden. Dies führt zu einer Vergrößerung des Aufgabentyps, der mit dem Balas - Algorithmus behandelbar ist, da seine Anwendung sich auf kleinere Teilaufgaben beschränkt.

Anwendungsfeld: Aufgabentypen, bei denen 0 - Setzung einzelner Variabler 0 - Setzung vieler Variabler erzwingt, also Fixkostenprobleme, dezentrale Versorgungsnetze, Projektauswahlprobleme unter Beachtung zeitlicher Reihenfolgen.

Beispiel: Fixkostenproblem 10 von Haldi in binärer Form:
Ohne Hierarchisierung wurde das Verfahren nach 20000 Iterationen erfolglos abgebrochen. Hierarchisierung mit den Gewichtungsfaktoren $g_i = 1$, $1 \leq i \leq 4$, $g_i = 1000$, $7 \leq i \leq 10$ beendete das Verfahren nach 3724 Iterationen. Vorschaltung einer Heuristik, die den Zielwert 16 erbrachte, und der mit LP gewonnene Schnitt

$$x_1 + x_2 + x_3 + x_4 + x_5 + x_6 \leq 4$$

mit der Gewichtung $g_{11} = 500$ beendete das Verfahren nach 1976 Iterationen.

Literatur:

Trauth, C. A.
Woolsey, R.E. Integer Linear Programming: A Study in Computational Efficiency, Management Science, Vol. 15, Jg. 1969, S. 481ff.

Solution of a Tinned Iron Purchasing Problem using Lagrangean Relaxation

B. Dorhout , Enschede

SUMMARY

The following problem is considered. A tin factory obtains its material from steel works. It consists of rectangular sheets of tinned iron of various sizes and thickness of tinfoil, with several other specifications, like aftertreatment, temper, etc. The prices per unit of volume of the sheets vary with thickness, thickness of tinfoil, and width. They also depend on discounts, which are given for large quantities of the same sheets. As a consequence, it is often profitable to the factory to order sheets of larger sizes and/or thicker tin layers than needed. This is allowed, provided that sheets of a certain size can only be replaced by sheets of one size. Leftover pieces can be sold as scrap.

The question arose which sizes and quantities should be ordered each month in order to minimize net total costs. (The way in which parts of tins are cut out of the sheets, and lengths and widths of these sheets are considered to be fixed.)

A first step in solving this problem is to consider separate problems for those sheets, which only differ in size or tin layers. In order to solve the resulting problems one must first generate all useful combinations of sizes and tinlayers which can be ordered. These problems can be formulated as follows. Define

$J =$ set of indices of sheets wanted

$I =$ set of indices of sheets with generated sizes and tinfoils

$K =$ set of quantity discount classes

$R_i =$ set of indices $j \in I \cap J$ of sheets that can be replaced by sheets with index $i, i \in I$

$S_j =$ set of indices $i \in I$ of sheets that can replace sheets with index $j, j \in J$

$\ell_k =$ lower bound of quantity, in tons, for discount class $k, k \in K$

$c_{ik} =$ price per ton of sheets with index i and quantity discount class k, $i \in I, k \in K$

$a_{ij} =$ quantity of sheets, in tons, with index i, needed to replace sheets with index $j, j \in R_i, i \in I$

$p_{ij} =$ price of scrap to be sold if sheets with index i replace sheets with index $j, j \in R_i, i \in I$ and the variables

$x_{ik} =$ quantity, in tons, of sheets with index i, to purchase in quantity discount class $k, i \in I, k \in K$ quantity

$$w_{ik} = \begin{cases} 1 & \text{if sheets with index } i \text{ are purchased in quantity discount class } k, \\ & i \in I,\ k \in K \\ 0 & \text{otherwise} \end{cases}$$

$$y_{ik} = \begin{cases} 1 & \text{if sheets with index } i \text{ replace sheets with index } j,\ j \in R_i, i \in I \\ 0 & \text{otherwise} \end{cases}$$

Then it is asked to minimize

$$\sum_{i \in I} \sum_{k \in K} c_{ik} x_{ik} - \sum_{i \in I} \sum_{j \in R_i} p_{ij} y_{ij} \qquad (1)$$

subject to

$$\sum_{k\in K} x_{ij} \geqslant \sum_{j\in R_i} a_{ij}y_{ij} \quad i\in I \tag{2}$$

$$x_{ik} \geqslant \ell_k w_{ik} \quad i\in I, k\in K \tag{3}$$

$$x_{ik} \leqslant \ell_{k+1} w_{ik} \quad i\in I, k\in K \tag{4}$$

$$\sum_{k\in K} w_{ik} \leqslant 1 \quad i\in I \tag{5}$$

$$\sum_{i\in S_j} y_{ij} = 1 \quad j\in J \tag{6}$$

$$x_{ik} \geqslant 0 \quad i\in I, k\in K \tag{7}$$

$$y_{ij} = 0 \text{ or } 1 \quad j\in R_i, i\in I \tag{8}$$

$$w_{ik} = 0 \text{ or } 1 \quad i\in I, k\in K \tag{9}$$

Applying Lagrangean relaxation to this problem with respect to (6) solve, with v = vector of multipliers $v_j, j\in J$: minimize

$$\sum_{i\in I}\sum_{k\in K} c_{ik}x_{ik} - \sum_{i\in I}\sum_{j\in R_i} p_{ij}y_{ij} + \sum_{j\in J} v_j - \sum_{i\in I}\sum_{j\in R_i} v_j y_{ij} \tag{10}$$

subject to (2),...,(5), (7),...,(9).
It can be seen easily that this problem is the same as to calculate

$$f(v) = \sum_{j\in J} v_j - \sum_{i\in I} \max_{k\in K} \max \tag{11}$$

$$(\max\{\sum_{j\in R_i}(v_j + p_{ij} - c_{ik}a_{ij})y_{ij} \mid \sum_{j\in R_i} a_{ij}y_{ij} \geqslant \ell_k, y_{ij} = 0 \text{ or } 1,\ j\in R_i\},$$

$$\max\{\sum_{j\in R_i}(v_j + p_{ij})y_{ij} \mid \sum_{j\in R_i} a_{ij}y_{ij} \leqslant \ell_k, y_{ij} = 0 \text{ or } 1,\ j\in R_i\} - c_{ik}\ell_k)$$

In order to solve the original problem we maximize the lower bound $f(v)$ for (1) by subgradient techniques. After a number of experiments we decided to use rules, proposed by CAMERINI ET AL. (1975). These always gave good results within 50 steps in $\mathbf{R}^{|J|}$.
From (11) it can be seen that the calculation of $f(v)$ in principle boils down to the solution of a great number of knapsack problems. A closer look at (11) learns that this number can be reduced drastically by calculating bounds to the values of the objective functions from the *LP*-relaxations of the knapsack problems. Moreover, it appears that in much cases resemblances between knapsack problems can be used to increase efficiency.
After each calculation of $f(v)$ a feasible solution to (2),...,(9) is easily obtained: if the solution of (11) indicates that sheet j should be replaced by more than one sheet i, the most profitable one of them is chosen; the demand for an unassigned sheet j is satisfied by the cheapest sheet $j\in S_i$.
The method described above was programmed in Turbo-PASCAL on a micro computer, with satisfying results. In our experiments the largest difference between the highest lower bound and the value of the objective function of the best solution found, never exceeded 0,7%.

REFERENCES

CAMERINI, P.M., L. FRATTA & F. MAFFIOLI, "On improving relaxation methods by modified gradient techniques", Math. Progr. Study 3 (1975), 26-34.

A STOCHASTIC ASSIGNMENT APPROACH TO RESOURCE-CONSTRAINED MULTI-PROJECT SCHEDULING

Andreas Drexl, Darmstadt

Zusammenfassung: Untersucht wird ein Mehrprojekt-Kapazitätsplanungsproblem, bei dem Vorgangsdauern und -kosten von der beanspruchten Ressource abhängen. Begonnene Vorgänge dürfen nicht unterbrochen werden. Es stellt sich die Frage, wann welcher Vorgang von welcher Kapazitätsart kostenminimal bearbeitet werden soll, wobei projektspezifische Reihenfolgebeziehungen, projektabhängige Anfangs- und Endtermine sowie Kapazitätsrestriktionen einzuhalten sind. Das Problem kann als binäres Optimierungsproblem formuliert und (allerdings nur für kleinere Problemstellungen) mit Branch & Bound - Methoden exakt gelöst werden [6]. Zur heuristischen Lösung beschreiben wir daher einen (sehr effizienten) stochastischen Zuordnungsansatz, mit dem auch Probleme praktischer Größenordnung lösbar sind.

Summary: We study the nonpreemptive resource-constrained multi-project scheduling problem in which activity durations as well as costs are a function of the assigned resource. Regarding projectspecific precedence relations, individual release dates and deadlines per project as well as resource restrictions, the question arises, how and when each activity should be scheduled. The problem may be formulated in terms of a zero-one program. Problems of smaller dimensions can be solved to optimality by branch & bound methods [6]. In this paper we present a (highly efficient) stochastic assignment algorithm, which allows to solve large real world problems approximately.

A. Description of the problem

Project scheduling problems with resource constraints have been studied extensively in the past [1, 2, 8, 10]. In traditional approaches activities need a prespecified amount of some resources. We consider the case where the resources are substitutional (i.e. may be assigned alternatively to the activities):

Each activity must be scheduled requiring only one of the doubly-constrained resources (limited per period and total availability). Activity durations as well as costs depend upon the assignment of each activity to a resource. Which activity should be scheduled by which resource at minimum overall cost regarding precedence relations, resource constraints, and individual release dates as well as deadlines per project?

Conceptually this problem may be either seen as a generalized assignment problem or as a modification of the multi-mode resource-constrained scheduling problem considered in [3, 12]. Our modification is on the one hand a restrictive one, because it does not consider the resources in the most general way; on the other hand it is a generalizing one, because it deals with the multi-project case with individual dates per project.

Although a number of different objectives may be incorporated in our model, this paper focuses on the time-cost tradeoff variant, where ativity costs (durations) appear in the objective function (constraints). The main reason is, that the resulting model is highly suitable for dealing with audit scheduling problems [4], which indeed represent the most important area, where our approach is applicable.

Our model is based on the following assumptions: Within the planning horizon there have to be scheduled $i = 1,2,\dots,I$ projects. Project i has the earliest start time (release date) PE_i and the latest finish time PL_i. The time interval (planning horizon) ranging from $0 := \min\{PE_i \mid i=1,2,\dots,I\}$ to $T := \max\{PL_i \mid i=1,2,\dots,I\}$ is divided into T periods of equal length. Project i consists of activities a_{ij}, $j = 1,2,\dots,J_i$; precedence relations between some of the activities are specified in advance. There are $k = 1,2\dots,K$ doubly-constrained resources, each resource being available with one unit per period; total availability of resource k is D_k time units. Each activity must be scheduled by one resource; each resource can only schedule one activity per period; activities once initiated cannot be interrupted (preempted) by another activity. Scheduling activity j of project i by resource k requires d_{ijk} time periods (duration); the corresponding scheduling costs are c_{ijk} units.

B. Formulation of the problem

We don't know in advance the resource k by which activity j of project i will be scheduled. Thus we get the maximum time interval, within which a_{ij} may be scheduled by any resource k, by taking $\delta_{ij} := \min\{d_{ijk} \mid k=1,2\dots,K\}$ and performing a traditional critical path analysis.

More precisely we determine the critical path earliest start time ES_{ij} of each activity with respect to PE_i by using δ_{ij} for all $j = 1,2,\dots,J_i$. Analogously we determine the critical path latest finish time LF_{ij} of each activity with respect to PL_i by using δ_{ij} once more. Thus $[ES_{ij}, LF_{ij}]$ represents the maximum time interval, within which a_{ij} may be scheduled without violating dates. Defining variables

$$x_{ijkt} := \begin{cases} 1, & \text{scheduling of } a_{ij} \text{ by k is finished in t} \\ 0, & \text{otherwise} \end{cases}$$

a binary program using the general framework given in [9] may be formulated:

$$\text{Minimize } Z(\underline{x}) = \sum_{i=1}^{I} \sum_{j=1}^{J_i} \sum_{k=1}^{K} c_{ijk} \sum_{t=ES_{ij}+d_{ijk}}^{LF_{ij}} x_{ijkt} \tag{1}$$

Subject to

$$\sum_{k=1}^{K} \sum_{t=ES_{ij}+d_{ijk}}^{LF_{ij}} x_{ijkt} = 1 \qquad \forall\ i,j \tag{2}$$

$$\sum_{k=1}^{K} \sum_{t=ES_{ih}+d_{ihk}}^{LF_{ih}} t\, x_{ihkt} \le \sum_{k=1}^{K} \sum_{t=ES_{ij}}^{LF_{ij}-d_{ijk}} (t-d_{ijk})\, x_{ijkt} \qquad (3)$$

$$\forall\; i,j \text{ and } [i,h] \in \Psi_{ij}$$

$$\sum_{i=1}^{I} \sum_{j=1}^{J_i} \sum_{q=t}^{t+d_{ijk}-1} x_{ijkq} \le 1 \quad \forall\; k,t \qquad (4)$$

$$\sum_{i=1}^{I} \sum_{j=1}^{J_i} \sum_{t=ES_{ij}+d_{ijk}}^{LF_{ij}} d_{ijk}\, x_{ijkt} \le D_k \quad \forall\; k \qquad (5)$$

$$x_{ijkt} \in \{0,1\} \quad \forall\; i,j,k,t \qquad (6)$$

In (1) $\underline{x}$ represents the vector of all binary variables, $Z(\underline{x})$ represents the objective function value for any feasible vector $\underline{x}$ with respect to (2)-(6). (2) are the activity completion constraints. (3) ensures that all precedence relations are maintained ([i,h] denotes the indices of a_{ih} and Ψ_{ij} denotes the set of immediate predecessors of [i,j]). (4) resp. (5) represent per period resp. total availability resource constraints.

C. A stochastic scheduling procedure

Traditional scheduling rules [7, 8] quite often do not find an (existing) feasible solution in the presence of time and resource restrictions. Therefore a more sophisticated method like the following should be used in this context. (A review of existing solution procedures can be found in [6]).

The algorithm successively constructs precedence and resource feasible partial schedules (solutions). "Partial" indicates that fewer than $\sum_{i=1}^{I} J_i$ activities have currently been scheduled (corresponds to "$\le$" instead of "=" in (2)) before terminating the solution process. Scheduling of one more activity is equivalent to augmenting the partial feasible solution.

Activities currently being unscheduled are selected as candidates, if all predecessors have been scheduled, and if the earliest start time is greater than or equal to the time t (:=0,1,...,T) of the simulation clock. Starting with t:=0 the simulation clock is increased successively. More formerly we obtain the set of candidates IJC as follows:

$IJ0 := \{[i,j] \mid a_{ij} \text{ currently is unscheduled}\}$

$IJ1 := \{[i,j] \mid a_{ij} \text{ currently is scheduled}\}$

$$IJC := \{[i,j] \in IJ0 \mid ES_{ij} \leq t, \forall_{ij} \in IJ1\} \tag{7}$$

Denoting with AK the <u>set of resources</u> which are <u>available</u> in t (assigned to an activity only until t-1; enough left-over capacity), the following <u>opportunity costs</u> may be calculated:

$$\omega_{ijk} := \max_{q \in AK} c_{ijq} - c_{ijk} \quad \forall \ [i,j] \in IJC \text{ and } k \in AK \tag{8}$$

ω_{ijk} compares the cost of scheduling a_{ij} by resource k with the worst-case consequence if k would be unavailable. In this sense it suggests itself to take ω_{ijk} in order to decide which available resource should be assigned to which candidate activity.

In the following we will take ω_{ijk} (slightly modified) as stochastic assignment probabilities. Taking ω_{ijk} corresponding to (8) all resources with highest scheduling cost would become an assignment probability of zero - a misleading consequence in the case of scarce resources and tight times. In order to overcome this difficulty we set

$\omega_{min} := \min\{\omega_{ijk} > 0 \mid \forall \ [i,j] \in IJC \text{ and } k \in AK\}$ and calculate

$$\Omega_{ijk} := (\omega_{ijk} + \omega_{min})^{\alpha} \quad \forall \ [i,j] \in IJC \text{ and } k \in AK \tag{9}$$

Taking Ω_{ijk} as <u>stochastic assignment probabilities</u> we get an arbitrarily large range of <u>stochastic scheduling rules</u> for $\alpha \geq 0$.

In general one does not know in advance the tightness of resources and dates; therefore one should start with an appropriate high value of α in order to hopefully get a near-optimum solution. Some trials failing to construct a feasible solution, α should be decreased and the solution process should be repeated once more.

There do exist a lot of possibilities for stochastically assigning resources to activities. For example a linear sum assignment problem [5] may be formulated taking Ω_{ijk} as cost coefficients, starting an algorithm and stopping before reaching optimality. Due to unavailability of appropriate stopping criteria for linear sum assignment algorithms we choose a pragmatic way of successively assigning resources to activities. Defining some other symbols

n	number of scheduled activities ($\lvert IJ1 \rvert$)
DK	array of length K (consumption of resource k)
FK	array of length k (availability date of resource k)

the stochastic assignment algorithm can be fully described as follows:

Stochastic Assignment Procedure

```
Initialize:  n := t := 0; DK(k) := FK(k) := 0 ∀ k;
IJ1 := Φ ; IJ0 := {[i,j]| i=1,2,...,I; j=1,2,...,J_i}.

Perform:
Determine ES_ij by traditional critical path analysis (using δ_ij).
M1: IJC := {[i,j] ∈ IJ0| ES_ij ≤ t, ψ_ij ∈ IJ1};
AK := {k| FK(k) ≤ t, DK(k) < D_k  ∀  k};
if |IJC| > 0 then
begin if |AK| > 0 then
  begin calculate Ω_ijk according to (9) ∀ [i,j] ∈ IJC and k ∈ AK (if
    DK(k) + d_ijk > D_k set Ω_ijk := 0);
    if Ω_ijk = 0 ∀ [i,j] ∈ IJC and k ∈ AK then goto M2;
    choose [β,γ] ∈ IJC and π ∈ AK randomly with probabiblity
    proportional to Ω_ijk;
    set n := n+1; DK(π) := DK(π) + d_βγπ ;
    IJ0 := IJ0 - [β,γ]; IJ1 := IJ1 ∪ [β,γ]; FK(π) := d_βγπ ;
    store the partial feasible schedule;
    update ES_ij for all successors of [β,γ];
                 I
    if   n  =    Σ  J_i    then stop else goto M1
                i=1
  end
  M2:  τ := max{0,min{FK(k)>0| ∀ k with DK(k) < D_k }};
  if  τ = 0 then stop else
  begin FK(k) := max{0,FK(k)-τ} ∀ k; set t := t + τ;
    update ES_ij  ∀ [i,j] ∈ IJ0;
    goto M1
  end
end
τ := min{ES_ij| [i,j] ∈ IJ0, ψ_ij ∈ IJ1};
FK(k) := max{0,FK(k)-(τ-t)}  ∀ k; set t := τ;
goto M1
```

Terminate: The procedure stops, if either all activities have been scheduled (feasible solution found) or if there is no resource with enough left-over capacity.

* * * * *

In some cases it may be interesting to restrict the set of candidate activities to those elements of IJC with minimum slack time. Denoting with LS_{ij} the latest start time of a_{ij} (calculated analogously to ES_{ij} and LF_{ij}) and with $S_{ij} := LS_{ij} - ES_{ij}$ the corresponding slack time we get an alternative activity candidate set as follows:

$$IJCS := \{[i,j] \in IJC \mid S_{ij} = \min\{S_{\beta\gamma} \mid [\beta,\gamma] \in IJC\}\} \tag{10}$$

Using IJCS instead of IJC in the above procedure may - in some cases - reveal an algorithmic variant with a better performance.

D. Computational results

The stochastic assignment procedure has been coded in FORTRAN, implemented on an IBM PC AT 02 and tested on 8 groups of problems, consisting of 10 problems each, each with a structure similar to the one used in [4, 11].

Table 1 summarizes the results. Each problem has been solved 50 times calling the above procedure using (a fairly large) $\alpha = 2,0$. In each group in at least one case the optimum solution (determined with the method of [6]) has been found regardless of using IJC or IJCS. The percentage deviation from the optimum solution is in the worst-case - once more for IJC and IJCS - less than 5%. The mean percentage deviation is in both cases quite good and does not exceed 1%. The CPU-time of the algorithm ranges from 0,04 to 0,14 sec on the average per execution.

Figure 1 demonstrates the effect of using different values for α. A problem out of the last group has been solved repeatedly using IJC until f = 100 feasible solutions had been found. Using $\alpha = 0,5$ the number of trials (executions of the algorithm) was 105 (case (a)). The cases (a)-(d) indicate that with increasing α (i) more trials are necessary in order to generate a prespecified number of feasible solutions and (ii) the percentage deviations of the objective function values from the optimal one are decreasing.

Thus we observe that the probability of finding a feasible solution is decreasing with increasing α - the quality of solutions found increasing as well.

E. Final remarks

As to the restricted computational experience the stochastic assignment approach seems to be a highly suitable method for approximately solving time-cost tradeoff problems of the type considered in this paper.

I	$J_i(\forall i)$	IJC (%) Min	IJC (%) Mean	IJC (%) Max	IJCS (%) Min	IJCS (%) Mean	IJCS (%) Max	CPU
1	10	0,0	0,5	3,5	0,0	0,4	3,3	0,04
	20	0,0	0,4	3,4	0,0	0,5	3,6	0,10
2	5	0,0	0,7	3,8	0,0	0,8	3,7	0,07
	10	0,0	0,8	3,9	0,0	0,8	3,8	0,11
3	4	0,0	0,8	4,2	0,0	0,8	3,6	0,09
	6	0,0	0,6	3,4	0,0	0,9	4,7	0,13
4	4	0,0	0,6	3,6	0,0	0,7	4,4	0,12
	5	0,0	0,7	4,5	0,0	0,6	3,8	0,14

Table 1: Summary of results

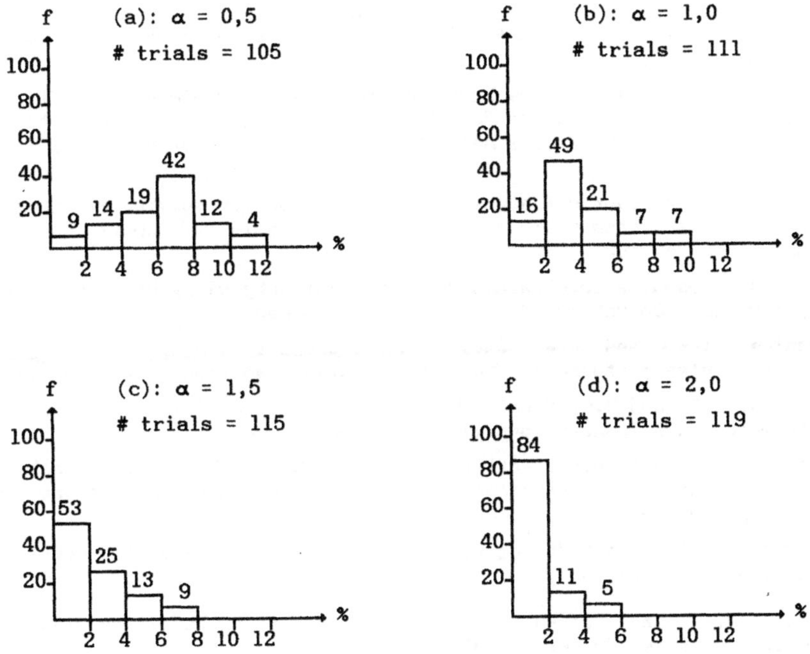

Figure 1

The structure of a specific problem may be characterized by the number of projects / activities, the number and tightness of the resources, the tightness of release dates / deadlines and the precedence structure. Depending upon this structure a variant of the above algorithm (using IJC or IJCS; choice of α) should be most suitable. Which variant to take for which problem structure cannot be suggested based on theoretical considerations. A large numerical study analogously to [8] should answer this question.

The fundamental idea of the algorithm, to resolve conflicts between activities (which are competing for scarce resources) by Monte-Carlo methods on the basis of opportunity costs, is a quite general one. It should be worth-while to investigate its usefulness for solving other hard combinatorial problems approximately, too.

References

[1] Baker, K.R.: Introduction to sequencing and scheduling, New York 1974.

[2] Bartusch, M.: Optimierung von Netzplänen mit Anordnungsbeziehungen bei knappen Betriebsmitteln, Diss. Aachen 1983.

[3] Blazewicz, J.; W. Cellary; R. Slowinski and J. Weglarz: Scheduling under resource constraints - deterministic models, Basel 1986 (Annals of Operations Research, Vol. 7).

[4] Bolenz, G. and R. Frank: Das Zuordnungsproblem von Prüfern zu Prüffeldern unter Berücksichtigung von Reihenfolgebedingungen - Ein Lösungsansatz der binären Optimierung. Zeitschrift für betriebswirtschaftliche Forschung 29 (1977), pp. 427-447.

[5] Derigs, U.: The shortest augmenting path method for solving assignment problems - motivation and computational experience. Annals of Operations Research 4 (1985/6), pp. 57-102.

[6] Drexl, A.: Resource-constrained multi-project scheduling by bounded enumeration. Symposium über Operations Research 1987 in Passau.

[7] Kurtules, I.S. and E.W. Davis: Multi-project scheduling: categorization of heuristic rules performance. Management Science 28 (1982), pp. 161-172.

[8] Kurtulus, I.S. and S.C. Narula: Multi-project scheduling: analysis of project performance. IIE Transactions 17 (1985), pp. 58-66

[9] Pritsker, A.A.B.; W.D. Watters and P.M. Wolfe: Multiproject scheduling with limited resources: a zero-one programming approach. Management Science 16 (1969), pp. 93-108.

[10] Radermacher, F.J.: Scheduling of project networks. Annals of Operations Research 4 (1985/6), pp. 227-252.

[11] Rohde, M.: Das Set-Partitioning-Problem: Wirtschaftliche Anwendungen und Algorithmen, Diss. FU Berlin 1978.

[12] Talbot, F.B.: Resource-constrained project scheduling with time-resource tradeoffs: the nonpreemptive case. Management Science 28 (1982), pp. 1197-1210.

CUTTING PLANES FOR THE SYMMETRIC TRAVELLING SALESMAN PROBLEM

Bernhard Fleischmann, Hamburg

We present a new class of cutting planes for the symmetric travelling salesman problem (TSP) which contains the known facetial cutting planes as special cases, i.e. the comb and clique tree inequalities [3], the path and wheelbarrow inequalities [1] and the star constraints [2] and, in addition, a large set of new cutting planes. Conditions for the cutting planes to be facetial are discussed. In many examples, these cutting planes are sufficient to define the optimal solution or to yield a very sharp lower bound, such as for Groetschel's 442-city-problem. Possible ways of using them in a dual cutting plane algorithm are presented, based on heuristic search procedures.

References

[1] G. Cornuéjols, J. Fonlupt, D. Naddef, The travelling salesman problem on a graph and some related integer polyhedra. Math. Progr. 33 (1985) 1 - 27.

[2] B. Fleischmann, A new class of cutting planes for the symmetric travelling salesman problem. To appear in: Math. Progr.

[3] M. Grötschel, M.W. Padberg, Polyhedral theory, in: E.L. Lawler et al. (eds), The travelling salesman problem (J. Wiley & Sons, 1985), 251-305.

SURVEY OF SOLVED AND OPEN PROBLEMS IN THE DEGENERACY PHENOMENON*)

Tomas Gal , Hermann-Josef Kruse , Peter Zörnig, Hagen

1. Introduction

Degeneracy is a phenomenon that may arise, e.g., in linear programming (LP for short), bottleneck LP, multiparametric LP, linear vectormaximization, etc. If it does arise then it certainly influences any vertex-searching method for mathematical models based on a system of linear inequalities and in some cases it leads to misinterpretation of optimal solutions.

Consider a vertex $x^o \in X$, $X := \{x \in \mathbb{R}^n \mid Ax \leq b,\ x \geq 0\}$, $A \in \mathbb{R}^{m \times n}$, $b \in \mathbb{R}^m$. Transforming X by means of slack variables $u_i = x_{n+i} \geq 0$, $i = 1, \dots, m$, into $\bar{X} := \{y \in \mathbb{R}^{m+n} \mid \bar{A}y = b,\ y \geq 0\}$, $y = \binom{x}{u}$, $\bar{A} = (A \mid I_m)$, a basis B of $\bar{A}$ is one-to-one assigned to x^o, provided that x^o is nondegenerate.

If $x^o \in X$ is degenerate, i.e., it is overdetermined (more than n hyperplanes pass through x^o), a set of bases $B^o := \{B_u^o \mid u = 1, \dots, U\}$, $U > 1$, is associated with x^o. Denote by $x_B^{(o)} \in \mathbb{R}^{m+n}$ a complete basic feasible solution of $\bar{A}y = b$, $y \geq 0$, i.e. $x_B^{(o)}$ and x^o are assigned uniquely to each other. Further denote by $\sigma \in \{1, \dots, m\}$ the degeneracy degree of x^o, i.e., σ is the number of basic variables with the value zero in $x_B^{(o)}$. In order to get an idea about the magnitude of the possible maximal number of bases that can be associated with a σ-degenerate vertex $x^o \in \mathbb{R}^n$ see Table 2.1 (below).

From the historical point of view, in 1976/77 the problem arose to find all or a part of neighbouring (adjacent) vertices of a given degenerate vertex $x^o \in X$ with minimal effort [10]. We tried then to approach this problem from a graph-theoretical point of view.

Assuming $X \neq \emptyset$, and, for simplicity, bounded, the set X defines obviously a convex polytope. Suppose there is no degenerate vertex $x \in X$. Then it is possible to represent any convex polytope by the so called graph of the polytope, $G'(X)$, [20].

*) Supported by the DFG Grant No. 335/1-1-II A3

Operations Research Proceedings 1987

The nodes of $G'(X)$ correspond to vertices of X (or: to bases of $\bar{X}$) and every edge of $G'(X)$ is one-to-one assigned to an edge (1-face) of X. Any basis B can be also represented by a simplex-type tableau (without the reduced costs row). "Travelling" from a vertex $x^i \in X$ (associated with the basis B_i) to another vertex $x^{i'} \in X$, $i \neq i'$, (associated with the basis $B_{i'}$) along an edge $(x^i, x^{i'}) \subset X$ corresponds to a unique basis-exchange B_i to $B_{i'}$ (or vice versa). In tableau representation this corresponds to a Gauss-Jordan elimination step with a positive pivot (because of $y \geq 0$). Denote this exchange by $B_i \overset{+}{\longleftrightarrow} B_{i'}$.

Suppose now that x^o is degenerate. Then the graph of the polytope, $G'(X)$, misses essential information about the degeneracy structure of x^o. We therefore introduced a so called (positive or proper) <u>degeneracy graph</u> (DG for short), $G_+^o := G_+^o(x^o)$, the nodes of which correspond to the bases B_u^o associated with x^o and the edges of which are defined as $\{B_u^o, B_{u'}^o\}$, $u, u' = 1, \ldots, U$, $u \neq u'$, iff $B_u^o \overset{+}{\longleftrightarrow} B_{u'}^o$. Embedding G_+^o into $G'(X)$ we obtain the so called <u>representation graph</u>, $G(X)$, of the polytope X. If $U = 1$ then $G'(X) = G(X)$.

Consider as an example a cube $X := \{x \in \mathbb{R}^3 \mid x_1 \leq 1, x_2 \leq 1, x_3 \leq 1, x_{1,2,3} \geq 0\}$ (Fig. 1.1a) and its graph representation (Fig. 1.1.b).

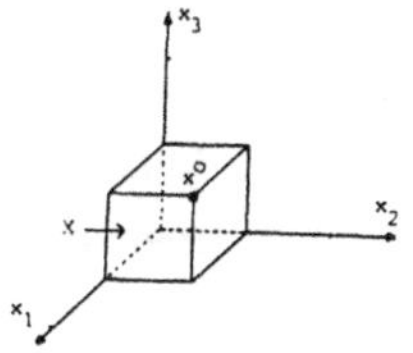

Fig. 1.1.a

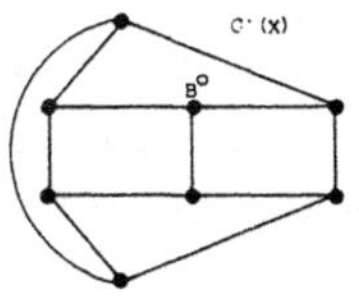

Fig. 1.1.b

Let us add the constraint $x_1 + x_2 + x_3 \leq 3$ which preserves X but makes x^o overdetermined (Fig. 1.2a). Symbolizing by a frame the DG G_+^o of x^o we obtain the representation graph $G(X')$ of the "new" cube $X' := \{x \in \mathbb{R}^3 \mid x_1 \leq 1, x_2 \leq 1, x_3 \leq 1, x_1 + x_2 + x_3 \leq 3, x_{1,2,3} \geq 0\}$ (Fig. 1.2b).

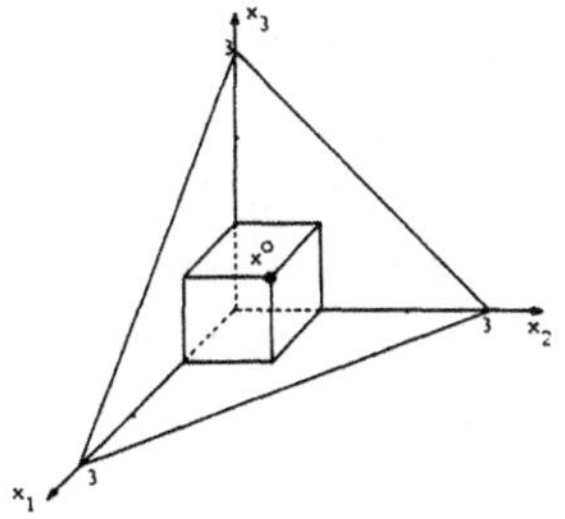

Fig. 1.2a

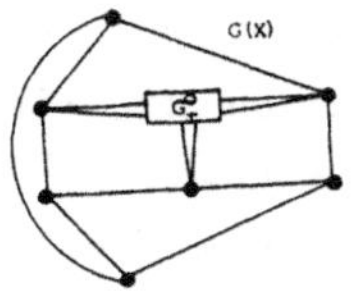

Fig. 1.2b

Let us have a closer look at a hypothetical graph G_+^o (Fig. 1.3). Hypothetical because G_+^o of x^o in the above cube X' would be a too simple graph with only 4 nodes. The nodes in frames are the so called transition nodes with the property that they connect G_+^o with an "outer" node, i.e., with a node belonging to G(X) but not to G_+^o. The remaining nodes of G_+^o are called internal nodes.

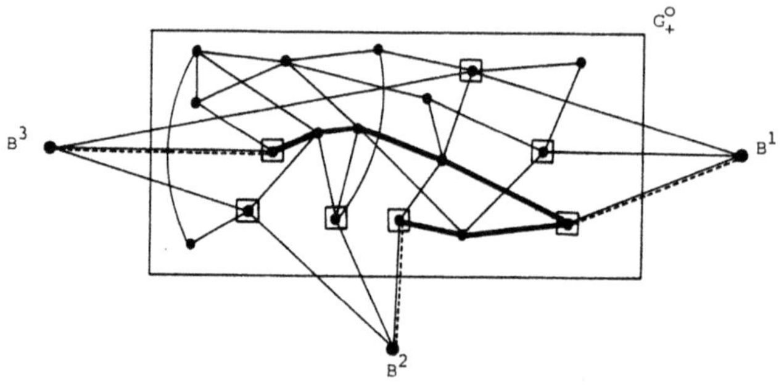

Fig. 1.3

In this paper we shall give a concise survey of results concerning the theory and some application of the DG's.

2. Some theoretical results

We proved [10, 12] that in G_+^o there exists a tree $\bar{G}^o \subset G_+^o$ that connects x^o with all its neighbours. In Fig. 1.3 such a tree is illustrated by bold lines. Thus having found such a tree the initial problem to find all neighbours of a degenerate x^o is solved.

In order to determine efficiently a tree $\bar{G}^o$ we elaborated a method called the N-tree-algorithm (N for neighbour) [15, 17]. It is a special tree-algorithm which is combined with a lexicographic selection rule. We found that the degree of the transition nodes might play a role in selecting the "best" starting node for the N-tree-algorithm. This might lead to the determination of a so called minimal tree $\bar{G}^o_{min}$ (a tree with minimal number of nodes). We conjectured that the greatest integer smaller or equal to $\frac{U}{\sigma + 1}$ is an upper bound on the number of nodes needed to form a minimal tree. A series of tests is running with a corresponding subroutine (find a transition node with the maximal degree as starting node). Recent computing results confirm the above conjecture and they show a possible direction for further improvement of the procedure [16]. We hope for new clues to determine $\bar{G}^o_{min}$ from further research into the theory of DG's.

We enlarged our research to DG's in which the edges can be defined via basis-exchange using negative pivots. Allowing negative pivots only we called the corresponding DG the <u>negative DG</u>, G_-^o. If any nonzero pivot is admissible we call the corresponding DG the <u>general DG</u>, G^o.

We found [27] that the general DG is always connected, while the proper or negative DG's may be unconnected.

Investigating the bounds of the number U of nodes B_u^o of B^o the upper bound is obviously

$$U_{max} = \binom{n+\sigma}{\sigma}$$

and we found ([27], p. 50) that the lower bound is

$$U_{min} = 2^{\sigma-1}(n-\sigma+2).$$

Currently a series of empirical tests is running in order to find $U \in [U_{min}, U_{max}]$ in real cases. The preliminary results show that U is not much larger than U_{min} even though U_{min} is itself an impressing number (Tab. 2.1).

Table 2.1: U_{min} associated with a σ-degenerate vertex $x^o \in \mathbb{R}^n$ for some selected values of σ and n in comparison with U_{max}

n	σ	$U_{min} = 2^{\sigma-1}(n-\sigma+2)$	$U_{max} = \binom{n+\sigma}{\sigma}$
5	3	16	56
10	5	112	3,003
50	5	752	$3.48\cdot10^{6}$ 1)
50	10	21,504	$1.62\cdot10^{17}$ 1)
50	40	$1.319\cdot10^{13}$ 1)	$5.99\cdot10^{25}$ 1)
100	30	$3.865\cdot10^{10}$ 1)	$2.61\cdot10^{39}$ 1)
100	50	$2.93\cdot10^{16}$ 1)	$2.01\cdot10^{40}$ 1)
100	80	$1.33\cdot10^{25}$ 1)	$3\cdot10^{52}$ 1)

We studied the general structure of DG's and we started with the construction principle for the case $\sigma \leq 2$ [17, 34]. The results are summarized in

Theorem 2.1
A graph G is a 2xn-DG (σxn, σ = 2, $x \in \mathbb{R}^n$) iff G isomorphic to a line graph

(*) $$L(K(p_1, \ldots, p_r))$$

---

1) These values are rounded

where r, $p_1, \ldots, p_r$ are positive integers with $r \geq 2$, $\sum_{i=1}^{r} p_i = n + 2$ ($p_1, p_2 > 1$ for $r = 2$).

The properties of a 2xn-DG are summarized in

Theorem 2.2
The 2xn-DG (*) has the following properties:
(i) the diameter is ≤ 2
(ii) the number of nodes is

$$\frac{1}{2}((n+2)^2 - \sum_{i=1}^{r} p_i^2)$$

(iii) the number of edges is

$$\sum_{i=1}^{r} p_i \binom{n+2-p_i}{2}$$

(iv) the connectivity is

$$2n + 2 - p_{r-1} - p_r.$$

For $\sigma = 1$ the DG's are complete graphs with n + 1 nodes ([27], p. 116). Further investigations into the structure of DG's with $\sigma > 2$ are currently being performed.

3. Degeneracy graphs and linear programming

3.1 Linear programming and simplex-cycling

Consider the linear program

$$\text{(LP)} \qquad \max_{x \in X} c^T x, \quad c \in \mathbb{R}^n$$

Speaking about the connection between (LP) and degeneracy the first idea comming to one's mind is simplex-cycling [4, 21]. Several anticycling methods have been developed during the past 35 years (see, e.g., [3, 5, 7, 33]) since Charnes [6] published the first known perturbation scheme in 1952. Anticycling devices are embedded into professional LP-software despite a discussion in the literature [18, 25, 26, 29, 32] on whether cycling appears in real-world applications or not. In our opinion more attention should be paid to the connection between the structure of the matrix (A|b) enlarged by the row $(c^T, 0)$ and cycling than to try to improve or invent new anticycling methods. In other words, (A|b) implies B^o and hence induces G_+^o. And in G_+^o there exist in general closed lines (circuits for short - see also Fig. 1.3). The question is then to identify the properties of G_+^o that cause simplex cycling.

We approached this problem from two points of view. Firstly, using known properties of the DG's, secondly, introducing new conceptions. For the first approach it was necessary to enlarge the notion of a DG by the so called LP-DG [35]. We then found that a circuite of G^o_+ is a simplex cycle iff the circuite can be enlarged to a so called star-shaped graph embedded into the LP-DG (Fig. 3.1); the depicted graph is based on the Beale cycling example [4].

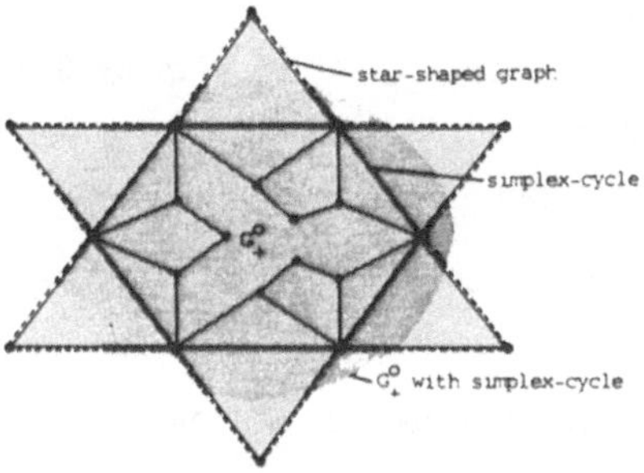

Fig. 3.1

The second approach is not exclusively based on the theory of the DG's. Given (A, b, c) and a circuite C of G^o_+; then it is possible to induce in a certain way a point-set M(C) such that C is a simplex-cycle iff $c \in M(C)$, where c stems from c^Tx [35]. We are currently trying to elaborate a method which solves the problem: given (A, b, c) and a degenerate vertex x^o. Determine whether $c \in M(C)$ for at least one C of G^o_+. If so, simplex-cycling occurs. We are in the very beginning of finding an efficient method to solve this problem.

3.2 Degeneracy in an optimal solution

Suppose that (LP) has an optimal solution with optimal degenerate vertex x^o. Let us select those nodes $\widetilde{B}^o_k$ of B^o which are associated with optimal bases. In general only a part $\widetilde{B}^o \subseteq B^o$ forms the set of optimal bases associated with x^o. The subgraph $\widetilde{G}^o \subset G^o$ induced by $\widetilde{B}^o$ is then called an o-DG (o for optimal). We studied various properties of $\widetilde{G}^o$.

We first investigated the number of nodes in $\widetilde{B}^o$. We proved [28] some results which are summarized in

Lemma 3.1

(i) There exists the case that $\widetilde{B}^o = B^o$

(ii) There is a triple (A, b, x^o) to which no objective function $z = c^Tx$ can be assigned such that $\widetilde{B}^o = B^o$.

(iii) There exists the case that one and only one basis of x^o is an optimal basis.

(iv) For the case that one and only one basis of x^o is an optimal basis the necessary condition is that $\tilde{G}_-^o$ has at least one isolated node.

Denote by $\tilde{G}_+^o$ and $\tilde{G}_-^o$ the positive and negative o-DG's, respectively.

We studied the connectivity properties of the o-DG's [28] and found - among other things - that a simultaneous dual degeneracy plays an important role in this connection which corresponds to very recent results by Greenberg [19]. Our results [28] are summarized in

Lemma 3.2

(i) Two optimal bases $\tilde{B}_k^o$, $\tilde{B}_{k'}^o \in \tilde{B}^o$, $k \neq k'$, for which $\tilde{B}_k^o \longleftrightarrow \tilde{B}_{k'}^o$, are always dual degenerate.

(ii) $\tilde{G}_+^o$ is trivial (i.e., any component of $\tilde{G}_+^o$ consists of exactly one node) iff there is no dual degeneracy

(iii) $\tilde{G}_-^o$ can be unconnected

3.3 Sensitivity analysis under degeneracy

Sensitivity analysis with respect to the right hand side ("RHS-ranging") or with respect to the objective function coefficients ("Cost-ranging") became nowadays a constituent part of commercial LP-software. However, in case of degeneracy the corresponding subprograms yield erroneous results [9, 13, 24].

Performing sensitivity analysis in a nondegenerate case means [11] considering a scalar parameter λ, e.g., in the RHS, and determining the parameter interval Λ such that for all $\lambda \in \Lambda$ the optimal basis associated with the optimal vertex $x \in X$ does not change (remains optimal). However, if $x^o \in X$ is a degenerate optimal vertex the set $\tilde{B}^o$ is assigned to x^o. What should then be invariant? In [9, 13, 24] it is suggested to determine Λ_k^o for each $\tilde{B}_k^o$ associated with x^o. Then the parameter interval is the union $\Lambda = \bigcup_{k=1}^{K} \Lambda_k^o$, K being the number of all bases in $\tilde{B}^o$. Changing the RHS implies, however, that $\tilde{B}_u^o \in \tilde{B}^o$ remains optimal.

We propose, therefore, the following formulation for sensitivity analysis with respect to the RHS b under degeneracy:

Determine Λ such that for all $\lambda \in \Lambda$ at least one basis $\tilde{B}_u^o \in \tilde{B}^o$ remains optimal.

With respect to the cost ranging the corresponding formulation reads:

Determine T (the overall parameter interval) such that for all $t \in T$ at least one $\tilde{B}^o_u \in \tilde{B}^o$ remains optimal, or equivalently, x^o remains the optimal vertex.

We are studying the economic impact of these formulations. An interesting contribution to these questions is found in [19]. The above "definitions" can easily be enlarged to the case with parameter-vectors.

3.4 Shadow prices under degeneracy

The determination of shadow prices under (primal) degeneracy is closely related to sensitivity analysis. It is known (see, e.g., [1, 2, 8, 9, 19, 23, 31]) that under primal degeneracy there does not exist the shadow price of a resource b_i. It has been proved that in such a case there exist two shadow prices, one for "buying one unit" and one for "selling one unit" of b_i. For details see [13]. Hence, the problem is not any more the existence of shadow prices under (primal) degeneracy but how to determine them. In [24] methods have been proposed, unfortunately, without complete proofs.

Therefore, we are currently investigating the questions [30]: If for instance the o-DG $\tilde{G}^o_-$ is not connected (see Lemma 3.2 (iii)) is it sufficient to consider only one component (maximal connected subgraph) of $\tilde{G}^o_-$? If yes, which of the components should be chosen? Or does it not matter which one?

3.5 Degeneracy and redundant constraints

Degeneracy is closely related to weakly redundant constraints. A weakly (w. for short) redundant constraint passing through $x^o \in X$ causes its overdetermination and thus degeneracy. Hence w. redundancy is a sufficient condition for degeneracy because the latter is not necessarily caused by w. redundancy only. Concerning the sources of degeneracy, compare also [19].

From a formal point of view it is known [22] that a w. redundant constraint can be omitted without influencing the set X. If degeneracy is caused by w. redundancy would then this circumstance not simplify sensitivity analysis and the determination of shadow prices?

We found, that in case of sensitivity analysis with respect to b and shadow prices it is not possible to omit the w. redundant constraints and then determine the shadow price or the parameter interval in the usual sense. We proved [14] that omitting the w. redundant constraints leads to false results in general. This is not true for sensitivity analysis with respect to c.

This is, incidentally, in contrast to formal methods for determining redundant constraints [22]. These methods aim to determine the redundancies in order to omit them. Hence, the number of constraints is reduced which reduces the CPU time for computing an optimal solution of (LP). However, omitting the w. redundant constraints may heavily influence the economic interpretation of the optimal solution, the determination of shadow prices and sensitivity analysis with respect to b.

References

[1] AKGÜL, M.: A Note on Shadow Prices in Linear Programming. J. of ORS 35, 1984, 425 - 431.

[2] AUCAMP, D. C., STEINBERG, D. I.: The Computation of Shadow Prices in Linear Programming. J. of ORS 33, 1982, 557 - 565.

[3] AVIS, D., CHVÁTAL, V.: Notes on Bland's Pivoting Rule. Math. Progr. 8, 1978, 24 - 34.

[4] BEALE, E. M. L.: Cycling in the Dual Simplex Algorithm. Naval Res. Log. Quart. 2, 1955, 269 - 276.

[5] BLAND, R. G.: New Finite Pivoting Rules for the Simplex Method. Math. of Oper. Res. 2, 1977, 103 - 107.

[6] CHARNES, A.: Optimality and Degeneracy in Linear Programming. Econometrica 20, 1952, 160 - 170.

[7] DANTZIG, G. B., ORDEN, A., WOLFE, P.: The Generalized Simplex Method for Minimizing a Linear Form under Linear Inequalities. Pacific J. of Math. 5, 1955, 183 - 195.

[8] EILON, A., FLAVELL, R.: Note on "Many-Sided Shadow Prices". OMEGA 2, 1974, 821 - 823.

[9] EVANS, J. R., BAKER, N. R.: Degeneracy and the (Mis-) Interpretation of Sensitivity Analysis in Linear Programming. Decision Sciences 13, 1982, 348 - 354.

[10] GAL, T.: Determination of all Neighbours of a Degenerate Extreme Point in Polytopes. Disc. Paper No. 17B, FernUniversität Hagen, 1978.

[11] GAL, T.: Postoptimal Analyses, Parametric Programming, and Related Topics. McGraw Hill,New York 1979.

[12] GAL, T.: On the Structure of the Set Bases of a Degenerate Point. JOTA 45, 1985, 577 - 589.

[13] GAL, T.: Shadow Prices and Sensitivity Analysis in Linear Programming under Degeneracy - A State-of-the-Art Survey. OR Spektrum 8, 1986, 59 - 71.

[14] GAL, T.: Degeneracy and Redundancy in Linear Programming. In Preparation, Fernuniversität Hagen, 1987.

[15] GAL, T., KRUSE, H.-J.: Ein Verfahren zur Lösung des Nachbarschaftsproblems. Oper. Res. Proceedings 1984, Springer Verlag 1985, 447 - 454.

[16] GAL, T., KRUSE, H.-J.: An Improved Method to Solve the Neighbourhood Problem under Degeneracy. Disc. Paper, FernUniversität Hagen, 1986.

[17] GAL, T., KRUSE, H.-J., ZÖRNIG, P.: New Developments in the Area of Degeneracy Graphs. Presented at the Joint National TIMS/ORSA Meeting, Los Angeles, April 1985.

[18] GASS, S. I.: Comments on the Possibility of Cycling with the Simplex Method. Oper. Res. 27, 1979, 848 - 852.

[19] GREENBERG, H.-J.: An Analysis of Degeneracy. Naval Res. Log. Quart. 33, 1986, 635 - 655.

[20] GRÜNBAUM, B.: Convex Polytopes. J. Wiley,London-New York 1969.

[21] HOFFMAN, A. J.: Cycling in the Simplex Algorithm. National Bureau of Standards Rep. No 2974, 1953.

[22] KARWAN, M. H., LOTFI, V., TELGEN, J., ZIONTS, S.: Redundancy in Mathematical Programming - A State-of-the-Art Survey. Springer Verlag, Berlin-Heidelberg-New York-Tokyo 1983.

[23] KNOLMAYER, G.: How Many Sided are Shadow Prices at Degenerate Primal Optima? OMEGA 4, 1976, 493 - 494.

[24] KNOLMAYER, G.: The Effects of Degeneracy on Cost-Coefficient Ranges and an Algorithm to Resolve Interpretation Problems. Decision Sciences 15, 1984, 14 - 21.

[25] KOTIAH, T. C. T., STEINBERG, D. I.: Occurrences of Cycling and other Phenomena arising in a Class of Linear Programming Models. Comm. of the Assoc. for Comput. Machinery 20, 1977, 107 - 112.

[26] KOTIAH, T. C. T., STEINBERG, D. I.: On the Possibility of Cycling with the Simplex Method. Oper. Res. 26, 1978, 374 - 376.

[27] KRUSE, H.-J.: Degeneracy Graphs and the Neighbourhood Problem. Lecture Notes in Economics and Mathematical Systems No 260, Springer Verlag, Berlin-Heidelberg-New York-Tokyo 1986.

[28] KRUSE, H.-J.: Über spezielle Teilgraphen von Entartungsgraphen. Disc. Paper No 121, FernUniversität Hagen, 1987.

[29] MAJTHAY, A.: On Degeneracy and Cycling with the Simplex Method. Disc. Paper No 41, Center for Econometrics and Decision Sciences, University of Florida 1981.

[30] PIEHLER, G.: Notes on Determination of Shadow Prices and Sensitivity Analysis under Degeneracy. In Preparation, Fernuniversität Hagen, 1987.

[31] STRUM, J. E.: Note on Two-Sided Shadow Prices. J. Account. Res. 7, 1969, 160 - 162.

[32] TELGEN, J.: A Note on a Linear Programming Problem that Cycled. COAL Newsletters 2, 1980, 8 - 11.

[33] WOLFE, P.: A Technique for Resolving Degeneracy in Linear Programming. J. of SIAM 11, 1963, 205 - 211.

[34] ZÖRNIG, P.: Strukturuntersuchungen an 2xn-Entartungsgraphen. Disc. Paper No 87, FernUniversität Hagen 1985.

[35] ZÖRNIG, P.: On Cycling of the Simplex-Algorithm in Linear Programming. In Preparation, Fernuniversität Hagen, 1988.

A shortest augmenting path algorithm for dense and sparse linear assignment problems

Roy Jonker
Ton Volgenant

The linear assignment problem (LAP) is useful as a relaxation for difficult combinatorial optimization problems (quadratic assignment, travelling salesman). Theoretical developments for the LAP can often be extended to problems as minimum cost flow and transportation.

The methods based on shortest paths are dual algorithms: dual feasibility exists and primal feasibility has to be reached. This is achieved by considering the LAP as a minimum cost flow problem, solved by steps that involve finding shortest paths on an auxiliary graph.

In our algorithm LAPJV the initialization is primarily aimed at reaching a high initial reduction. It consists of three procedures: (1) reduction of columns in a standard way, (2) reduction transfer from unassigned to assigned rows; although it has no direct net effect on the reduction sum, a higher reduction sum may be obtained afterwards when unassigned rows are reduced; (3) augmenting reduction of unassigned rows. An attempt is made to find augmenting paths starting in unassigned rows, to which at the same time reduction is transferred. In the process assigned columns remain so, but rows may become assigned, unassigned or reassigned.

Augmentation starts by finding an alternating path, a sequence of, alternately, row and column indices. A shortest path algorithm is adapted to save many rows scans. Indirect comparison gives that LAPJV is about twice as fast as HUNG & ROM [4]. We left primal simplex methods out of consideration, as these are outperformed by other methods. We made comparisons both on dense and on sparse problems. The computation times in milliseconds (Fortran codes on a CDC Cyber 750 with OPT =2) are averages for 10 full density problems.

As usual with pure Hungarian methods, ASSCT [2] turns out to be very sensitive to cost range. Only for small ranges the use of pointer techniques makes ASSCT competitive. LAPJV [5] is less sensitive to the cost range and clearly faster than ASSIGN [1] and than LSAP [3], Dorhout's improved version of Tomizawa's algorithm. The code of moderate size and the required memory is small. The algorithm is also suited for sparse problems.

cost range	problem size	ASSCT	LSAP	ASSIGN	LAPJV
1-100/10000	50	51/145	32/31	22/28	15/25
	100	149/1085	168/168	114/134	64/103
	150	283/3562	535/453	256/410	179/259
	200	420/9689	1363/919	520/779	322/456

REFERENCES

[1] D.P. BERTSEKAS (1981). A New Algorithm for the Linear Assignment Problem. *Mathematical Programming 21*, 152-171.

[2] G. CARPANETO, P. TOTH (1980). Algorithm 5458 (Solution of the Assignment Problem). ACM Transactions on Mathematical Software 6, 104-111.

[3] B. DORHOUT (1975). Experiments with Some Algorithms for the Linear Assignment Problem. Report BW 39, Mathematisch Centrum, Amsterdam.

[4] M.S. HUNG, W.O. ROM (1980). Solving the Assignment Problem by Relaxation. Operations Research 28, 969-982.

[5] R. JONKER, A. VOLGENANT (1987) A shortest augmenting path algorithm for dense and sparse linear assignment problems. Computing 38, 325-340.

Approximation Algorithms for Scheduling Unrelated Parallel Machines

Jan Karel Lenstra
David B. Shmoys
Éva Tardos

We consider the following problem. There are m parallel machines and n independent jobs. Each job is to be assigned to one of the machines. The processing of job j on machine i requires time p_{ij}. The objective is to find a schedule that minimizes the makespan.

Our main result is a polynomial which constructs a schedule that is guaranteed to be no longer than twice the optimum. We also present a polynomial approximation scheme for the case that the number of machines is fixed. Both approximation results are corollaries of a theorem about the relationship of a class of integer programming problems and their linear programming relaxations. In particular, we give a polynomial method to round the fractional extreme points of the linear program to integral points that nearly satisfy the constraints.

In contrast to our main result, we prove that no polynomial algorithm can achieve a worst-case ratio less than 3/2 unless $P=NP$. We finally obtain a complexity classification for all special cases with a fixed number of processing times.

1980 Mathematics Subject Classification: 90B35, 68Q25, 68R05.
Key Words & Phrases: scheduling, parallel machines, approximation algorithm, worst case analysis, linear programming, integer programming, rounding.
Note: The complete paper has appeared in the Proceedings of the 28th annual IEEE Symposium on the Foundations of Computer Science, Computer Society Press of the IEEE, Washington, D.C. (1987), pp. 217-224. It has been submitted for journal publication.

COMBINATORIAL IMPROVEMENTS OF THE 1-TREE BOUND FOR THE TRAVELING SALESMAN PROBLEM

Franz Rendl, Graz

1-trees are used as relaxation for the symmetric Traveling Salesman Problem. We propose to strengthen this relaxation by adding additional constraints of the following type: select a set of nonadjacent vertices and require the 1-tree to contain exactly two edges adjacent to any of those vertices. This leads to the intersection of a graphical matroid with a partition matroid. We propose an algorithm for this problem (more efficient than general matroid intersection) and provide computational results to indicate the improvement over the general 1-tree relaxation.

Local Search for Constrained Routing Problems

M.W.P. Savelsbergh, Amsterdam

Distribution management is becoming more and more the subject of mathematical research. On the operational level, two problems prevail: the routing of capacitated vehicles through a collection of points to pick up or deliver goods, the vehicle routing problem, and the scheduling of vehicles to meet time or precedence constraints imposed upon their routes, the vehicle scheduling problem.

Only recently researchers started to emphasize the development of methods to solve real life problems. Because real-life problems tend to be large-scale we look at approximation algorithms. Among those, iterative improvement procedures have always played an important role. We develop local search algorithms for routing problems with various side constraints. The presented algorithms are based on the k-exchange concept. We examine side constraints that arise when customers have time windows, when there are both collections and deliveries on a route, and when there are precedence relations between customers. The presence of side constraints introduces feasibility problems. Testing whether an exchange results in a feasible tour no longer takes constant time, but requires $O(n)$ time, where n is the number of vertices. This is due to fact that although the tour is only locally changed, it can affect the complete tour. We show that by applying a lexicograph search strategy and maintaining a set of properly chosen global variables, the testing effort can be reduced to constant time.

The Stochastic Knapsack Problem

Leen Stougie , Amsterdam

Whereas the deterministic knapsack problem is *NP*-hard, a stochastic version of it shown to be polynomially solvable by a list policy.

1980 Mathematics Subject Classification: 90C15, 90B99
Keywords & Phrases: stochastic knapsack problem, list policy, computational complexity

The Stochastic Knapsack Problem

In the literature various stochastic versions of *NP*-hard scheduling problems have been shown to be solvable in polynomial time by simple list scheduling rules [Weiss 1982], [Pinedo 1982], [Derman et al. 1978] and [Pinedo 1983]. Here we will show that the same phenomenon occurs for the *knapsack problem*, the deterministic model of which is *binary NP-hard. The stochastic result is straightforward from the equivalence between the knapsack problem and the scheduling problem in which the weighted number of late jobs on a single machine is minimized.*

We consider the following model. Let n items have requirement coefficient $\mathbf{a}_1,\mathbf{a}_2,\ldots,\mathbf{a}_n$ that are independent *exponentially* distributed random variables with rates $\lambda_1,\lambda_2,\ldots,\lambda_n$. The profit coefficients $\mathbf{c}_1,\mathbf{c}_2,\ldots,\mathbf{c}_n$ are independent random variables with distributions that have finite means. Finally, the capacity $\mathbf{b}$ of the knapsack is random and has distribution F and a density function f. The *stochastic knapsack problem* is formulated as

$$\max E\{\sum_{j=1}^{n}\mathbf{c}_j\mathbf{x}_j \mid \sum_{j=1}^{n}\mathbf{a}_j\mathbf{x}_j\leq\mathbf{b},\ \mathbf{x}_j\in\{0,1\},\ (j=1,\ldots,n)\}.$$

We are interested in finding optimal policies within the following two classes of policies, defined according to [Pinedo 1983].

The first class of policies is the class of *static list* policies. A list policy arranges all items in a list, and consider the items for inclusion in the knapsack according to that list. An item is included if its requirements added to the total requirement of the items already included does not violate the knapsack capacity. In a static list policy the order of the items is *prefixed*, and hence does not depend on particular realizations of the parameter values. Of course, the outcome of the policy does depend on realizations of values.

The second class of policies is the class of *dynamic* policies. In a dynamic policy, the decision maker is allowed to determine his actions at any moment in the selection procedure, taking into account all the information that has become available up to that moment.

Theorem 1. *The optimal static list for solving the stochastic knapsack problem considers the items in decreasing order of $\lambda_j E\mathbf{c}_j$. If inclusion of an item considered violates the capacity constraint this item is rejected, otherwise it is included.*

Proof. Consider first the case of two items. We compare the values of the objective function when the items are ordered 1.2 and 2.1. Let Z_1, and Z_2 denote the respective values. Then

$$Z_1 = Ec_1 Pr\{a_1 \leq b\} + Ec_2 Pr\{a_1 + a_2 \leq b\}$$
$$= Ec_1(1 - Pr\{a_1 > b\}) + Ec_2(1 - Pr\{a_1 + a_2 > b\})\}$$
$$= Ec_1 + Ec_2 - Ec_1 \int_0^\infty e^{-\lambda_1 x} dF(x) - Ec_2 \int_0^\infty (\lambda_1/\lambda_2))e^{-\lambda_2 x} dF(x)$$
$$+ Ec_2 \int_0^\infty (\lambda_2/(\lambda_1 - \lambda_2))e^{-\lambda_1 x} dF(x)$$

with a similar expression for Z_2. Hence,

$$Z_1 - Z_2 = (\lambda_1 Ec_1 - \lambda_1 - \lambda_2))e^{-\lambda_1 x} dF(x)$$

where

$$K(x) = (e^{-\lambda_1 x} - e^{-\lambda_2 x})/(\lambda_2 - \lambda_1) > 0 \text{ for } x > 0.$$

Hence, $Z_1 \geq Z_2$ when $\lambda_1 Ec_1 \geq \lambda_2 Ec^2$. We next extend this argument to problems with n items. Compare the sequence $j_1, ..., j_{i-1}, j_{i+1}, j_i, hj_{i+2}, ..., j_n$ and $j_{i+2}, ..., j_n$ is the same in the two sequences. Hence, we need only compare the expected profit of j_i and j_{i+1} in the two sequences. Conditional on the capacity requirement of the first $i-1$ items, the amount of capacity still available for j_i and j_{j+1} is the same in the two sequences. Therefore, comparing the expected additional profit given upon inclusion of j_i and j_{i+1} in the two interchange that puts an item with a higher value of $\lambda_j Ec_j$ first increases the expected profit. □

We notice that this policy runs in time $O(\log n)$, required for ordering the item. Hence this stochastic knapsack problem is solved in polynomial time.

THEOREM 2. *The static list policy given in Theorem 1 is also optimal within the class of dynamical policies for the stochastic knapsack problem.*

PROOF. There are n decision points at which the decision maker must determine the item to be included next. That the static policy is also optimal within the class of dynamic policies can be shown by induction on the number of items still to be considered for inclusion. The assertion clearly holds when only one item remains to be considered. Assume the static policy of Theorem 1 is optimal in the class of dynamic policies when $k-1$ items remain to be considered. When k items are to be considered for inclusion, one item must be chosen to be included first provided that inclusion of it does not violate the capacity constraint. This static policy considers the items in decreasing $\lambda_j Ec_j$-order. Thus the choice to be made when k items remain to be possibly included is a choice between static policies. The static policy of Theorem 1 is therefore optimal in the class of dynamic policies. □

REFERENCES.

C. DERMAN, G. LIEBERMAN and S. ROS (1978), A renewal decision problem, *Management Sci.* 24, pp. 554-561.

M. PINEDO (1982), Computational complexity of stochastic scheduling problems, in M.A.H. Dempster, J.K. Lenstra, A.H.G. Rinnooy Kan(eds.), *Deterministic and stochastic scheduling,* D. Reidel, Dordrecht.

M. PINEDO (1983), Stochastic scheduling with release dates and due dates, *Operations Research* 31, pp. 559-572.

G. WEISS (1982), Multiserver stochastic scheduling, in M.A.H. Dempster, J.K. Lenstra, A.H.G. Rinnooy Kan (eds.), *Deterministic and stochastic scheduling,* D. Reidel, Dordrecht.

A Simulation Tool for the Performance Evaluation of Parallel Branch and Bound Algorithms

H.W.J.M. Trienekens, Rotterdam

Parallel Computation offers a challenging opportunity to speed up the time consuming enumerative procedures taht are necessary to solve hard combinatorial problems. The performance of such a parallel branch and bound algorithm cannot be evaluated by simply executing the algorithm on a few parallel systems. Among the difficulties encountered are the noise produced by the other users on the system, the limited variation in parallellism (the number and kind of the processors in the system is strictly bounded) and the waste of resources involved: most of the time, the outcomes of all computations are already known and the only issue of interest is when these outcomes are produced.

We will describe a way to simulate the execution of parallel branch and bound algorithms on arbitrary parallel systems in such a way that the memory and cpu requirements are very reasonable. The use of simulation only has minor consequences for the formulation of the algorithm.

Report 8720/A, Econometric Institute, Erasmus University Rotterdam

Minimizing Makespan on Unrelated Parallel Machine

S.L. van de Velde , Amsterdam

There are m parallel machines available for processing n independent jobs. A machine can process one job at a time and a job, once started, has to be finished without preemption. Processing job j on machine i requires time p_{ij}. The objective is to find a schedule of minimum length.

Three solution approaches are presented. The first is an approximation algorithm, based on iteratively solving Lagrangian subproblems. This yields a lower bound as well as a reduction in problem size. Only minor modifications in this algorithm are necessary to convert it into a multiplier adjustment method for lower bound calculations. These aspects are employed in the second approach, which is a branch-and bound algorithm of the depth first type. Its branching rule is based on dual costs for assigning jobs to machines, as they resulted from the first algorithm. A powerful truncated version of this branch-and-bound algorithm is conceded by this branching rule. A condition is given under which the solution of the truncated branch-and-bound algorithm is recognized to be optimal. If this cannot be decided, an increased lower bound is returned. Computational results show that the first procedure outperforms existing approximation algorithms. The optimization algorithm handles problems of moderate size effectively. The truncated version is capable of considerably improving the approximate solution. It finds solutions within one percent from the optimum for large problems.

A HEURISTIC FOR SCHEDULING PROBLEMS ESPECIALLY FOR SCHEDULING FARM OPERATIONS

Peter Wijngaard , Eindhoven

To obtain a schedule for farm operations with techniques like linear programming or dynamic programming is a time-consuming task. It asks for a lot of computer-memory and therefore, only small problems can be solved. But, practice asks for better solvers which can be used for large problems.
Therefore, we have developed a new combinatorial algorithm which has the following properties:

1) it is fast: an admissible schedule is derived in very short time;
2) it makes less use of computer-memory: the program for the algorithm fits on a micro-computer.

Due to the properties, the algorithm is a heuristic. It is based on a network. In this network, each stage is a moment in time and every state of this stage stands for an amount of material (e.g. crops, soil). The link between states of different stages represents the decision to process material or not.
The heuristic is based on the network-search technique hill-climbing. It makes use of a heuristic evaluation function. The procedure is as follows: of a certain parent state n, each possible decision is evaluated. A decision will give a decrease or increase of material and involve a quantity of actual costs. These actual costs are the sum of timeliness costs of materials, overtime costs and extra costs (costs due to extra operations). Of the remaining amount of material, an expectationof costs is made with the heuristic evaluation function. The best decision is the one with the minimal sum of actual and expected costs. This decision will lead to a new parent state at stage n+1. The procedure will be continued until all the materials are processed or until the planningshorizon is derived.
Due to the strong heuristic evaluation function, the solution differs not much from the optimal solution ($\pm$ 4 %, derived by dynamic programming for a number of test-instances). Contrary, the decrease of calculation time is considerable ($\pm$ 10^6 %).
The algorithm makes use of farm management data, climatological and meteorological data. It will be used to calculate schedules for farm business, but it can also be applied in an industrial environment (e.g. the agribusiness or road construction).
A decision support system which uses the algorithm is developed. This system works on a micro-computer under the operating system MS-DOS. The system can perform planning, simulation and the combination of these two. This system will be tested thoroughly for the harvest on a large farm in the IJssel Lake Polders.

BERICHTE AUS DEN DGOR-ARBEITSGRUPPEN

Aus den ständigen Arbeitsgruppen der DGOR berichten während der Tagung:

Arbeitsgruppe:	**Berichterstatter:**
OR im Gesundheitswesen	K. Heidenberger, Nürnberg
Netzplantechnik und Projektmanagement	J. Schwarze, Braunschweig
Praxis der linearen Optimierung	H. Weber, Hannover

VERZEICHNIS DER AUTOREN UND REFERENTEN

GAL 612
Prof. Dr. Dr. T. Gal, FernUniversität Gesamthochschule,
Fachbereich Wirtschaftswissenschaft, Feithstr. 140,
D-5800 Hagen

GAUL 338,405
Prof. Dr. W. Gaul, Universität Karlsruhe, Institut für
Entscheidungstheorie und Unternehmensforschung,
D-7500 Karlsruhe

VAN GELDER 367
Drs. H. van Gelder RA, MSc., North-Holland Consulting bv.,
Boekelermeerweg 23, NL-1851 LS Heiloo

GELDERS 42
Prof. Dr. Ir. L.F. Gelders, Katholieke Universiteit Leuven,
Afd. Industrieel Beleid, Celestijnenlaan 30A, B-3030 Leuven
(Heverlee)

GERKEN 272
Dipl.-Math. J. Gerken, Universität Bremen, Fachbereich
Elektrotechnik/Physik, Postfach 330440, D-2800 Bremen 33

GIESBERTS 166
P. Giesberts, Technische Universiteit Eindhoven, Faculteit der
Bedrijfskunde, Postbus 513, NL-5600 MB Eindhoven

GRAD 501
Prof. Dr. J. Grad, Univerza Edvarda Kardelja V Ljubljani,
Ekonomska fakulteta Borisa Kidrica, Kardeljeva ploscad 17,
YU-61109 Ljubljana

GRIEM 358
W. Griem, Mannesmann, Datenverarbeitung, D-4030 Ratingen 4

GRÜNWALD 232
Dipl.-Ing. H.J. Grünwald, Philips International B.V.,
Corporate ISA, VN-402, Postbus 218, NL-5600 MD Eindhoven

GÜNTHER 116
Dr. H.O. Günther, Universität Mannheim, Lehrstuhl f.
Betriebswirtschaftslehre und Unternehmungsforschung,
D-6800 Mannheim

Dr. S. Indihar, University of Maribor, VEKS, Razlagova Ul. 14, Postni Predal 180, YU-62000 Maribor

Prof. Dr. J. Jahn, Universität Erlangen-Nürnberg, Institut für Angewandte Mathematik, Martensstr. 3, D-8520 Erlangen

Dr. W. Jammernegg, Universität Graz, Institut für Statistik, Ökonometrie und Operations Research, A-8010 Graz

Dr. Ch. Jansson, Universität Kaiserslautern, FB Mathematik, Erwin-Schrödinger-Straße, D-6750 Kaiserslautern

Prof. Dr. M. Jarke, Universität Passau, Lehrstuhl für Dialogorientierte Systeme, D-8390 Passau

P.D. Dr. U. Jensen, Universität Hohenheim, Institut für Angewandte Mathematik und Statistik, D-7000 Stuttgart 70

Dr. R. Jonker, Koninklijke/Shell-Laboratorium, P.O. Box 3003, NL-1003 AA Amsterdam

Prof. Dr. L.C.M. Kallenberg, Rijksuniversiteit Leiden, Subfaculteit Wiskunde en Informatica, P.O. Box 9512, NL-2300 RA Leiden

W.C.M. Kallenberg, University of Twente, P.O. Box 217, NL-7500 AE Enschede

E.M.F. Kalvelagen, Vrije Universiteit, Econometric Institute, de Boelelaan 1105, NL-1081 HV Amsterdam

Dr. F.L. Klapp, Mannesmann-Röhrenwerke, D-4330 Mühlheim

NAAMANI 390
Drs. O. Naamani, Operations Research Dept., ABN Bank,
P.O. Box 669, NL-1000 EG Amsterdam

NIEKAMP 260
E.G. Niekamp, Erasmus Universiteit Rotterdam, Postbus 1738,
NL-3000 DR Rotterdam

OHMANN 87
R. Ohmann, Universität Osnabrück, Inst. f. BWL und
Wirtschaftsinformatik, Luisenstr. 16, D-4500 Osnabrück

VAN OMMEREN 490
Drs. J.C.W. van Ommeren, Dept. of Econometrics, Vrije
Universiteit, P.O. Box 7161, NL-1007 MC Amsterdam

OOMES 525
V. Oomes, University of Twente, Department of Management
Studies, P.B. 217, NL-7500 AE Enschede

PAESSENS 290
Prof. Dr. H. Paessens, Fachhochschule Flensburg, Fachgebiet
Wirtschaftsinformatik, Kanzleistr. 91-93, D-2390 Flensburg

PIETRZAK 170
Dipl.-Ing. R. Pietrzak, Institut für Fertigungstechnik und
Spanende Werkzeugmaschinen, Schloßwender Str. 5,
D-3000 Hannover 1

VAN DER PLAS 291
Drs. C.J. van der Plas, Nationale Nederlanden, Prinses
Beatrixlaan 15, NL-2595 AK The Hague

POPP 364
Prof. Dr. W. Popp, Universität Bern, Betriebswirtschaftliches
Institut, Sennweg 2, CH-3012 Bern

PROTH 167
J.M. Proth, INRIA-Lorraine, Chateau du Montet, Rue du Doyen
Roubault, F-54500 Vandoeuvre les Nancy

VAN PUTTEN 362
Drs. C. van Putten, Nederlandse Philips Bedrijven B.V., Centre
for Quantitative Methods, P.O. Box 218, NL-5600 MD Eindhoven

RADERMACHER 235
Prof. Dr. Dr. F.J. Radermacher, Universität Passau, Lehrstuhl für Informatik und OR, Postfach 2540, D-8390 Passau

REESE 115,198
Dr. J. Reese, Fernuniversität Gesamthochschule, Fachbereich Wirtschaftswissenschaft, Feithstr. 140, D-5800 Hagen

REINDERS 236
Ir. M.P. Reinders, Landbouwuniversiteit, Vakgroep Wiskunde, De Dreijen 8, NL-6703 BC Wageningen

REMMERSWAAL 291
Drs. J.J. Remmerswaal, Algemene Bank Nederland NV, P.O. Box 669, NL-1000 EG Amsterdam

RENDL 624
Dr. F. Rendl, Technische Universität Graz, Institut für Mathematik, Kopernikusgasse 24, A-8010 Graz

RENTZ 349
Prof. Dr. O. Rentz, Universität Karlsruhe, Inst. für Industriebetriebslehre und Industrielle Produktion (IIP), D-7500 Karlsruhe 21

REYMAN 515
G. Reyman, University of Technology, P.O. Box 513, NL-5600 MB Eindhoven

RIDDER 491
Dr. A. Ridder, Rijksuniversiteit Leiden, Subfaculteit Wiskunde en Informatica, P.O. Box 9512, NL-2300 RA Leiden

RITTER 66
Prof. Dr. K. Ritter, Technische Universität, Inst. f. Angew. Mathematik u. Statistik, Arcisstr. 21, D-8000 München 2

ROBRADE 96
A. Robrade, Universität der Bundeswehr, Betriebliche Logistik und Organisation, Holsteinhofweg 85, D-2000 Hamburg 70

ROMMELFANGER 206,592
Prof. Dr. H. Rommelfanger, Universität Frankfurt, Institut für Statistik und Mathematik, Mertonstr. 17, D-6000 Frankfurt

Batch number: 10844827

Printed by Printforce, the Netherlands